CONTAMINANTS OF EMERGING ENVIRONMENTAL CONCERN

SPONSORED BY
Emerging Contaminants of Concern Task Committee of the
Environmental Council

Environmental and Water Resources Institute (EWRI)
of the American Society of Civil Engineers

EDITED BY
Alok Bhandari
Rao Y. Surampalli
Craig D. Adams
Pascale Champagne
Say Kee Ong
R. D. Tyagi
Tian C. Zhang

Published by the American Society of Civil Engineers

Library of Congress Cataloging-in-Publication Data

Contaminants of emerging environmental concern / sponsored by Emerging Contaminants of Concern Task Committee of the Environmental Council [and] Environmental and Water Resources Institute (EWRI) of the American Society of Civil Engineers ; edited by Alok Bhandari ... [et al.].
    p. cm.
Includes bibliographical references and index.
ISBN 978-0-7844-1014-1
1. Pollutants. I. Bhandari, Alok. II. Environmental Council of the States (U.S.) Emerging Contaminants of Concern Task Committee. III. Environmental and Water Resources Institute (U.S.)

TD174.C66 2009
628.5'2--dc22   2008048522

American Society of Civil Engineers
1801 Alexander Bell Drive
Reston, Virginia, 20191-4400

www.pubs.asce.org

Preface

The 21st century has unfolded the widespread occurrence of a new category of contaminants which have attracted the attention of citizens, scientists and engineers, researchers, state and federal agencies, environmental groups, industrial and commodity groups, and regulators. These contaminants are predominantly unregulated anthropogenic chemicals that occur in air, soil, water, food, and, human/animal tissues in trace concentrations, are persistent in the environment, and are capable of perturbing the physiology of target receptors. These chemicals are considered to be contaminants of emerging environmental concern (CoEECs).

The ASCE's Technical Committee on Hazardous, Toxic, and Radioactive Waste Management identified the need to collect and present the latest information on the occurrence and fate of CoEECs in natural and engineered systems. The committee envisioned to prepare an easy-to-read book that would serve as a reference for practicing professionals and be equally effective as a text in undergraduate or graduate courses.

This book report is organized by types of commonly occurring and widely studied CoEECs. Chapter 1 introduces the topic of the book report and presents the need to understand the characteristics and environmental occurrence of these chemicals. Chapter 2 discusses analytical chemistry methods for sampling, separation, purification, identification and quantification of CoEECs in environmental samples. Chapter 3 discusses pharmaceuticals, while Chapter 4 talks about personal care products. Chapters 5 and 6 present information about antibiotics and hormones, respectively. Chapter 7 discusses the occurrence and fate of phthalate plasticizers and their degradation products in natural and engineered systems. Chapters 8 and 9 focus on surfactants and their derivatives, and fire retardants. Chapter 10 discusses pesticides, several of which are considered to be CoEECs at trace concentrations. Chapter 11 focuses on nanomaterials, a category of CoEECs whose health and environmental implications are just beginning to be evaluated. Finally, Chapter 12 talks about organisms capable of degrading CoEECs and the molecular biology tools used to study these organisms.

The editors acknowledge the hard work and patience of all authors who have contributed to this book.

- AB, RYS, CDA, PC, SKO, RDT, TCZ

Contributing Authors

Craig D. Adams, *University of Kansas, Lawrence, KS*

Shankha K. Banerji, *University of Missouri – Columbia, MO*

S. Barnabe, *Pulp and Paper Center, University of Quebec at Trois-Rivières, QC*

I. Beauchesne, *INRS, Universite du Quebec, Quebec, QC*

Alok Bhandari, *Iowa State University, Ames, IA*

Satinder K. Brar, *INRS, Universite du Quebec, Quebec, QC*

Pascale Champagne, *Queens University, Kingston, ON*

Supreeda Homklin, *Chulalongkorn University, Bangkok*

Keith C.K. Lai, *University of Texas, Austin, TX*

Warisara Lertpaitoonpan, *Iowa State University, Ames, IA*

Tawan Limpiyakorn, *Chulalongkorn University, Bangkok*

Say Kee Ong, *Iowa State University, Ames, IA*

Bala Subramanian, *INRS, Universite du Quebec, Quebec, QC*

Rao Y. Surampalli, *U.S. Environmental Protection Agency, Kansas City, KS*

R.D. Tyagi, *INRS, Universite du Quebec, Quebec, QC*

Mausam Verma, *Dalhousie University, Halifax, NS*

Kang Xia, *Mississippi State University, Mississippi State, MS*

Song Yan, *INRS, Universite du Quebec, Quebec, QC*

Tian C. Zhang, *University of Nebraska-Lincoln, Omaha, NE*

Contents

CHAPTER 1

Introduction

Alok Bhandari

1.1 Background

Over the past century our industries have produced thousands of chemicals to enhance our overall quality of life. These compounds have allowed us to increase agricultural productivity, improve animal health, and boost human longevity. The improper use and disposal of some of these chemicals, however, has also resulted in adverse human health impacts and environmental problems. The societal need to ensure the quality of soil and water environments was first recognized in the mid-20th century. Soon thereafter, we began regulating our coastal waters and surface streams for sediments and oxygen-consuming organic material. Then came Rachel Carson's *Silent Spring* which introduced the public to the environmental and human health impacts of a wide array of persistent and toxic anthropogenic chemicals. Regulations were reinforced and maximum contaminant levels (MCLs) were established to minimize human exposure to chemicals characterized as acutely toxic or carcinogenic. The National Pollutant Discharge Elimination System (NPDES) ensured that these contaminants were not discharged into surface water from point-sources such as industries and municipal wastewater treatment plants (WWTPs).

The late 20th century produced great advancements in trace chemical analysis including separation and identification methods such as solid phase extraction, gas-chromatography/mass-spectrometry (GC/MS), and liquid-chromatography/mass-spectrometry (LC/MS). These technologies allowed scientists to separate compounds from environmental media and identify them at unprecedented levels of parts per trillion (ppt or ng/L) and lower. Suddenly, a broad range of previously undetected micropollutants became apparent in media as diverse as food, tissue, breast milk, and water samples collected from rivers, lakes, aquifers, municipal water treatment plants and WWTPs. Thus, the dawn of the 21st century has brought with it the knowledge of widespread, trace-level environmental occurrence of a generally unregulated group of chemicals that have been variously termed as 'emerging chemicals of concern', 'contaminants of concern', 'contaminants of emerging concern', 'micro-constituents', 'unregulated contaminants', 'persistent organic pollutants', 'pharmaceuticals and

1

personal care products', or 'contaminants of emerging environmental concern (CoEECs)'.

CoEECs are usually unregulated, but persistent anthropogenic chemicals that are discharged into the environment or generated therein at low concentrations. These are potentially harmful compounds whose effects on ecology and human health are poorly understood because of their recent discovery in the environment. CoEECs enter the environment through a variety of domestic, industrial and agricultural activities. Some CoEECs are used by humans as pharmaceuticals or personal care products and enter the hydrological cycle through discharges from municipal wastewater treatment plants or on-site septic systems. Others are used in the animal agriculture for growth promotion or disease control and enter the environment when animal waste is applied on agricultural fields. Still others are used in industrial surfactants, plasticizers, or flame retardants and released via industrial discharges. These compounds are mobile and persistent in the environment and occur in air, water, sediments, and tissues of ecological receptors at concentrations of ppt to low parts-per-billion (ppb, μg/L).

Scientists from the U.S. Geological Survey (USGS) were among the first to report a widespread occurrence of CoEECs at targeted sites in streams across the United States (Kolpin et al., 2002). USGS's Toxic Substances Hydrology Program has since conducted other nation-wide field investigations and have recently reported on the presence of these compounds in groundwater and sources of municipal drinking water (Barnes et al. 2008; Focazio et al., 2008). The most frequently detected compounds in groundwater in human and animal waste sources included N,N-diethyltoluamide, (DEET, insect repellent), bisphenol-A (plasticizer), tri(2-chlroethyl)phosphate (flame retardant), sulfamethoxazole (antibiotic) and 4-octylphenol monoethoxylate (surfactant metabolite). When untreated sources of municipal drinking water were tested, the five most frequently detected CoEECs in surface water included cholesterol, metolachlor (herbicide), cotinine (nicotine metabolite), β-sitosterol (natural plant sterol), and 1,7-dimethylxanthine (caffeine metabolite).

Robust data on ecological and human health effects of exposure to CoEECs at environmentally relevant concentrations is still lacking. However, extensive research is being conducted on this topic and early results appear to suggest some ecological concern. Frogs exposed to the anti-bacterial agent, triclosan, have been found to show reduced activity and startle response (Fraker and Smith, 2004). Residues of the anti-inflammatory, diclofenac, have been implicated with declines in vulture populations in South Asia (Oaks et al., 2004). Fish collected downstream from municipal wastewater treatment plants discharging effluent containing a variety of estrogenic chemicals have revealed signs of altered sex ratios, reduced gonad size, gonadal intersex, disrupted ovarian and testicular histopathology and vitellogenin induction, that are typical of exposure to estrogenic chemicals (Vajda et al., 2008). Intersex gonads and high vitellogenin concentrations were recently reported in male fish collected from four river basins of the Southeastern United States indicating exposure to EDCs and estrogenic chemicals (Hinck et al., 2008).

The antiepileptic drug, carbamazepine, has been shown to impact the survivability of the non-biting midge *Chironomus riparius* but it had no effects on other organisms such as the oligochaete *Lumbriculus variegates* and the freshwater snail *Potamopyrgus antipodarum* (Oetken et al., 2005). Other researchers have found little or no ecological impact of some CoEECs. No adverse impact of exposure of the antidepressant, fluoxetine, was observed on zooplankton at environmentally relevant concentrations (Laird et al., 2007). The antibiotic, tylosin, was shown to pose little or no risk for aquatic macrophytes (Brain et al., 2005). Exposure to the surfactant derivative, 4-nonylphenol, at or above the EPA toxicity-based chronic exposure level produced no changes in morphological endpoints such as the gonadosomatic and hepatosomatic indices, secondary sexual characteristics and histopathology in male fathead minnows (Schoenfuss et al., 2008).

1.2 Types of CoEECs

Several classes of anthropogenic compounds can be classified as CoEECs. In its national reconnaissance studies of US streams, groundwater and raw drinking water, the USGS has focused on the subset of CoEECs summarized in Table 1.1 (USGS, 2008). These include human and veterinary antibiotics, sex and steroidal hormones, household and industrial chemicals and other human pharmaceuticals. The latter category includes lifestyle medicines analgesics, anti-inflammatories, stimulants, antacids, antidepressants, antihypertensives, antidiabetics, anticoagulants, antianxiety, antiasthmatics, antihyperlipidimics, and antianginals. This book report includes specific chapters focusing on pharmaceuticals (chapter 3), personal care products (chapter 4), antibiotics (chapter 5), hormones (chapter 6), plasticizers (chapter 7), surfactants (chapter 8), fire retardants (chapter 9), pesticides (chapter 10), and nanomaterials (chapter 11). Also discussed are analytical methods used for the separation, clean-up, and identification of CoEECs in environmental samples (chapter 2) and molecular biology techniques for CoEEC degrading organisms (chapter 12).

1.3 Future Challenges

Modern analytical tools and techniques have allowed detection of CoEECs at minute concentrations. Continued development and refinement of techniques for separation and quantitation of trace level CoEECs in complex environmental matrices is necessary to improve our understanding of factors affecting the fate and transport of these contaminants. This knowledge will lead to the development of models that can accurately predict the behavior of these contaminants in the soil and water environment.

Table 1.1. CoEECs targeted in USGS's National Reconnaissance Study (USGS 2008)

Antibiotics (Human and Veterinary)		Other Human Pharmaceuticals	
Tetracyclines	*Sulfonamides*	*Prescription*	*Non-Prescription*
chlortetracycline	sulfachlorpyridazine	cimetidine	acetaminophen
doxycycline	sulfadimethoxine	dehydronifedipine	ibuprofen
oxytetracycline	sulfamerazine	digoxygenin	codeine
tetracycline	sulfamethazine	digoxin	caffeine
	sulfamethiazole	diltiazem	1,7-dimethylxanthine
Fluoroquinolones	sulfamethoxazole	fluoxetine	cotinine
ciprofloxacin	sulfathiazole	gemfibrozil	
enrofloxacin		metformin	
norfloxacin		paroxetine	
sarafloxacin	*Other Antibiotics*	ranitidine	
	carbadox	salbutamol	
Macrolides	lincomycin	warfarin	
erythromycin-H_2O	trimethoprim		
roxithromycin	virginiamycin		
tylosin			

Sex and Steroidal Hormones		
Biogenics	*Pharmaceuticals*	*Sterols*
17β-estradiol (E2)	17α-ethynylestradiol	cholesterol
17α-estradiol (E1)	mestranol	3β-coprostanol
estrone	19-norethisterone	stigmastanol
estriol	equilenin	
testosterone	equilin	
progesterone		
cis-androsterone		

Household and Industrial Chemicals			
Insecticides	*PAHs*	*Plasticizers*	*Surfactant Derivatives*
carbaryl	anthracene	bis(2-ethylhexyl)adipate	nonylphenol-monoethoxylate
chlorpyrifos	benzo(*a*)pyrene	ethanol-2-butoxyphosphate	nonylphenol diethoxylate
cis-chlordane	fluoranthene	bis(2-ethylhexyl)phthalate	octylphenol-monoethoxylate
diazinon	naphathalene	diethylphthalate	octylphenol-diethoxylate
dieldrin	phenanthrene	triphenyl phosphate	*p*-nonylphenol
lindane	pyrene		
methyl parathion			
N,N-diethyltoluamide			
Antioxidants	*Fire Retardants*	*Others*	
butylatedhydroxyanisole	tri(2-chloroethyl)-phosphate	Acetophenone	
butylatedhydroxytoluene	tri(dichlorisopropyl)-phosphate	bisphenol-A	
2,6-di-tert-butylphenol		1,2-dichlorbenzene	
2,6-di-tert-butyl-p-benzoquinote		*p*-cresol	
5-methyl-1H-benzotriazole		phenol	
		phthalic anhydride	
		tetrachloroethylene	
		triclosan	

The ecotoxicological significance of CoEECs at concentrations in which they have been reported to occur in the environment remains largely unknown. Future research should focus on the effects of long-term, low-level exposures to CoEECs on human health and ecosystem response. Environmental exposures and ecosystem responses need to be studied at all scales. Synergistic and antagonistic effects from exposure to multiple contaminants of concern need to be understood by the scientific and regulatory communities.

Ensuring the absence of trace-level CoEECs from potable water will require future water treatment plants to adopt advanced technologies such as activated carbon, advanced oxidation processes, membrane filtration, and reverse osmosis (Snyder et al., 2007). However, since it is not yet certain that exposure at these concentrations poses a credible human health risk, robust cost-benefit analyses would need to be performed before industry-wide adoption of these expensive technologies for drinking water treatment. Nonetheless, these technologies will be indispensable in situations where municipal water is directly or indirectly augmented with treated wastewater.

Future challenges for reducing human exposure to CoEECs also include the management of combined sewer overflows, onsite septic systems, and landfills which can introduce considerable amounts of CoEECs into the soil and water environment. Agricultural practices such as land-application of animal wastes on farms, and waste management at confined animal feeding operations require improvement to minimize the release of contaminants of concern into soil and water. Best management practices need to be developed for managing urban and agricultural non-point sources of emerging contaminants.

Finally, the scientific community, practicing engineers, regulators, federal and state governments, affected industries, and citizens should work together to craft the necessary framework for the proper use and disposal of these chemicals such that expensive treatment of affected soil, water or air is minimized.

References

Barnes, K.K.; Kolpin, D.W.; Furlong, E.T.; Zaugg, S.D.; Meyer, M.T.; Barber, L.B. (2008) 'A national reconnaissance of pharmaceuticals and other organic wastewater contaminants in the United States – I. Groundwater.' Science of the Total Environment, 402:192-200.

Brain, R.A.; Bestari, J.; Snaderson, H.; Hanson, M.L.; Wilson, C.J.; Johnson, D.J.; Subket, P.K.; Solomon, K.R. (2005) 'Aquatic microcosm assessment of the effects of tylosin on Lemna gibba and Myriophyllum spicatum.' Environmental Pollution, 133:389-401.

Focazio, M.J.; Kolpin, D.W.; Barnes, K.K.; Furlong, E.T.; Meyer, M.T.; Zaugg, S.D.; Barber, L.B.; Thurman, M.E. (2008) 'A national reconnaissance of pharmaceuticals and other organic wastewater contaminants in the United States –

II. Untreated drinking water sources.' Science of the Total Environment, 402:201-216.

Fraker, S.L.; Smith, G.R. (2004) 'Direct and interactive effects of ecologically relevant concentrations of organic wastewater contaminants on Rana pipiens tadpoles.' Environmental Toxicology, 19:250-256.

Hinck, J.E.; Blazer, V.S.; Denslow, N.D.; Echols, K.R.; Gale, R.W.; Wieser, C.; May, T.W.; Ellersieck, M.; Coyle, J.J.; Tillitt, D.E. (2008) 'Chemical contaminants, health indicators, and reproductive biomarker responses in fish from rivers in the Southeastern United States.' Science of the Total Environment, 390:538-557.

Kolpin, D.W.; Furlong, E.T.; Meyer, M.T.; Thurman, E.M.; Zaugg, S.D.; Barber, L.B.; Buxton, H.T. (2002) 'Pharmaceuticals, hormones, and other organic wastewater contaminants in U.S. streams, 1999-2000: A national reconnaissance. Environmental Science and Technology, 36:1202-1211.

Laird, B.D.; Brain, R.A.; Johnson, D.J.; Wilson, C.J.; Sanderson, H.; Solomon, K.R. (2007) 'Toxicity and hazard of a mixture of SSRIs to zooplankton communities evaluated in aquatic microcosms.' Chemosphere, 69:949-954.

Oaks, J.L.; Gilbert, M.; Virani, M.Z.; Watson, R.T.; Meteyer, C.U.; Rideout, B.A.; Shivaprasad, H.L.; Ahmed, S.; Chaudhry, M.J.I.; Arshad, M.; Mahmood, S.; Ali, A.; Khan, A.A. 'Diclofenac residues as the cause of vulture population decline in Pakistan.' Nature, 427:630-633.

Oetken, M.; Nentwig, G.; Loffler, D.; Ternes, T.; Oehlmann, J. (2005) 'Effects of pharmaceuticals on aquatic invertebrates. Part I. The antiepileptic drug carbamazepine.' Archives of Environmental Contamination and Toxicology, 49:353-361.

Schoenfuss, H.L.; Bartell, S.E.; Bistodeau, T.B.; Cediel, R.A.; Grove, K.J.; Zintek, L.; Lee, K.E.; Barber, L.B. (2008) 'Impairment of the reproductive potential of male fathead minnows by environmentally relevant exposures to 4-nonylphenol.' Aquatic Toxicology, 86:91-98.

Snyder, S.A.; Wert, E.C.; Lei, H.; Westerhoff, P.; and Yoon, Y. (2007) 'Removal of EDCs and Pharmaceuticals in Drinking and Reuse Treatment Processes.' American Water Works Association Research Foundation, Denver, Colorado.

USGS (2008) 'Target compounds for national reconnaissance of emerging contaminants in US streams' URL: toxics.usgs.gov/regional/contaminants.html (accessed 09/10/2008).

Vajda, A.M.; Barber, L.B.; Gray, J.L.; Lopez, E.M.; Woodling, J.D.; Norris, D.O. (2008) 'Reproductive disruption in fish downstream from an estrogenic wastewater effluent.' Environmental Science and Technology, 42:3407-3414.

CHAPTER 2

Analytical Methods for Environmental Samples

Kang Xia

2.1 Introduction

During recent decades, chemicals representing active ingredients in pharmaceuticals and personal care products (PPCPs) have emerged as a new class of environmental contaminants due to concerns of their negative hormonal and toxic impacts on various organisms at trace levels (Daughton and Ternes, 1999). Because of their frequent usage, a variety of human use PPCPs are discharged daily into WWTPs via excretion with urine and feces as parent compounds, conjugated compounds, or metabolites, and through washing or direct disposal. Those PPCPs that can not be completely degraded during wastewater treatment processes enter the environment via WWTP effluent and biosolids (Keller et al., 2003).

With the rapid development of sophisticated and sensitive analytical methods and instruments, more and more PPCPs have been detected in a variety of environment samples (Halling-Sørensen et al., 1998; Daughton and Ternes, 1999; Kolpin et al., 2002; Peck 2006). PPCPs include a wide spectrum of compounds with large differences in chemical properties, making it a daunting task to describe in detail the analytical method for thousands of registered PPCPs and their possible metabolites in different environmental matrices. The objective of this chapter is to provide a comprehensive summary of the published analytical methods during the past decade on some commonly used PPCPs by classifying them into neutral, acidic, basic, and zwitterionic groups (Tables 2.1-2.4). The chapter begins with discussion on sample handling, preparation, and cleanup for water, soil/sediment/biosolids, and biological samples, followed by a section summarizing instrumentation for analysis. Some examples on detections of representative PPCPs in environmental samples are demonstrated at the end of the chapter.

7

Table 2.1 Characteristics of representative neutral PPCPs detected in environmental samples

Compound	Use	Structure	Solubility (mg L^{-1}) (pH = 6)
Carbamazepine Log K_{ow} = 2.83 MW = 236.27	antiepileptics		78
Loratadine Log K_{ow} = 4.60 MW = 382.88	Allergy treatment		0.05
Triclosan Log K_{ow} = 4.20 MW = 289.54	Antibacterial agents		4.1
Trichlocarban Log K_{ow} = 4.50 MW = 315.58			0.7
17α-ethinylestradiol Log K_{ow} = 3.83 MW = 296.40	Synthetic hormone		2
Musk xylene Log K_{ow} = 3.46 MW = 297.26	Synthetic musk fragrances		0.02

Table 2.1 Characteristics of representative neutral PPCPs detected in environmental samples (continued)

Compound	Use	Structure	Solubility (mg L^{-1}) (pH = 6)
Galaxolide Log K_{ow} = 4.77 MW = 258.40	Synthetic musk fragrances		9
Octydimethyl-*p*-aminobenzoic acid Log K_{ow} = 4.72 MW = 277.40			2
Oxybenzone Log K_{ow} = 3.34 MW = 228.24	UV blocking agents		210
Octocrylene Log K_{ow} = 5.47 MW = 361.48			0.2
DEET Log K_{ow} = 2.44 MW = 191.27			1,000
Bayrepel Log K_{ow} = 2.22 MW = 229.32	Insect repellents		2,500

Table 2.1 Characteristics of representative neutral PPCPs detected in environmental samples (continued)

Compound	Use	Structure	Solubility (mg L^{-1}) (pH = 6)
Diazepam Log K_{ow} = 2.99 MW = 284.74	Tranquilizer		20
Caffeine Log K_{ow} = 1.31 MW = 194.19	Psychoactive stimulant		3,700
Glibenclamide Log K_{ow} = 2.41 MW = 494.00	Anti-diabetic agent		16
Polybrominated diphenyl ethers (PBDEs) log K_{ow} ∃ 5.74 log K_{ow} = 0.621(#Br) + 4.12 (Braekevelt et al., 2003)	Flame retardants	 (m + n = 10)	not water soluble

2.2 Sample Handling, Preparation, Extraction, and Clean-up

Detailed protocols for water and solid sample collection were described in the "Standard Practice for Sampling Water" (ASTM, 1980) and the "Test Methods for Evaluating Solid Waste (SW-846)" (EPA, 1996), respectively. The biological sample collection procedures can be found in the "Guidance for Assessing Chemical Contaminant Data for Use in Fish Advisories" (EPA, 2000). Prior to sample collection, the following major parameters need to be specified in a sampling plan: sampling site, target analytes, sampling times, type of sample matrix, method of sampling, number of replicates (EPA, 2000).

Table 2.2 Characteristics of representative acidic PPCPs detected in environmental samples

Compound	Use	Structure	Solubility (mg L^{-1}) (pH = 6)
Clofibric acid pK$_a$ = 3.2 MW = 214.65			339,000
Gemfibrozil pK$_a$ = 4.8 MW = 250.33	Lipid regulator		480
Bezafibrate pK$_a$ = 3.3 MW = 361.82			7,600
Atorvastatin pK$_a$ = 4.3 MW = 558.64			300
Ibuprofen pK$_a$ = 4.4 MW = 206.28			2,000
Salicylic acid pK$_a$ = 3.0 MW = 138.12	Anti-inflammatory agent		1,000,000
Fenoprofen pK$_a$ = 4.2 MW = 242.27			4,100

Table 2.2 Characteristics of representative acidic PPCPs detected in environmental samples (continued)

Compound	Use	Structure	Water solubility (mg L^{-1}) (pH = 6)
Naproxen $pK_a = 4.8$ MW = 230.26			510
Ketoprofen $pK_a = 4.2$ MW = 254.28			9.9
Tolfenamic acid $pK_a = 3.7$ MW = 261.70	Anti-inflammatory agent		440
Diclofenac $pK_a = 4.2$ MW = 296.15			980
Indomethacin $pK_a = 4.0$ MW = 357.79			3,100
Furosemide $pK_a = 3.0$ MW = 330.74	Diuretic agent		6,900

Table 2.2 Characteristics of representative acidic PPCPs detected in environmental samples (continued)

Compound	Use	Structure	Solubility (mg L^{-1}) (pH = 6)
Amoxicillin $pK_a = 2.4$ MW = 365.40	Gram-positive and gram-negative antibiotic		300
Tetracycline $pK_a = 4.5$ MW = 444.43	Broad-spectrum antibiotic		3,100

To minimize contamination of samples, use of personal care products containing analytes of interests is discouraged during sample collection, processing, and analysis. Gloves must be used at all time when handling samples. Careful cleaning of samplers and containers prior to sampling, proper selection of container materials, correct sampler preservation, and adequate sample holding time are crucial for assessing occurrence, fate, and behavior of trace levels of PPCPs in the environment (EPA, 1996; EPA, 2000).

2.2.1 Aqueous Samples

Aqueous samples containing < 1% solids are often filtered first through solvent pre-washed glass fiber filters (pore size < 1 μm) or 0.45 μm cellulose filters to remove particulates. If both acid and basic analytes are of interests, the filtrate is separated into two equal portions, with one portion acidified to pH 2 using concentrated HCl or H_2SO_4 and the second basified to pH 10 using concentrated NH_4OH. Whenever possible, stable isotopically labeled analogs of analytes of interest are spiked into their respective acidified or basified portion. The acidified portion should be stabilized with Na_4EDTA if analytes of interests (e.g., tetracycline) form complexes with metal ions such as Ca^{2+} or Mg^{2+} in the matrix of interest (EPA, 2007). The samples need to be stored in the dark at < 6°C before arrival at the laboratory. The samples need to be kept frozen at < -10°C if they are not extracted and analyzed right away upon arrival at the laboratory. The frozen aqueous samples need to be extracted within 7 days of collection, and analyzed within 40 days of extraction. In order to prevent transformation of target analytes before analysis, extraction of analytes using appropriate solid phase extraction (SPE) cartridges at the collection sites right after spiking the pH-adjusted aqueous samples with stable isotopes is highly recommended (McArdell et al., 2007). The target analytes loaded on the dried SPE cartridges are more stable compared to in aqueous phase. It is because the SPE process may able to remove some components that otherwise can interact or transform target analytes during the transport and storage of aqueous samples. In addition, the dried target analytes-loaded SPE cartridges are easy to store and transport.

Table 2.3 Characteristics of representative basic PPCPs detected in environmental samples

Compound	Use	Structure	Solubility (mg L^{-1}) (pH = 6)
Acebutolol pK$_a$ = 9.1 MW = 336.43	β-blockers for treating disorders of cardiovascular system		98,000
Metoprolol pK$_a$ = 9.2 MW = 267.36			615,000
Ranitidine pK$_a$ = 8.4 MW = 314.40	Anti-ulcer agents		110,000
Cimetidine pK$_a$ = 7.1 MW = 252.34			14,000
Omeprazole pK$_a$ = 4.7 pK$_a$ = 8.4 MW = 345.42			41
Lansoprazole pK$_a$ = 3.6 pK$_a$ = 8.9 MW = 369.36			4.8

Table 2.3 Characteristics of representative basic PPCPs detected in environmental samples (continued)

Compound	Use	Structure	Solubility (mg L^{-1}) (pH = 6)
Pheniramine pK$_a$ = 9.4 MW = 240.34	antihistamines	CH–CH$_2$–CH$_2$–NMe$_2$, Ph (pyridine ring)	240,000
Cyclizine pK$_a$ = 8.0 MW = 266.38		CHPh$_2$ (piperazine ring), Me	16,000
Thonzylamine pK$_a$ = 8.8 MW = 286.37		CH$_2$–CH$_2$–NMe$_2$, N–CH$_2$, OMe (pyrimidine ring)	20,000
Methylparaben pK$_a$ = 8.3 MW = 152.15	Preservatives	C–OMe (O double bond), HO (benzene ring)	64,000
Benzylparaben pK$_a$ = 8.2 MW = 228.24		C–O–CH$_2$–Ph (O double bond), HO (benzene ring)	1,200
Methadone pK$_a$ = 9.1 MW = 309.45	Opioid used for pain relief	NMe$_2$, Ph, O, Me–CH–CH$_2$–C–C–Et, Ph	8,700

Table 2.3 Characteristics of representative basic PPCPs detected in environmental samples (continued)

Compound	Use	Structure	Solubility (mg L^{-1}) (pH = 6)
Oxycodone pK$_a$ = 7.6 MW = 315.36	Opioid used for pain relief		4,700
Acetaminophen pK$_a$ = 9.9 MW = 151.16	Analgesic, antitussive, antipyretic agent		10,000
Erythromycin pK$_a$ = 8.2 MW = 747.95	Macrolide antibiotic		19,000
Amlodipine pK$_a$ = 9.0 MW = 408.88	Diuretic agent		2,100
Diphenhydramine pK$_a$ = 8.8 MW = 255.35	Antihistamine used in cold and cough medicine	Ph$_2$CH$-$O$-$CH$_2-$CH$_2-$NMe$_2$	19,000

Table 2.3 Characteristics of representative basic PPCPs detected in environmental samples (continued)

Compound	Use	Structure	Solubility (mg L^{-1}) (pH = 6)
Fluoxetine pK$_a$ = 10.1 MW = 309.33	selective serotonin reuptake inhibitor (antidepressant)		5,300
Sertraline pK$_a$ = 9.5 MW = 306.23			2,000

Because of their trace levels in environmental samples, target PPCPs need to be extracted from their matrix, separated from interference components, and enriched via various extraction and clean-up methods. For aqueous samples, the most widely used method for target analyte extraction, clean-up, and concentration is solid phase extraction (SPE), although, traditional liquid/liquid extraction (LLE) methods are still used by some researchers occasionally. Compared to LLE, SPE methods provide better recoveries, create less phase separation problems, consume less solvent, and can be operated off line or on-line when coupled directly with chromatographic system (Berrueta et al., 1995).

In SPE, an aqueous sample is passed through a solid sorbent normally packed into a cartridge or imbedded into disks made of Teflon or glass fiber. During this process, analytes of interest can be retained onto the solid phase and be separated from other sample components that pass through. The target analytes retained on the solid phase are then eluted with another solvent, concentrated down, and finally analyzed on instruments. Similar to the packing materials used in liquid chromatography columns, there are four basic types of SPE sorbent (stationary phase): reversed phase, normal phase, ion exchange, and adsorption (Table 2.5). Recently, various other type of sorbents, for example, those involving antigen-antibody interactions have recently been developed as SPE stationary materials with high extraction specificity for single analytes or classes of compounds (Mikaela and Pilar, 2005). Higher specificity for target analytes means more efficient sample extraction and clean-up for trace level analytes in complex environmental samples.

Table 2.4 Characteristics of representative zwitterionic PPCPs detected in environmental samples

Compound	Use	Structure	Solubility (mg L^{-1}) (pH = 6)
Ciprofloxacin pK$_a$ = 6.03 (-COOH) pK$_a$ = 8.38 (=NH$_2^+$) MW = 331.34			1,300
Norfloxacin pK$_a$ = 6.18 (-COOH) pK$_a$ = 8.38 (=NH$_2^+$) MW = 319.33	Fluoroquinolone antibiotics		990
Ofloxacin pK$_a$ = 5.23 (-COOH) pK$_a$ = 7.39 (≡NH$^+$) MW = 361.37			190

Up to date, SPE with reversed phase, normal phase, and ion exchange stationary phases are commonly used for extraction and clean-up of PPCPs in aqueous samples by following the general steps: conditioning the SPE with appropriate solvent → passing aqueous sample though SPE stationary phase at certain flow rate → wash the stationary phase with appropriate solvent → elute the analytes of interest. Pre-conditioning of SPE opens up reaction sites on the stationary phase as well as removes impurities. It is important not to allow the SPE stationary phase to dry before sample addition. During the sample addition step, an adequate flow rate is essential for achieving maximum retention of analytes to the stationary phase. Depending on the nature of functional group(s) on an analyte and the chemistry of a SPE stationary phase, the pH of a sample needs to be adjusted before sample addition to SPE cartridge or disks. In order to remove interference compounds and leaving analytes retained on the stationary phase, washing solvent needs to be stronger than the sample matrix but weaker than the eluant solvent that is used to collect the analytes off the stationary phase. Again, the flow rate of the eluant solvent should be slow enough to ensure best recovery. Representative SPE extraction and clean-up methods for PPCPs in aqueous phase are summarized in Table 2.6.

Table 2.5 Characteristics of common SPE packing materials and their target analytes*

| SPE phase | Polarity | | | | Stationary phase and analytes interactions |
	Stationary phase	Sample matrix	Analytes	Eluant solvent	
Reversed phase	nonpolar	polar to moderately polar	mid to nonpolar	nonpolar	van der Waals forces
Normal phase	polar	mid to nonpolar	polar	polar	hydrogen bonding, π-π interactions, dipole-diple interactions, dipole-induced dipole ineractions
Ion exchange	charged	having a pH at which functional groups on both analytes and stationary phase are charged	charged	having a pH that neutralizes functional groups either on the anlytes or on the stationary phase	electrostatic attraction
Adsorption	vary				Hydrophobic and hydrophilic interactions may apply depends on which stationary phase is used and the nature of analytes.

* Based on Supelco publication (Bulletin 910)

Table 2.6 SPE extraction and clean-up methods for selected PPCPs in aqueous samples

Analytes	SPE material	Pre-conditioning solvent	Washing solvent	Eluent solvent	References
Neutral analytes					
carbamazepine	60 mg Oasis HLB	2 mL *n*-hexane → 2 mL acetone → 10 mL methanol → 10 mL non-contaminated groundwater (pH adjusted to 10)	2 mL 5% methanol in 2% aqueous NH$_4$OH	4 x 1 mL methanol	Vieno et al., 2006
	Lichrolut EN	6 mL methanol → 6 mL water	none	3 mL methanol → 3 mL ethyl acetate, pooled	Castiglioni et al., 2005
caffeine diazepam glibenclamid propyphenazone nifedipine omeprazole oxyphenbutazone phenylbutazone 17α-ethinylestradiol	500 mg Isolute C$_{18}$	none	none	3 x 1 mL methanol	Ternes et al., 2001
	Oasis MCX	6 mL methanol → 3 mL water → 3 mL water acidified to pH 2	none	2 mL methanol → 2 mL 2% ammonia solution in methanol → 2 mL0.2% NaOH in methanol, pooled	Castiglioni et al., 2005

Table 2.6 SPE extraction and clean-up methods for selected PPCPs in aqueous samples (continued)

Analytes	SPE material	Pre-conditioning solvent	Washing solvent	Eluent solvent	References
		Acidic analytes			
clofibric acid gemfibrozil bezafibrate ibuprofen salicylic acid fenoprofen naproxen ketoprofen tolfenamic acid diclofenac indomethacin	150 mg Oasis MAX (Waters Corp.)	4 mL methanol → 10 mL H_2O (pH = 3)	5 mL methanol/sodium acetate	10 mL 2% formic acid in methanol	Lee et al., 2005
	60 mg Oasis MCX (Waters Corp.)	2 mL n-hexane → 2 mL acetone → 10 mL methanol → 10 mL non-contaminated groundwater (pH adjusted to 2)	none	4 x 1 mL acetone	Lindqvist et al., 2005
	60 mg Oasis HLB	2 mL EtOAC-acetone (50:50, v/v) → 2 mL Methanol → 3 mL water	1 mL methanol-water (10:90, v/v)	6 mL EtOAC-acetone (50:50, v/v)	Öllers et al., 2001
	Oasis MCX	6 mL methanol → 3 mL water → 3 mL water acidified to pH 2	none	2 mL methanol → 2 mL 2% ammonia solution in methanol → 2 mL0.2% NaOH in methanol, pooled	Castiglioni et al., 2005
	200 mg Strata X	3 x 2 mL water → 3 x 2 mL water (pH=3)	none	3 x 2 mL methanol	Hilton and Thomas, 2003

Table 2.6 SPE extraction and clean-up methods for selected PPCPs in aqueous samples (continued)

Analytes	SPE material	Pre-conditioning solvent	Washing solvent	Eluent solvent	References
Basic analytes					
acebutolol metoprolol	60 mg Oasis HLB	2 mL n-hexane $\rightarrow$ 2 mL acetone $\rightarrow$ 10 mL methanol $\rightarrow$ 10 mL non-contaminated groundwater (pH adjusted to 10)	2 mL 5% methanol in 2% aqueous NH$_4$OH	4 x 1 mL methanol	Vieno et al., 2006
	60 mg Oasis HLB	2 × 3mL methanol $\rightarrow$ 2 × 3mL water	none	6 mL methanol	Nikolai et al., 2006
acebutolol metoprolol ranitidine omeprazole cimetidine lansoprazole loratadine	Oasis MCX	6 mL methanol $\rightarrow$ 3 mL water $\rightarrow$ 3 mL water acidified to pH 2	none	2 mL methanol $\rightarrow$ 2 mL 2% ammonia solution in methanol $\rightarrow$ 2 mL0.2% NaOH in methanol, pooled	Castiglioni et al., 2005
acetaminophen	C$_{18}$/SDB-XC disks	10 mL CH$_2$Cl$_2$ $\rightarrow$ 10 mL acetone $\rightarrow$10 mL methanol $\rightarrow$10 mL water	none	10 mL methanol $\rightarrow$ 10 mL acetone $\rightarrow$ 10 mL CH$_2$Cl$_2$, pooled	Zhang et al., 2007

Table 2.6 SPE extraction and clean-up methods for selected PPCPs in aqueous samples (continued)

Analytes	SPE material	Pre-conditioning solvent	Washing solvent	Eluent solvent	References
methadone oxycodone codeine morphine	200 mg Oasis HLB	2 mL n-heptane → 2 mL acetone → 3 x 2 mL methanol → 4 x 2 mL water (pH=7)	none	4 x 2 mL acetone, pooled	Hummel et al., 2006
erythromycin (macrolides)	60 mg Oasis HLB	6 mL acetone → 6 mL methanol → 6 mL water (pH=6)	none	3 x 2 mL methanol, pooled	Miao et al., 2004
Zwitterionic analytes					
ciprofloxacin norfloxacin ofloxacin	60 mg Oasis HLB	2 mL n-hexane → 2 mL acetone → 10 mL methanol → 10 mL non-contaminated groundwater (pH adjusted to 10)	2 mL 5% methanol in 2% aqueous NH_4OH	4 x 1 mL methanol	Vieno et al., 2006
	60 mg Oasis HLB	6 mL aceton → 6 mL methanol → 6 mL 50 mM Na_2EDTA	none	3 x 2 mL methanol, pooled	Miao et al., 2004

2.2.2 Soil, Sediment, and Biosolids Samples

Compared to aqueous samples, solid samples have more complicated matrixes containing a variety of organic and inorganic components that may chemically and physically interact with target analytes, making it more challenging to completely recover target analytes out of the matrix. Commonly used extraction methods are solvent extraction (SE) or ultrasonic solvent extraction (USE), Soxhlet extraction, accelerated solvent extraction (ASE), also known as pressurized liquid extraction (PLE), and microwave assisted solvent extraction (MASE). Once target analytes are extracted, the above described SPE extraction and clean-up procedures for aqueous samples can be used to clean-up interfering large molecules such as proteins, lipids, or humic substances out of the extracts before instrumental analysis (Göbel et al., 2005). Gel permeation chromatography (GPC) also known as size exclusion chromatography (SEC) and activated silica gel, aluminum oxides, or Florisil columns are also commonly used for cleaning-up of extracts of solid samples (La Guardia et al., 2001; Liu et al., 2004; Zeng et al., 2005; Kinney et al., 2006).

Because SE methods are laborious, subject to problems associated with using large volumes solvent, and time consuming, they are not methods of choice by many researchers for PPCPs analysis. In recent years, ASE has replaced traditional Soxhlet extraction which consumes large quantities of extracting solvents. Since the extraction using ASE is normally performed under high pressure (up to 2×10^7 pa) and sometimes under high temperature (up to 200°C), faster and more efficient extraction using much less solvent can be achieved. In addition, a gradient of multiple solvents can be utilized for ASE to extract compounds with varying hydrophobicities. The ASE method is used in most of the published methods to date for extraction of PPCPs from solid samples (Table 2.7).

2.2.3 Biological Samples

Up to date, published analytical methods of PPCPs in biological samples are limited. However with increasing concern of environmental exposures of various organisms to PPCPs, there is an urgent need for analytical method development for this class of compounds in biological samples. Similar to soil, sediment, and biosolids samples, the ASE method is the most common extraction method for solid biological samples (Table 2.8). Before extraction, the moisture in a solid biological sample is removed by homogenizing the sample with hydromatrix or sodium sulfate. For biological samples, it is important to remove the lipids and other high molecular weight compounds from the extractants before instrumental analysis because they can cause interferences during detection. The general approach employed for lipids removal is to use deactivated alumina packed at the outlet of ASE extraction cells (Draisci et al., 1998) or gel permeation chromatography (GPC) using Bio Beads coupled with activated silica gel column (Nakata, 2005). Florisil and Strata-NH2/alumina have also been used for lipids clean up (Covaci et al., 2003; Osemwengie and Steinberg, 2003).

2.3 Sample Analysis

After sample extraction and clean-up, target analytes are analyzed most commonly using chromatographic techniques, in which analytes are first separated on a stationary phase with the aid of a mobile phase, and are subsequently detected on a detector. Gas chromatography (GC) or high performance liquid chromatography (HPLC) are commonly used chromatographic instruments for separation of PPCPs.

Table 2.7. Extraction and clean-up methods for selected PPCPs analysis in soil, sediment and biosolids samples.

Analytes	Matrix	Pre-treatment	Extraction	Clean-up	References
HHCB, AHTN	digested sludge	homogenize with hydromatrix	ASE with CH_2Cl_2 at 60°C, 1.4×10^7 Pa	Activated silica solvent exchanged to hexane	Simonich et al., 2000
		freeze-dried	Soxhlet extraction with CH_2Cl_2 for 72 h	silica gel and alumina and solvent exchanged to hexane	Zeng et al., 2005
		Freeze-dried	Ultrasonic extraction with methanol and acetone	RP-C18 SPE	Ternes et al., 2005
NP1EO	biosolids	freeze-dried, sieved through 2mm	ASE with CH_2Cl_2 at 100°C, 6.9×10^6 Pa	Size exclusion, solvent exchanged to hexane, activated silica eluted sequencially with hexane, hexane/CH_2Cl_2 (6:4), acetone. Solvent exchanged to toluene	La Guardia et al., 2001
NP2EOs					
NP					
PBDEs	sediment	air dry and ground by hand	Soxhlet extraction with n-hexane/acetone (50:50 v/v) for 4 h	alumina (5 % deactivated with water) column and solvent exchanged to hexane	Allchin et al., 1999
benzalkonium chlorides	sediment	freeze dry	ASE with acetonitrile/water (6:4) at 120°C, 3.4×10^5 Pa	cleanup and concentration using an automated SPE system with PLRP-s cartridges eluted with acetonitrile/water (6:4)	Ferrer and Furlong, 2002

Table 2.7. Extraction and clean-up methods for selected PPCPs analysis in soil, sediment and biosolids samples (continued).

Analytes	Matrix	Pre-treatment	Extraction	Clean-up	References
ciprofloxacin norfloxacin	digested sludge	Dried at 60°C for 72 h, ground to < 0.5 mm	ASE with 50 mM aqueous phosphoric acid (pH=2) and acetonitrile (1:1, v/v) at 100°C and 1.0×10^7 Pa	cleanup with MPC disk cartridge	Golet et al., 2002
azithromycin clarithromycin roxithromycin tylosin erythromycin	digested sludge	Freeze dry	USE with methanol then acetone, pooled ASE with methanol–water (50/50, v/v)	Extract was diluted with water, then cleaned-up using SPE (Oasis HLB)	Göbel et al., 2005
carbamazapine diphenhydramine fluoxetine triclosan triclocarban	biosolids	none	ASE with water/isopropyl aclcolhol (50:50, v/v) at 120°C and then re-extracted with water/isopropyl aclcolhol (20:80, v/v) at 200°C, 1.4×10^7 Pa	Extract was diluted with water then cleaned-up using SPE (PSDVB packing material) and Florisil SPE cartridge	Kinney et al., 2006
caffeine 17β-estradiiol ibuprofen ketoprofen musk ketone naproxen triclosan	sediment	Dried at 50°C and homogenized using blender	MASE using methylene chloride:methanol (2:1, v/v)	Slica gel packed with anhydrous sodium sulfate and activated copper granules	Rice and Mitra, 2007

Table 2.7. Extraction and clean-up methods for selected PPCPs analysis in soil, sediment and biosolids samples (continued).

Analytes	Matrix	Pre-treatment	Extraction	Clean-up	References
bezafibrate clofibric acid diclofenac fenoprofen fenoprop gemfibrozil ibuprofen indomethacin ketoprofen naproxen	sediment	Autoclaved at 131°C for 2 h	USE with 45 mL of acetone/acetic acid (20/1,v/v) → 3 x 45 mL ethyl acetate, pooled	Extract is diluted with 500 mL uncontaminated groundwater and cleaned-up with Oasis MCX	Löffler and Ternes, 2003
clarithromycin erythromycin oleandomycin roxithromycin sulfadiazine sulfamethazine sulfamethoxazole trimethoprim			USE with 2 x 45 mL methanol → 45 mL acetone → 45 mL ethyl acetate	Extract is diluted with 500 mL uncontaminated groundwater and cleaned-up with Lichrolute EN and Lichrolute C18	

Mass spectrometer (MS) is the most preferred detector to be coupled with GC or HPLC for many trace level PPCPs analysis due to its high selectivity and sensitivity and its capability of providing structural information. For PPCPs with halogens, electron capture detector is also often coupled with GC to provide sensitive detection without the high cost associated with purchasing a MS. However, unlike MS, structural confirmation can not be achieved using ECD. For HPLC analysis, in addition to MS, florescence detector (FLD) or ultraviolet detector (UVD) are sometimes used. FLD is in general an order of magnitude more sensitive than UVD, but it only works for fluorescent compounds. When FLD is used for detection of analytes without fluorophore functional group, preliminary treatment of samples with reagents that form fluorescent derivatives is required. Both detectors are not capable of providing structural information. Which chromatographic technique is used and what detector is chosen for analysis of target analytes largely depend on the chemical and physical properties of the compounds, as well as the desired sensitivity and specificity.

Table 2.8. Extraction and clean-up methods for selected PPCPs analysis in biological samples.

Analytes	Matrix	Pre-treatment	Extraction	Clean-up	References
HHCB AHTN	fish	homogenize with hydromatrix	ASE with ethylacetate–hexane (1:5, v/v) at 80°C, 1.0×10^7 Pa	deactivated alumina packed at outlet of the ASE extraction cell	Draisci et al., 1998
	marine mammal	ground with sodium sulfate	Soxhlet extraction with CH_2Cl_2/hexane (8:1, v:v) for 7 h	gel permeation chromatography (GPC) using Bio-beads S-X3 packed glass column with CH_2Cl_2/hexane (1:1, v:v) as mobile phase, then activated silica gel packed glass column with hexane as mobile phase	Nakata, 2005
NP1EO NP2EOs NP	fish, bird	homogenize with sodium sulfate	Soxhlet extraction with CH_2Cl_2/methanol (7:3 v/v) for 24 h	aminopropyl silica cartridges	Hu et al., 2005
PBDEs	fish	homoginized and dehydrated with anhydrous sodium sulphate	Soxhlet extraction with n-hexane/acetone (50:50 v/v) for 4 h	alumina (5% deactivated with water) column and solvent exchanged to hexane	Allchin et al., 1999
		homogenized and mixed with hydromatrix	ASE extraction with CH_2Cl_2 (1.0×10^7 Pa and 100°C)	deactivated silica gel column and solvent exchange to hexane	Xia et al., 2008
perfluoro-alkanesulfonate perfluoro-carboxylate	fish	homogenize in 2-3 mL of 0.25M sodium carbonate and 1 mL of 0.5M TBAS	liquid extraction twice for 10 min with 5 mL MTBE	filter through 0.2 μm nylon filter	Moody et al., 2001

Table 2.8. Extraction and clean-up methods for selected PPCPs analysis in biological samples (continued).

Analytes	Matrix	Pre-treatment	Extraction	Clean-up	References
acetaminophen atenolol cimetidine codeine 1,7-dimethylxanthine lincomycin trimethoprim thiabendazole caffeine sulfamethoxazole metoprolol propranolol diphenhydramine diltiazem carbamezepine tylosin fluoxetine norfluoxetine sertraline erythromycin clofibric acid warfarin miconazole ibuprofen gemfibrozil	fish fillet	none	Homogenize 1 g tissue with 8 mL acetonitrile/methanol (50:50, v:v), centrifuge at 16,000 rpm for 40 min at 4°C, dry supernatant under N2 at 45°C, reconstitute with 1 mL mobile phase.	Filter the final 1-mL solution through Pall Acrodisc hydrophobic Teflon Supor membrane syringe filter (0.2 μm pore size)	Ramirez et al., 2007
triclosan benzophenone-3 (BP-3) 4-methylbenzylidene camphor (4-MBC) ethylhexyl methoxy cinnamate (EHMC)	fish filet	homogenize with Na₂SO₄	liquid extraction with CH₂Cl₂/cyclohexane 1:1 (v:v)	gel permeation chromatography using a Biobeads S-X3 column and CH₂Cl₂/cyclohexane 35:65 (v:v) as mobile phase, followed by silica chromatography	Balmer et al., 2005; Buser et al., 2006

Table 2.8. Extraction and clean-up methods for selected PPCPs analysis in soil, sediment and biosolids samples (continued).

Analytes	Matrix	Pre-treatment	Extraction	Clean-up	References
sulfonamides, benzimidazoles, levamisole, nitroimidazoles, tranquillisers, fluroquinolones	pig kidney, chicken meat egg	homogenize with acetonitrile then add Na₂SO₄	pass extract through Bond Elut SCX cartridge, rinse the reservoir with acetonitrile, wash the cartridge sequentially with acetone, methanol and acetonitrile. elute the cartridge with acetonitrile/35% ammonia (95:5, v/v) into a test tube to obtain the B1 fraction, replace the test tube and elute with 5 mL of methanol/35% ammonia (75:25, v/v) to collect the B2 fraction before analysis of both fraction		Stubbings et al., 2005

2.3.1 Instrumentation Consideration

In GC, a sample is vaporized and injected onto the inlet of a chromatographic column. An inert gaseous mobile phase passing through the column transports the analytes through the column to the detector (Skoog and Leary, 1997). GC is best suited for neutral compounds with sufficient volatility (volatile and semi-volatile compounds). Analytes with basic or acidic functional group(s) in their structure tend to have low volatility and need to be derivatized, most often via acetylation, methylation, silylation, or pentafluorobenzyl bromide derivatization processes, to increase their volatility prior to GC analysis. Factors such as temperature, inert gas flow, type and thickness of column stationary phase, column length and diameter affect the separation of analytes on the column. The chemical nature of the stationary phase affects the partition ratio of an analyte to the stationary phase and, therefore its retention time. Table 2.9 lists some common stationary phases for GC. The polarity of the stationary phase should match that of the target analytes. The principle of "like dissolves like" applies to the partition between analytes and the GC column stationary phase. When the polarity match is good, the order of analytes elution through the column is largely determined by the boiling point of the analytes (Skoog and Leary, 1997).

For analysis of basic or acidic PPCPs, HPLC is often a better choice than GC because in HPLC a liquid mobile phase is used to transport analytes through the column to the detector. A liquid sample can be injected directly onto the column without vaporization. Analytes are separated due to differences in their partitioning between the mobile phase and the stationary phase. Flow rate and composition of mobile phase and column size and length can also affect analytes separation. Similar to GC stationary phase, methods for separation of analytes on an HPLC column largely depend on the polarity of analytes (Figure 2.1). For solutes having molecular weights greater than 10^4, exclusion method is often used. For low molecular weight ionic compounds, ion exchange method is preferred. Small polar but nonionic compounds

Table 2.9 Common GC column stationary phases (Restek, 2008)

Structure and Composition	Property	Common Trade Names	Applications
100% dimethyl polysiloxane CH_3 $\mid$ $-Si-O$ $\mid$ CH_3 100%	non-polar	Rtx-1, TR-1, DB-1, PE-1, ZB-1, BP1 SPB-1	solvents, petroleum products, pharmaceutical samples, waxes
6% cyanopropylphenyl, 94% dimethyl polysiloxane $C\equiv N$ $\mid$ $(CH_2)_3$ CH_3 $\mid$ $\mid$ $-Si-O$ $-Si-O$ $\mid$ $\mid$ Ph CH_3 6% 94%	slightly polar	Rtx-624, DB-624, Rtx-1301, DB-1301, SPB-1301, HP-1301, HP-624, AT-324,	volatile compounds, insecticides, residue solvents in pharmaceutical products
14% cyanopropylphenyl, 86% dimethyl polysiloxane $C\equiv N$ $\mid$ $(CH_2)_3$ CH_3 $\mid$ $\mid$ $-Si-O$ $-Si-O$ $\mid$ $\mid$ Ph CH_3 14% 86%	intermediately polar	Rtx-1701, TR-1701, DB-1701, CP-Sil19, BP10	pesticides, aroclors, alcohols, oxygenates
50% cyanopropylphenyl, 50% dimethyl polysiloxane $C\equiv N$ $\mid$ $(CH_2)_3$ CH_3 $\mid$ $\mid$ $-Si-O$ $-Si-O$ $\mid$ $\mid$ Ph CH_3 50% 50%	polar	Rtx-225, DB-225, HP-225, AT-225, BP-225, CP Sil43, PE-225	fatty acid methylesters (FAMEs), carbohydrates
90% biscyanopropyl, 10% cyanopropylphenyl $C\equiv N$ $C\equiv N$ $\mid$ $\mid$ $(CH_2)_3$ $(CH_2)_3$ $\mid$ $\mid$ $-Si-O$ $-Si-O$ $\mid$ $\mid$ $(CH_2)_3$ Ph $\mid$ $C\equiv N$ 90% 10%	very polar	Rtx-2330	FAMEs, cis/trans and dioxin isomers, rosin acids

Table 2.9 Common GC column stationary phases (continued)

Structure and Composition	Property	Common Trade Names	Applications
5% dipheyl, 95% dimethyl polysiloxane Ph — Si—O / Ph — 5%　　CH₃ — Si—O / CH₃ — 95%	non-polar	Rtx-5, DB-5, SPB-5, HP-5, AT-5, BP-5, CP Sil8, PE-2	flavors, aromatic hydrocarbons
35% dipheyl, 65% dimethyl polysiloxane Ph — Si—O / Ph — 35%　　CH₃ — Si—O / CH₃ — 65%	intermediately polar	Rtx-35, DB-35, SPB-35, HP-35, AT-35, BPX-35, PE-11	pesticides, aroclors, amines, nitrogen containing herbicides
20% dipheyl, 80% dimethyl polysiloxane Ph — Si—O / Ph — 20%　　CH₃ — Si—O / CH₃ — 80%	slightly polar	Rtx-20, SPB-20, AT-20, PE-7	volatile compounds, alcohols
65% dipheyl, 35% dimethyl polysiloxane Ph — Si—O / Ph — 65%　　CH₃ — Si—O / CH₃ — 35%	intermediately polar	Rtx-65TG, TAP-CB	triglycerides, rosin acids, free fatty acids
50% phenyl-50%methyl polysiloxane Ph — Si—O / CH₃ — 100%	intermediately polar	Rtx-50, DB-17, DB-608, SP-2250, SPB-50, AT-50, CP Sil24, CB, PE-17	triglycerides, phthalate esters, steroids, phenols

Table 2.9 Common GC column stationary phases (continued)

Structure and Composition	Property	Common Trade Names	Applications
trifluoropropylmethyl polysiloxane CF_3 | C_2H_4 | —Si—O | CH_3	selective for lone pair electrons	Rtx-200, DB-210, AT-210	solvents, freons, drugs, ketones, alcohols
Carbowax® PEG H H | | —C—C—O | | H H	polar	Stabilwax®, DB-Wax, Carbowas 20M, DB-waxetr, Supelcowax-10, Carbowas PEG 20M, HP-20M, InnoWax, HP-Wax, AT-Wax, BP-20, CP Wax52, PE-CW	FAMEs, flavors, acids, amines, solvents, xylene

are best handled by partition methods using reversed phase stationary phase or normal phase stationary phase (Skoog and Leary, 1997). Adsorption method is often chosen for separating nonpolar compounds, however, for analysis of those compounds, GC is a better choice.

As shown in Table 2.10, for trace level analysis of PPCPs in environmental samples, MS is the most preferred detector coupled GC or HPLC. During detection, a mass spectrum is obtained by converting components of a sample into rapidly moving gaseous ions, separating them on the basis of their mass-to-charge ratios (m/z), and detecting the ions on an ion detector (Skoog and Leary, 1997). MS is capable of providing information about the qualitative and quantitative composition of analytes in complex mixtures, the structures of analytes, and isotopic ratios of atoms in samples.

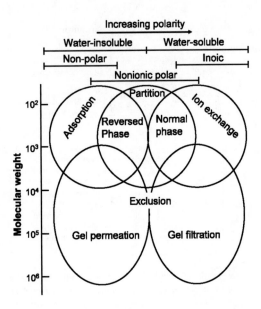

Figure 2.1 Interrelationship between analyte properties and the appropriate LC chromatographic separation method (Sauders., 1975, with permission from Van Nostrand Reinhold, New York).

For MS coupled with GC, electron ionization and chemical ionization are used to produce ions. During electron ionization, electrons are emitted from a heated filament and accelerated by a potential of approximately 70 eV. Charged positive ions (radical cations) are formed when the energetic electrons approach neutral molecules closely enough to cause them to lose electrons by electrostatic repulsion (Skoog and Leary, 1997). Subsequent relaxation of the radical cations takes place by extensive fragmentation, giving a large number of positive ions of various masses that are in general less than that of the molecular ion. The mass spectrum generated under defined conditions is highly reproducible for a specific analyte and, therefore, can be used for compound identification against extensive collections of standardized EI mass spectra (e.g. NIST/EPA/NIH Mass Spectral Library 2005). During chemical ionization, gaseous atoms of the sample are ionized by collision with ions produced by electron bombardment of an excess of reagent gas, for example, methane. The resulting ions formed are usually $[M+H]^+$ due to proton transfer reaction, $[M-1]^+$ due to hydride transfer reaction, or $[M+29]^+$ due to $C_2H_5^+$ transfer reaction. Other reagents such as propane, isobutene, and ammonia are sometimes used for chemical ionization, resulting in different spectrum with a given analyte (Skoog and Leary, 1997).

Table 2.10. Instruments used for analysis of PPCPs in environmental samples

Analytes	Chromatography	Mobile Phase	Column	Detection	References
clofibric acid bezafibrate ibuprofen furosemide hydrochlorothiazide sulphamethoxazole 17β-Estradiol estrone 17α-ethinylestradiol		A: 0.05% TEA in water (pH 8) B: acetonitrile		negative ESI-MS/MS in multiple reaction monitoring (MRM) mode	
amoxycillin atenolol ciprofloxacin demethyl diazepam diazepam enalapril lincomycin methotrexate oleandomycin ofloxacin omeprazole oxytetracycline ranitidine salbutamol tilmicosin carbamazepine clarithromycin cyclophosphamide erythromycin spiramycin tylosin	HPLC	A: 0.1% formic acid in water (pH 2) B: acetonitrile	3 μm, 2 mm x 50 mm Luna C8	positive ESI-MS/MS in MRM mode	Castiglioni et al., 2005

Table 2.10. Instruments used for analysis of PPCPs in environmental samples (continued)

Analytes	Chromato-graphy	Mobile Phase	Column	Detection	References
bezafibrate diclofenac fenoprop ibuprofen ketoprofen naproxen	HPLC	A: acetonitril B: 10 mM NH₄OH	5 μm, 2.1 mm x 50 mm Zorbax XDB-C18	negative ESI-MS/MS in MRM mode	Vieno et al., 2005
acebutolol metoprolol metoprolol sotalol ciprofloxacin norfloxacin ofloxacin carbamazepine	HPLC	A: acetonitrile B: 1% acetic acid		positive ESI-MS/MS in MRM mode	Vieno et al., 2006
caffeine 4-Aminoantipyrine propyphenazone diazepam glibenclamide nifedipine omeprazole oxyphenbutazone phenylbutazone	HPLC	A: 90% 20 mM ammonium aetate (pH 5.7) + 10% CH₃CN B: 40% 20 mM ammonium aetate (pH 5.7) + 60% CH₃CN	5 μm, 125 x 33 mm Merck LiChrospher 100RP-C18 (end capped)	positive ESI-MS/MS in MRM mode	Ternes et al., 2001
parabens 4-octylphenol 4-nonylphenol triclosan estrone 17β-Estradiol	GC (analytes derivatized with pentafluoropropionic acid anhydride (PFPA)	helium gas	Restek Rtx-5Sil MS (30 m, 0.25 mm I.D., 0.25 μm thickness)	EI-MS in ion monitoring (SIM) mode	Lee et al., 2005
clofibric acid ibuprofen salicylic acid gemfibrozil fenoprofen naproxen ketoprofen tolfenamic acid diclofenac-Na indomethacin	GC (analytes derivatized with N-t-butyldimethyl silyl-N-methyl-trifluoroacetamide (MTB-STFA) with 1% TBDMSCI				

Table 2.10. Instruments used for analysis of PPCPs in environmental samples (continued)

Analytes	Chromato-graphy	Mobile Phase	Column	Detection	References
carbamazepine ibuprofen clofibric acid ketoprofen naproxen diclofenac	GC GC analytes derivatized with diazomethane	helium gas	Rtx-5MS 30 m, 0.25 mm I.D., 0.25 μm thickness	EI-MS in SIM mode	Öllers et al., 2001
erythromycin sulfamethoxazole acetyl- sulfamethoxazole trimethoprim mefenamic acid lofepramine propranolol dextropropoxyphene diclofenac tamoxifen	HPLC	A: 40 mM ammonium acetate (adjust to pH 5.5 with fomic acid) B: methanol C: water	5μm, 250mm × 2mm Luna C18	positive ESI- /MS/MS in consecutive reaction monitoring (CRM) mode	Hilton and Thomas, 2003
clofibric acid				negative ESI- /MS/MS in CRM mode	
paracetamol				positive ESI- /MS/MS in SIM mode	
ibuprofen				negative ESI- /MS/MS in SIM mode	
clofibric acid ibuprofen acetaminophen caffeine chlorophen naproxen triclosan carbamazepine Estrone 17β-estradiol 17α-ethinylestradiol	GC (analytes derivatized with N,O- bis(trimethyls ily) trifluoroaceta mide (BSTFA)		HP-5MS 30 m, 0.25 mm I.D., 0.25 μm thickness	EI-MS in SIM mode	Zhang et al., 2007

Table 2.10. Instruments used for analysis of PPCPs in environmental samples (continued)

Analytes	Chromato-graphy	Mobile Phase	Column	Detection	References
benzoylecgonine codeine dihydrocodeine hydrocodone methadone morphine oxycodone tramadol bromazepam diazepam medazepam nordiazepam oxazepam temazepam-1 temazepam-2 carbamazepine primidone doxepin verapamil	HPLC	A: acetonitrile B: 10 mM ammonium formate in water (adjust to pH 4 with formic acid)	4 μm, 150 × 3 mm, Synergi Polar-RP 80 Å column	positive ESI-MS/MS in MRM mode	Hummel et al., 2006
triclosan 2,4-dichlorophenol 2,4,6-trichlorophenol	GC (analytes derivatized with N-methyl-N-(tert-butyldimethyl silyl)trifluoro acetamide (MTBSTFA)	helium gas	HP-5 MS 30m×0.2 5mm I.D., 0.25 μm thickness	EI-MS-MS in SIM mode	Morales et al., 2005
estrone ethynylestradiol estradiol estriol	GC analytes derivatized with penta-fluorobenzoyl		HP-5 MS 15m×0.2 5mm I.D., 0.25 μm thickness	negative CI-MS in SIM mode (reagent gas: methane)	Xiao et al., 2001
	GC (analytes derivatized with pentafluorobe nzyl bromide (PFBBR) and N-trimethylsilyl imidazole (TMSI))	helium gas	DB5-XLB 60m×0.2 5mm I.D., 0.25 μm thickness	negative CI-MS/MS in SIM mode (reagent gas: methane)	Fine et al., 2003

Table 2.10. Instruments used for analysis of PPCPs in environmental samples (continued)

Analytes	Chromato-graphy	Mobile Phase	Column	Detection	References
macrolides	HPLC	A:acetonitrile B: 20 mM ammonium acetate (0.05% formic acid, pH 5)	3μm, 2.1 × 50 mm Genesis C18 column	positive ESI-MS/MS in SRM mode	Miao et al., 2004
quinolones quinoxaline dioxides sulfonamides		A acetonitrile B: 20 mM ammonium acetate (0.1% formic acid, pH 4.0)	3μm, 2.1 × 150 mm Genesis C18 column		
tetracyclines		A:acetonitrile B: 20 mM ammonium acetate (0.1% formic acid and 4 mM oxalic acid)	3μm, 2.1 × 50 mm Genesis C18 column		
phenazone, propyphenazone dimethyl-aminophenazone	GC (analytes derivatized with MTBSTFA)	helium gas	HP-5 MS 30m×0.2 5mm I.D., 0.25 μm thickness	EI-MS in SIM mode	Zühlke et al., 2004
synthetic fragrance compounds	GC	helium gas	DB-5 MS 15m×0.2 5mm I.D., 0.25 μm thickness	EI-MS in SIM mode	Simonich et al., 2000
			HP-5 30m×0.3 2mm I.D., 0.25 μm thickness	EI-MS in full scan mode	Zeng et al., 2005
octylphenol nonylphenols nonylphenol monoethoxylates nonylphenol diethoxylates	GC	helium gas	DB-5 MS 60m×0.3 2mm I.D., 0.25 μm thickness	EI-MS in SIM mode	La Guardia et al., 2001

Table 2.10. Instruments used for analysis of PPCPs in environmental samples

Analytes	Chromato-graphy	Mobile Phase	Column	Detection	References
PBDEs	GC	helium gas	Restek Rtx-CLPestic ides 30m×0.2 5mm I.D., 0.25 μm thickness	ECD or EI-MS SIM mode	Xia et al., 2008
perfluoro-alkanesulfonates sperfluoro-carboxylates		A: 10 mM ammonium acetate in water B: 10 mM ammonium acetate in methanol		negative ESI-MS-MS in MRM mode	
amoxycillin atenolol ciprofloxacin demethyl diazepam diazepam enalapril lincomycin methotrexate oleandomycin ofloxacin omeprazole oxytetracycline ranitidine salbutamol tilmicosin carbamazepine clarithromycin cyclophosphamide erythromycin spiramycin tylosin	HPLC	A: 0.1% formic acid in water (pH 2) B: acetonitrile	4 μm , 2.1 mm x 50 mm, Genesis C8	positive ESI-MS/MS in MRM mode	Moody et al., 2001

Table 2.10. Instruments used for analysis of PPCPs in environmental samples (continued)

Analytes	Chromatography	Mobile Phase	Column	Detection	References
bezafibrate diclofenac fenoprop ibuprofen ketoprofen naproxen	HPLC	A: acetonitril B: 10 mM NH$_4$OH	5 μm, 2.1 mm x 50 mm	negative ESI-MS/MS in MRM mode	Vieno et al., 2005
acebutolol metoprolol metoprolol sotalol ciprofloxacin norfloxacin ofloxacin carbamazepine	HPLC	A: acetonitrile B: 1% acetic acid	Zorbax XDB-C18	positive ESI-MS/MS in MRM mode	Vieno et al., 2006
caffeine 4-Aminoantipyrine propyphenazone diazepam glibenclamide nifedipine omeprazole oxyphenbutazone phenylbutazone	HPLC	A:90% 20 mM amm. acetate (pH 5.7) + 10% CH$_3$CN B:40% 20 mM amm. aetate (pH 5.7) + 60% CH$_3$CN	5 μm, 125 x 33 mm Merck LiChrospher 100RP-C18 (end capped)	positive ESI-MS/MS in MRM mode	Ternes et al., 2001
parabens 4-octylphenol 4-nonylphenol triclosan estrone 17β-Estradiol	GC (analytes derivatized with pentafluoropropionic acid anhydride (PFPA)	helium gas	Restek Rtx-5Sil MS 30 m, 0.25 mm I.D., 0.25 μm thickness	EI-MS in ion monitoring (SIM) mode	Lee et al., 2005
clofibric acid ibuprofen salicylic acid gemfibrozil fenoprofen naproxen ketoprofen tolfenamic acid diclofenac-Na indomethacin	GC (analytes derivatized with N-t-butyldimethyl silyl-N-methyl-trifluoroaceta mide (MTB-STFA) with 1% TBDMSCI				

Table 2.10. Instruments used for analysis of PPCPs in environmental samples (continued)

Analytes	Chromato-graphy	Mobile Phase	Column	Detection	References
carbamazepine	GC		Rtx-5MS 30 m, 0.25 mm I.D., 0.25 μm thickness	EI-MS in SIM mode	Öllers et al., 2001
ibuprofen	GC				
clofibric acid	(analytes	helium gas			
ketoprofen	derivatized				
naproxen	with				
diclofenac	diazomethane)				
erythromycin sulfamethoxazole acetyl-sulfamethoxazole trimethoprim mefenamic acid lofepramine propranolol dextropropoxyphene diclofenac tamoxifen	HPLC	A: 40 mM ammonium acetate (adjust to pH 5.5 with fomic acid) B: methanol C: water	5μm, 250mm × 2mm Luna C18	positive ESI-/MS/MS in consecutive reaction monitoring (CRM) mode	Hilton and Thomas, 2003
clofibric acid				negative ESI-/MS/MS in CRM mode	
ibuprofen				negative ESI-/MS/MS in SIM mode	
paracetamol				positive ESI-/MS/MS in SIM mode	
clofibric acid ibuprofen acetaminophen caffeine chlorophen naproxen triclosan carbamazepine Estrone 17β-estradiol 17α-ethinylestradiol	GC (analytes derivatized with *N,O*-bis(trimethylsil y) trifluoroaceta mide (BSTFA)		HP-5MS 30 m, 0.25 mm I.D., 0.25 μm thickness	EI-MS in SIM mode	Zhang et al., 2007

Table 2.10. Instruments used for analysis of PPCPs in environmental samples (continued)

Analytes	Chromato-graphy	Mobile Phase	Column	Detection	References
benzoylecgonine codeine dihydrocodeine hydrocodone methadone morphine oxycodone tramadol bromazepam diazepam medazepam nordiazepam oxazepam temazepam-1 temazepam-2 carbamazepine primidone doxepin verapamil	HPLC	A: acetonitrile B: 10 mM ammonium formate in water (adjust to pH 4 with formic acid)	4 μm, 150 × 3 mm, Synergi Polar-RP 80 Å column	positive ESI- MS/MS in MRM mode	Hummel et al., 2006
triclosan 2,4-dichlorophenol 2,4,6-trichlorophenol	GC (analytes derivatized with N- methyl-N- (tert- butyldimethyl silyl)trifluoro acetamide (MTBSTFA)	helium gas	HP-5 MS 30m×0.2 5mm I.D., 0.25 μm thickness	EI-MS-MS in SIM mode	Morales et al., 2005
estrone ethynylestradiol estradiol estriol	GC (analytes derivatized with penta- fluorobenzoyl)	helium gas	HP-5 MS 15m×0.2 5mm I.D., 0.25 μm thickness	negative CI-MS in SIM mode (reagent gas: methane)	Xiao et al., 2001
	GC (analytes derivatized with pentafluorobe nzyl bromide (PFBBR) and N- trimethylsilyl imidazole (TMSI))		DB5- XLB 60m×0.2 5mm I.D., 0.25 μm thickness	negative CI-MS/MS in SIM mode (reagent gas: methane)	Fine et al., 2003

Table 2.10. Instruments used for analysis of PPCPs in environmental samples (continued)

Analytes	Chromato-graphy	Mobile Phase	Column	Detection	References
macrolides	HPLC	A: acetonitrile B: 20 mM ammonium acetate (0.05% formic acid, pH 5)	3µm, 2.1 × 50 mm Genesis C18 column	positive ESI-MS/MS in SRM mode	Miao et al., 2004
quinolones quinoxaline dioxides sulfonamides		A: acetonitrile B: 20 mM ammonium acetate (0.1% formic acid, pH 4.0)	3µm, 2.1 × 150 mm Genesis C18 column		
tetracyclines		A: acetonitrile B: 20 mM ammonium acetate (0.1% formic acid and 4 mM oxalic acid)	3µm, 2.1 × 50 mm Genesis C18 column		
phenazone, propyphenazone dimethyl-aminophenazone	GC (analytes derivatized with MTBSTFA)	helium gas	HP-5 MS 30m×0.2 5mm I.D., 0.25 µm thickness	EI-MS in SIM mode	Zühlke et al., 2004
synthetic fragrance compounds	GC	helium gas	DB-5 MS 15m×0.2 5mm I.D., 0.25 µm thickness	EI-MS in SIM mode	Simonich et al., 2000
			HP-5 30m×0.3 2mm I.D., 0.25 µm thickness	EI-MS in full scan mode	Zeng et al., 2005

Table 2.10. Instruments used for analysis of PPCPs in environmental samples (continued)

Analytes	Chromato-graphy	Mobile Phase	Column	Detection	References
octylphenol nonylphenols nonylphenol monoethoxylates nonylphenol diethoxylates	GC	helium gas	DB-5 MS 60m×0.3 2mm I.D., 0.25 μm thickness	EI-MS in SIM mode	La Guardia et al., 2001
PBDEs	GC	helium gas	Restek Rtx-CLPestic ides 30m×0.2 5mm I.D., 0.25 μm thickness	ECD or EI-MS SIM mode	Xia et al., 2008
perfluoro-alkanesulfonates sperfluoro-carboxylates	HPLC	A: 10 mM ammonium acetate in water B: 10 mM ammonium acetate in methanol	4 μm, 2.1 mm x 50 mm, Genesis C8	negative ESI-MS-MS in MRM mode	Moody et al., 2001

For MS coupled with HPLC, electrospray ionization (ESI) and atmosphere pressure chemical ionization (APCI) are commonly used for ionization of analytes that are introduced into the MS with HPLC eluent. Both ionization methods generate mainly "pseudo molecular ions" (loss or gain of a proton, or other adduct) and can be operated in positive and negative ion mode. Detailed description on mechanisms of ESI and APCI can be found elsewhere (Hoffmann and Stroobant, 2007; Watson and Sparkman, 2007). Because ESI transfers analyte ions from solution into the gas phase, whereas APCI ionizes analytes in the gas phase, analytes occurring as ions in solution may be best analyzed by ESI, while non-ionic analytes may be well suited for APCI (Reemtsma, 2003), The ionization-continuum diagram developed by Thurman et al (2001) as shown in Figure 2.1 illustrates that ESI works well in ionic regions and also works well on polar analytes where there is acidity or basicity in solution. Whereas APCI works on both polar and non-polar species but usually does not work on ionic species in solution.

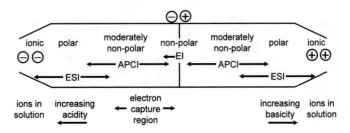

Figure 2.2 Ionization–continuum diagram showing the selection of atmospheric pressure chemical ionization (APCI), electrospray ionization (ESI), and electro impact (EI) ionization based on analyte chemical properties (Thruman et al., 2001). (permission from American Chemical Society).

To overcome complex matrices and to achieve trace level detection limits, tandem mass spectrometry is predominantly used for analysis of PPCPs in environmental samples (Table 2-10). Tandem mass spectrometry (often abbreviated as MS/MS) is a hyphenated technique involves coupling one mass spectrometer to a second. In this technique, the first spectrometer serves to isolate the molecular ions of interest. These ions are then introduced one at a time into a second mass spectrometer, where they are fragmented to give a series of mass spectra, which can provide sensitive qualitative and quantitative information of target analytes (Skoog and Leary, 1997).

2.3.2 Evaluating and Overcoming Matrix Effects

Due to the complex nature of environmental samples, the matrix effect is often encountered during analysis of trace levels of PPCPs. The matrix effect could be caused by those interfering components that can not be separated by an analytical column (Hao et al., 2007). Those co-eluted compounds can create overlapping peaks if detected by detectors that can not discriminate them or they may exhibit similar ions in the MS or MS/MS detection, resulting in false identification or overestimation of target compounds. This phenomenon is more common in GC-MS and GC/MS/MS analysis. The matrix effect could also be a result from the interactions between the target analytes and those co-extracted or co-eluted matrix compounds, resulting in signal suppression or less frequently enhancement due to their effect on ionization of target analytes. This kind of matrix effect can result in under- or over-quantitation of target analytes and is frequently observed in LC/MS and LC/MS/MS analysis (Hao et al., 2007).

Hernando et al. (2004) reported signal loss of up to 17% and 13% for LC-ESI-MS/MS analysis of clofibric acid and betaxolol, resepectively, in spiked river water. The signal loss went up to 60% and 49% for the two compounds, respectively, in spiked wastewater influent samples (Figure 2.3). Quintana and Reemtsma (2004) observed a clear tendency of decreasing signal suppression with increasing retention time for acidic PPCPs, a result of non-specific matrix effects of moderately polar matrix components. Löffler and Ternes (2003) observed 51%, 134%, and 106% signal enhancement in negative LC–APCI–MS/MS analysis of ketoprofen, naproxen and bezafibrate, respectively, in analytes spiked-sediment samples. Ohlenbusch et al. (2002) also demonstrated that Ca^{2+} in the sample can cause significant signal enhancement (300%) in direct infusion ESI-MS/MS analysis of naphthoic acids in groundwater. Ikonomou et al. (1990) suggested, among many factors contributing to the matrix effect, that organic compounds present in the sample in concentrations exceeding $10^{-5}M$ may compete with the analyte for access to the droplet surface for gas phase emission. Ohlenbusch et al. (2002) suggested that Ca^{2+} ions can increase the conductivity of a solution and, therefore, increase the formation of fine droplets during spray process, resulting in production of higher signal intensities.

To overcome the matrix effects during analysis of trace level of PPCPs in complicated environmental samples, first approach to take is to implement more selective and extensive extraction and clean-up procedures. A thorough extraction and clean-up before instrumental analysis would reduce the amount of matrix components that are introduced to instruments during analysis. For example, a hydrophobic matrix may be extracted by a two-step extraction procedure, in which a first SPE with C18 material under neutral conditions, while polar acidic analytes remain in the aqueous phase for a second extraction with a polar polymeric phase at acidic pH (Reemtsma, 2003). More selective sorbents such as immunosorbents (Pichon et al., 1999), molecular imprinted polymers (Gilar et al., 2001), and restricted access materials (Petrovic et al., 2002) have shown great promise as packing material for SPE during extraction and clean-up to improve the selectivity of target analytes. The disadvantage of extensive extraction and clean-up procedures is that it is time consuming and the possibility of losing analytes increases with increasing consecutive extraction and clean-up steps. In addition, this strategy can fail if the physico-chemical properties of the target analytes and the matrix components are similar.

Sample clean-up during chromatography by using a pre-analytical column in the chromatographic system or improved chromatographic separation efficiency on an analytical column can also be options to separate matrix components from target analytes (Pascoe et al., 2001; Petrovic et al., 2002). In GC/MS or GC/MS/MS analysis, using a non-interfered fragment ion or, when there are isomers or chiral compounds can sometime overcome the matrix effect. In LC/MS or LC/MS/MS, dilution of samples before injection or reduction of postcolumn eluent flow rate sometimes proved to be an effective approach to significantly reduce matrix effect in cases when the pre-concentration of matrix components during sample preparation magnified matrix effect (Hernando et al., 2004; Kloepfera et al., 2005). However, when target analytes are at trace levels, dilution of a sample many decrease analytical sensitivity and would not be a method of choice.

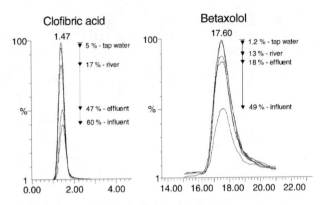

Figure 2.3. Extracted MRM chromatograms for clofibric acid and betaxolol in spiked aqueous samples showing signal suppression (%) in LC/MS/MS analysis (Hernando et al., 2004). With permission from *J. Chromatogr. A.*

If the above mentioned strategies fail to eliminate the matrix effect, several calibration techniques compensating for matrix effects should be applied: external matrix-matched standards (Pfeifer et al., 2002); isotopically labeled internal standards (Halden and Paull, 2004); standard addition calibration (Quintana and Reemtsma 2004); and postcolumn addition of internal standard (Kaufmann and Butcher, 2005).

External matrix matched calibration often uses standards with the same or similar matrix composition as the analyzed sample to establish a calibration curve to be used for calculation of an unknown sample (Pfeifer et al., 2002; Schlüsener et al., 2003; Miao et al., 2004). However, this approach requires the availability of uncontaminated sample matrix to be used and also assumes minimum matrix variation from sample to sample. In reality, obtaining uncontaminated sample might be impossible under certain circumstances and significant matrix variation among samples may often exist, especially for biological samples (Ito and Tsukada, 2001).

Isotopically labeled internal standards are used for correction of the matrix effect because they have almost the same chemical properties and the same retention times as non-labeled substances (Halden and Paull, 2004). Isotopically labeled target analytes are normally added to the samples immediately after sample collection. Using of isotopically labeled internal standards can not only compensate the losses of target analytes during extraction and clean-up steps but also correct matrix effect during analysis. However, their use is expensive, especially in a multicomponents analysis, where a separate internal standard for each analyte is required. Most often isotopically labeled analytes are not commercially available. One or few isotopically labeled compound with similar characteristics of a group target analytes is often used as surrogate/internal standards for multi-residue analysis of PPCPs in environmental samples (Kasprzyk-Hordern et al., 2007). However, this approach may not reliable compensate matrix effects for all analytes.

Standard addition calibration method can correct for sensitivity losses caused by matrix components, however, it cannot avoid a loss of sensitivity (Reemtsma, 2001). In standard addition calibration, the calibration standards are added to a sample solution under investigation at serial dilutions. Calibration curves are then prepared for the quantitation of the analytes in the sample (Quintana and Reemtsma 2004). This approach requires three to four analysis of each sample and each sample has to be calibrated and quantified separately. This method becomes impractical for monitoring projects in which large sets of samples are analyzed. To reduce the number of the analysis, two runs of analysis per sample (single-point standard addition) are sometimes applied instead of obtaining a calibration curve for each sample (Ito and Tsukada, 2001). The concentration of analyte in the sample is calculated using the following equation:

$$C_u = \frac{C_s I_u}{I_s - I_u}$$

where C_u is the concentration of target analyte in the sample solution, C_s is the concentration of target analyte spiked into the sample solution, I_u and I_s are the signal intensities of target analyte in the sample solution and in the spiked sample solution, respectively.

In the postcolumn addition method, an internal standard that is structurally similar to target compounds is injected at a constant flow rate into the effluent from the analytical column directly into MS detector (Kaufmann and Butcher, 2005). In this method, the response of a target anlyte is affected by matrix effect to the same extent as the spiked internal standard which appears at the same retention time, and hence variations of the spike response can be used to correct the peak area of an analyte. This method has been proven to be reliable and less labor intensive then other calibration methods (Kaufmann and Butcher, 2005).

2.4 Conclusion

The quality of data from detection of PPCPs at trace levels in environmental samples depends largely on proper sample handling and preparation, thorough sample extraction and clean-up to remove as much matrix interfering components as possible, and sensitive and robust analytical instruments. When analyzing environmental samples, matrix effect is the most challenging problem to overcome. More research is needed to develop selective stationary phases used in SPE and analytical columns. Careful selection of specific operating parameters, like the choice of mobile phase and type of additives may generate better separation between target analytes and matrix compounds and optimize ionization of target analytes at the MS detector. For large scale surveillance projects, the use of on-line extraction, clean-up, and detection systems would significantly increase the workflow and cut down the cost. Development of multi-residue analytical protocol that can qualify and quantify PPCPs as well as their metabolites will be essential for our understanding of the occurrence, fate, and environmental impacts those compounds.

References

Allchin, C. R., R. J. Law, and S. Morris. 1999. Polybrominated diphenyl ethers in sediments and biota downstream of potential sources in the UK. Environ. Pollut. 105:197-207.

ASTM. 1980. Standard Practice for Sampling Water. ASTM Annual Book of Standards, ASTM 1916 Race St., Philadelphia, PA 19103-1187.

Balmer, M. E., H. Buser, M. D. Müller, and T. Poiger. 2005. Occurrence of Some Organic UV Filters in Wastewater, in Surface Waters, and in Fish from Swiss Lakes. 39:953 – 962.

Berrueta, L. A., B. Gallo, and F. Vicente. 1995. A review of solid phase extraction: Basic principles and new developments. Chromatographia. 40:474-483.

Braekevelt, E., S. A. Tittlemier, and G. T. Tomy. 2003. Direct measurement of octanol-water partition coefficients of some environmentally relevant brominated diphenyl ether congeners. Chemosphere. 51: 563-567.

Buser, H., M. E. Balmer, P. Schmid, and M. Kohler. 2006. Occurrence of UV Filters 4-Methylbenzylidene Camphor and Octocrylene in Fish from Various Swiss Rivers with Inputs from Wastewater Treatment Plants. Enviorn. Sci. Technol. 40:1427 – 1431.

Castiglioni, S., R. Bagnati, D. Calamari, R. Fanelli, and E. Zuccato. 2005. A multiresidue analytical method using solid-phase extraction and high-pressure liquid chromatography tandem mass spectrometry to measure pharmaceuticals of different therapeutic classes in urban wastewaters. J. Chromatogr. A. 1092:206-215.

Covaci, A., S. Voorspoels, and J. de Boer . 2003. Determination of brominated flame retardants, with emphasis on polybrominated diphenyl ethers (PBDEs) in environmental and human samples—a review. Environ. Intern. 29:735-756.

Daughton, C. G, and T. A. Thernes. 1999. Pharmaceuticals and personal care products in the environment: Agents of subtle changes? Environ. Health Persp. 107:907-938.

Draisci, R., C. Marchiafava, E. Ferretti, L. Palleschi, G. Catellani, and A. Anastasio. 1998. Evaluation of musk contamination of freshwater fish in Italy by accelerated solvent extraction and gas chromatography with mass spectrometric detection. J. Chromatogr. A. 814:187-197.

EPA. 1996. Test Methods for Evaluating Solid Waste (SW-846). http://www.epa.gov/epaoswer/hazwaste/test/main.htm

EPA. 2000. Guidance for Assessing Chemical Contaminant Data for Use in Fish Advisories. Volume 1: Fish Sampling and Analysis. 3rd Ed. EPA 823-B-00-007.

EPA. 2007. Method 1694: Pharmaceuticals and personal-care productes in water, soil, sediment, and biosolids by HPLC/MS/MS. EPA Office of Water. EPA-821-R-08-002.

Ferrer, I., and E. T. Furlong. 2002. Accelerated solvent extraction followed by on-line solid-phase extraction coupled to ion trap LC/MS/MS for analysis of benzalkonium chlorides in sediment samples. Anal. Chem. 74:1275-1280.

Fine, D. D., G. P. Breidenbach, T. L. Price, and S. R. Hutchins. 2003. Quantitation of estrogens in ground water and swine lagoon samples using solid-phase extraction, pentafluorobenzyl/trimethylsilyl derivatizations and gas chromatography–negative ion chemical ionization tandem mass spectrometry. 1017:167-185.

Gilar, M., E. S. P. Bouvier, and B. J. Compton. 2001. Advances in sample preparation in electromigration, chromatographic and mass spectrometric separation methods. J. Chromatogr. A. 909:111-135.

Golet, E. M., A. Strehler, A. C. Alder, and W. Giger. 2002. Determination of Fluoroquinolone Antibacterial Agents in Sewage Sludge and Sludge-Treated Soil Using Accelerated Solvent Extraction Followed by Solid-Phase Extraction. Anal. Chem. 74:5455-5462.

Göbel, A., A. Thomsen, C. S. McArdell, A. C. Alder, W. Giger, N. Theiß, D. Löffler and T. A. Ternes. 2005. Extraction and determination of sulfonamides, macrolides, and trimethoprim in sewage sludge. J. Chromatogr. A. 1085: 179-189.

Halling-Sørensen, B, S. N. Nielsen, P. F. Lanzky, F. Ingerslev, H. C. H. Lηzhρft, and S. E. Jρgensen. 1998. Occurrence, fate and effects of pharmaceutical substances in the environment – A review. Chemosphere. 36:357-393.

Hao, C. X. Zhao, and P. Yang. 2007. GC-MS and HPLC-MS analysis of bioactive pharmaceuticals and personal-care products in environmental matrices. Trends in Anal. Chem. 26:569-580.

Hernando, M.D., M. Petrovic, A.R. Fernández-Alba, and D. Barceló. 2004. Analysis by liquid chromatography–electrospray ionization tandem mass spectrometry and acute toxicity evaluation for β-blockers and lipid-regulating agents in wastewater samples. J. Chromatogr. A. 1046:133-140.

Hilton, M. J., and K. V. Thomas. 2003. Determination of selected human pharmaceutical compounds in effluent and surface water samples by high-performance liquid chromatography–electrospray tandem mass spectrometry. J. Chromatogr. A. 1015:129-141.

Hoffmann, E., and V. Stroobant. 2007. Mass Spectrometry: Principles and Applications. 3rd ed. Wiley-Interscience.

Halden, R. U., and D. H. Paull. 2004. Analysis of Triclocarban in Aquatic Samples by Liquid Chromatography Electrospray Ionization Mass Spectrometry. Environ. Sci. Technol. 38:4849 – 4855.

Hu, J., F. Jin, Y. Wan, M. Yang, L. An, W. An, and S. Tao. 2005. Trophodynamic Behavior of 4-Nonylphenol and Nonylphenol Polyethoxylate in a Marine Aquatic Food Web from Bohai Bay, North China: Comparison to DDTs. Environ. Sci. Technol. 39:4801 – 4807.

Hummel, D., D. Löffler, G. Fink, and T. A. Ternes. 2006. Simultaneous Determination of Psychoactive Drugs and Their Metabolites in Aqueous Matrices by Liquid Chromatography Mass Spectrometry. Environ. Sci. Technol. 40:7321-7328.

Ikonomou, M. G., A. T. Blades, and P. Kebarle. 1990. Investigations of the electrospray interface for liquid chromatography/mass spectrometry. Anal. Chem. 62:957-967.

Ito, S., and K. Tsukada. 2001. Matrix effect and correction by standard addition in quantitative liquid chromatographic-mass spectrometric analysis of diarrhetic shellfish poisoing toxins. J. Chromatogr. A. 943:39-46.

Kasprzyk-Hordern, B., R. M. Dinsdale, and A. J. Guwy. 2007. Multi-residue method for the determination of basic/neutral pharmaceuticals and illicit drugs in surface water by solid-phase extraction and ultra performance liquid chromatography–positive electrospray ionisation tandem mass spectrometry. J. Chromatogr. A. 1161:132-145.

Kaufmann, A., and P. Butcher. 2005. Segmented post-column analyte addition; a concept for continuous response control of liquid chromatography/mass spectrometry peaks affected by signal suppression/enhancement. Rapid Comm. in Mass Spectro. 19:611 – 617.

Keller, H., K. Xia, and A. Bhandari. 2003. Occurrence and degradation of estrogenic nonylphenol and its precursors in Northeast Kansas wastewater treatment plants. Practice Per. Hazard. Toxic. Radioactive. Waste Manag. 7:203-213.

Kinney, C. A., E. T. Furlong, S. D. Zaugg, M. R. Burkhardt, S. L. Werner, J. D. Cahill, and G. R. Jorgensen. 2006. Survey of organic wastewater contaminants in biosolids destined for land application. Environ. Sci. Technol. 40:7207-7215.

Kloepfera, A., J. B. Quintanaa, T. Reemtsma. 2005. Operational options to reduce matrix effects in liquid chromatography–electrospray ionisation-mass spectrometry analysis of aqueous environmental samples. J. Chromatogr. A. 1067:153-160

Kolpin, D. W., Furlong, E. T., Meyer, M. T., Thurman, E. M., Zaugg, S. D., Barber, L. B., and Buxton, H. T. 2002. Pharmaceuticals, hormones, and other organic wastewater contaminants in U.S. streams, 1999-2000: a national reconnaissance. Environ. Sci. Technol. 36:1202-1211.

La Guardia, M. J., R. C. Hale, E. Harvey, and T. M. Mainor. 2001. Alkylphenol ethoxylate degradation products in land-applied sewage sludge (biosolids). Environ. Sci. Technol. 35:4798-4804

Lambropoulou, D. A., and T. A. Albanis. 2007. Methods of sample preparation for determination of pesticide residues in food matrices by chromatography-mass spectrometry-based techniques: a review. Anal Bioanal. Chem. 389:1663-1683.

Lee, H. B., T. E. Peart, and M. L. Svoboda. 2005. Determination of endocrine-disrupting phenols, acidic pharmaceuticals, and personal-care products in sewage by solid-phase extraction and gas chromatography–mass spectrometry. J. Chrom. A. 1094:122-129.

Lindqvist, N., T. Tuhkanen, and L. Kronberg. 2005. Occurrence of acidic pharmaceuticals in raw and treated sewages and in receiving waters. Water. Res. 39:2219-2228.

Liu, R., J. L. Zhou and A. Wilding. 2004. Microwave-assisted extraction followed by gas chromatography–mass spectrometry for the determination of endocrine disrupting chemicals in river sediments. J. Chrom. A. 1038:19-26.

Löffler, D., and T. A. Ternes. 2003. Determination of acidic pharmaceuticals, antibiotics and ivermectin in river sediment using liquid chromatography–tandem mass spectrometry. J. Chromatogr. A. 1021:133-144.

Miao, X. S., F. Bishay, M. Chen, and C. D. Metcalfe. 2004. Occurrence of Antimicrobials in the Final Effluents of Wastewater Treatment Plants in Canada. Environ. Sci. Technol. 38:3533-3541.

McArdell, C. S., Alder, A. C., Göbel, A., Löffler, D., Suter, M. J. F., and Terns, T. A. (2007) *in* Human Pharmaceuticals, Hormones and Fragrances: The Challenge of Micropollutants in Urban Water Management. Editor(s): Thomas Ternes, Adriano Joss. IWA Publishing. New York.

Mikaela, N., and M. M. Pilar. 2005. Development and evaluation of C18 and immunosorbent solid-phase extraction methods prior immunochemical analysis of chlorophenols in human urine. Anal. Chim. Acta. 533:67-82.

Moody, C. A., W. C. Kwan, J. W. Martin, D. C. G. Muir, and S. A. Mabury. 2001. Determination of Perfluorinated Surfactants in Surface Water Samples by Two Independent Analytical Techniques: Liquid Chromatography/Tandem Mass Spectrometry and 19F NMR. Anal. Chem. 73:2200 – 2206.

Morales, S., P. Canosa, I. Rodríguez, E. Rubí and R. Cela. 2005. Microwave assisted extraction followed by gas chromatography with tandem mass spectrometry for the determination of triclosan and two related chlorophenols in sludge and sediments. J. Chromgrah. A. 1082:128-135.

Nakata, H. 2005. Occurrence of Synthetic Musk Fragrances in Marine Mammals and Sharks from Japanese Coastal Waters. Environ. Sci. Technol. 39:3430 – 3434.

Nikolai, L. N., E. L. McClure, S. L. MacLeod, and C. S. Wong. 2006. Stereoisomer quantification of the blocker drugs atenolol, metoprolol, and propranolol in wastewaters by chiral high-performance liquid chromatography–tandem mass spectrometry. J. Chromatogr. A. 1131:103-109.

Ohlenbusch, G., C. Zwiener, R. U. Meckenstock, and F. H. Frimmel. 2002. Identification and quantification of polar naphthalene derivatives in contaminated groundwater of a former gas plant site by liquid chromatography–electrospray ionization tandem mass spectrometry. J. Chromatogr. A. 967:201-207.

Öllers, S., H. P. Singer, P. Fässler, and S. R. Müller. 2001. Simultaneous quantification of neutral and acidic pharmaceuticals and pesticides at the low-ng/l level in surface and waste water. J. Chromatogr. A. 911:225-234.

Osemwengie, L. I. and S. Steinberg. 2003. Closed-loop stripping analysis of synthetic musk compounds from fish tissues with measurement by gas chromatography–mass spectrometry with selected-ion monitoring. J. Chromatogr. A. 993:1-15.

Pascoe, R., J. P. Foley, and A. I. Gusev. 2001. Reduction in Matrix-Related Signal Suppression Effects in Electrospray Ionization Mass Spectrometry Using On-Line Two-Dimensional Liquid Chromatography. Anal. Chem. 73: 6014 – 6023.

Petrovic, M., S. Tavazzi, and D. Barcelo. 2002. Column-switching system with restricted access pre-column packing for an integrated sample cleanup and liquid chromatographic–mass spectrometric analysis of alkylphenolic compounds and steroid sex hormones in sediment. J. Chromatogr. A. 971:37-45.

Petrović, M., M. D. Hernando, M. S. Díaz-Cruz, and D. Barceló. 2005. Liquid chromatography–tandem mass spectrometry for the analysis of pharmaceutical residues in environmental samples: a review. J. chromatogr. A. 1067:1-14.

Peck, A. 2006. Analytical methods for the determination of persistent ingredients of personal care products in environmental matrices. Anal. Bioanal. Chem. 386, 907-939.

Pfeifer, T., J. Tuerk, K. Bester, and M. Spiteller. 2002. Determination of selected sulfonamide antibiotics and trimethoprim in manure by electrospray and atmospheric pressure chemical ionization tandem mass spectrometry. Rapid. Commu. in Mass Spectro.16:663-669.

Pichon, V., M. Bouzige, C. Miège, and M. Hennion. 1999. Immunosorbents: natural molecular recognition materials for sample preparation of complex environmental matrices. TrAC-Trends in Anal. Chem. 18:219-235.

Quintana, J. B., and T. Reemtsma. 2004. Sensitive determination of acidic drugs and triclosan in surface and wastewater by ion-pair reverse-phase liquid chromatography/tandem mass spectrometry. Rapid. Comm. Mass Sectro. 18:765-774.

Ramirez, A. J., M. A. Mottaleb, B. W. Brooks, and C. K. Chambliss. 2007. Analysis of Pharmaceuticals in Fish Using Liquid Chromatography-Tandem Mass Spectrometry. Anal. Chem. 79, 3155-3163.

Reemtsma, T. 2001. The use of liquid chromatography-atomspheric pressure ionization-mass spectrometry in water analysis – Part II: Obstacles. Trends in Analy. Chem. 20:533-542.

Reemtsma, T. 2003. Liquid chromatography-mass spectrometry and strategies for trace-level analysis of polar organic pollutants. J. Chromatogr. A. 1000:477-501.

Restek. 2008. Restek GC column selection guide. http://www.restek.com/guide_gccolsel_sect2.asp (accessed April 2008).

Rice, S. L., and S. Mitra. 2007. Microwave-assisted solvent extraction of solid matrices and subsequent detection of pharmaceuticals and personal care products (PPCPs) using gas chromatography–mass spectrometry. Analytica Chimica Acta. 589:125-132.

Richter, B. E., B. A. Jones, J. L. Ezzell, N. L. Porter, N. Avdalovic, and C. Pohl. 1996. Accelerated Solvent Extraction: A technique for sample preparation. Anal. Chem. 68:1033-1039.

Saunders, D. L. 1975. In Chromatography : a laboratory handbook of chromatographic and electrophoretic methods. E. Heftmann, ed. 3rd ed. Van Nostrand Reinhold, New York. pp81.

Schlüsener, M. P., M. Spiteller, and K. Bester. 2003. Determination of antibiotics from soil by pressurized liquid extraction and liquid chromatography–tandem mass spectrometry. J. Chromatogr. A. 1003:21-28.

Simonich, S. L., W. M. Begley, G. Debaere, and W. S. Eckhoff. 2000. Trace analysis of fragrance materials in wastewater and treated wastewater. Environ. Sci. Technol. 34:959-965.

Skoog, D. A., and J. J. Leary. 1997. Principles of Instrumental Analysis. 5th ed. Brooks Cole.

Stubbings, G., J. Tarbin, A. Cooper, M. Sharman, T. Bigwood, and P. Robb. 2005. A multi-residue cation-exchange clean up procedure for basic drugs in produce of animal origin. Anal. Chim. Acta. 547:262-268.

Supelco. 1998. Guide to solid phase extraction. http://www.sigmaaldrich.com/ Graphics/Supelco/objects/4600/4538.pdf (accessed on March 3 2008).

Ternes, T. A., M. Bonerz, and T. Schmidt. 2001. Determination of neutral pharmaceuticals in wastewater and rivers by liquid chromatography–electrospray tandem mass spectrometry. J. Chromatogr. A. 938:175-185.

Ternes, t. A., M. Bonerz, N. Herrmann, D. Löffler, E. Keller, B. B. Lacida and A. C. Alder. 2005. Determination of pharmaceuticals, iodinated contrast media and musk fragrances in sludge by LC tandem MS and GC/MS. J. Chromatogr. A. 1067:213-223.

Thurman, E. M., I. Ferrer, and D. Barceló. 2001. Choosing between atmospheric pressure chemical ionization and electrospray ioninaton interfaces for the HPLC/MS analysis of pesticides. Anal. Chem. 73:5441-5449.

Vieno, N. M., T. Tuhkanen, and L. Kronberg. 2005. Seasonal variation in the occurrence of pharmaceuticals in effluents from a sewage treatment plant and in the recipient water. Environ. Sci. Technol. 39:8220-8226.

Vieno, N. M., T. Tuhkanen, and L. Kronberg. 2006. Analysis of neutral and basic pharmaceuticals in sewage treatment plants and in recipient rivers using solid phase extraction and liquid chromatography–tandem mass spectrometry detection. J. Chromatogr. A. 1134:101-111.

Watson J. T., and O. D. Sparkman. 2007. Introduction to Mass Spectrometry: Instrumentation, Applications, and Strategies for Data Interpretation. 4th ed. Wiley.

Xia, K., M. B. Luo, C. Lusk, K. Armbrust, L. Skinner, R. Sloan. 2008. Polybrominated diphenyl ethers (PBDEs) in biota representing different trophic levels of the Hudson River, New York: from 1999 to 2005. Environ. Sci. Technol. (in press).

Xiao, X., D. V. McCalley, J. McEvoy. 2001. Analysis of estrogens in river water and effluents using solid-phase extraction and gas chromatography–negative chemical ionisation mass spectrometry of the pentafluorobenzoyl derivatives. 923:195-204.

Zeng, X., G. Sheng, Y. Xiong, and J. Fu. 2005. Determination of polycyclic musks in sewage sludge from Guangdong, China using GC–EI-MS. Chemosphere. 60:817-823.

Zhang, S., Q. Zhang, S. Darisaw, O. Ehie, and G. Wang. Simultaneous quantification of polycyclic aromatic hydrocarbons (PAHs), polychlorinated biphenyls (PCBs), and pharmaceuticals and personal care products (PPCPs) in Mississippi river water, in New Orleans, Louisiana, USA. Chemosphere. 66:1057-1069.

Zrostlíková, J., J. Hajlová, J. Poustka, and P. Begany. 2002. Alternative calibration approaches to compensate the effect of co-extracted matrix components in liquid chromatography–electrospray ionisation tandem mass spectrometry analysis of pesticide residues in plant materials. J. Chromatogr. A. 973:13-26.

Zühlke, S., U. Dünnbier, T. Heberer. 2004. Detection and identification of phenazone-type drugs and their microbial metabolites in ground and drinking water applying solid-phase extraction and gas chromatography with mass spectrometric detection. 1050:201-209.

CHAPTER 3

Pharmaceuticals

Craig D. Adams

3.1   Introduction

Many pharmaceuticals enter the environment as a result of human use, veterinary use, and through industrial discharges. Pharmaceuticals are, by definition and design, biological active molecules that can have profound influence on human and other animal biochemical processes. Risks posed by these pharmaceuticals are largely uncertain.

Sources of pharmaceuticals in the environment include municipal sewage outfalls, hospital wastewater, pharmaceutical industry discharges, and run-off from animal agriculture. Pharmaceuticals emanating from animal agriculture are primarily antibiotics and hormonal compounds (covered in detail in Chapters 5 and 6, respectively). Additional sources include distributed sewage treatment systems.

Pharmaceuticals from all of these sources may be transported through the environment into groundwater or surface water used as a drinking water supply, thereby posing a potential risk to humans (Lindberg *et al.*, 2007). Further, pharmaceuticals may pose a significant risk in the environment, for example, due to hormonal effects or by supplying pressure to formation of antibiotic resistance (Kummerer, 2003).

To estimate and understand the risks associated with a particular pharmaceutical requires knowledge of toxicity and exposure. The toxicity of various pharmaceuticals at trace (e.g., sub-μg/L, or ng/L) concentrations is the subject of study. The second component of risk involves exposure of humans and of animals in the environment, to particular pharmaceuticals. Exposure assessment has many facets including generation and discharge from a variety of sources, fate in sewage treatment plants, transport and degradation in the environment, and removal in drinking water treatment. Various studies and analyses have been conducted to estimate each of these facets. Estimates of behaviors (e.g., sorption to biosolids) can be made from simple parameters such as K_{OW}, though studies (discussed below) show that such a simple approach often is insufficient to accurately predict physicochemical behaviors of pharmaceuticals.

Endocrine disruption effects of hormonal pharmaceuticals, due to their high potency at extremely low concentrations, are of particular concern for humans and animals. Antibiotics are also especially important because of their potential to form and promote antibiotic-resistance for human pathogens, and their potential to significantly impact natural microbial consortia (Kümmerer, 2001). Cytostatics are important due to their potential for mutagenicity, carcinogenicity, and embryotoxicity, as well as their heavy metal (platinum) content (Kümmerer, 2001). Other classes of pharmaceuticals, such as analgesics and psychopharmacologicals, may also be highly important due to their strength and common use.

Relatively few detailed studies have been conducted on the fate, effects, and occurrence of pharmaceuticals in the environment and in treatment processes, as compared with pesticides, for example, which have been studied extensively. Using computational software based on chemical structures to predict environmental behavior of pharmaceuticals, therefore, has great value in that a comprehensive set of data can be generated. However, such estimates must be used with caution, and primarily for screening purposes unless a correlation has been validated for a given compound or class of compounds. Detailed studies at laboratory-, pilot- and full-scale are needed to fully assess the fate of pharmaceuticals in natural and engineered systems, and to predict exposure of both humans and aquatic organisms.

Pharmaceuticals are generally classified by their therapeutic use. For humans, key classes of pharmaceuticals include:

- Analgesics
- Antibiotics and antimicrobials
- Anticonvulsant/antiepileptics
- Antidiabetics
- Antihystamines
- Antipsychotics, antidepressants, and antianxiety drugs
- Beta-blockers (β-blockers)
- Cytostatics and antineoplastics
- Estrogens and hormonal compounds
- Lipid-regulators
- Stimulants
- X-ray contrast media

Analgesics are pain-relief drugs that include narcotic analgesics, non-narcotic analgesics, and non-steroidal anti-inflammatory drugs (NSAID). Common nonnarcotic analgesics include acetaminophen and aspirin. Narcotic analgesics include codeine, methadone, morphine, and oxycodone. Common NSAIDs include diclofenac, fenoprofen, ketoprofen, mefenamic acid, indomethacin, naproxen, and ibuprofen.

Antibiotics and *antimicrobials* are used to fight bacterially-related disease. Major classes of antibiotics include fluoroquinolones (e.g., sarafloxacin, enrofloxacin, norfloxacis, ciprofloxacin), sulfonamides (e.g., sulfamethizole, sulfamerazine, sulfamethazine), macrolides (e.g., erythromycin, azithromycin, roxithromycin, clarithromycin, tylosin), tetracyclines (e.g., tetracycline, oxytetracytcline, chlorotetracycline, doxycycline), fenicoles (e.g., chloramphenicol, florfenicol, thiamphenicol), nitroimidazoles (e.g., benznidazole, metronidazole), β-lactams (e.g., penicillins, cefradine, carbapenems, cephalosporins), and aminoglycosides (e.g., streptomycin) (Ikehata *et al.*, 2006). Due to the special environmental importance of this class of pharmaceutical, antibiotics are covered in detail in this book in Chapter 5.

Anticonvulsants (including *antiepileptics*) are used to control seizures. Important antiepileptics include dilantin, primidone, and carbamazepine (Ikehata *et al.*, 2006).

Antidiabetics help control blood sugar levels. A commonly used antidiabetics is metformin (Kolpin *et al.*, 2002).

Antihystamines are used to block the histamine production in the body. Common examples include cimetidine and ranitidine.

Antipsychotics, antidepressants, and *antianxiety drugs* are used to treat a wide range of psychotic and depression disorders. Common examples include paroxetine and fluoxetine buspirone, meprobamate, and diazepam (Kolpin *et al.*, 2002; Westerhoff *et al.*, 2005).

Beta-blockers (β-blockers) are used for treating cardiovascular disease including hypertension, arrhythmias, and coronary artery disease by blocking the physiological effects of norepinephrine and epinephrine at the β-adrenargic receptors in the human body (Ikehata *et al.*, 2006). Common *β-blockers* include atenolol, celiprolol, metoprolol, propranolol, sotalol (Ikehata *et al.*, 2006).

Cytostatics and *antineoplastics* are used for treating cancers. Common examples include anthracyclines (e.g., aclarubicin, daunorubicin, doxorubicin, epirubicin, belomycin, idarubicin, pirarubicin), antimetabolites (e.g., azathioprine, cytarubine, 5-fluorouracil, methotrexate), and alkylating agents (cyclophosphamide, ifosfamid, and melphalan) (Ikehata *et al.*, 2006).

Estrogens and hormonal compounds are an especially important class of pharmaceuticals due to their common use and powerful impact on the human and other animal endocrine systems impacting general development, sexual development, and metabolism. The most important natural estrogens include estriol (E3), estradiol (E2), and estrone (E1) which are excreted from humans and other animals. A common synthetic estrogen is 17α-ethinylestradiol (EE2) which is used for contraception by women. Other natural hormones include progesterone and testosterone (Kolpin *et al.* 2002). Hormonal compounds are covered in detail in

Chapter 6. In addition to estrogens and natural hormones, many xenobiotic chemicals also affect the endocrine systems of humans and other animals. These endocrine disrupting chemicals (EDCs) include phthalates, bisphenol A, alkylphenol ethoxylates, chloro-*s*-triazine herbicides, and other compounds. Various EDCs are covered in detail in Chapters 4, 6, and 7 in this book.

Lipid regulators are used to control disorders such as hypercholesterolemia. Common lipid regulators include bezafibrate, clofibrate, fenofibrate, and gemfibrozil (Ikehata *et al.*, 2006).

Stimulants may be narcotic or non-narcotic. The most common stimulant is caffeine and is commonly found in surface waters (Kolpin *et al.*, 2002).

X-ray contrast media are non-therapeutic, however, and are used to allow x-ray visualization of structures within the human body. These compounds include diatrizoate, iomeprol, iopamidol, iopentol, iothalamic acid, and ioxithalamic acid (Ikehata *et al.*, 2006).

3.2   Properties

Important processes for pharmaceuticals in the environment include partitioning onto soils and sediments, hydrolysis, and biodegradation. For sewage and drinking water treatment, key partitioning processes include sorption to biosolids and to activated carbon, respectively. Important transformation processes are hydrolysis, and chemical oxidation by chlorine, ozone, or chlorine dioxide.

Table 3.1 lists selected properties for 106 pharmaceuticals (and natural hormones) organized into thirteen. Some natural hormones (e.g., estrone (E1), estradiol (E2), and estriol (E3)) were included in the table (and the discussions in this chapter) due to their importance and for comparison with synthetic pharmaceutical hormonal compounds (e.g., 17α-ethinylestradiol (EE2), and mestranol). Log K_{OW} values (which describe the lipophilic/hydrophilic balance) are also provided in Table 3.1 along with CAS number, molecular weight, aqueous solubility (estimated using EPI Suite (v. 3.2)), and acid dissociation constants (pK_a).

Log K_{OW} values in Table 3.1 are both calculated values (by the author using KOWWIN (v. 1.65)), and experimental values (when available) listed in KOWWIN. Of the 78 compounds for which experimental log K_{OW} values were included in KOWWIN, 66 (or 84%) of the log K_{OW} values were within ±0.5 units. The greatest deviations were for oxytetracycline ($\Delta_{exp-calc}=1.97$) and procarbazine ($\Delta_{exp-calc}=-1.21$). Both computational and experimental values are subject to error which may be due to unusual structural features or to complex chemistry (e.g., strong complexation to metals as with tetracyclines), respectively.

Table 3.1 Properties of selected pharmaceuticals (and selected natural hormones) organized by class.

	CAS	MWt	pK	KOWWIN (v3.06) log KOW (exp)	KOWWIN (v3.06) log KOW (calc) ++	WSKOWWIN v1.36 Aqueous solubility (mg/L)
ANTIBIOTICS						
fluoroquinolones						
sarafloxacin	98105-99-8	385.36	5.6, 8.2 †			
enrofloxacin	93106-60-6	359.40	3.85, 6.19, 7.59, 9.86 ‡		0.7	
moxifloxacin	151096-09-2		6.4, 9.5 †			
norfloxacin	70458-96-7	319.34	3.11, 6.10, 8.60, 10.56 ‡	-1.03	-0.31	1.78E+05
ofloxacin	82419-36-1	361.38	6.0, 8.2 †	-0.39 +	-0.2	2.83E+04
ciprofloxacin	93107-08-5	331.35	3.01, 6.14, 8.70, 10.58 ‡			1.15E+04
sulfonamides						
sulfamethizole	144-82-1	270.33	1.86, 5.29 ‡	0.54 +	0.41 ++	6292
sulfamerazine	127-58-2	264.30	2.06, 6.9 ‡	0.14 +	0.21 ++	1.49E+04
sulfamethazine	1981-58-4	278.33	2.07, 7.49 ‡		1.23 ++	1466
sulfadimethoxine	122-11-2	310.33	2.13, 6.08 ‡	1.63 +	1.17 ++	433.1
sulfamethoxazole	723-46-6	253.28	2.1, 5.7 ξ	0.89 +	0.48 ++	3942
sulfapyridine	144-83-2	249.29		0.35 +	0.53 ++	1.20E+04
sulfathiazole	72-14-0	255.31		0.05 +	0.72 ++	2.00E+04
macrolides						
erythromycin (H₂O)	114-07-8	733.95	8.9 ‡	3.06 *	2.48 ++	0.5168
azithromycin	83905-01-5	749.00	8.7, 9.5 †	4.02 *	3.24 ++	0.06204
roxithromycin	80214-83-1	837.07	9.17 ‡		2.75 ++	0.01887
clarithromycin	81103-11-9	747.97	8.9 †	3.16 *	3.18 ++	0.342
tylosin	1401-69-0	916.12	7.1, 7.73 †	1.63 *	1.05 ++	0.5065
tetracyclines						
tetracycline	60-54-8	444.44		-1.3 *	-1.33 ++	3877
oxytetracycline	79-57-2	460.44	3.3, 7.3, 9.1, †	-0.9 *	-2.87 ++	1399
chlortetracycline	64-72-2	450.88	3.33, 7.55, 9.33 ‡		-3.6 ++	8.49E+04
demeclocycline	64-73-3	450.88	3.37, 7.36, 9.44 ‡		-3.6 ++	8.50E+04
doxycycline	10592-13-9	414.42	3.02, 7.97, 9.15 ‡		-2.08 ++	7275
minocycline	10118-90-8	457.49	2.8, 5.0, 7.8, 9.5 †	0.05 **	-0.42 ++	59.3
fenicoles						
chloramphenicol	56-75-7	323.13	5.5 ‡	1.14 +	0.92 ++	388
thiamphenicol	15318-45-3	356.22		-0.27 +	-0.33 ++	9660
nitroimidazoles						
benznidazole	22994-85-0	260.25		0.91 **	1.22 ++	1411
metronidazole	443-48-1	171.16	2.5 †	-0.02 +	0 ++	2.57E+04
β-lactams						
penicillins						
amoxicillin	26787-78-0	365.41	2.4, 7.4, 9.6 †	0.87 **	0.97 ++	3433
clavulanic acid	58001-44-8	199.16			-2.04 ++	1.00E+06
cefradine	38821-53-3	349.41				2805
aminoglycosides						
streptomycin	57-92-1	581.58			-7.53 ++	1.00E+06
amikacin	37517-28-5	585.61	6.7, 8.4, 9.7 †		-8.78 ++	1.00E+06
other						
triclosan	3380-34-5	289.55	8, 7.9 ξ	4.76 #	4.66 ++	4.62
trimethoprim	738-70-5	290.32	3.23, 6.76 ‡	0.91 +	0.73 ++	2334
trimethoprim						
vancomycin	1404-90-6	1449.28	7.1 ‡		-0.84 ++	1.90E-05
lincomycin	859-18-7	406.54		0.56 +	0.29 ++	927
spectinomycin	21736-83-4	332.36	6.8, 8.8		-0.82 ++	4.05E+05
tinidazole	19387-91-8	247.27		-0.35 +	0.01 ++	1.99E+04

Table 3.1 (cont'd)

					KOWWIN (v3.06)				WSKOWWIN v1.36
	CAS	MWt	pK		log KOW (exp)		log KOW (calc)	++	Aqueous solubility (mg/L)
HORMONAL COMPOUNDS									
estriol (E3)	50-27-1	288.39	10.4	ξ	2.45	+	2.81	++	441
estradiol (E2)	50-28-2	272.39	10.4	ξ	4.01	+	3.94	++	82
estrone (E1)	53-16-7	270.37	10.3	ξ	3.13	+	3.43	++	146.8
17α-ethinylestradiol (EE2)	57-63-6	296.41	10.5	ξ	3.67	+	4.12	++	116.4
mestranol	72-33-3	310.44					4.68	++	3.5
19-norethisterone	68-22-4	298.43			2.97	+	2.99	++	118
equilenin	517-09-9	266.34					3.93	++	32.2
equilin	474-86-2	268.36					3.35	++	98.5
progesterone	57-83-0	314.47		ξ	3.87	+	3.67	++	5
testosterone	58-22-0	288.43	17.4	ξ	3.32	+	3.27	++	67.8
cis-androsterone	53-41-8	290.45			3.69	+	3.07	++	31.9
ANALGESICS									
nonnarcotic analgesics									
acetaminophen	103-90-2	151.17	9.7	ξ	0.46	**	0.27	++	3.03E+04
aspirin	50-78-2	180.16	3.5	†	1.19	+	1.13	++	5296
narcotic analgesics									
codeine	76-57-3	299.37	8.2, 10.6	†	1.19	##	1.28	++	1.22E+04
methadone	76-99-3	309.46			3.93	+	4.17	++	48.48
morphine	57-27-2	285.35			0.89	##	0.72	++	2.64E+04
oxycodone	76-42-6	315.37					0.66	++	2.76E+04
non-steroidal anti-inflamatory drug (NSAID)									
aminopyrine	58-15-1	231.30	5	†	1	+	0.6	++	4191
diclofenac	15307-79-6	318.14	4.2	ξ	0.7	**	0.57	++	2425
fenoprofen	31879-05-7	242.28	4.5	†			3.9	++	30.1
ketoprofen	22071-15-4	254.29	4.5	†	3.12	**	3	++	120
mefenamic acid	61-68-7	241.29			5.12	+	5.28	++	1.12
indomethacin	53-86-1	357.80	4.5	†	4.27	+	4.23	++	3.11
naproxen	22204-53-1	230.27	4.5	ξ	3.18	+	3.1	++	145
ibuprofen	15687-27-1	206.29	4.5	ξ	3.97	##	3.79	++	41.1
CYTOSTATICS/ANTINEOPLASTICS									
anthracyclines									
aclarubicin	57576-44-0	497.51							35.6
antimetabolites									
azathioprine	446-86-6	277.26	8.2		0.1	+	-0.09	++	272
cytarubine	147-94-4	243.22	4.3, 4.22	†	-2.51	+	-2.46	++	1.76E+05
5-fluorouracil	51-21-8	130.08			-0.89	+	-0.81	++	2.59E+04
methotrexate	59-05-2	454.45	3.8, 4.8, 5.6	†	-1.85	+	-1.28	++	2600
alkylating agents									
busulfan	55-98-1	246.29	- -	†	-0.52	+	-0.68	++	6.90E+04
chlorambucil	305-03-3	304.22	5.8	†			4.44	++	4.61
cyclophosphamide	50-18-0	261.09			0.63	+	0.97	++	5943
ifosfamid	3778-73-2	261.09			0.86	+	0.97	++	3781
melphalan	148-82-3	305.20					0.39	++	45.7
other									
carboplatin	41575-94-4	369.24					-0.46	++	1.17E+04
dactinomycin	50-76-0	1255.45	11.9	†			-0.91	++	0.00167
procarbazine	671-16-9	221.30	6.8, 8.8	†	0.08	+	1.29	++	1418
pyrimethamine	58-14-0	248.72	7.3	†	2.69	+	2.41	++	121.3

Table 3.1 (cont'd)

	CAS	MWt	pK			KOWWIN (v3.06) log KOW (exp)			KOWWIN (v3.06) log KOW (calc) ++		WSKOWWIN v1.36 Aqueous solubility (mg/L)
ANTICONVULSANTS/ANTIEPILEPTICS											
gabapentin	60142-96-3	171.24	3.68, 10.7	†	-1.1	**	-1.37	++			4491
primidone	125-33-7	218.26			0.91	+	0.73				5874
carbamazepine	298-46-4	236.28	<2	ξ	2.45	θ	2.25	++			17.66
oxcarbazepine	28721-07-5	252.27	10.7	†			1.11				202.8
phenytoin	57-41-0	252.27	8.3	†	2.47	+	2.16				178.6
valproate	99-66-1	144.22	4.6	†	2.75	**	2.96				894.6
BETA-BLOCKERS (β-BLOCKERS)											
acetbutotol	37517-30-9	336,43	9.2	†	1.71	+	1.19				259
atenolol	29122-68-7	266.34	9.6	†	0.16	+	-0.03				685.2
betaxolol	63659-18-7	307.44			2.81	θθ	2.98				450.7
celiprolol	56980-93-9	379.50			1.92	**	1.93				93.92
metoprolol	37350-58-6	267.37	--		1.88	+	1.69				4777
nadolol	42200-33-9	309.41	9.67	†	0.81	**	1.17				2.24E+04
propranolol	525-66-6	259.35	9.5	†	3.48	##	2.6				228
sotalol	3930-20-9	272.36	8.2, 9.8	†	0.24	‡‡	0.37				5513
ANTIHYSTAMINES											
cimetidine	51481-61-9	252.34	6.8		0.57	+	0.4				1.05E+04
ranitidine	66357-35-5	314.41			0.27	+	0.29				2.47E+04
ANTI-DIABETICS											
metformin	657-24-9	129.17					-2.64				1.00E+06
LIPID-REGULATORS											
bezafibrate	41859-67-0	361.83	3.6	†			4.25				1.22
clofibrate	882-09-7	214.65	3	†	2.57	+	2.84				582.5
fenofibrate	42017-89-0	318.76					4				9.11
gemfibrozil	25812-30-0	250.34	4.7	ξ			4.77				4.96
atorvastatin	134523-00-5	558,66	4.46	†			6.36				0.00112
ANTIPSYCHOTICS/ANTIDEPRESSANTS											
paroxetine	78246-49-8	365.84					2.1				79.91
fluoxetine	54910-89-3	255.36			4.05	††	4.23				78.24
buspirone	36505-84-7	385.51			2.63	αα	12.31				21.35
meprobamate	57-53-4	218.25	<2	ξ	0.7	+	0.98				8877
diazepam	439-14-5	284.75	2.4, 1.5	ξ	2.82		2.7				58.78
X-RAY CONTRAST MEDIA											
acetrizoic acid	85-36-9	556.87					2.26				3.62
diatrizoate	117-96-4	613.92	1, 3.4	†			1.37				8.885
iopromide	73334-07-3	791.12	<2,>13	†							23.75
STIMULANTS											
caffeine	58-08-2	194.19	6.1	ξ	-0.07	+	0.16				2632
ILLICIT DRUGS											
heroine (diacetylmorphine)	561-27-3	369.42			1.58	##	1.8				2152
methamphetamine	537-46-2	149.24			2.07	+	2.22				1.33E+04
LSD	50-37-3	323.44			2.95	+	2.26				2.098
cocaine	50-36-2	303.36			2.3	+	2.17				1298

†† Adlard et al. (1995) from KOWWIN
Avdeef et al., (1996) from KOWWIN (v. 3.06)
‡‡ Burgot (1990) from KOWWIN (v. 3.06)
θ Dal Pozzo et al. (1989) from KOWWIN (v. 3.06)
+ Hansch et al. (1995) from KOWWIN (v. 3.06)
θθ Recanatini (1992) from KOWWIN (v. 3.06)

αα Takacs-Novak (1995)
‡ Qiang and Adams (2004)
† Ternes and Joss (2006)
ξ Vanderford and Snyder (2006)
** Sangster (1993) from KOWWIN (v. 3.06)
* McFarland et al. (1997) from KOWWIN (v. 3.06)
++ KOWWIN v. 3.06

Experimental log K_{OW} values are shown in rank order in Figure 3.1. Hormonal compounds, analgesics and lipid-regulators dominated the highest hydrophilic/lipophilic ratio group ($3 < \log K_{OW} < 12$) (Table 2). While distributed over a wider range, compounds with the lowest log K_{OW} values (i.e., $-9 < \log K_{OW} < 1$) include cytostatics/antineoplastics, antidiabetics, antihystamines, illicit drugs, and stimulants (Table 3.2).

Log K_{OW} is a valuable parameter often used to predict or interpret the partitioning behavior (and toxicity) of compounds. Care must be used, however, in not misinterpreting, or overinterpreting, what a log K_{OW} may indicate. For example, estimated aqueous solubility itself may not correlate well with log K_{OW} for predicted aqueous solubility versus experimental K_{OW} (e.g., $r<0.46$ for the compounds in Table 3.1 and Figure 3.2).

The partition coefficient (K_D) onto various solids is sometimes predicted from log K_{OW}, although these correlations may often falter. For example, Tolls (2001) demonstrated that prediction of even organic carbon normalized partition coefficients (K_{OC}) may be underestimated by log K_{OW}. The primary reason for less than perfect correlations is that many mechanisms other than sorption or partitioning to an organic phase (e.g., hydrogen bonding, surface complexation, and cation exchange) also play an important role in partitioning behaviors of pharmaceuticals between aqueous and solid (or organic liquid) phases.

Pharmaceuticals generally exhibit very low volatility as exhibited by low Henry's Law constants (H). For example, for the compounds listed in Table 3.1, H ranged from $7(10^{-36})$ to $1(10^{-5})$ atm·m^3·mol^{-1} (calculated by the author using EPI Suite software (v. 3.2); data not shown).

The toxicity of pharmaceuticals is not addressed in detail in this chapter. However, pharmaceuticals are, by nature and design, biologically active. A comprehensive review of human and animal toxicology of pharmaceuticals is provided by Mückter (2006). Of most concern with respect to toxicity are the hormonal compounds, cytostatics/antineoplastics compounds, antimicrobials, and X-ray contrast media.

3.3 Occurrence and Behavior in Natural Systems

When looked for in the environment, pharmaceuticals are regularly observed in surface water, drinking water, and wastewater throughout the world (Kolpin et al., 2002; Sprague and Battaglin, 2005; Conn et al., 2006).

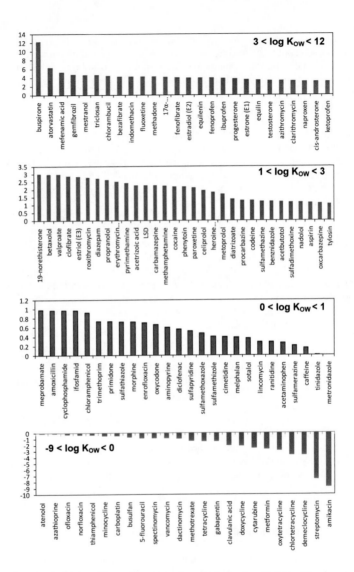

Figure 3.1. Ordered log K_{OW} values for selected pharmaceuticals and hormones. The *higher range* would have lower relative hydrophobicity and is dominated by lipid regulators, hormonal compounds, and other pharmaceuticals. The *intermediate* range is a mix of pharmaceuticals. The *low range* would tend to have higher relative hydrophilicity and is dominated by antibiotics cytostatics/antineoplastics, and a common x-ray contrast media (iopromide).

Table 3.2 Sorting of classes of pharmaceuticals (and selected natural hormones) into highest to lowest log K_{OW} bins. (n= number of pharmaceuticals within the class)

	log K_{OW}				
	3 - 12	1 - 3	0 - 1	-9 - 12	n
lipid-regulators	100	0	0	0	11
hormonal compounds	82	18	0	0	14
analgesics	50	14	36	0	4
antipsychotics/antidepressants	40	40	20	0	35
antibiotics	9	17	34	40	5
cytostatics/antineoplastics	8	15	23	54	13
x-ray contrast media	0	100	0	0	6
beta-blockers (β-blockers)	0	75	13	13	1
anticonvulsants/antiepileptics	0	67	17	17	2
antihystamines	0	0	100	0	8
illicit drugs	0	0	100	0	4
stimulants	0	0	100	0	1
anti-diabetics	0	0	0	100	2

3.3.1 Surface Water and Groundwater

The occurrence of pharmaceuticals in surface water and groundwater has been reported by several research groups (e.g., Coppage and Braidech, 1976; Stan et al.,1994; Holm, 1995; Heberer et al., 1997; Seiler et al., 1999; Ternes and Hirsch, 2000; Sacher et al., 2001; Ternes, 2001; Snyder et al., 2003; Williams et al., 2003; Drewes et al., 2003; Clara et al., 2004; Snyder et al., 2004; Anderson et al., 2004; Heberer et al., 2004; Mückter et al., 2004; Buerge et al., 2006; Giger et al., 2006; Fono et al., 2006; Haggard et al., 2006; Hummel et al., 2006; Peschka et al., 2006; Rabiet et al., 2006; Loraine and Pettigrove, 2006; Kim and Carlson, 2006; Kim et al., 2007). One important study was conducted for surface waters by Kolpin et al. (2002) on the occurrence of pharmaceuticals, hormonal compounds, and other organic wastewater contaminants. The study examined a wide suite of organic compounds in 139 streams and rivers in 30 states across the United States in 1999 and 2000. The study found that 80% of the streams tested had organic wastewater contaminants. The results of the Kolpin et al. (2002) study that were specifically related to pharmaceuticals (and selected natural hormones) are presented in Table 3.3. The results show (not surprisingly) that the stimulant, caffeine, was the most commonly detected drug, which was found in 71% of surface waters tested with a median concentration of 100 ng·L^{-1} (Table 3). Of the natural hormonal compounds, estriol (E3), followed by estradiol (E2), estrone (E1), progesterone, and testosterone, were most common. The synthetic estrogen used for birth control, 17α-ethinylestradiol (EE2), was second only to estriol (E3) in occurrence, and was found in 16% of surface water samples analyzed with a median concentration of 73 ng·L^{-1}.

Table 3.3 Occurrence of selected pharmaceuticals (and selected natural hormones) in surface water, treated sewage, and drinking water.

COMPOUND	CLASS	CAS	Surface Water Det. Freq (%)	Med (ng/L)		Drinking Water Det. Freq (%)	Med (ng/L)		Treated Sewage [ε] Greatest Med (ng/L)	Max (ng/L)
caffeine	stimulants	58-08-2	70.6	100	λ					
acetaminophen	nonnarcotic analgesics	103-90-2	23.8	110	λ					
estriol (E3)	hormonal compounds	50-27-1	21.4	19	λ					
erythromycin (H₂O)	antibiotics - macrolides	114-07-8	21	100	λ				270	6,000
lincomycin	antibiotics - other	859-18-7	19.2	60	λ					
17α-ethinylestradiol (EE2)	hormonal compounds	57-63-6	15.7	73	λ					
clofibrate	lipid-regulators	882-09-7	15	600	ξξ	ND	ND	ξξ		
tylosin	antibiotics - macrolides	1401-69-0	13.5	40	λ					
sulfamethoxazole	antibiotics - sulfonamides	723-46-6	12.5	150	λ				1,400	2,000
trimethoprim	antibiotics - other	738-70-5	12.5	150	λ				550	1,900
estradiol (E2)	hormonal compounds	50-28-2	10.6	160	λ					
codeine	narcotic analgesics	76-57-3	10.6	200	λ					
ibuprofen	NSAID	15687-27-1	8	5850	ξξ	7	734	ξξ	3,100	27,300
triclosan	antibiotics - other	3380-34-5	8	734	ξξ	7	49	ξξ		
estrone (E1)	hormonal compounds	53-16-7	7.1	27	λ					
roxithromycin	antibiotics - macrolides	80214-83-1	4.8	50	λ				40	1,700
progesterone	hormonal compounds	57-83-0	4.3	110	λ					
gemfibrozil	lipid-regulators	25812-30-0	3.6	48	λ				920	5,500
testosterone	hormonal compounds	58-22-0	2.8	116	λ					
ciprofloxacin	antibiotics - fluoroquinolones	93107-08-5	2.6	20	λ				170	860
chlortetracycline	antibiotics - tetracyclines	64-72-2	2.4	420	λ					
sulfamethazine	antibiotics - sulfonamides	1981-58-4	1.2	220	λ					
sulfadimethoxine	antibiotics - sulfonamides	122-11-2	1.2	60	λ					
tetracycline	antibiotics - tetracyclines	60-54-8	1.2	110	λ					
oxytetracycline	antibiotics - tetracyclines	79-57-2	1.2	340	λ					
norfloxacin	antibiotics - fluoroquinolones	70458-96-7	0.9	120	λ					
diclofenac									1,500	28,400

λ   Kolpin et al (2002)
ξξ Loraine and Pettigrove (2006)
ε   Alder et al., (2006)

Non-narcotic, narcotic and non-steroidal anti-inflammatory drugs (NSAID) analgesics were also common with 24% of surface waters containing acetaminophen (median concentration = 110 ng·L⁻¹), and 11% containing codeine (median concentration = 200 ng·L⁻¹). Loraine and Pettigrove (2006) also found 8% of surface waters sampled contained ibuprofen (median concentration 5,850 ng·L⁻¹). Many antibiotics were detected in the study with 13 different antibiotics found with a frequency of at least 1%. Most commonly observed antibiotics/antimicrobials were erythromycin (H₂O), lincomycin, tylosin, sulfamethoxazole, and trimethoprim with frequencies ranging from 12.5 to 21%, and median concentrations of from 40 ng·L⁻¹ for tylosin to 150 ng·L⁻¹ for trimethoprim (Table 3). Triclosan, a common antibiotic used in hand soaps, was found by Loraine and Pettigrove (2006) in 8% of samples at a median concentration of 734 ng·L⁻¹. Finally, the lipid regulators, clofibrate and gemfibrozil, were found in 15 and 4% of surface water samples, respectively, with median concentrations of 600 and 48 ng·L⁻¹, respectively (Kolpin et al, 2001; Loraine and Pettigrove, 2006).

Another study by Sprague and Battaglin (2005) studied pharmaceuticals and related organic compounds in streams and groundwater in Colorado (USA). The study detected pharmaceuticals including caffeine, cotinine, and triclosan, primarily in urban streams.

3.3.2 *Municipal and Industrial Wastewater*

Occurrence and/or removal data for pharmaceuticals in wastewaters has been published by a variety of groups (e.g., Andreozzi *et al.*, 2003; Ashton *et al.*, 2004; Metcalfe *et al.*, 2004; Karthikeyan and Meyer, 2006; Nakada *et al.*, 2006; Vieno *et al.*, 2007). Studies on pharmaceutical concentrations in wastewater include Karthikeyan and Meyer (2006) who monitored antibiotic pharmaceuticals in municipal sewage treatment plants in Wisconsin (USA) (Karthikeyan and Meyer, 2006). Out of 21 antibiotic compounds tested, the study found six antibiotics present with frequencies of 80, 70, 45, 40, and 10% for tetracyclines and trimethoprim, sulfamethoxazole, erythromycin (H_2O), ciprofloxacins, and sulfamethazine, respectively. The study also noted concentrations in the low $\mu g \cdot L^{-1}$ range, and that small and large sewage treatment plants had similar antibiotic profiles.

A review by Alder *et al.* (2006) tabulated concentrations in treated sewage of 22 pharmaceuticals including analgesics, lipid regulators, X-ray contrast media, antidepressants, antiepileptics, β -blockers, and antibiotics for 8 countries (France, Greece, Italy, Sweden, Germany, UK, Canada, and USA). Some pharmaceuticals had particularly high concentrations such as ibuprofen and diclofenac with maximum concentrations of 27,300 and 28,400 $ng \cdot L^{-1}$, respectively (Table 3). The review by Alder *et al.* (2006) also examined usage patterns of pharmaceuticals and their occurrence in hospital sewage, and cited the occurrence of antibiotics, antipsychotics, cytostatics/antineoplastics, and X-ray contrast media at concentrations in the $\mu g \cdot L^{-1}$ range.

3.3.3 *Soils and Sediments*

Only limited study of the occurrence of pharmaceuticals in soils and sediments systems have been reported. The primary reason for this lack of data has been the lack of appropriate instrumentation and methods to allow accurate measurement of pharmaceuticals at low concentration in complex matrices. With the growing prevalence of sensitive triple-quadruple mass spectrometers in environmental research facilities, rapidly expanding amounts of data on the occurrence of pharmaceuticals in soils and sediments in anticipated.

Work by Tolls (2001) compiled log K_{OC} and log K_D values for a variety of antibiotics in soils (Table 3.4). Log K_{OC} correlated well with log K_D for both clayey and non-clayey soils (α=0.05) for the paired data from Tolls (2001) (Figure 3.2).

Table 3.4 Log K_{OW}, log K_d, and log K_{OC} for selected pharmaceuticals.

Compounds	CAS	MWt	log K_{OW} (calc by KOWWIN)	log K_{OC} (calc by PCKOCWIN)	Sediments/ soils log K_{OC} (L/kg) [oo]	Sediments/ soils log K_d (L/kg) [oo]
fluoroquinolones						
enrofloxacin	93106-60-6	359.40	0.7		4.2-5.9	2.4-3.8
ofloxacin	82419-36-1	361.38	-0.2	1.918	4.6	2.5
ciprofloxacin	93107-08-5	331.35		1.55	4.7	2.6
sulfonamides						
sulfamethazine	1981-58-4	278.33	1.23	2.782	1.8-2.3	0.6-3.1
sulfathiazole	72-14-0	255.31	0.72	2.975	2.3	0.69
macrolides						
tylosin	1401-69-0	916.12	1.05	1	2.7-3.9	0.9-2.1
tetracyclines						
tetracycline	60-54-8	444.44	-1.33	1.76		2.6-3.2
oxytetracycline	79-57-2	460.44	-2.87	1.99	4.4-5.0	2.5-3.0
fenicoles						
chloramphenicol	56-75-7	323.13	0.92	1		0.2-0.4
metronidazole	443-48-1	171.16	0	1	1.6-1.7	0.54-0.67

δδ Tolls (2001)

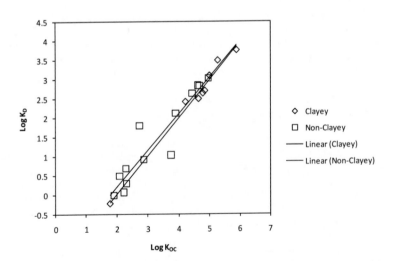

Figure 3.2 Correlation between log K_{OC} and log K_D for clayey and non-clayey soils from Tolls (2001) for a suite of antibiotics listed in Table 3.4. Correlations are significant at $\alpha=0.05$.

3.3.4 Air and Atmosphere

Due to the physiochemical properties required for their use, pharmaceuticals generally have very low volatility and vapor pressure. This is expressed in Henry's law constants (i.e., linear partition coefficients between the aqueous and gaseous phases). For example, for the compounds listed in Table 1, the range of Henry's Law constants was from $6.60(10^{-36})$ (for the aminoglycoside, amikacin) to $9.70(10^{-06})$ (for the anticonvulsant, valproate). These very low values show that volatilization of pharmaceuticals and their gas-phase occurrence will be minimal in most typical conditions.

3.4   Fate and Transformation in Engineered Systems

A growing body of work is providing insight into the behavior and fate of pharmaceuticals in water and wastewater treatment systems. Key processes with respect to pharmaceuticals in engineered treatment systems includes partitioning and transformation processes. Important partitioning processes for pharmaceuticals in engineered systems include adsorption (e.g., onto activated carbon or biosolids) as well as membrane separations. Many transformation processes are also possible including aerobic and anaerobic biodegradation, hydrolysis, and chemical oxidation with chlorine, ozone, or advanced oxidation.

3.4.1 Water Treatment

Sorption - In water treatment, partitioning of pharmaceuticals onto activated carbon is most likely. Activated carbon is typically used in one of two manners. First, powdered activated carbon (PAC) is applied near the front end of a treatment plant, and pharmaceuticals (and other organics) adsorb to the carbon surface through a variety of mechanisms as the PAC passes through the plant. The PAC is removed on granular (typically sand/anthracite) filters. PAC has a very large internal pore structure consisting of micropores ($d_{micropore} < 2$ nm), mesopores (2 nm $< d_{mesopore} < 50$ nm), and macropores (50 nm $< d_{macropore}$) (Patni *et al.*, 2008). Greater than 99% of the surface area of a PAC is in the internal pore space of the solid. Because a pharmaceutical must diffuse through the pore to reach internal adsorption sites, the adsorption process may take days to come to equilibrium. However, the contact time for PAC and a pharmaceutical in a water treatment plant is typically on the order of hours so that only partition equilibrium is achieved.

Another type of activated carbon used in water treatment is granular activated carbon (GAC). GAC is larger than PAC and, hence, does not stay suspended in water as PAC does in a water treatment plant. Instead, GAC is used on GAC-capped filters or in post-filter GAC contactors. In general, GAC adsorption capacities approach those in equilibrium with the higher influent concentration of a pharmaceutical.

The effectiveness of both PAC and GAC for the removal of pharmaceuticals is highly dependent on a wide variety of parameters. Lower *temperatures*, as might be seen in a wintertime scenario, can cause a significantly higher adsorption capacity for a carbon than at higher temperatures potentially corresponding to summertime temperatures. The *specific surface area* (m^2/g) of a carbon has a direct effect on the amount of a pharmaceutical that can be adsorbed on a carbon for a given mass (and volume) of carbon applied in a process. The *type* and *activation method* of a carbon has a major effect on both the *pore-size distribution* of the carbon as well as the surface functional groups and *point of zero charge* (PZC) for the surface of the carbon. *Pore-size distribution* is important because larger pharmaceuticals can only access surface area with pores that have a greater diameter than the diameter of the pharmaceutical itself.

Natural organic matter (NOM) (especially humic substances including humic and fulvic acids) can strongly adsorb to activated carbon. In the process, NOM can block pores making their internal surface area inaccessible to subsequent sorption of a pharmaceutical. Furthermore, NOM can competitively adsorb with pharmaceuticals on carbons by tying up adsorption surface sites that would otherwise have been available for pharmaceutical adsorption.

pH can also have a profound effect on the capacity of a carbon for a particular pharmaceutical. pH affects the speciation of both the pharmaceutical and the surface of the carbon. For example, if the pH of a solution is such that the pharmaceutical is in a neutral form, it may have a greater propensity to adsorb to a carbon with a neutral surface. If, on the other hand, the pH is such that the pharmaceutical is in an anionic form, then its polarity is much greater and it may have a much lower propensity to adsorb (thereby reducing the observed sorption capacity). At lower pH, some pharmaceuticals may become cationic increasing the likelihood of adsorption based on ionic interactions between a cationic pharmaceutical and an anionic surface functional group.

Several studies have examined the adsorption of pharmaceuticals on activated carbons (e.g., Adams *et al.*, 2002; Westerhoff *et al.*, 2005; Snyder *et al.*, 2006). Adams *et al.* (2002) determined the percent removal of 7 antibiotics on commercial PAC with a contact time of 4 hours to simulate typical conditions in a conventional water treatment plant. The results showed that for PAC dosages of 10 mg/L or greater, average removals were approximately 60 percent or greater from the filtered Missouri River water. PAC dosages this high are rarely used, however, with lower PAC dosages on the order of 1-2 mg/L being commonly used to improve the taste and odor of the product water. Thus, only limited removal of these antibiotics would be expected under typical application conditions for PAC in a water treatment plant.

Westerhoff *et al.* (2005) examined the removal of a large suite of pharmaceuticals and personal care products with two common coal-based PACs (Acticarb AC800 and Calgon WPM) in filtered surface water. WPM provided a slightly better removal of the pharmaceuticals than the AC800 (Table 3.5). The effect of PAC dosage can be clearly seen by the often dramatic difference in removals of pharmaceuticals between typical PAC dosages used for taste and odor control (e.g., 1 mg/L), and the much higher dosages used for synthetic organic chemical removal (e.g., 5-20 mg/L) (Table 3.5). For example, the percent removal for diclofenac was 0% for 1 mg/L of PAC and 92% for 20 mg/L of PAC. Clearly, generalizations regarding the expected removals of pharmaceuticals in treatment plants using PAC must be made with care.

Table 3.5 Removal of selected pharmaceuticals by PAC treatment with Acticarb AC800 PAC at a dosage of 5 mg/L in filtered surface water with a contact time of 4 hours.

Class	Pharmaceutical	Percent removal by PAC[α]				log KOW (calc)[++]
		5 mg/L AC800	1 mg/L WPM	5 mg/L WPM	20 mg/L WPM	
antibiotics	triclosan	98				4.66
antipsychotics/antidepressants	fluoxetine	95				4.23
hormonal compounds	estradiol (E2) (natural)	94	62	97	97	3.94
antibiotics	trimethoprim	93				0.73
hormonal compounds	progesterone	93	47	91	91	3.67
hormonal compounds	17α-ethinylestradiol (EE2)	88	55	97	97	4.12
hormonal compounds	testosterone	88	47	97	99	3.27
anticonvulsants/antiepileptics	carbamazepine	80				2.25
stimulants	caffeine	78				0.16
antipsychotics/antidepressants	diazepam	73				2.7
analgesics	acetaminophen	67				0.27
lipid-regulators	gemfibrozil	49				4.77
analgesics	diclofenac	44	0	44	92	0.57
analgesics	naproxen	47				3.1
antipsychotics/antidepressants	meprobamate	29	8	44	94	0.98
analgesics	ibuprofen	21	2	35	80	3.79
antibiotics	sulfamethoxazole	16				0.48
x-ray contrast media	iopromide	14				0

α Westerhoff et al. (2005)
++ Calc by author using KOWWIN v. 3.02

In Figure 3.3, the removal data from Table 3.5 are plotted versus log K_{OW} for a 5 mg/L dosage of AC800 PAC with a contact time of 4 hours to simulate a typical treatment process (Westerhoff et al., 2005). The data showed that at this dosage, pharmaceutical removals of between 14 and 98 percent were achieved. There is no significant correlation ($\alpha=0.05$) for the overall data between adsorption on PAC and log K_{OW} (for the neutral species). However, if the data for protonated bases (cationic), deprotonated acids (anionic), and caffeine (with its heterocylic nitrogen) are excluded, the data correlate relatively well ($\alpha=0.05$) (Figure 3.3). The lack of correlation for the outliers can be understood based on molecular structure and ionization of the compounds. The three compounds furthest below the correlation line are gemfibrozil, naproxen, and ibuprofen are all (anionic) deprotonated acids at neutral pH (Figure 3.3). Thus, the lack of correlation is due to the log K_{OW} being for the more lipophilic neutral species, and the (lower) observed sorptive removal being for the more hydrophilic ionized species. Similarly, two of the outliers above the correlation line are for (cationic) protonated bases (Figure 3.3). The greater than expected removal based solely on lipophilicity is likely due to electrostatic sorptive mechanisms between the cationic sorbates and the solid surface (in addition to the non-specific interations). More detailed analysis of the physicochemical factors affecting the correlation between adsorption and log K_{OW} may be found in Westerhoff *et al.* (2005).

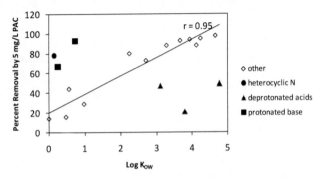

Figure 3.3 Correlation between log K_{OW} (calculated by author using KOWWIN) and percent removal by PAC (AC800) treatment from Westerhoff *et al.* (2005) from Table 3.5.

Another potential adsorbent in a water treatment process are the coagulation and softening solids. Work by Westerhoff *et al.* (2005) also examined the removal of caffeine, estrogens (E1, E2, E3, EE2), androstenedione, progesterone, fluoxetine, hydrocodone, and erythromycin-H_2O, as well as other xenobiotic organic compounds on precipitation solids associated with alum and iron salt coagulation, and with lime softening at pH 11. All of these pharmaceuticals and hormones were removed less than 10 percent except for erythromycin-H_2O which was removed by 33%. Work by Adams *et al.* (2002) for 7 antibiotics also showed negligible removal on these same types of solids.

Overall, both PAC and GAC have the ability for highly effective removal of pharmaceuticals in drinking water treatment. However, PAC dosages greater than those typically used for taste-and-odor control are required for effective pharmaceutical removal in water treatment. For both PAC and GAC, proper selection of the carbon type (and consideration of operating conditions) is critical to obtain effective removals.

Transformation – Common oxidants used in drinking water treatment include chlorine, chloramines, chlorine dioxide, permanganate, and ozone, with chlorine and ozone being most common as primary oxidants. Second-order rate constants are tabulated in Table 3.6 for ozone, free chlorine (HOCl/OCl⁻), and chlorine dioxide. From these selected data, it is apparent that ozone tends to be most reactive followed by chlorine and then chlorine dioxide. Westerhoff *et al.* (2005) studied the removal of selected pharmaceuticals in natural waters and found significant removal of most compounds with a (relatively high) chlorine dosage of 5 mg/L. Notable exceptions to the high removals were caffeine, dilantin, fluoxetine, iopromide, meprobamate, and testosterone.

Table 3.6 Second-order rate constants for selected pharmaceuticals (and selected natural hormones) for ozone, free chlorine, and chlorine dioxide.

				k (L/mol/s)		
		CAS	MWt	Ozone	HOCl/OCl-	CLO2
17α-ethinylestradiol (EE2)	hormonal compounds	57-63-6	296.41	7.00E+09 β	122 β	2.00E+05 β
amoxicillin	antibiotic - β-lactams	26787-78-0	365.41	6.00E+06 ζζ		
roxithromycin	antibiotic - macrolides	80214-83-1	837.07	4.50E+06 β	10-100 β	220 β
sulfamethoxazole	antibiotic - sulfonamides	723-46-6	253.28	2.50E+06 β	700-1900	6.70E+03 β
diclofenac	NSAID	15307-79-6	318.14	1.00E+06 β	"fast" β	1.00E+04 β
lincomycin	antibiotic - other	859-18-7	406.54	4.00E+05 ζ		
carbamazepine	anticonvulsants/antiepileptics	298-46-4	236.28	3.00E+05 β	"No Reaction" β	<0.015 β
spectinomycin	antibiotic - other	21736-83-4	332.36	3.40E+04 ζ		
bezafibrate	lipid-regulators	41859-67-0	361.83	5.90E+02 β	"No Reaction" β	<0.01 β
ibuprofen	NSAID	15687-27-1	206.29	9.6 β	"No Reaction" β	<0.01 β
diazepam	antipsychotics/depressants	439-14-5	284.75	0.75 β		
iopromide	x-ray contrast media	73334-07-3	791.12	<0.8 β		
sulfamethizole	antibiotic - sulfonamides	144-82-1	270.33		130-1600 δ	
sulfamerazine	antibiotic - sulfonamides	127-58-2	264.30		210-2300 δ	
sulfamethazine	antibiotic - sulfonamides	1981-58-4	278.33		190-4500 δ	
sulfadimethoxine	antibiotic - sulfonamides	122-11-2	310.33		18,300-19,400 δ	
sulfathiazole	antibiotic - sulfonamides	72-14-0	255.31		580-8800 δ	

β Huber et al. 2003
ζζ Ikehata et al. (2006)
ζ Qiang and Adams (2004)
θ Dal Pozzo et al., (1989) from KOWWIN (v. 3.06)
δ Chamberlain and Adams (2006)

At pH levels below about 7.6, free chlorine exists in the more reactive hypochlorous acid (HOCl) versus the less reactive (OCl-). Westerhoff *et al.* (2005) noted, however, that many of the pharmaceuticals had less than 20% difference in chloridative removals between pH 5.5 versus 6.8-8.2 (with lower pH always providing greater removals).

Ozone also can be highly effective for removing selected pharmaceuticals during drinking water treatment. Ternes *et al.* (2002) found that under typical drinking water treatment conditions, carbamazepine and diclofenac were completely removed (>99%) while clofibric acid was removed 77%.

It must be remembered that transformation byproducts have the potential to be as biologically active as parent compounds. In such cases, removal of the parent compounds does not necessarily diminish the toxicity or risk associated with the parent. For example, Shah *et al.* (2006) found that the chlorination byproducts of carbodox formed during drinking water disinfection retained the biologically active N-oxide groups.

3.4.2 Municipal Wastewater Treatment

Sorption - As in drinking water, major processes for pharmaceuticals in wastewater treatment include partitioning reactions and transformation reactions. Particularly important partitioning reactions in municipal wastewater (sewage) treatment include sorption to primary and secondary biosolids, as well as membrane separations.

Sorption of pharmaceuticals generally follows linear relationships between the solid and aqueous phases. Commonly the linear partition coefficient (K_D) is used to describe or model the sorption reaction because the very low pharmaceutical concentrations in wastewater tend to promote linear (rather than Langmuir-shaped) isotherms.

In a municipal wastewater treatment plant, key residuals (sludge) streams include primary ($1°$) and secondary ($2°$) sludge from their respective clarifiers. Primary clarifiers remove a majority of the total suspended solids (and associated biochemical oxygen demand (BOD)) entering a wastewater treatment plant. The primary effluent is then treated in secondary (or biological) treatment often consisting of an aeration basin followed by the secondary clarifier.

As compared with primary sludge, secondary sludge tends to have very high microorganism concentrations and less fatty content. In primary sludge, partitioning of a pharmaceutical into organic (lipophilic) layer may dominate as an adsorption mechanism (Ternes *et al.*, 2004). This might favor the adsorption of neutrally-charged pharmaceuticals in the pH range of the wastewater (e.g., 7 to 8). In secondary (biological) sludge, other adsorption mechanisms may play a more dominant role including ion exchange, surface complexation, and/or hydrogen bonding. Because each mechanism is highly charge dependent, solution pH can play an important role as it affects the speciation of functionalities of both the pharmaceutical and the biosolids. Other factors can also be critically important such as total dissolved solids, calcium content, temperature, and other factors.

For any given suspended solids (SS) concentration, the percent sorption of a pharmaceutical on primary or secondary sludge may be modeled if the linear partition coefficient is known:

$$Percent\,adsorbed = \frac{mass_{solids}}{mass_{solids} + mass_{aqueous}}$$

$$= \frac{\dfrac{q(mg/g)}{C(mg/L)} SS(g/L)}{1 + \dfrac{q(mg/g)}{C(mg/L)} SS(g/L)}$$

$$= \frac{\left(K_D(L/kg) \cdot SS(g/L) \cdot (0.001\,kg/g)\right)}{1 + \left(K_D(L/kg) \cdot SS(g/L) \cdot (0.001\,kg/g)\right)} \tag{1}$$

where q is solids concentration, C is aqueous concentration, SS is suspended solids concentration, and K_D is the linear partition coefficient. Thus, for any given suspended solids (SS) concentration, the percent sorption of a pharmaceutical on primary or secondary sludge may be modeled if the linear partition coefficient is known (Figure 3.4). Log K_D values for selected pharmaceuticals on primary and secondary sludge are tabulated in Table 3.7. Comparison of Log K_D values for primary versus secondary sludge demonstrate that generalizations may be difficult regarding which sludge is likely to have the greatest sorption potential, and hence, contain the highest concentrations of a given pharmaceutical.

Table 3.7 Log K_D values for selected pharmaceuticals on primary and secondary sludge. Log K_{OW} values and aqueous solubility are also presented for comparison.

Compound	Class	Aqueous solubility (mg/L)	log KOW (exp)	log KOW (calc)	Primary log Kd (L/gSS)	Secondary log Kd (L/gSS)
17α-ethinylestradiol (EE2)	hormonal compounds	116.4	3.67	4.12	2.4	2.5
diazepam	antipsychotics/antidepressants	58.78	2.82	2.7	1.6	1.3
ifosfamid	cytostatics/antineoplastics	3781	0.86	0.97	1.3	0.15
diclofenac	analgesics	2425	0.7	0.57	2.7	1.2
cyclophosphamide	cytostatics/antineoplastics	5943	0.63	0.97	1.7	0.0024
norfloxacin	antibiotics	1.78E+05	-1.03	-0.31	2.5	37
ciprofloxacin	antibiotics	1.15E+04			2.6	26

ββ Golet et al. (2003)
** Sangster (1993) from KOWWIN (v. 3.06)
+ Hansch et al., 1995 from KOWWIN (v. 3.06)

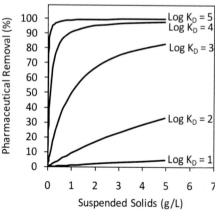

Figure 3.4 Percent pharmaceutical removal for various suspended solids (SS) concentrations and log linear partition coefficients.

Transformation – Removal of pharmaceuticals in sewage treatment plants may be via biological transformation, chemical oxidation (e.g., during chlorine or ozone disinfection), or by hydrolysis.

Studies of *biodegradation* of pharmaceuticals in sewage treatment plants shows that a wide range of biodegradation can occur due both to differences in the biodegradability of specific pharmaceuticals, and due to the nature and operation of the biological treatment process itself. For example, it has been shown that the solids retention time (SRT; also known as sludge age) is a critical parameter with respect to the degree of biodegradation that can be achieved within a class of pharmaceuticals (e.g., Ternes *et al.*, 2004). Depending on the treatment objective in an activated sludge process, sludge age can vary from 2 to 4 days for BOD removal processes, to 8 to 12 days for nitrification, to 10 to 15 days for denitrification, to 14 to 20 days for biological phosphorus removal (Tchobanoglous, 2003; Ternes *et al.*, 2004). Ternes *et al.* (2004) demonstrated that sludge age requirements for removal of compounds such as sulfamethoxazole and ibuprofen require only 2 to 5 days, while 5 to 15 days are required for 17α-ethinylestradiol, and greater than 20 days are required for carbamazepine and diazepam removal. Similarly, Andersen *et al.* (2003) found that for activated sludge plants that were practicing nitrification and/or denitrification, significant removal of 17α-ethinylestradiol was achieved in full scale treatment. Batt *et al.* (2006) reported similar effects regarding the effectiveness of extended sludge age (used for nitrification) on the removal of iopromide and trimethoprim.

About half of existing activated sludge processes in the United States are primarily designed for BOD removal only with corresponding shorter sludge ages. New activated sludge processes in the United States, however, tend to be designed with longer sludge ages to provide for at least nitrification (Daigger, 2007).

The reason for this dependency on sludge age for pharmaceutical removal may have several reasons. First, greater sludge age correlates linearly with greater mixed liquor suspended solids concentrations for a given basin volume (V), flow rate (Q), yield (Y), and biochemical oxygen demand (S_0-S) removal:

$$\text{Sludge age(d)} = \frac{V(m^3)}{Q(m^3/d) \cdot Y(kg/kg) \cdot (S_0 - S(kg/m^3))} \cdot \text{MLSS}(kg/m^3) \qquad (2)$$

Due to their low concentrations, removal of pharmaceuticals (A) will tend to follow second-order kinetics (or pseudo-first order kinetics at a constant suspended solids (or biomass) concentration) as shown in:

$$\frac{dA}{dt} = -\frac{U_{max}[Biomass][A]}{Y(K_C +[A])}$$

$$= -\frac{U_{max}[Biomass][A]}{Y(K_C +[A])} \tag{3}$$

$$= -\left(\frac{U_{max}}{YK_C}\right)[Biomass][A]$$

$$= -k_{biodeg}[Biomass][A]$$

where dA/dt is the removal rate of the pharmaceutical (A), U_{max} is the specific growth rate, Y is the yield, and K_C is the half-saturation coefficient. Thus, greater sludge age corresponds to proportionately greater MLSS (Equation 2), and also to proportionately greater pharmaceutical removal rates (Equation 3). Other potential causes of a greater pharmaceutical removal at a greater sludge age include more diversity of the bacterial consortia present in the sludge, as well as more enzyme diversity (Ternes et al., 2004).

Second-order biological rate constants compiled from Ternes et al. (2004) are tabulated in Table 3.8. From these values, half-lives may be calculated based for any given suspended solids concentration. If the suspended solids concentration were 2,000 mg/L, for example, corresponding half lives for various pharmaceuticals can range from less than one hour to many days (Table 3.8).

EPI Suite (v. 3.2) software was used by the author to estimate the removal of various pharmaceuticals in a standard sewage treatment plant (STP) (presumed to be based on BOD removal rather than nutrient removal with longer sludge ages). The removals predicted by this software ranged from 3 to 93% (Table 3.8). Interestingly, almost all of the predicted removal was based on sorption to biosolids as opposed to biodegradation (Table 3.8). No volatilization was estimated due to very low Henry's Law constants (Table 3.8). These estimates are provided for informational purposes only. The estimates are not validated, and are expected to contain considerable error.

Chemical oxidation of pharmaceuticals in sewage treatment plants may also occur during wastewater disinfection with chlorine or ozone. For example, work by Bedner and MacCrehan (2005) investigated the transformation of acetaminophen using simulated wastewater disinfection using free chlorine (hypochlorite) at neutral pH of a wide range of reaction times (2 to 90 minutes). Acetaminophen concentrations studied were 150 and 1,500 µg/L, and the chlorine dosage was 4 mg/L as Cl_2. The study showed that 88% of the acetaminophen was transformed within 1 hour of contact time, and that various chlorination byproducts could be identified (e.g., 1,4-benzoquinone, chloro-4-acetamidophenol, and dichloro-4-acetamidophenol). Dechlorination with sulfite is commonly used to prevent discharge or residual chlorine, and was seen by Bedner and MacCrehan (2005) to cause further transformation of the some oxidation byproducts.

Table 3.8. Second-order biological rate constants from Ternes *et al.* (2004), and calculated half-lives (hours) assuming a 2,000 mg/L mixed liquor suspended solids concentration. Also shown are estimated overall removals (from EPI Suite (v. 3.2)) along with the relative biodegradation and sludge sorption percentages (calculated by author).

		KOWWIN log KOW (calc)	Biological k (CAS) (L/gSS/d)†	Half life (hours)κκ	Secondary log Kd (L/gSS) ββ	Removal in STP (%)κ Total	Bio-deg	Sludge sorp-tion
atorvastatin	lipid-regulators	6.36				93.21	0.77	92.43
mefenamic acid	NSAID	5.28				81.16	0.70	80.50
gemfibrozil	lipid-regulators	4.77	6.4-9.6	0.9-1.3		69.10	0.62	68.48
triclosan	antibiotics - other	4.66				68.67	0.61	68.06
mestranol	hormonal compounds	4.68				65.05	0.59	64.46
chlorambucil	alkylating agents	4.44				52.84	0.50	52.34
indomethacin	NSAID	4.23	<0.3	>28		43.65	0.43	43.22
bezafibrate	lipid-regulators	4.25	2.1-3.0	2.8-4.0		42.58	0.42	42.16
fluoxetine	antipsychotics/antidepressants	4.23				32.40	0.34	32.06
azithromycin	antibiotics - macrolides	3.24	<0.1	>83		30.99	0.33	30.66
estradiol (E2)	hormonal compounds	3.94				30.52	0.32	30.20
fenofibrate	lipid-regulators	4.00	7.2-10.8	0.8-1.2		30.06	0.32	29.74
ibuprofen	NSAID	3.79	21-35	0.2-0.4		28.72	0.31	28.40
equilenin	hormonal compounds	3.93				26.97	0.30	26.68
methadone	narcotic analgesics	4.17				26.97	0.30	26.68
fenoprofen	NSAID	3.90	10-14	0.6-0.8		25.72	0.29	25.43
progesterone	hormonal compounds	3.67				24.50	0.28	24.22
cis-androsterone	hormonal compounds	3.07				18.10	0.22	17.90
17α-ethinylestradiol (EE2)	hormonal compounds	4.12			2.5	17.51	0.22	17.29
propranolol	beta-blockers (β-blockers)	2.60				12.58	0.18	12.40
equilin	hormonal compounds	3.35				10.04	0.16	9.88
testosterone	hormonal compounds	3.27				9.54	0.16	9.39
naproxen	NSAID	3.10	1.0-1.9	4.4-8.3		7.55	0.14	7.41
clarithromycin	antibiotics - macrolides	3.18	<0.4	>21		7.30	0.14	7.17
estrone (E1)	hormonal compounds	3.43				6.96	0.13	6.83
ketoprofen	NSAID	3.00				6.85	0.13	6.72
erythromycin (H₂O)	antibiotics - macrolides	2.48	<0.1	>83		6.23	0.13	6.11
19-norethisterone	hormonal compounds	2.99				5.44	0.12	5.32
LSD	illicit drugs	2.26				5.29	0.12	5.17
betaxolol	beta-blockers (β-blockers)	2.98				4.36	0.11	4.25
valproate	anticonvulsants/antiepileptics	2.96				4.24	0.11	3.93
diazepam	antipsychotics/antidepressants	2.70			1.3	4.42	0.11	4.31
roxithromycin	antibiotics - macrolides	2.75	<0.2	>42		4.05	0.11	3.94
pyrimethamine	other cytostatics	2.41				3.77	0.11	3.66
buspirone	antipsychotics/antidepressants	2.31				3.52	0.11	3.42
clofibrate	lipid-regulators	2.84	0.3-0.8	10-28		3.31	0.10	3.20
phenytoin	anticonvulsants/antiepileptics	2.16				3.01	0.10	2.91

Bedner and MacCrehan (2006) also studied chlorination of fluoxetine and metoprolol (cationic pharmaceuticals) under typical sewage disinfection conditions. They found that both compounds were rapidly converted to byproducts (e.g., neutral chloramines), that were stable relative to reduction during wastewater dechlorination.

Qiang *et al.* (2006) studied the removal of sulfonamide antibiotics with chlorine and chloramines from a lagoon slurry, and found that only negligible removal was achieved with common disinfection dosages of chlorine. Thus, a wide range of transformation and byproduct formation reaction may be anticipated for chlorine use disinfection of treated wastewater. Additional study is warranted due to the common used of chlorine for the disinfection of wastewater.

Ozone is sometimes used for wastewater disinfections, and is recently being discussed as a viable means for controlling pharmaceuticals in wastewater. Huber *et al.* (2005) found that ozone dosages as low as 1 mg/L applied to secondary treated municipal sewage can effectively remove many pharmaceuticals. Huber *et al.* (2005) also found that an ozone dosage of 2 mg/L could remove 90 to 99% of macrolides (e.g., azithromycin, erythromycin-H_2O, clarithromycin, and roxithromycin) and sulfonamides (e.g., sulfadiazine, sulfathiazole, sulfapyridine, and sulfamethoxazole) from all wastewaters tested. Ozone could be less effective for more ozone-recalcitrant pharmaceuticals (e.g., ibuprofen, diazepam, and iopromide (from Table 3.6)), or for pharmaceuticals that might adsorb more strongly to biosolids (and, thereby, be partially protected from ozone oxidation).

Hydrolysis is another potential abiotic degradation pathway for pharmaceuticals in sewage treatment plants. For example, related work in lagoon slurry by Loftin and Adams (2004) studied the hydrolysis of antibiotics including tylosin, oxytetracycline, and lincomycin. The work found that oxytetracycline degraded rapidly (except at low temperature), while tylosin and lincomycin were relatively recalcitrant. They also noted a strong temperature and pH dependence on the hydrolysis rates.

3.4.3 Solids Treatment and Disposal

A wide range of pharmaceuticals are observed in biosolids from wastewater treatment plants. For example, Kinney *et al.* (2006) found median concentrations of carbamazapine (an antiepileptic), diphenhydramine (an antihystamine), and fluoxetine (an antidepressant) at concentrations of 68, 340, and 370 µg/kg C, respectively, in nine biosolids residuals.

Carballa *et al.* (2007) studied the removal of various pharmaceuticals under both mesophilic and thermophilic digestion of municipal biosolids, and found similar results for the two temperatures and classes of microorganisms. Generally, minimal removals were observed for some pharmaceuticals (e.g., carbamazepine, diclofenac, and iopromide) while other pharmaceuticals were significantly removed (e.g., naproxen, sulfamethoxazole, roxithromycin, and 17α-ethinylestradiol). The study reported that no effect of either temperature or SRT was observed.

3.5   Conclusion

Pharmaceuticals play a vital role in providing the public with cures and relief from a wide range of ailments and disease. The modern era of medicine, however, has also brought with it massive manufacture, sales, and use of pharmaceuticals. Much of these pharmaceuticals reach the environment (and drinking water sources) through excretion or through improper disposal. The most common pharmaceuticals (or natural hormones) observed include an over-the-counter stimulant (caffeine), an over-the-counter analgesic (acetaminophen), a natural and a synthetic hormonal compound (estriol and 17α-ethyinylestradiol, respectively), a lipid-regulator (clofibrate), and various antibiotics (erythromycin, lincomycin, tylosin, sulfamethoxazole, and trimethoprim). With this large range of different pharmaceuticals in the environment (including drinking water sources), and due to the generally low concentrations (e.g., less than 1000 ng/L), the effects on human and environmental health are difficult to predict or measure.

However, due to the specifically-designed biological activity of pharmaceutical compounds, knowledge and minimization of human and environmental exposure to pharmaceuticals is certainly a sensible and rational objective. Examination of the fate and behavior of pharmaceuticals on an individual basis is justified for key pharmaceuticals that are heavily used or that pose a particularly significant health effect on humans or other animals. It is also of great benefit to be able to predict (albeit with some uncertainty) the behavior of pharmaceuticals in the environment, in water treatment and in wastewater treatment.

Many simple approaches to predicting fate (e.g., use of K_{OW} to predict partitioning to biosolids) certainly provide a reasonable first screening tool as long as it is kept clear that many compounds may deviate in their behavior from the prediction due to other chemical properties (e.g., ionization and speciation). This chapter provided an overview of both modeling and experiment data of key partitioning and transformation processes in natural systems, water, wastewater, and residuals treatment.

Modeling plays a valuable role in testing our understanding of systems, in screening for specific behaviors, and in ranking potential exposure of humans and the environment to individual or classes of pharmaceuticals. Much more experimental research is needed on the fate and effects of pharmaceuticals, as well as on the development and validation of fate and effects models, in order to appropriately protect the public and environmental health.

References

Adams, C., Wang Y., Loftin K., Meyer M. (2002) "Removal of Antibiotics from Surface and Distilled Water in Conventional Water Treatment Processes," J. Environmental Engineering, 128:3, 253-260.

Alder, A., Bruchet, A., Carballa, M., Clara, M., Joss, A., Löffler, D., McArdell, C., Miksch, K., Omil,F., Tuhkanen, T., Ternes, T. (2006) Consumption and Occurrence. In Human Phamaceuticals, Hormones and Fragrances. The Challenge of Micropollutants in Urban Water Management. T. Ternes and A. Joss (eds.) London.

Andersen, H., Siefrist, H., Halling-Sorensen, B., Ternes, T. (2003) "Fate of Estrogens in a Municipal Sewage Treatment Plant," Environmental Science and Technology, 37, 4021-4026.

Anderson, P., D'Aco, V., Shanahan, P., Chapra, S., Buzby, M., Cunningham, V., Duplessie, B., Hayes, E., Mastrocco, F., Parke, N., Rader, J., Samuelian, J., Schwab, B. (2004) "Screening Analysis of Human Pharmaceutical Compounds in U.S. Surface Waters," Environmental Science and Technology, 38, 838-849.

Andreozzi, R., Raffaele, M., Nicklas, P. (2003) "Pharmaceuticals in STP Effluents and their Solar Photodegradation in Aquatic Environment," Chemosphere, 50, 1319-1330.

Ashton, D., Hilton, M., Thomas, K. (2004) "Investigating the environmental transport of human pharmaceuticals to streams in the United Kingdom," Science of the Total Environment, 333, 167-184.

Batt, A., Kim, S., Aga, D. (2006) "Enhanced Biodegradation of Iopromide and Trimethoprim in Nitrifying Activated Sludge," Environmental Science and Technology, 40, 7367-7373,

Bedner, M., MacCrehan, W. (2005) "Transformation of Acetaminophen by Chlorination Produces the Toxicants 1,4-Benzoquinon and N-Acetyl-p-benzoquinone Imine.", Environmental Science and Technology, 40, 516-522.

Bedner, M., MacCrehan, W., (2006) "Reactions of the amine-containing drugs fluoxetine and metoprolol during chlorination and dechlorination processes used in wastewater treatment," Chemosphere, 65, 2130-2137.

Buerge, I., Buser, H-R., Poiger, T., Muller, M. (2006) "Occurrence and Fate of Cytostatic Drugs Cyclohosphamide and Ifosfamide in Wastewater and Surface Waters," Environmental Science and Technology, 40, 7242-7250.

Carballa, M., Omil, F., Ternes, T., Lema, J. (2007) "Fate of Pharmaceutical and Personal Care Products (PPCPs) during Anaerobic Digestion of Sewage Sludge," Water Research, 41, 2139-2150.

Chamberlain, E., Adams, C. (2006) "Oxidation of Sulfonamides, Macrolides, and Carbadox with Free Chlorine and Monochloramine," Water Research, 40, 2463-2592. Clara et al., 2004

Conn, K., Barber, L., Brown, Gregory K.; Siegrist, R. (2006) "Occurrence and Fate of Organic Contaminants during Onsite Wastewater Treatment," Environmental Science and Technology, 40, 7358-7366.

Coppage, D., Braidech, T. (1976) "River Pollution by Anticholinesterase Agents," Water Research, 10, 19-24.

Daigger, G. (2007) Personal communication.

Drewes, J., Heberer, T., Rouch, T., Reddersen, K. (2003) "Fate of Pharmaceuticals during Ground Water Recharge," Ground Water Monitoring and Remediation, 23, 64-72.

EPI Suite (v. 3.2) http://www.epa.gov/opptintr/exposure/pubs/episuite.htm

Fono, L., Kolodziej, E., Sedlak, D. (2006) "Attenuation of Wastewater-Derived Contaminants in an Effluent-Dominated River," Environmental Science and Technology, 40, 7257-7262.

Giger, W., Schaffner, C., Kohler H-P. (2006) "Benzotriazole and Tolyltriazole as Aquatic Contaminants. 1. Input and Occurrence in Rivers and Lakes," Environmental Science and Technology, 40, 7186-7192.

Golet, E., Xifra, I., Siegrist, H., Alder, A., Giger, W. (2003) "Environmental Exposure Assessment of Fluoroquinolone Antibacterial Agents from Sewage to Soil," Environmental Science and Technology, 37, 3243-3249.

Haggard, B., Galloway, J., Reed Green, W., Meyer, M. (2006) "Pharmaceuticals and Other Organic Chemicals in Selected North-Central and Northwestern Arkansas Streams," Journal of Environmental Quality, 35, 1078-1087.

Heberer, T., Dunnbier, U., Reilich, C., Stan, H. (1997) "Detection of drugs and drug metabolites in groundwater samples of a drinking water treatment plant," Fresenius Environmental Bulletin, 6, 438-443.

Heberer, T., Mechlinski, A., Fanck, B., Knappe, A., Massmann, G., Pekdeger, A., Fritz, B. (2004) "Field studies on the fate and transport of pharmaceutical residues in bank filtration," Ground Water Monitoring and Remediation, 24, 70-77.

Holm, J., Ruegge, K., Bjerg, P., Christensen, T. (1995) "Occurrence and distribution of pharmaceutical organic compounds in the groundwater down gradient of a landfill (Grindsted, Denmark)," Environmental Science and Technology, 29, 1415-1420.

Huber, M., Gobel, A., Joss, A., Hermann, N., Loffler, D., McArdell, C., Ried, A., Siegrist, H., Ternes, T., von Gunten, G. (2005) "Oxidation of Pharmaceuticals during Ozonation of Municipal Wastewater Effluents: A Pilot Study," Environmental Science and Technology, 39, 42900-4299.

Huber, M., Canonica, S., Park, G.-Y., von Gunten, U. (2003) "Oxidation of Pharmaceuticals during Ozonation and Advanced Oxidation Processes," Environmental Science and Technology, 37, 1016-1024.

Hummel, D., Loffler, D., Fink, G., Ternes, T. (2006) "Simultaneous Determination of Psychoactive Drugs and their Metabolites in Aqueous Matrices by Liquid Chromatography Mass Spectrometry," Environmental Science and Technology, 40, 7321-7328.

Ikehata, K., Naghashkar, N., Gamal El-Din, M. (2006) "Degradation of Aqueous Pharmaceuticals by Ozonation and Advanced Oxidation Processes: A Review," Ozone Science and Engineering, 28, 353-414.

Karthikeyan, K., Meyer, M. (2006) "Occurrence of Antibiotics in Wastewater Treatment Facilities in Wisconsin, USA.," The Science of the Total Environment, 361, 196-207.

Kim, S-C., and Carlson, K. (2006) "Occurrence of Ionophore Antibiotics in Water and Sediments of a Mixed-Landscape Watershed," Water Research, 40, 2549-2560.

Kim, S., Cho, J., Kim, I., Vanderford, B., Snyder, S. (2007) "Occurrence and Removal of Pharmaceuticals and Endocrine Disruptors in South Korean Surface, Drinking, and Waste Waters," Water Research, 41, 1013-1021.

Kinney, C., Furlong, E., Zaugg, S., Burkhardt, M., Werner, S., Cahill, J., Jorgensen, G. (2006) "Survey of Organic Wastewater Contaminants in Biosolids Destined for Land Application," Environmental Science and Technology, 20, 7207-7215.

Kolpin, D., Furlong, E., Meyer, M., Thurman, E. M., Zaugg, S., Barber, L., Buxton, H. (2002) "Pharmaceuticals, Hormones, and Other Organic Wastewater Contaminants in U.S. Streams 1999-2000: A National Reconnaissance." Environmental Science and Technology, 36, 1202-1211.

Kummerer, K. (2001) "Chapter 1: Introduction: Pharmaceuticals in the Environment. In Pharmaceuticals" in the Environment: Sources, Fate, Effects and Risks, Springer, New York, NY.

Kummerer, K. (2003) "Significance of Antibiotics in the Environment," Journal of Antimicrobial Chemotherapy, 52, 5-7.

Lindberg, R., Bjorklund, K., Rendahl, P., Johansson, M., Tysklind, M., Andersson, B. (2007) "Environmental Risk Assessment of Antibiotics in the Swedish Environment with Emphasis on Sewage Treatment Plants," Water Research, 41, 613-619.

Loftin, K., Adams, C., Surampalli, R. (2004) "The Fate and Effects of Selected Veterinary Antibiotics on Anaerobic Swine Lagoons," Water Environment Federation's 77th Annual Technical Exhibition and Conference (WEFTEC), Washington, D.C., USA (October, 2004).

Loraine, G., Pettigrove, M. (2006) "Seasonal Variations in Concentrations of Pharmaceuticals and Personal Care Products in Drinking Water and Reclaimed Wastewater in Southern California," Environmental Science and Technology, 40, 687-695.

Metcalfe, C., Miao, X.-S., Hua, W., Letcher, R., Servos, M. (2004) "Pharmaceuticals in the Canadian Environment," Pharmaceuticals in the Environment (2nd Edition), 67-90.

Mueckter, H. (2006) "Do Drug Residues in Drinking Water Endanger Our Health?," Berichte aus Siedlungswasserwirtschaft, 37, 39-40.

Nakada, N., Tanishima, T., Shinohara, H., Kiri, K., Takada, H. (2006) "Pharmaceutical Chemicals and Endocrine Disruptors in Municipal Wastewater in Tokyo and their Removal During Activated Sludge Treatment," Water Research, 40, 3297-3303.

Qiang, Z., Macauley, J., Mormile, M., Surampalli, R., Adams, C. (2006) "Treatment of Antibiotics and Antibiotic Resistant Bacteria in Swine Wastewater with Free Chlorine," J. Agriculture and Food Chemistry, 54, 8144-8154.

Qiang, Z., Adams, C., Surampalli, R. (2004) "Determination of Ozonation Rate Constants for Lincomycin and Spectinomycin," Ozone Science and Engineering, 26, 1-13.

Patni, A., Ludlow, D., Adams, C. (2008) "Characteristics of Ground Granular Activated Carbon for Rapid Small Scale Column Tests," J. Environmental Engineering, 131, 216-221.

Peschka, M., Eubeler, J., Knepper, T. (2006) "Occurrence and Fate of Barbiturates in the Aquatic Environment," Environmental Science and Technology, 40, 7200-7206.

Rabiet, M., Togola, A., Brissaud, F., Seidel, F., Budzinski, H., Elbaz-Poulichet, F. (2006) "Consequences of Treated Water Recycling as Regards Pharmaceuticals and Drugs in Surface and Ground Waters of a Medium-sized Mediterranean Catchment," Environmental Science and Technology, 40, 5282-5288.

Sacher, F., Lange, R., Brauch, H-J., Blankenhorn, I. (2001) "Pharmaceuticals in Groundwaters Analytical Methods and Results of a Monitoring Program in Baden-Wurttemberb, Germany," Journal of Chromatography A, 938, 199-210.

Seiler, R., Zaugg, S., Thomas, J., Howcroft, D. (1999) "Caffeine and Pharmaceuticals as Indicators of Waste Water Contamination in Wells," Ground Water, 37, 405-410.

Shah, A., Kim, J-H., Huang, C-H. (2006) "Reaction Kinetics and Transformation of Carbadox and Structurally-Related Compounds with Aqueous Chlorine," Environmental Science and Technology, 40, 7228-7235.

Snyder, S., Westerhoff, P., Yoon, Y., Sedlak, D. (2003) "Pharmaceuticals, Personal Care Products, and Endocrine Disruptors in Water: Implications for the Water Industry," Environmental Engineering Science, 20, 449-469.

Snyder, S., Leising, J., Westerhoff, P., Yoon, Y., Mash, H., Vanderford, B. (2004) "Biological and Physical Attenuation of Endocrine Disruptors and Pharmaceuticals: Implications for Water Reuse," Ground Water Monitoring and Remediation, 24, 1-11.

Snyder, S., Adham, S., Redding, A., Cannon, F., DeCarolis, J., Oppenheimer, J., Wert, E., Yoon, Y. (2006) "Role of Membranes and Activated Carbon in the Removal of Endocrine Disruptors and Pharmaceuticals," Desalination, 202, 156-181.

Sprague, L., Battaglin, W. (2005) "Wastewater Chemicals in Colorado's Streams and Ground Water," USGS Fact Sheet 2004-3127 (January, 2005).

Stan, H., Heberer, T., Linkerhaegner, M. (1994) "Occurrence of Clofibric Acid in the Aquatic System. Is the Use in Human Medical Care the Source of the Contamination of Surface, Groundwater, and Drinking Water?," Vom Wasser, 83 57-68.

Takacs-Novak, K., Szasz, Gy., Budvari-Barany, Zs., Jozan, M., Lore, A. (1995) "Relationship Study Between Reversed Phase HPLC Retention and Octanol/water Partition among Amphoteric Compounds," Journal of Liquid Chromatography, 18, 807-825.

Tchobanoglous, G., Stensel, D. (2003) Wastewater Engineering - Treatment and Reuse (4th edition), McGraw-Hill, New York.

Ternes, T., Joss, A., Siegrist, H. (2004) "Scrutinizing Pharmaceuticals and Personal Care Products in Wastewater Treatment," Environmental Science and Technology, 38, 393A-399A.

Ternes, T., Hirsch, R. (2000) "Occurrence and Behavior of X-ray Contrast Media in Sewage Facilities and the Aquatic Environment," Environmental Science and Technology, 34, 2741-2748.

Ternes, T., Meisenheimer, M., McDowell, D., Sacher, F., Brauch, H-J., Haist-Gulde, B., Preuss, G., Wilme, U., Zulei-Seibert, N. (2002) "Removal of Pharmaceuticals During Drinking Water Treatment," Environmental Science and Technology, 36, 3855-3863.

Ternes, T. (2001) "Pharmaceuticals and metabolites as contaminants of the aquatic environment," ACS Symposium Series 791(Pharmaceuticals and Personal Care Products in the Environment), 39-54.

Tolls, J. (2001) "Sorption of Veterinary Pharmaceuticals in Soils: A Review," Environmental Science and Technology, 35, 3397-3406.

Vanderford, B., Snyder, S. (2006) "Analysis of Pharmaceuticals in Water by Isotope Dilution Liquid Chromatography/Tandem Mass Spectrometry," Environmental Science and Technology, 40, 7312-7320.

Vieno, N., Tuhkanen, T., Kronberg, L. (2007) "Elimination of Pharmaceuticals in Sewage Treatment Plants in Finland," Water Research, 41, 1001-1012.

Westerhoff, P., Yoon, Y., Snyder, Sh., Wert, S. (2005) "Fate of Endocrine-Disruptor, Pharmaceutical, and Personal Care Product Chemicals during Simulated Drinking Water Treatment Processes" Environmental Science and Technology, 39, 6649-6663.

Williams, R., Johnson, A., Smith, J., Kanda, R. (2003) "Steroid Estrogens Profiles along River Stretches Arising from Sewage Treatment Works Discharges," Environmental Science and Technology, 37, 1744-1750.

CHAPTER 4

Personal Care Products

Pascale Champagne

4.1 Introduction

Personal care products (PCPs) are a class of compounds which consist of consumer products marketed for direct use primarily on the human body. These products exclude prescription and over the counter pharmaceutical products which are intended to have documented physiological effects (Daughton and Ternes, 1999), and with the exception of neutraceuticals and food supplements, PCPs are generally not intended for ingestion. There are thousands of chemicals that are constituents of PCPs. These are diverse and are used as active ingredients or preservatives in cosmetics; skin care, dental care and hair care products; soaps and cleansers; insect repellents; sunscreen agents; fragrances; and flame retardants. Many of these PCPs are used in very large quantities, and often at dosages and frequencies larger than are recommended.

The active ingredients in a number of PCPs are considered bioactive chemicals, which inherently implies that they have the potential to affect the flora and fauna of soil and aquatic receiving environments. In some cases, bioactive ingredients are first subjected to metabolism by the consumer and excreted metabolites and parent compounds can then be subjected to further transformations in receiving environments. Some metabolites can be converted back to their parent compounds. The numbers and effects of these chemical compounds can be amplified via further biological and physicochemical transformation processes, which can result in the formation of other modified structures such as chains of metabolites, some of which are also considered to be bioactive (Daughton, 2004). These bioactive parent compounds, metabolites and transformation products can be continuously introduced to receiving environments as complex mixtures (Daughton and Ternes, 1999).

Personal care products differ from pharmaceuticals in that large quantities can be directly introduced into receiving environments (air, surface and ground water, sewage, sludges and biosolids, landfills, soils) through regular use, such as showering, bathing, spraying, excretion or the disposal of expired or used products. Because of this uncontrolled release, they can bypass possible treatment systems.

Many treatment systems have not been designed for the treatment or removal of these compounds, particularly those present in micro-concentrations. Hence many of these compounds are eventually discharged into receiving environments, particularly aquatic environments. Through their continuous introduction into receiving environments, even active parent compounds of PCPs or metabolites that are generally considered to have a low persistence can display the same exposure potential as persistent pollutants, since their transformation or removal rates, which are typically more rapid than those of persistent pollutants, can be compensated by their replacement rates (Daughton and Ternes, 1999). As a result, PCPs are referred to as pseudo-persistent contaminants (Barceló and Petrovic, 2007).

From the perspective of industrial ecology, exposure to the bioactive parent compounds or metabolites in PCPs is an example of consumption-related environmental impacts, that is, impacts that occur in the consumption or use phase of the product life cycle (Wormuth et al., 2005). The environmental fates and effects of many of the broad class of chemical constituents found in PCPs are poorly understood, although considerable persistence and bioaccumulation in non-target aquatic organisms have been reported (Ternes et al., 2004), some exhibiting negative hormonal and toxic effects on a number of aquatic organisms at concentrations as low as µg/kg (Daugton and Ternes, 1999). To date, few studies that have considered whether the interactive effects of combined low concentrations from various classes of PCPs, and whether their chemical constituents, metabolites and transformation products have any significance with respect to ecological function. Of particular concern is the fact that subtle immediate effects could occur and go undetected in the flora and fauna of receiving environments, but long-term cumulative effects could result in irreversible changes, by the time they are recognized (Daughton and Ternes, 1999). Data are also limited regarding the effects of unexpected exposure on human health.

Although they are produced and sold in large quantities on an annual basis, with annual production exceeding 1×10^6 tonnes worldwide (Richardson et al., 2005), compared to pharmaceutical products, concerns and interest regarding the potential fate and effects of PCPs on receiving air, aquatic and soil environments only emerged in the late 1990's. An increase in activity has been possible in this research area, in part, as a result of the development of more sophisticated and sensitive analytical instruments, which have allowed for more PCPs to be detected at trace levels in various environmental matrices (Xia et al., 2005). Hence, in recent years, a wide range of PCPs has been detected ranging from ng/kg to g/kg scales in a variety of environmental matrices (Daughton and Ternes, 1999; Snyder et al., 2001a; Koplin et al., 2002; Boyd et al., 2004; Chen et al., 2004; Jjemba, 2006; Moldovan, 2006; Jones-Lepp and Stevens, 2007; Zhang et al., 2007).

This chapter will outline some of the issues and concerns associated with the broad class of chemical constituents derived from PCPs, through consideration of their properties and potential ecological effects; their source and occurrence in natural systems; and their fate in treatment systems and receiving environments.

4.2   Properties

Personal care products include a broad class of chemical compounds, their metabolites and transformation products. Table 4.1 lists some of the more common PCPs of concern to receiving environments with their respective solubility and log K_{ow} values, and categorizes them in terms of their primary functional classes or uses. Most PCPs are more polar than traditional contaminants and the majority have acidic or basic functional groups (Shon et al., 2007). They are also considered to be micro-pollutants present at trace level concentrations (<1 ug/L), which is often below the detection limits of current analytical approaches. These properties create unique challenges for both analytical detection and removal processes (Snyder et al., 2003b). In addition, the pseudo-persistent nature of these compounds, characterized by their often high transformation or removal rates compensated by their continuous introduction into the environment (Barcelo, 2003), has the potential to have significant ecological effects on receiving environments. However, the ecotoxicolotical effects of chronic low-concentration exposures to the majority of these compounds is not clearly known or understood, hence most of these compounds are not currently regulated. In fact, for most emerging contaminants, detection, occurrence, risk assessment and ecotoxicological data are not yet available and it is therefore difficult to predict their actual risk (Shon et al., 2007).

As the majority of emerging contaminants of concern found in PCPs are released through activities such as showering, bathing, washing or disposed of down the drain, water is an extremely important medium for transporting these compounds to and within receiving environments. The effectiveness of removal mechanisms for PCPs (sorption, biodegradation, volatilization and photooxidation) are largely dependent on the chemical structure of the compounds and their physico-chemical properties such as solubility, lipophilicity and acidity. According to Carballa et al. (2005a), PCPs can be divided into 3 main groups: lipophilic (high K_{ow}), neutral (non-ionic) and acidic (hydrophilic and ionic) compounds.

Solubility refers to the maximum quantity of solute that dissolves in water at a specified temperature, where an increase in dissolution results in the breaking up of solute-solute and solvent-solvent intermolecular bonds and the subsequent formation of solute-solvent intermolecular bonds. The log K_{ow} reflects the portioning of the organic compounds between natural organic phases and water in addition to being an indicator of the lipophilicity of a compound. A high K_{ow} are typically associated with hydrophobic compounds which are more soluble in octanol than in water, while a low K_{ow} typically indicates a compound that is soluble in water. In theory, an increase in the C/O and H/O ratios implies the substitution of the oxygen atom with halogens, H or N, which results in a lower the K_{ow}. From an ecological perspective, this would facilitate the transfer of the polar compound onto cells and enhance the bioaccumulation of the compound (Jjimba, 2006). The partitioning coefficient (K_{ow}) can also affect the sorption of the compound on soils, sediments minerals and dissolved organic material, where low K_{ow} compounds have a reduced affinity for sorption, leading to the enhanced bioavailability of the compound in the environment

Table 4.1. Selected common personal care product compounds of concern and their uses.

Compound (synonym)	Use	CAS Number	Chemical Formula	Chemical Structure	Molecular Weight (g/mol)	Water Solubility (mg/L)	Log Kow
Acetophenone (methylphenyl ketone)	Fragrance	98-86-2	CH₃COC₆H₅		120.2	6130	1.6
Benzalkonium chloride (alkylbenzyl dimethylammonium chloride)	Surfactant	63449-41-2	C₆H₅CH₂N(CH₃)₂RCl (R=C₈H₁₇ to C₁₈H₃₇)		340.0	> 1000	n/a
Benzenesulfonic acid (sodium dodecylbenzenesulfonate)	Surfactant	25155-30-0	CH₃(CH₂)₁₁C₆H₄SO₃Na		348.5	800	0.45
Benzyl salicylate (benzyl-2-hydroxybenzoate)	Sunscreen	118-58-1	2-(HO)C₆H₄CO₂CH₂C₆H₅		228.2	25	4.3
Biphenylol (2-hydroxybiphenyl)	Disinfectant	90-43-7	C₆H₅C₆H₄OH		170.2	700	3.1
Bisphenol A (2,2-Bis(4-hydroxyphenyl)propane)	Disinfectant	80-05-7	(CH₃)₂C(C₆H₄OH)₂		228.3	300	3.4
Butylparaben (butyl-4-hydroxybenzoate)	Preservative Anti-oxidant	94-26-8	HOC₆H₄CO₂(CH₂)₃CH₃		194.2	207	3.6

Compound (synonym)	Use	CAS Number	Chemical Formula	Chemical Structure	Molecular Weight (g/mol)	Water Solubility (mg/L)	Log K_{ow}
Butylated hydroxyanisole (BHA) (tert-butyl-4-methoxyphenol)	Preservative Anti-oxidant	25013-16-5	$(CH_3)_3CC_6H_3(OCH_3)OH$		180.2	213	3.5
Cashmeran (DPMI) (6,7-dihydro-1,1,2,3,3-pentamethyl-4(5H)-indanone)	Fragrance	33704-61-9	$C_{14}H_{22}O$		206.3	0.17	4.9
Celestolide (ADBI) (6-tert-butyl-1,1-dimethylindan-4-yl methyl ketone)	Fragrance	13171-00-1	$C_{17}H_{24}O$		244.4	0.015	6.6
Chloropentabromocyclohexane (1,2,3,4,5-pentabromo-6-chlorocyclohexane)	Flame Retardant	87-84-3	$C_6H_6Br_5Cl$		513.1	0.055	4.7
Chloroxylenol (4-chloro-3,5-xylenol)	Disinfectant Antiseptic	88-04-0	$ClC_6H_2(CH_3)_2OH$		156.6	250	3.3
Clorophene (2-benzyl-4-chlorophenol)	Disinfectant Antiseptic	120-32-1	$C_6H_5CH_2C_6H_3(Cl)OH$		218.7	149	4.2

Compound (synonym)	Use	CAS Number	Chemical Formula	Chemical Structure	Molecular Weight (g/mol)	Water Solubility (mg/L)	Log K_{ow}
Chlorocresol (4-Chloro-m-cresol)	Disinfectant Antiseptic	50-59-7	$ClC_6H_3(CH_3)OH$		142.6	3830	3.1
DEET (N,N-diethyl-m-toluamide)	Insect Repellent	134-62-3	$C_{12}H_{17}NO$		191.3	912	2.2
Ethyl hydrocinnamate (ethyl 3-phenylpropionate)	Sunscreen	2021-28-5	$C_6H_5CH_2CH_2COOC_2H_5$		178.2	220	2.7
Eusolex 2292 (2-ethylhexyl-p-methoxycinnamate)	Sunscreen	5466-77-3	$CH_3OC_6H_4CH=CHCO_2CH_2CH(C_2H_5)(CH_2)_3CH_3$		290.4	0.16	5.8
Eusolex 4360 (Benzophenone)	Sunscreen	119-61-9	$(C_6H_5)_2CO$		182.2	137	3.2
Eusolex 6007 ((2-ethylhexyl 4-(dimethylamino)benzoate)	Sunscreen	21245-02-3	$(CH_3)_2NC_6H_4CO_2CH_2CH(C_2H_5)(CH_2)_3CH_3$		277.4	0.0053	5.8

Compound (synonym)	Use	CAS Number	Chemical Formula	Chemical Structure	Molecular Weight (g/mol)	Water Solubility (mg/L)	Log K_{ow}
Eusolex 6300 (4-methylbenzylidene camphor)	Sunscreen	38102-62-4	$C_{18}H_{22}O$		254.4	0.013	5.9
Eusolex HMS - Homosalate (3,3,5-trimethylcyclohexyl salicylate)	Sunscreen	118-56-9	$C_{16}H_{22}O_3$		262.4	Insoluble	6.2
Eusolex OCR - Octocrylene (2-ethylhexyl 2-cyano-3,3-diphenylacrylate)	Sunscreen	6197-30-4	$C_6H_5)_2C=C(CN)CO_2CH_2CH(C_2H_5)(CH_2)_3CH_3$		361.5	Insoluble	6.9
Galaxolide (HHCB) (1,3,4,6,7,8-hexahydro-4,6,6,7,8,8-hexamethyl-cyclopenta[γ]-2-benzopyran)	Fragrance	1222-05-5	$C_{18}H_{26}O$		258.4	1.75	5.9
Hydrocinnamic Acid (3-Phenylpropionic acid)	Sunscreen	501-52-1	$C_6H_5CH_2CH_2COOH$		150.17	5900	1.8
Methyl dihydrojasmonate (cyclopentaneacetic acid, 3-oxo-2-pentylmethyl ester)	Fragrance	24851-98-7	$C_{13}H_{22}O_3$		226.3	Insoluble	3.0

Compound (synonym)	Use	CAS Number	Chemical Formula	Chemical Structure	Molecular Weight (g/mol)	Water Solubility (mg/L)	Log K_{ow}
Methylparaben (methyl 4-hydroxybenzoate)	Preservative Anti-Oxidant	99-76-3	$HOC_6H_4CO_2CH_3$		152.2	2500	2.0
Methyl cinnamate (methyl 3-phenylpropenoate)	Sunscreen	103-26-4	$C_6H_5CH=CHCOOCH_3$		162.2	387	2.6
Musk ambrette (2-tert-butyl-4,6-dinitro-5-methylanisole)	Fragrance	83-66-9	$C_{12}H_{16}N_2O_5$		268.3	Insoluble	4.2
Musk ketone (4-tert-butyl-2,6-dimethyl-3,5-dinitroacetophenone)	Fragrance	81-14-1	$C_{14}H_{18}N_2O_5$		294.3	1.9	4.3
Musk moskene (1,1,3,3,5-pentamethyl-4,6-dinitroindan)	Fragrance	116-66-5	$C_{14}H_{18}N_2O_4$		278.3	Insoluble	5.4
Musk tibetene	Fragrance	145-39-1	$C_{13}H_{18}N_2O_4$		266.3	Insoluble	5.2

Compound (synonym)	Use	CAS Number	Chemical Formula	Chemical Structure	Molecular Weight (g/mol)	Water Solubility (mg/L)	Log K_{ow}
Musk xylene (5-tert-butyl-2,4,6-trinitro-m-xylene)	Fragrance	81-15-2	$C_{12}H_{15}N_3O_6$		297.2	0.49	4.5
4-Nonylphenol (p-nonylphenol)	Surfactant	104-40-5	$CH_3(CH_2)_8C_6H_4OH$		220.4	6.4	6.0
4-Octylphenon (p-octylphenol)	Surfactant	1806-26-4	$CH_3(CH_2)_7C_6H_4OH$		206.3	2.9	5.9
Oxybenzone (2-hydroxy-4-methoxybenzophenone)	Sunscreen	131-57-7	$HOC_6H_3(OCH_3)COC_6H_5$		228.2	68.6	3.8
Pentabromotoluene (2,3,4,5,6-pentabromotoluene)	Flame Retardant	87-83-2	$C_6Br_5CH_3$		486.6	0.00094	7.0
Phantolide (AHMI) (acetyl-1,1,2,3,3,6-hexamethylindan)	Fragrance	15323-35-0	$C_{17}H_{24}O$		244.3	0.027	6.7
Surfynol 104 (2,4,7,9-Tetramethyl-5-decyne-4,7-diol)	Surfactant	128-86-3	$(CH_3)_2CHCH_2C(CH_3)(OH)C{\equiv}C(CH_3)(OH)CH_2CH(CH_3)_2$		226.4	1700	2.8

Compound (synonym)	Use	CAS Number	Chemical Formula	Chemical Structure	Molecular Weight (g/mol)	Water Solubility (mg/L)	Log K_{ow}
TBPP (tris[2,3-dibromo-1-propyl] phosphate)	Flame Retardant	126-72-7	$C_9H_{15}Br_6O_4P$		697.6	8	4.3
TCEP (tris[2-carboxyethyl]phosphine hydrochloride)	Flame Retardant	51805-45-9	$C_9H_{15}O_6P \cdot HCl$		286.7	7820	1.6
Tergitol NP10 (polyethylene glycol nonylphenyl ether)	Surfactant	9016-45-9	$(C_2H_4O)n(C_{15}H_{24}O)$			>1000	n/a
Tetrabromo-o-cresol (2-Methyl-3,4,5,6-tetrabromophenol)	Disinfectant Antiseptic	576-55-6	$CH_3C_6Br_4OH$		423.7	insoluble	5.6
Tetrabromobisphenol A (4,4'-isopropylidenebis[2,6-dibromophenol])	Flame Retardant	79-94-7	$(CH_3)_2C[C_6H_2(Br)_2OH]_2$		543.87	0.001	7.2
Tetrabromophthalic anhydride (3,4,5,6-tetrabromophthalic anhydride)	Flame Retardant	632-79-1	$C_8Br_4O_3$		463.7	0.019	5.6
Tonalide (AHTN) (1-[5,6,7,8-tetrahydro-3,5,5,6,8,8-hexamethyl-2-naphthyl]ethan-1-one)	Fragrance	21145-77-7	$C_{18}H_{26}O$		258.4	1.25	5.7

Compound (synonym)	Use	CAS Number	Chemical Formula	Chemical Structure	Molecular Weight (g/mol)	Water Solubility (mg/L)	Log K_{ow}
Traseolide (ATII) (1-(2,3-dihydro-1,1,2,6-tetramethyl-3-(1-methylethyl)-1H-inden-5-yl)ethan-1-one)	Fragrance	68140-48-7	$C_{18}H_{26}O$		258.4	0.085	8.1
Tris(2-chloroethyl) phosphate	Flame Retardant	115-96-8	$(ClCH_2CH_2O)_3P(O)$		285.5	7000	1.4
Triclocarban (1-(4-chlorophenyl)-3-(3,4-dichlorophenyl)urea)	Disinfectant Antiseptic	101-20-2	$Cl_2C_6H_3NHCONHC_6H_4Cl$		315.6	Insoluble	4.9
Triclosan (5-chloro-2-(2,4-dichlorophenoxy)phenol)	Disinfectant Antiseptic	3380-34-5	$C_{12}H_7Cl_3O_2$		289.5	10	4.8
Triphenyl phosphate	Flame Retardant	115-86-6	$(C_6H_5O)_3PO$		326.3	4.6	1.9

(Jjimba, 2004). From Table 4.1, it can be seen that PCPs with a range of solubilities are released into the environment. However, the solubility did not correlate with the bioavailability, the proportion of PCPs excreted nor the concentrations of the compounds in the aquatic environment (Jjimba, 2006). It should be noted that studies by Tolls (2001) have indicated that the K_{ow} may not be a good descriptor of the behavior of PCPs in the environment.

4.2.1 Functional Classes

Large quantities of PCPs such as food supplements, fragrances, skin and hair care products, insect repellents, cleaning products, flame retardants and homeopathic products are produced and sold in large quantities worldwide each year. The chemical constituents found PCPs are often grouped into functional classes depending on their use: fragrances, sun-screen agents and photoinitiators, disinfectants/antiseptics and insect repellents, surfactants, preservatives and anti-oxidants, fire retardants and nutraceuticals (Table 4.1).

4.2.1.1 Fragrances

Fragrances are used extensively in PCPs such as soap, perfumes, detergents and shampoos, and other PCPs. This class of compounds includes acetophenone and synthetic musks (nitro, polycyclic and macrocyclic musks). Synthetic musks were first identified in aquatic samples in the early 1980's and comprise a series of structurally similar chemicals, which are designed to imitate the scent of the natural product from Asian musk deer. As a result of their widespread use, they are ubiquitous, persistent and bioaccumlative pollutants, and some forms are considered to be highly toxic (Daughton and Ternes, 1999). The global distribution of synthetic musks use in PCPs include candles, air fresheners, and aroma therapy (41%); perfumes, cosmetics and toiletries (25%); soaps, shampoos and detergents (34%) (Xia et al, 2005). Nitro musks and polycyclic musks are the most commonly used synthetic musks, where a worldwide production of 7600 Mg was reported in 1996 (Rimkus, 1999). Polycyclic musks, which are substituted indanes and tetralins, are the primary musks used today, accounting for almost two-thirds of the global market, while the less expensive nitro musks, which are nitrated aromatics, account for approximately one third of global production (Daughton and Ternes 1999)

Polycyclic musks, have been reported to act as inhibitors of multixenobiotic resistance in aquatic organisms (Luckenbach et al., 2004). Recent studies regarding the environmental risk assessment of exposure to polycyclic musks show that their concentrations in the aquatic environment could exceed predicted no-effect concentrations (PNEC) (Matamoros and Bayona 2006). Galaxolide and tonalide, a tetralin derivative, are polycyclic musks fragrances that are not readily biodegradable and are generally classified as persistent. They are considered to be lipophilic with log K_{ow} values around 5.5-6 (Carballa et al., 2005a) and, as such, are likely to accumulate in sediments, sewage sludge, and also in aquatic fauna (Fromme et al., 2001; Rimkus 1999; Gatermann et al, 1999). These compounds have also been found in human adipose tissue and in breast milk (Zehringer and Herrman (2001); Muller et al, 1996; Rimkus and Wolf 1996). Hence, these compounds have a high potential for bioaccumulation and biomagnifications (Liebig et al., 2005). Galaxolide is included in the US EPA's list of high production volume chemicals targeting chemicals produced or imported in the US at more than 450 Mg/yr (Xia et al 2005).

The use of nitro musks has been reduced considerable and is being phased out in many parts of the worlds because of their persistence and related toxicity concerns. Amino musks are the transformation products of nitro musks and are considered to be toxicologically significant, where amino musk compounds often show greater toxicity than parent nitro musk compounds (Daughton and Ternes, 1999).

4.2.1.2 Sunscreen Agents and Photoinitiators

A number of PCPs contain UV filters and photoinitiators. They are extensively used in sunscreen products and cosmetics to protect the skin from damage by blocking out harmful UV radiation, where some filters protect against UVB irradiation at 280-315 nm, while others protect against UVA irradiation at wavelengths in the range of 315-400 nm (Cuderman and Heath, 2007). Two or more compounds are typically used to protect against UVA and UVB, and higher SPF values implies a higher concentration of UV filter in the product (Steinberg, 2000). UV filters (e.g. benzophenone) are also used to prevent UV light from damaging scents and colors in products such as perfumes and soaps, particularly products packaged in clear glass or plastic. They are also used as additives in plastics and packaging as UV blockers and/or flavor agents. (Cuderman and Heath, 2007).

There are two basic types of UV filters, inorganic and organic. Inorganic filters (TiO2, ZnO) primarily reflect, scatter and absorb UV light, whereas organic filters absorb UV light (Balmer et al., 2005; Steinberg, 2000). Organic filters include benzophenone, methylbenzylidene camphor, methyl phenylpropionate, ethyl phenylpropionate, octyl methoxycinnamate, benzophenone and oxybenzone. Due to the nature of their use and application, a number of sunscreen products including HCA, methyl phenylpropionate, ethyl phenylpropionate, octyl methoxycinnamate, benzophenone, and oxybenzone have been reported in drinking water and wastewater samples. Methylbenzylidene camphor has been reported to bioconcentrate in lake species (Daughton and Ternes, 1999). Oxybenzone and octyl methoxycinnamate have been shown to be weakly estrogenic (Schlumpf et al., 2004). Nagtegall et al. (1997) reported that several sunscreens bioconcentrate in fish. For the most part environmental toxicity is unknown (Lorraine and Pettigrove, 2006).

4.2.1.3 Disinfectants, Antiseptics and Insect Repellents

Disinfectants, antiseptics and insect repellents are a class of pesticide compounds used in a wide array of PCPs as general antibacterial, antifungal or insecticidal agents in toothpastes, cosmetics, lotions, soaps, plastic kitchenware and toys. These compounds are often substituted phenolics such as biphenylol, 4-

chlorocresol, clorophene, bromophene, 4-chloroxylenol, and tetrabromo-*o*-cresol and triclosan, as well as others such as the insecticide DEET added in insect repellents. These are found as active ingredients in PCP formulations at percentage volumes of < 1-20 % (Daughton and Ternes, 1999).

Of these compounds, triclosan is perhaps the most commonly used in PCPs. It is an antimicrobial agent found in many had soaps (0.1-0.3%) (Thomas and Foster, 2005) and a bactericide added in detergents, dishwashing detergents, launder soaps, deodorants, cosmetics, lotions, creams, toothpastes and mouthwashes and footwear. It is also used as a preservative and disinfectant in many consumer products including medical skin creams (Singer et al., 2002) and has more recently been introduced as a slow release product (Microban) incorporated in a wide variety of plastic products (Okumura and Nishikawa, 1996). Triclosan is often formulated with chloroxylenol (Perencevich et al., 2001). Although triclosan was long perceived as a toxicant with non-specific mechanism(s) of action, such as gross membrane disruption, it is now thought that triclosan blocks lipid biosynthesis by specifically inhibiting the enzyme enoyl-acyl carrier protein reductase (McMurray et al., 1998). Levy et al. (1999) found that triclosan acts as a potent site-directed enzyme inhibitor by mimicking its natural substrate. Given its mechanism of action, bacteria could develop resistance and change in microbial diversity.

Clorophene is commonly used bactericide and fungicide added in "down-the-drain" PCPs such as disinfectant solutions and soaps (Xia et al., 2005). It acts as a preservative in cosmetic products to prevent the growth of microorganisms (Cuderman and Heath, 2007). Biphenylol is a bactericide and virucide added in dishwashing detergents, soaps, general surface disinfectants in hospitals, nursing homes, veterinary hospitals, commercial laundries, barbershops, and food processing plants (Xia et al., 2005).

4.2.1.4 Preservatives and Anti-Oxidants

Parbens (alkyl-*p*-hydroxybenzoates) and butylated hydroxyanisole (BHA) are common and widely used anti-microbial preservatives in cosmetics, skin creams, tanning lotions, toiletries, neutraceuticals and food (up to 0.1 % wt/wt). The acute toxicity of these compounds is generally considered to be very low (Daughton and Ternes, 1999). However, Routledge et al. (1998) reported that methyl, ethyl and butyl parabens displayed weak estrogenic activity in several assays. In particular, butylparaben showed the most competitive binding to the rat estrogen receptor at concentrations of one to two orders of magnitude higher than that of nonylphenol (Daughton and Ternes, 1999). Similarly, BHA is a known endocrine disruptor (Lorraine and Pettigrove, 2006). Hence, the continual introduction of this class of PCPs into receiving aquatic environments leads to the question of risk to aquatic organisms.

4.2.1.5 Surfactants

Surfactants are surface-active agents that lower the surface tension of a liquid, allowing easier spreading. They are typically organic compounds that consist of both hydrophobic and hydrophilic groups, which allows them to be semi-soluble in both organic and aqueous solvents. Surfactants are widely added to cleaning products used in a range of household applications including detergents, toothpastes, shampoos, shaving creams, contact lens solutions and bubble bath (Xia et al., 2005; Shon et al., 2007).

Surfactants can be ionic or non ionic, and in some cases they will congregate together and form micelles. Ionic surfactants include sodium deoxycholate and sodium dodecylbenzyenesulfonate added to detergents, as well as benzalkonium chloride added to contact lens solutions (Xia et al., 2005). Non ionic surfactants such as alklyphenon ethoxylates, nonylphenol, octylphenol, alkylphenol carboxylates, Surfynol 104 (2,4,7,9-tetramethyl-5-decyn-4,7-diol) and Tergitol are extensively used in PCPs. In 1995, more than 200 000 Mg/yr alklyphenon ethoxylates, almost half of the world's annual production , were produced in the US. (Xia et al., 2005). Tergitol NP10 is a polymer of ethylene oxide and nonylphenol that has been shown to disrupts cell membrane integrity and to deionize the cells. It is also contained in most spermicidal lubricants and gels (nonoxynol-9), hair dyes and nail treatments (Wilson et al., 2003).

4.2.1.6 Flame Retardants

Flame retardants are often included in the class of PCPs and include compounds such as alkylphenone ethoxylates, nonylphenol, octylphenol and alkylphenol carboxylates. These compounds are used as additives in flexible polyurethane foam, coatings for furniture and textiles, plastics for electrical and electronic equipment, wire, cable insulation, electrical connectors, automobiles, as well as construction and building materials. In 1998, the distribution of 1.14 million Mg consumed on a global sale included: Al, Mg and N based (56%), Br based (23%), P based (15%) and Cl based (6%) (Xia et al., 2005). The global market demand for PBDEs was 67,440 Mg in 2001, of which 83% was distributed to the North and South Americas (Hites, 2004). The worldwide market growth for flame retardants is currently estimated to be 4% per year (Xia et al., 2005).

4.2.1.7 Nutraceuticals and Herbal Remedies

In the last two decades, there has been an explosive rise in the popularity of bioactive food supplements (nutraceuticals), as well as herbal and natural remedies as products of alternative medicine. The majority of these are not regulated and are available over the counter, and many can have significant physiological effects. Although a number of nutraceuticals and herbal remedies have either proven or suspected biologic activity, because a given botanical compound usually has an array of particular compounds that in combination elicits a targeted response, these products are difficult to standardize, which makes the regulation of these compounds particularly challenging (Daughton and Ternes 1999). Some of these nutraceuticals and herbal remedies are so effective that the medical community is concerned about their abuse/misuse.

Several nutraceutical are consumed and enter receiving environment extensively. It is believe that because these are naturally occurring compound, they should be more easily degraded and pose less of an ecotoxicological risk to organisms in receiving environments. However, other factors that need to be considered include the extent to which these compounds are consumed. In addition, the fact that many of these natural compounds originate from isolated parts of the world, their use and introduction into foreign environments renders them effectively anthropogenic to new receiving environments (Daughton and Ternes, 1999). These considerations have yet to be addressed in the literature.

4.2.2 Effects of Personal Care Products on Receiving Environments

Similarly to pharmaceuticals, their human metabolites and estrogens, PCPs are often not completely removed by wastewater treatment systems. Hence, receiving environments including surface waters through direct effluent discharge, as well as soils and groundwater from the land application of wastewater treatment sludges and biosoilds, are continuously exposed to PCPs compounds and their transformation products. In this respect, it is assumed that PCPs could act as persistent compounds, which could lead to a multigenerational exposure for the organisms in receiving environments (Ferrari et al., 2003). Aquatic organisms, in particular, are captives in their environment from which they cannot avoid continual exposure (Daughton and Ternes, 1999). Concerns regarding the uncontrolled release of PCP and their potentially harmful effects have grown in the last decade (Lishman et al., 2006). Hence, as a result of the pseudo-persistent nature of PCPs compounds released into receiving environments, a number of research initiatives have emerged to examine the potential short-term and long-term ecological effects of individual and combinations of the active chemicals, metabolites and transformation products derived from PCPs.

Daughton (2004) described three factors that, in combination, could enhance the adverse effects from chemical stressors in organism in receiving environments. These chemical compounds would (1) possess structural stability in a particular environment (persistence or long environmental half-lives); (2) be lipophilic and, as such, more amenable to passively crossing cellular membranes leading to the accumulation and concentration the compound in lipids and fat, as well as, bioconcentration leading to bioaccumulation via the food chain; and (3) possess acute or chronic toxicity properties in target and non-target species.

The physico-chemical characteristics such as solubility, log K_{ow} and pK_a are used in pharmacokinetic studies in clinical settings and their use has been transplanted in predicting the behavior of pharmaceuticals, estrogens, PCPs, as well as their metabolites and transformation products in environmental assessments. For example, Kasim et al. (2004) reported that log K_{ow} values of 1.72 or greater were associated with higher bioavailability of pharmaceuticals in clinical settings and that the compound was likely to bioaccumulate in the environment. However, studies by Tolls (2001) indicated that the log K_{ow} may not be a good descriptor of the behavior of PCPs in the environment. Similarly, Jjimba (2006) noted that the solubility, log K_{ow} and pK_a values compared in his study did not correlate well with environmentally significant factors such as the proportion of compounds released or their concentrations in the environment. It was further noted that current approaches in assessing the risks from PPCPs in the environment generally focused on acute toxicity. These findings underscored the need for further research regarding the behavior of PCPs in the environment and with consideration for the ecotoxicity of these compounds.

It is reasonable to assume that PCPs have significant impacts on natural biotic communities. For example widely used antimicrobial agents such as those found in hand soaps and toothpastes are typically designed to kill or to inhibit the growth of a broad spectrum of undesirable microbial species, which could potentially cause unintended impacts on sensitive co-existing non-target organisms within the exposed environment. The ability to predict these potential collateral effects on receiving environments is currently very limited. Moreover, individual chemical compounds potentially can interact synergistically or antagonistically with other chemical compounds that may also be present in the receiving environment (Wilson et al., 2003). An ecotoxicity potential assessment was proposed by Jjimba (2006) that would take into account the fact that a variety of PCPs and other contaminants are often present at very low concentrations over prolonged periods of time, as well varying biological activity.

However, to date, the ecotoxicological effects of PCPs are even less well known and understood than their environmental distribution. The majority of PCPs have ill-defined biochemical mechanisms of action, and many interact with multiple non-therapeutic receptors resulting in potentially adverse effects in non-targeted receptors, in a number of organisms (Richardson et al, 2005). Daughton and Ternes (1999) identified the major threats resulting from exposure to PCPs as subtle, continual but unperceivable effects, which accumulate slowly such that major changes are undetected until the cumulative effects finally amount to irreversible change. Cumulative effects are defined as general, diffuse and ill-defined alterations impairment or inhibition of enzyme systems, protein turnover, metabolism and cytotoxic repair; leading to reduced fitness, gradual degeneration or atrophy of tissues and organs, reduced growth, accelerated aging, impaired immunologic systems, impaired reproduction, higher incidence of disease, and impaired adaptation, survival and succession (Daughton and Ternes 1999). Changes related to community composition and organization, as a result of the presence of PCPs can also be incurred.

The widespread addition of triclosan to numerous household products has resulted in significant releases of this compound to the environment in the last 40 years. A number of studies have investigated the effects of Triclosan on receiving environments, particularly aquatic ecosystems (Heath and Rock, 2000; Chuanchuen et al., 2001; Kolpin et al, 2002; Singer et al., 2002; Bester, 2003; Wilson et al., 2003; Canesi et al., 2007; DeLorenzo and Fleming, 2008; Dussault et al., 2008). Triclosan can act as an antimicrobial agent either by membrane disruption or via enoyl-ACP reductase inhibition (Loraine and Pettigrove, 2005). Whether or not sublethal exposures to triclosan could lead to microbial resistance has been the subject of discussion (Heath and Rock, 2000; Chuanchuen et al., 2001). Early studies demonstrated that triclosan could be photolized to 2,8-dichlorodibenzodioxin or be methylated to a more bioaccumulative form in natural waters (Bester, 2003). Triclosan has been reported to be present in wastewater treatment effluents and surface waters at concentrations high enough to affect algal speciation (Kolpin et al., 2002; Singer et al., 2002; Wilson et al., 2003). Wilson et al., 2003 demonstrated a consistent reduction of algal genus diversity as the concentrations of triclosan (0.015 μg/L-1.5 μg/L) were experimentally increased. The algal bioassays revealed that triclosan could significantly modify algal community structure in vitro, suggesting that there could be a strong potential for corresponding effects on the structure and function of natural stream ecosystems receiving this compound. The possible modes of action of triclosan were investigated in the marine bivalve *Mytilus galloprovincialis* Lam by Canesi et al. (2007). Results demonstrated that triclosan could act on a kinase-mediated cell signalling, lysosomal membranes and redox balance in different systems/organs of mussels. In mussel hemocytes (immune cells), *in vitro*, short-term exposures to triclosan in the low μM range were observed to reduce lysosomal membrane stability and induce the extracellular release of lysosomal hydrolytic enzymes (Canesi et al., 2007). Dussault et al. (2008) noted that that benthic invertebrates species *Chironomus tentans* and *Hyalella azteca* were most sensitive to triclosan, when compared to the lipid regulator atorvastatin, the antiepliptic drug carbamezepine and the synthetic hormone 17α-ethinylestradiaol. Triclosan has also been found to bioconcentrate in fish and in human breast milk (Adolfsson-Eric et al., 2002)

Polycyclic musks have also been found in a various receiving environments including the aquatic food chain, as well as in fatty tissues and breast milk (Fromme et al., 2001; Kannan et al., 2005). The bioconcentration/bioaccumulation potential for nitro musks has been reported to compare to that of more persistent organohalogens. In addition, nitro and amino nitro musks have shown very high acute aquatic toxicity in aquatic receiving environments (Daughton and Ternes, 1999).

Alkylbenzene sulfonates were the most commonly employed surfactants in commercial products until 1965, when it was determined that they could resist biodegradation. As a result, they were replaced by linear alkylsulfonate, which is biodegradable (Shon et al., 2007). Surfactants are widely used despite some risks to aquatic environments presented by certain types of detergents and metabolic degradation products. For instance, certain degradation products from the widely used alkylphenol polyethoxylate surfactants have been shown to be estrogenic and to bioaccumulate (Snyder et al., 2001b; Snyder et al., 2001c). A study by Wilson et al. (2003) investigating the effect of Tergitol NP10 at concentrations ranging between 5 µg/L-500 µg/L on natural freshwater algal populations, demonstrated a consistent reduction of algal genus diversity as the concentration for these compounds was experimentally increased.

As a result of dermal application followed by swimming activities, common sunscreen ingredients such as 2-phenylbenzimidazole-5-sulfonic acid and 2-phenylbenzimidazole, are widely released into recreational waters. These compounds have been shown to cause DNA breakage when exposed to UV-B (Daughton and Ternes, 1999). In addition, nutraceuticals could present risks to non-target species in receiving environments because their usage serves to redistribute and extend their normal occurrence in the environment. This could result in higher concentrations in surface waters than would normally be detected at their geographic sites of origin (Daughton and Ternes, 1999).

Given the current widespread influx of PCPs and other contaminants in receiving environments, organisms are most likely exposed to complex mixtures of compounds of which PCPs represent only a fraction. Unfortunately, very little is known regarding the effects of these mixtures on the organisms of receiving ecosystems. The breakdown products and conjugates of many PCPs may have chemical structures similar to those of the parent compounds and may also have effects on organisms sensitive to mode of action of the parent compound (Wilson et al., 2003). Acute toxicity, the major type of investigations used with nontarget species, is only one of many possible ecotoxicological points of concern. The simple extrapolation of effects from higher concentrations does not necessarily provide a predictive value for lower concentrations. Hence, further investigations must include chronic toxicity studies (Oetken et al., 2005), which will contribute important information regarding the risks posed by these contaminants in receiving environments. The toxicological data that currently exist for nontarget species are almost exclusively focused on antibiotics or endocrine disruptors (Daughton and Ternes, 1999).

The reaction of these compounds when mixed at various concentrations in the receiving environments is largely unknown. Mixtures of organic contaminants have been found to have significant direct effects on vertebrates even at low-level concentrations (Harris et al., 2001; Sohoni et al., 2001). In higher-level organisms exposed to multiple contaminants, additive effects are likely to occur, particularly with chemicals that are designed to cross cell membranes (Harris et al., 2001; Sohoni et al., 2001). DeLorenzo and Fleming (2008) performed a study on the marine phytoplankton species *Dunaliella tertiolecta*, using different mixtures of PPCPs including triclosan. They found that, as an individual compound under acute toxicity testing, triclosan had the highest impact on phytoplankton compared to individual pharmaceuticals: simvastatin, clofibric acid, diclofenac, carbamazepine and fluoxetine. They also reported that in mixtures, triclosan demonstrated additive toxicity effects. It was concluded that the presence of PPCP mixtures could lead to a decrease in the toxicity threshold for phytoplankton populations and that negative effects on phytoplankton populations could in turn impact nutrient cycling and food availability for higher trophic levels (DeLorenzo and Fleming, 2008).

Clearly the effect of the discharge of PCPs on receiving environments is a complex issue that remains to be addressed by ecotoxicologists. Comprehensive risk assessments will not be possible without considering the simultaneous presence of other contaminants along with PCPs. In determining ecological risk, consideration will need to be given to both additive and synergistic effects. Although the concentration of one PCP many be relatively low, the additive effects of multiple PCPs sharing a like mode of action must be considered as these could be significant (Daughton and Ternes, 1999). The nature of these relationships should be explored further to determine whether threshold concentrations exist at which no significant ecological effects of these chemicals can be observed. However, these studies will likely be very complex because parent compounds can be transformed into other biologically active compounds via various physical, chemical and biological processes. To better predict exposure risks, data on consumer behaviours, including frequencies of product application and amount of product applied are essential, in addition to physico-chemical properties of the compounds of concern (Wormuth et al. 2005).

4.2.3 Analysis of Personal Care Products

To date, no single analytical method is sufficient to provide comprehensive PCP monitoring. However, a number of analytical methods have been developed for the extraction and detection of PCPs from various environmental matrices (Simonich et al., 2000; Golet et al., 2001; LaGuardia et al., 2001; Petrovic et al, 2001; Hyotylainene and Hartonen, 2002; Patterson et al, 2002; |Covaci et al., 2003; Snyder et al., 2003a; Petrovic et al., 2002; Laguna et al., 2004; Richardson, 2004; Xia et al., 2005; Trenholm et al., 2006). Although some studies have shown that PCPs can be detected using advanced analytical techniques, these techniques are still limited to the identification and quantification of particular compounds within the matrix, while whether these compounds can have ecotoxicological impacts cannot yet be ascertained directly (Snyder et al., 2003a).

Since PCPs represent broad classes of chemical compounds, the analytical methods employed in their analysis are also quite varied. However, most analytical instruments are not able to directly detect PCPs at the trace levels at which they are present. In general, sample analyses involve some or all of the following general procedures : multiple extractions, large sample volumes, sophisticated sample clean up, and derivatization prior to analysis. The general procedure is outlined in Table 4.2 for environmental samples in solid and liquid matrices. An extraction step is used to concentrate target compounds to a detectable level. Conventional extraction techniques such as liquid-liqued (LLE), solid phase (SPE) and accelerated solvent (ASE) have been used to extract PCPs from liquid samples. However, SPE is by far

Table 4.2. General procedure for the analysis of personal care products in environmental samples for solid and liquid matrices

Solid Matrices (bisolids, sediments, sludges, soils)	Liquid Matrices (leachates, surface waters, wastewaters, wetlands)
Sample pretreatment	
• Griding • Homogenization • Lyophilization	• Filtration with glass filter • pH adjustment based on compound of interest
Extraction *(Target compound-defined solvent)*	
• ASE • MASE • PLE • Sohlet extraction • SPE • SPME	• LLE • SPE • ASE
Clean up **Optional*	
• Solvent exchange • Column chromatography (packing material target compound dependent) o Alumina o Carbon o Florisil o Modified silica gel	• Solvent exchange • Column chromatography (packing material target compound dependent) o Alumina o Carbon o Florisil o Modified silica gel
Concentration *(drying of extractant and redissolution in smaller volume of target compound defined solvent)*	
• Rotary evaporation • Vacuum evaporation • Drying with nitrogen gas	• Rotary evaporation • Vacuum evaporation • Drying with nitrogen gas
Analytical Instrumentation and Detectors *(derivitization required prior to analysis)*	
• GC/MS • CG/MS-SIM • GC/MS/MS • HPLC/UVvis • LC/MS-ESI	• GC/MS • GC/MS-SIM • GC/MS/MS • NCI-GC/MS • HPLC/DAD-MS • HPLC/FLD • HPLC/FLD/UV • HPLC/UV • LC/MS/MS

ASE = accelerated solvent extraction; CC = capillary column; DAD = diode array detection; ESI = electron spray ionization; FLD = fluorescence detector; GC = gas chromatography; HPLC = high performance liquid chromatography; LC = liquid chromatography; LLE = liquid-liquid extraction; MS = mass spectrometry; NCI = negative chemical ionization; PLE = pressurized liquid extraction; SIM = select ion monitoring; SPE = solid phase extraction; SPME = solid phase micro-extraction; UV = ultraviolet detector; UVvis = ultraviolet visible detector

the most commonly used (Snyder et al., 2003a). Among the extraction methods for solid samples, ASE has been reported to provide the most efficient and rapid extractions (Richter et al., 1996) and a gradient of multiple solvents can be used for to extract compounds of varying hydrophobicities (Xia et al., 2005). Following extraction, a cleanup step is often required to remove interferences from the sample matrix, particularly for environmental samples in solid matrices. These may include analyses such as gel permeation chromatography to remove large molecular weight compounds, or packed columns that can fractionate the extract base on polarity (Khim et al., 1999a; Khim et al., 1999b; Snyder et al., 1999; Snyder et al., 2001a; Snyder et al., 2001c). The most commonly used packing materials are modified silica, Florisil, alumnia and different types of carbon (Xia et al., 2005). The sample is then dried and redissolved in a smaller volume of solvent to increase the concentration of the target compounds and improve its detection. Finally, depending on the nature of the target compounds, PCPs in the concentrated samples are identified and quantified using a variety of analytical instruments. Specific analytical approaches that have been employed in the detection of selected PCPs are listed in Table 4.3. With the growing concern of EDCs, pharmaceuticals and PCPs in environmental matrices, there is a need for robust, rapid and sensitive analytical methods that can screen for a wide variety of compounds simultaneously (Trenholm et al., 2006).

4.3   Occurrence and Behavior in Natural Systems

A number of PCPs have been detected over a wide range of concentrations in a variety of receiving environments (Halling-Sorensen et al., 1998; Daughton and Ternes, 1999; Kolpin et al., 2002; Guenther et al., 2002; Snyder et al., 2003b; Richarson, 2003), as well as in animal tissues, human blood, and breast milk (Rimkus et al., 1994; Rimkus et al., 1999; Snyder et al., 2001a; Adolfsson-Erici et al., 2002; Hites, 2004) due to the bioaccumulation of some of these compounds. Some PCPs appear to have considerable persistence in the environment which is likely due to the fact that PCPs are used and, therefore, released to the environment on a continual basis. In the absence of very short half-lives, even exposures to non-persistent compounds can be constant and have a significant effect on organisms in receiving environments after extended periods of time (Daughton and Ternes, 1999).

4.3.1 Sources

As has been previously noted, large quantities of PCPs are consumed each year worldwide through the use of skin and hair care products, dental care products, cosmetics, personal scents and deodorants, soaps, sunscreen agents, hair styling products, cleaning products, nutraceuticals and herbal remedies. Little is known about the transport, occurrence and fate of many of the PCPs that are released into the global environment by human activities (Kolpin et al., 2002). It is also unknown how these compounds react and/or interact when introduced as mixtures of varying concentrations in receiving environmental matrices. The release of nutraceuticals and herbal remedies to receiving environments is a growing area of concern as these compounds could have the same potential fate as pharmaceuticals. However, their usage rates could be much higher than pharmaceuticals, as they are currently unregulated in most countries and can be taken without a prescription.

Table 4.3. Analytical methods for selected personal care products in environmental samples

Functional Class of Compounds	Compounds	Extraction Methods	Analytical Methods	Matrix	Reference
	Bisphenol A	n/a	GC/MS	Leachate effluent	Behnisch et al., 2001[b]
	Bisphenol A	n/a	HPLC	Sewage effluent, surface water	Fawell et al., 2001[b]
	Bisphenol A	n/a	GC/MS	Wastewater	Nasu et al., 2001[b]
	Triclosan	ASE	GC/MS-SIM	Wastewater	Kenda et al., 2003
	Bisphenol A, clorophene, triclosan	SPE	GC/MS-SIM	Stormwater	Boyd et al., 2004
	Triclosan	SPE	GC/MS-SIM	Wastewater	Lishman et al., 2006
	Triclosan	SPE	GC/MS	Drinking water, reclaimed water	Loraine and Pettigrove, 2006
	Triclosan	SPE	GS/MS-SIM	Surface water	Moldovan, 2006
Disinfectant/ Antiseptic	Triclosan	SPE	LC/MS/MS	Wastewater	Trenholm et al., 2006
	Triclosan	SPE	LC/MS/MS	Surface water, synthetic water	Westerhoff et al., 2006
	Triclosan	SPE	LC/MS/MS	Surface water, synthetic water	Yoon et al., 2006; 2007
	Biosol, biphenylol, p-chloro-m-xylenol, p-chloro-m-cresol, clorophene, triclosan,	SPE	GS/MS-SIM	Wastewater	Yu et al., 2006
	Clorophene, triclosan	SPE	GS/MS	Recreational waters, wastewater	Cuderman and Heath, 2007
	Bisphenol A, triclosan	SPE	GC/MS	Wastewater	Nakada et al., 2007
	Triclosan	MASE	GC/MS-SIM	Spiked soil, natural sediment	Rice and Mitra, 2007
	Bisphenol A, triclosan	SPE	GS/MS-SIM	Surface water	Zhang et al., 2007
	Triclosan	LLE	NCI-GC/MS	Surface water	Dussault et al., 2008
	Polybrominated diphenylethers (PBDEs)	Soxhlet	GC/MS-SIM	Sediment	Sellstrom et al., 1998[c]; Allchin et al., 1999[c]; Covaci et al., 2003[c]
Flame Retardants	Polybrominated diphenyl ether (PBDEs)	n/a	GC/MS	Synthetic water	Rahman et al., 2001[b]
	Tetrabromobisphenol A Tetrachlorobisphenol A	n/a	HPLC/UVvis	Sediment	Voordeckers et al., 2002[b]
	TCEP	SPE/LLE	GC/MS/MS	Wastewater	Trenholm et al., 2006
	TCEP	SPE	LC/MS/MS	Surface water, synthetic water	Westerhoff et al., 2006
	TCEP	SPE	LC/MS/MS	Surface water, synthetic water	Yoon et al., 2006
	Musk ketone, musk xylene	n/a	GC	Surface water, fish	Yamagishi et al., 1983[b]
Fragrances	Galaxolide	n/a	HPLC/UV	Synthetic water	Rimkus, 1999[b]
	Musk ketone, Musk xylene	LLE	GC/MS/MS	Liquid sewage	Berset et al., 2000[c]
	Galaxolide, tonalide	SPE	GC/MS	Wastewater	Simonich et al., 2000[c]
		ASE	GC/MS	Solid sludge	

Functional Class of Compounds	Compounds	Extraction Methods	Analytical Methods	Matrix	Reference
	Cashmeran, celestolide, galaxolide, musk ambrette, musk ketone, musk moskene, musk, tibetene, phantolide, tonalide, traseolide	ASE	GC/MS	Wastewater	Kenda et al., 2003
	Galaxolide, tonalide	Soxhlet	GC/MS-SIM	Digested sewage sludge	Steven et al., 2003[a]
	Galaxolide, tonalide	SPME	GC/MS	Wastewater	Carballa et al., 2004; 2005a; 2005b
	Galaxolide, tonalide	PLE	GC/MS	Biosolids, sludge	Kinney et al., 2006[a]
	Celestolide, galaxolide, phantolide, tonalide, traseolide	LLE	GC/MS	Wastewater	Lishman et al., 2006
	Galaxolide, methyl-dihydrojasmonate, tonalide	SPE	GC/MS	Wetland, wastewater	Matamoros and Bayona, 2006
	Galaxolide, tonalide	SPE	GC/MS	Surface water	Moldovan et al., 2006
	Galaxolide, tonalide	PLE	GC/MS-SIM	Biosolids	Osemwengie, 2006[a]
	Galaxolide, musk ketone	SPE/LLE	GC/MS/MS	Wastewater	Trenholm et al., 2006
	Galaxolide, musk ketone	SPE	GC/MS/MS	Surface water, synthteric water	Westerhoff et al., 2006
	Galaxolide, musk ketone	SPE/LLE	GC/MS/MS	Surface water, synthteric water	Yoon et al., 2006
	Galaxolide, tonalide	SPME	GC/MS	Biosolids, sludge	Carballa et al., 2007a; 2007b
	Galaxolide, methyl-dihydrojasmonate, tonalide	SPE	GC/MS	Wetland, wastewater	Matamoros et al., 2007
	Musk ketone	MASE	GC/MS-SIM	Spiked soil, natural sediment	Rice and Mitra, 2007
	Galaxolide, tonalide	SPE	GC/MS	Wastewater, groundwater	Ternes et al., 2007
Insect repellent	DEET	SPE	GC/MS	Drinking water, reclaimed water	Loraine and Pettigrove, 2006
	DEET	SPE	LC/MS/MS	Wastewater	Trenholm et al., 2006
	DEET	SPE	LC/MS/MS	Surface water, synthetic water	Westerhoff et al., 2006
	DEET	SPE	LC/MS/MS	Surface water, synthetic water	Yoon et al., 2006; 2007
Preservative	BHA	SPE	GC/MS	Drinking water, reclaimed water	Loraine and Pettigrove, 2006
Sunscreen	Benzophenone, hydrocinnamic acid, octyl methoxy cinnamate	SPE	GC/MS	Drinking water, reclaimed water	Loraine and Pettigrove, 2006
	Oxybenzone	SPE	LC/MS/MS	Wastewater	Trenholm et al., 2006
	Oxybenzone	SPE	LC/MS/MS	Surface water, synthetic water	Yoon et al., 2006; 2007
	Benzophenone-3, ethylhexyl dimethyl PABA, ethlyhexyl methoxycinnamate, homosalate, 4-methylbenzylidene	SPE	GC/MS	Recreational waters, wastewater	Cuderman and Heath, 2007

Functional Class of Compounds	Compounds	Extraction Methods	Analytical Methods	Matrix	Reference
	camphor, octocrylene hydrocinnamic acid, oxybenzone	SPE	GC/MS-SIM	Wetland, wastewater	Matamoros et al., 2007
	Nonylphenol, nonylphenol polyethoxylates, octylphenol	n/a	GC/MS- silica CC	Drinking water	Carlile et al., 1996[b]
	Nonylphenol polyethoxy carboxylates	SPE	GC/MS	Sewage effluent, surface water	Field and Reed, 1996[b]
	Nonylphenol ethoxylate	n/a	HPLC/FLD	Synthetic water	Fielding et al., 1998[b]
	4-nonylphenol	n/a	HPLC/FLD	Fathead Minnows	Giesy et al., 2000[b]
	Nonylphenol, nonylphenol polyethoxylates, octylphenol	n/a	HPLC/FLD	Wastewater effluent, surface water	Snyder et al., 2000[b]
	Nonylphenol, nonylphenol polyethoxylates	n/a	HPLC/DAD-MS	Sewage water, surface water	Sole et al., 2000[b]
	Nonylphenol	n/a	GC/MS	Leachate effluent	Behnisch et al., 2001[b]
	Nonylphenol ethoxylate	n/a	GC/MS/MS	Sewage effluent, surface water	Fawell et al., 2001[b]
	Nonylphenol, octylphenol	n/a	NCI-GC/MS	Drinking water, surface water	Kuch and Ballschmiter, 2001[b]
Surfactants	Nonylphenol, nonylphenol polyethoxylates	ASE	GC/MS	Biosolids	La Guardia et al., 2001[c]
	Nonylphenol, nonylphenol ethoxylate	n/a	GC/MS	Wastewater	Nasu et al., 2001[b]
	4-nonylphenol, 4-tert-octylphenol	n/a	GC/MS	Fish	Tsuda et al., 2001[b]
	Benzalkonium chlorides	ASE/SPE	LC/MS-ESI	Sediment	Ferrer and Furlong, 2002[c]
	Nonylphenol, nonylphenol polyethoxylates	SPE	HPLC/FLD/UV	Wastewater	Keller et al., 2003[c]
	Surfynol	SPE	GC/MS	Drinking water, reclaimed water	Loraine and Pettigrove, 2006
	Nonylphenol, octylphenol	SPE	GC/MS	Wastewater	Nakada et al., 2007
	Nonylphenol	ASE/SPE	GC/MS, GC/MS/MS	Septic, groundwater Soil	Stanford and Weinberg, 2007

ASE = accelerated solvent extraction; CC = capillary column; DAD = diode array detection; ESI = electron spray ionization; FLD = fluorescence detector; GC = gas chromatography; HPLC = high performance liquid chromatography; LC = liquid chromatography; LLE = liquid-liquid extraction; MS = mass spectrometry; NCI = negative chemical ionization; PLE = pressurized liquid extraction; SIM = select ion monitoring; SPE = solid phase extraction; SPME = solid phase micro-extraction; UV = ultraviolet detector; UVvis = ultraviolet visible detector; [a]Jones-Lepp and Stevens, 2007; [b]Snyder et al., 2003a; [c]Xia et al., 2005

Figure 4.1 indicates the sources and pathways of PCPs in urban environments. the anthropogenic sources (oval shape) of PCPs are from households and hospitals where the products are utilized, ingested or applied, as well as industries where the PCPs are manufactured or employed in various PCP formulations. The primary cycle leading to PCP persistence in the environment is illustrated in Figure 1 (bold solid line). PCPs primarily enter wastewater treatment plants via sanitary sewers after their regular use during showering, bathing or washing, or through down the drain product disposal. Treated wastewater effluents have been shown to contain a variety of PCPs with treatment levels varying depending primarily on the design, operation and availability of the treatment processes (Ternes, 1998). PCPs that are not removed during various wastewater treatment processes enter receiving surface waters as via wastewater treatment plant discharges. Following the discharge of treated wastewater into the receiving water, residual PCPs are diluted and mixed with residuals that can be present in both surface waters and groundwater seepages (Ellis, 2006). Conventional water treatment processes are often not sufficient for removing residual PCPs from natural waters and require additional oxidation, activated carbon or membrane filtration treatments to achieve non-detectable concentrations for these compounds (Ternes et al., 2002). Compounds that are not degraded in surface waters and are not amenable to treatment during water treatment processes are then re-introduced into the urban water cycle via drinking water, which perpetuates the pseudo-persistent cycle of these compounds.

Household, hospital and industrial sources of PCPs can also be released via direct liquid discharge to receiving waters (thin dashed line), through swimming activities, stormwater runoff and sewer discharges, as well as industrial wastewater discharges from the manufacture, processing and distribution of PCPs. Entry from these sources typically pass directly to receiving surface waters, and are not subject to treatment. Reclaimed water is currently being employed for non-potable uses and has not been utilized for the purpose of direct reuse as a drinking water supply. To date, contributions from these pathways are generally considered to be negligible. However, population growth and drought cycles are limiting the availability of freshwater many arid parts of the world (Loraine and Pettigrove, 2006). As such, the increasing reuse of reclaimed domestic wastewater for irrigation could also lead to the release of PCPs via stormwater runoff from the irrigated areas that would not normally be a source of PCP (Ellis, 2006). In addition, the presence of PCPs in water supplies may become a more significant issue of concern as water reuse becomes more intensive in arid regions around the world (Loraine and Pettigrove).

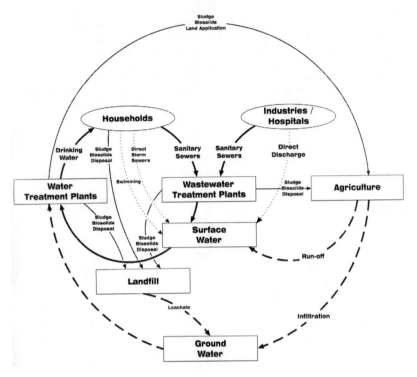

Figure 4.1 Sources and pathways of PCPs in urban environments.

Alternatively, PCPs can enter receiving environments through solid waste disposal routes (thin solid lines), in particular the disposal of unused products in refuse destined for landfills, as well as the landfilling or agricultural land application of water and wastewater treatment process sludges and biosolids. Because of the high lipophilicity of some PCPs compounds, these can significantly sorb onto sludges and sediments (Ternes et al., 2004). PCPs associated with land applied sludges and biosolids can then mobilize in the soil and leach to the groundwater or enter surface water through runoff (Jjimba, 2002; Pedersen et al., 2003). PCPs can also leach into groundwater receiving environments via improperly contained landfill leachate. Losses from agricultural and landfill sources (bold dashed line) pass directly to surface water and groundwater receiving environments, and are not subject to treatment.

Some PCPs, primarily fragrances, have also been detected in air samples. The most prevalent have been found to be the polycyclic musks which ranged in concentration from a few pg/m^3 to hundreds of pg/m^3 (Daughton and Ternes, 1999). However, to date, little research has focused on the potential short-term and long-term effects of continuous atmospheric releases of low concentrations of PCPs on humans and other organisms.

4.3.2 Surface Water and Groundwater

A number of fragrances, disinfectants and antiseptics, sunscreen agents, flame retardants and pesticides have been detected in surface water, groundwater and reclaimed water samples. Musks are considered to be non-biodegradable, with the exception of the reduction of nitro musks to amino derivatives, and have lent themselves to detection in a number of monitoring studies throughout the world. They are also highly lipophilic and which can lead to their bioconcentration and bioaccumulation. Concerns have been expressed regarding ecotoxicity of these PCPs in aquatic organisms (Daughtone and Ternes, 1999). In the early 1980's, Yamagishi et al. (1981; 1983) completed one of the first comprehensive monitoring studies of surface waters in Japan during which they identified musk xylene and musk ketone in freshwater fish, marine shellfish, river waters and sewage samples. Musk xylene was detected in 100% of the samples collected both upstream and downstream of wastewater treatment plants in concentrations ranging from 1 to 23 ng/L, while musk ketone was also found in a similar range of concentrations, but was generally not detected in samples upstream of wastewater treatment facilities. Muller et al. (1996) identified galaxolide, tonalide and celestolide at average concentrations of 136 ng/L, 75 ng/L and 3.2 ng/L, respectively, in water samples of the Glatt River in Switzerland. In the same study, the nitro musks tibetene, ambrette, moskene, ketone, and xylene were also detected at concentrations of 0.04, <0.03, 0.08, 8.3, and 0.62 ng/L, respectively. Heberer et al. (1999) noted concentrations of the polycyclic musks galaxolide, tonalide, and celestolide above 10 ug/L in surface water samples in and around Berlin (Germany). These waters received relatively high effluent discharges from waterwater treatment facilities and were considered to be contaminated despite receiving treated wastewaters. Galaxolide and tonalide were detected in concentrations ranging between 172-300 ng/L and 78-100 ng/L respectively, in water samples from the Some River in Romania (Moldovan, 2006). These concentrations were attributed to the discharge of untreated wastewaters at certain locations.

Triclosan is an antimicrobial agent found in a number of PCPs such as soaps (0.1-0.3%) and a it has been detected in a number of water samples. Singer et al., (2002) reported triclosan concentrations ranging between 11 and 98 ng/L in Swiss Rivers, Lindstrom et al. (2002) noted concentrations up to 74 ng/L in Swiss Lakes and concentrations ranging from 38-57 ng/L were detected in the Somes River in Romania (Moldovan 2006). Boyd et al. (2004) sampled the Orleans and London stormwater canals of New Orleans (Louisiana, U.S.) which are used to drain a portion of the city's stormwater directly into the Mississippi River or Lake Pontchartrain. Results from their 6-month monitoring study indicated triclosan and bisphenol A, a disinfecting agent, concentrations ranging ND-29 ng/L and 1.9-158 ng/L, respectively. Canal concentrations were generally found to increase with periods of increasing rainfall which was attributed to contamination due to cross flows with sanitary sewers in the aging New Orleans sewer system (Englande et al., 2003). In the same study, Boyd et al. (2004) also detected triclosan in Lake Pontchartrain samples, with a median concentration of 4.6 ng/L, which was lower than the concentrations of the canal waters (15 and 15.2 ng/L) and attributed to possible removal and degradation processes taking place in the surface water (Lindsrom et al., 2002; Tixier et al., 2002; Zhang and Huang, 2003). Bisphenol A was also detected in the Lake samples with a median concentration of 19 ng/L which was comparable to the concentrations of the canal waters (15.6 and 56 ng/L). These concentrations were likely due to the discharge from stormwater canals into Lake Pontchartrain. Other surface water monitoring studies noted bisphenol A concentrations ranging from 70-4000 ng/L in surface waters of Portugal (Azevedo et al., 2001), 9-776 ng/L in the Elbe River in Germany (Heemken et al., 2001), and 20-150 ng/L in the Tama River of Japan (Takahashi et al., 2003).

Sunscreen agents used in PCPs generally enter receiving environments through direct input into surface waters as a result of recreational activities such as swimming. The use of sunscreens is most prevalent in regions in which high-intensity sunshine is experienced for only a limited period of the year (Cuderman and Heath, 2007). Cuderman and Heath (2007) collected water samples from 17 recreational water sources (seawater, pools, lakes, rivers) in Slovania. They noted that the most commonly detected sunscreen agent was benzophenone-3 which was detected in concentrations ranging from 11-400 ng/L. Loraine and Pettigrove (2006) conducted a survey of raw and treated drinking water from 4 water filtration plants in San Diego County (California, U.S.) which indicated the occurrence of several polar organic pharmaceuticals and personal care products including sunscreen agents (hydrocinnamic acid (4.99-20.3 µg/L), octyl methoxycinnamate (0.56-5.61 µg/L), benzophenone (0.36-0.79 µg/L), triclosan (7.34 µg/L), surfynol 104 (0.326-0.818 µg/L), BHA (3.49-3.52 µg/L) and DEET (0.131 µg/L). A number of these were also found in the finished drinking water, with mean concentrations of 10.0 µg/L, 0.26 µg/L, 0.45 µg/L, 0.734 µg/L, 0.161 µg/L and 3.45 µg/L noted for hydrocinnamic acid, benzophenone, octyl methoxycinnamate, triclosan, surfynol 104 and BHA, respectively. Most of the PCPs found in the raw water were also found in the finished water, with the exception of DEET. Hence it was concluded that conventional water treatment plant processes may not be adequate to completely remove PCPs. The occurrence and concentrations of these compounds were in surface waters were found to be highly seasonally dependent, where higher concentrations were detected during the dry season and a number of PCPs were only detected during the dry season. Loraine and Pettigrove (2006) also examined reclaimed wastewaters, and noted mean concentrations of 22.3 µg/L, 0.993 µg/L, 1.43 µg/L, 1.08 µg/L and 1.31 µg/L for hydrocinnamic acid, benzophenone, triclosan, surfynol 104 and DEET, respectively, again indicating the limitations of current conventional wastewater treatment practices.

4.3.3 Wastewaters

The use of fragrances, sunscreens, preservatives, surfactants and anti-microbial agents PCPs, as well as nutraceuticals and herbal remedies is widespread and continues to grow. Many of these compounds are directly discharged to sanitary sewage systems, and many of these compounds are not readily amenable to treatment in conventional wastewater treatment processes, hence, a number of these compounds can find their way into receiving waters. A number of studies are available which have investigated the occurrence and removal of pharmaceutical and estrogenic compounds in wastewater treatment facilities due to their potential impact on receiving environments. Only on the last decade have the potential ecotoxicolgical effects of PCPs on receiving environments and the limitations of conventional wastewater treatment facilities for the effective removal of these compounds been recognized. Conventional municipal sewage treatment facilities were not traditionally designed to remove complex anthropogenic chemical mixtures with structures and transformation mechanisms of biological action that are foreign to biodegradation systems (Daughton, 2004). As such a number of recent studies have focused on the occurrence and fate of PCPs in wastewater treatment facilities.

In their early comprehensive monitoring study, Yamagishi et al. (1981; 1983) of surface waters completed in Japan, musk xylene was noted in 100% of samples and musk ketone was in 80% of the 74 samples analyzed. Concentrations in wastewater treatment facility effluents ranged from 25-36 ng/L for musk xylene and from 140-410 ng/L for musk ketone. Another early study of antiseptics and acid drugs in sewage and surface waters was conducted by Ternes (1998). Fourty-nine wastewater treatment facilities were surveyed in Germany where biphenylol and chlorophene were consistently detected in treatment facility influents and effluents. They noted concentrations as high as 2.6 µg/L for biphenylol and 0.71 µg/L for chlorophene in the influent and reported that the removal of clorophene from the effluent was less extensive than for biphenylol, where surface waters with concentrations similar to that of the effluents were observed. Gatermann et al. (1999) examined polycyclic and nitro musks in the environment. They found that in wastewater treatment facility influent, concentrations of musk xylene and musk ketone were 150 and 550 ng/L, respectively, while in the effluent, their concentrations dropped to 10 and 6 ng/L, respectively. They also detected the amino derivatives of musk xylene and musk ketone: 4-amino (34 ng/L) and 2-amino (10 ng/L) musk xylenes and 2-amino (250 ng/L) musk ketone. Although these amino derivatives could not be detected in the influent, their concentrations in the effluent increased, indicating an significant transformation of the parent nitro musks. It was concluded that the amino derivatives had the potential to be dischcarged in treated wastewater effluents at concentrations higher than those of the parent nitro musks.

In one of the first PPCP surveys in the U.K., Kenda et al. (2003) investigated 18 PPCPs including the antiseptic agent triclosan, 6 polycyclic musks (galaxolide, tonalide, cashmeran celestolide, phantolide, traseolide) and 5 nitro musks (musk ambrette, musk xylene, musk moskene, musk tibetene, musk ketone) in 6 wastewater treatment facilities in the U.K. Triclosan was detected in all influent wastewater samples as well as in all effluent samples, with concentrations as high as 3100 ng/L. Triclosan removal efficiencies at different wastewater treatment facilities was found to be highly variable and removal rates ranged from complete removal to completely ineffective, with average removal efficiencies of 95.6%. The concentrations of musks detected in the wastewater influents crude sewage were generally low with the exception of galaxolide, tonalide, musk xylene and musk ketone which ranged from 7.8-19.2 μg/L, 2.2-8.1 μg/L, <10-4700 ng/L and <10-2900 ng/L, respectively, which account for 95% of European usage (Kenda et al., 2003). Other musks compounds were also detected at lower concentrations and frequencies: traesolide (<10-2900 ng/L), phantolide (<10-100 ng/L), celestolide (<10-440 ng/L), cashmeran (<10-400 ng/L). Treated effluent samples indicated significant removal of galaxolide (39-83%) and tonalide (53-96%). Carballa et al. (2004) examined the fate of 13 PPCPs in a municipal wastewater treatment facility in Galicia (Spain). They noted that among the PPCPs considered, compounds with significant influent concentrations included the polycyclic musks galaxolide and tonalide, which were detected at concentrations in the ranges of 2.1-3.4 μg/L and 0.9-1.7 μg/L, respectively.

In a study examining the occurrence and reduction of 18 PPCPs at a 12 municipal wastewater treatment facilities discharging along the Thames River (Canada), Lishman et al. (2006) investigated the PCPs triclosan, as well as the 5 polycyclic musks celestolide, phantolide, traseolide, galaxolide and tonalide. These were detected at concentrations of 1.93 μg/L, 0.0372 μg/L, 0.0220 μg/L, 0.131 μg/L, 1.701 μg/L and 6.87 μg/L, respectively. Effluent concentrations were found to be on the order of 0.106 μg/L, 0.02 μg/L, ND, 0.047 μg/L, 0.876 μg/L and 0.298 μg/L for each of the compounds, respectively. A comparison between Canadian values and those of European studies indicated that polycyclic musk concentrations in Canadian treated wastewater effluents were generally 5-10 times lower. Matamoros and Bayona (2006) examined the removal efficiency and elimination rates of 11 PPCPs including galaxolide, tonalide and methyl-dihydrojasmonate in subsurface horizontal flow constructed wetlands partially treating wastewater generated by 200 inhabitants in the municipality of Les Franqueses del Valles (Spain). They reported average concentrations of these fragrances in the ranges of 0.23-1.32 μg/L, 0.14-0.45 μg/L and 6.88-38.71 μg/L.

Yu et al. (2006), examined the occurrence and biodegradability for 18 PPCPs in a Baltimore (U.S.) wastewater treatment facility. These included a number of phenolic antiseptic agents, triclosan, clorophene, biphenylol, biosol, 4-chloro-m-cresol, PCMS, which have similar chemical structures within similar concentration ranges, with average concentrations of 800, 750, 900, 250, 600 and 400 ng/L, respectively. Nakada et al. (2007) investigated the removal efficiencies of 24 pharmaceutically active compounds during sand filtration and ozonation in a water treatment facility that serves 460,000 people and treats 1.7×10^5 m^3 of wastewater, daily. Influent concentrations ranging from 316-784 ng/L, 123-359 ng/L, 505-1470 ng/L, 39.9-251 ng/L, 193-11,800 ng/L were noted for triclosan, thymol, nonylphenol, octylphenol, bisphenol A, respectively.

A great deal of work has been conducted recently to identify the occurrence and fate of these trace organic contaminants in wastewater treatment systems. To date, much of this work has focused on natural and synthetic hormones and antibiotics, but other compounds such as PCPs have been reported as well. Further research is required to expand the database of these for these compounds and establish effective removal strategies.

4.3.4 Soils, Sediments, Sludges and Biosolids

In many countries, the primary approaches for the management and disposal of sludges and biosolids from wastewater treatment processes involve incineration, landfilling or land application onto municipal and agricultural lands as a soil amendment or fertilizer. A significant concern arises when considering that biosolids and sludges applied onto land can contain high concentrations of contaminants including PCPs sorbed to the biosolid or sludge matrix. These can subsequently concentrate in soils with time and provide a source of pollutants that could eventually enter surface waters and groundwater via leaching and runoff processes (Xia et al., 2005). The available data regarding concentrations of PCPs in sludges and biosolids, are generally limited and these have primarily focused on fragrances and antimicrobial agents or disinfectants. Various fragrances have been detected at concentrations ranging from 1.5 to 147 ug/kg in biosolids from the U.S., Switzerland, and the Netherlands (Berset et al., 2000; Difrancesco et al., 2004). Stevens et al. (2003), reported that the most abundant synthetic musks were galaxolide and tonalide at concentrations of 27 mg/kg and 4.7 mg/kg, respectively, found in sewage sludge from 14 wastewater treatment facilities surveyed in the UK. Similar levels were found by Kupper et al. (2004) with concentrations of 20 mg/kg and 7 mg/kg, for Galoxilide and tonalide, respectively, and those detected by Osemwengie (2006) with concentrations of 18 mg/kg and 4.0 mg/kg, respectively. Nonylphenol, polyexthoxylates and nonylphenol have been detected in U.S. biosolids at concentrations as high as 981 mg/kg and 1380 mg/kg, respectively (La Guardia et al., 2001; Keller et al., 2003; Xia and Pillar, 2003). Fire retardants such as brominated diphenylethers have been reported at concentrations ranging from 32-4890 ug/kg in biosolids from a number of wastewater treatment facilities in the US and the Netherlands (Hale et al., 2001; De Boer et al., 2003).

The occurrence of PCPs in sediments is also of concern. With the persistence of these compounds in surface water environments, surface water sediments are continuously exposed to these compounds which can be sorbed onto the sediment matrix and accumulated over extended periods of time. Fooken (2004) measured polycyclic musk fragrances in streams in the German state of Hessen, and reported levels as high as 13 mg/kg total solids in suspended matter and 3.2 mg/kg dry weight in sediments, even though the water concentrations were only in the ng/L range. Winkler et al. (1998) measured musks in 31 particulate matter and water samples from the Elbe River in Germany. Concentrations of musk ketone, galaxolide, tonalide and celestolide in particulate matter samples, were in the range of 4-22 ng/g, 148-736 ng/g, 194-770 ng/g and 4-43 ng/g, respectively. Concentrations were generally found to be lower in the 31 water samples with concentrations of 2-10 ng/L, 36-152 ng/L, 24-88 ng/L and 2-8 ng/L for musk ketone, galaxolide, tonalide and celestolide, respectively.

The current knowledge with respect to the fate of PCPs in solid matrices is generally hindered by a lack of quantitative data regarding their occurrences in these environmental compartments (Cunningham et al., 2006; Difrancesco et al., 2004). Only in the last decade has there been a focus on the potential contamination of sludges and biosolids, and receiving amended soils with PCPs removed and often concentrated during water and wastewater treatment processes. Data concerning mixtures of pharmaceuticals, personal care products and other contaminants of concern are typically limited although individual subclasses have been identified in solid samples (Ternes et al., 2002; La Guardia et al., 2004). The primary barrier in the collection of these data is the limited number of analytical methods capable of reliably detecting the diversity of contaminants, including PCPs, present at trace concentrations in the complex solid matrices (La Guardia et al., 2004; Morales-Munoz et al., 2005). The differences in the extraction efficiencies of analytical methods for these compounds may not solely be attributed to the chemical structure, polarity or class of these compounds, but also to matrix effects which may limited the ability to more efficiently extract some of the target compounds from natural solids (Rice and Mitra, 2007). Jones-Lepp and Stevens (2007) stated that the analytical challenges to a completed analysis of sewage sludges and biosolids included overcoming the large negative surface charges and the interstitial spaces that provide multiple active sites for charged compounds.

4.4   Fate and Transformation in Engineered Systems

As previously discussed, PCPs have been detected in soils, sludges, biosolids, wastewater effluents, surface waters and groundwater supplies around the world. The pseudo-persistence of some PCPs in the environment could make them as important as currently regulated contaminants from an environmental perspective. This coupled with their occurrence at trace levels creates unique challenges for analytical detection in a variety of matrices and the assessment of removal performance by biosolids, water and wastewater treatment processes. To prevent PCPs form further entering the environment, there is an urgent need to document effective treatment techniques. A thorough understanding of transformation mechanisms for PCPs is the key foundation for developing better treatment approaches.

4.4.1 Surface Waters and Water Treatment

Following the discharge of treated wastewaters into the receiving surface waters, residual PCPs are diluted and mixed with compounds from surface water discharges, runoff and groundwater seepages. In surface waters, direct phototranformation has been proposed as a major reduction pathway (Tixier et al., 2002). Conventional water treatment processes appear to be insufficient in removing raw water residuals and require advanced treatment processes such as oxidation, activated carbon and adsorptive processes, or membrane filtration to achieve non-detectable concentrations (Ternes et al., 2002).

In studies involving surface waters, it was shown that triclosan could be photolized to 2,8-dichlorodibenzodioxin or methylated to a more bioaccumulative form (Bester, 2003). Phototransformation was also noted as the most likely removal pathway for triclosan in Romanian surface waters (Muldovan, 2006) and along the Rhine River (Tixier et al, 2002).

Westerhoff et al. (2005) investigated the treatment of 3 drinking water supplies spiked with 62 endocrine disrupting compounds (EDCs) and PPCPs, as well as one model water with a NOM isolate and spiked with 49 EDCs and PPCPs. These included 6 PCPs, namely DEET, oxybenzone, TCEP, triclosan, galaxolide and musk ketone. The removal of PCPs using coagulation was not found to be effective, where < 20% removal of PCPs were reported with ferric chloride and alum, respectively, DEET (0%, 6%), oxybenzone (0%, 0%), TCEP (0%, 0%), triclosan (0%, 13%), galaxolide (15%, 18%) and musk ketone (0%, 18%). Compared to coagulation, adsorption onto PAC was found to significantly improved the removal of these PCPs, where removals of 0%, 93%, 71%, 93%, 63% and 73% were observed for DEET, oxybenzone, TCEP, triclosan, galaxolide and musk ketone, respectively. This was attributed to the hydrophobic characteristics of PAC which allows it to interact with nonpolar organics (Shon et al., 2007). It was found that the octanol-water partitioning coefficients could serve as a reasonable indicator of the feasibility of compound removal under controlled PAC test conditions with the exception of compounds that were protonated or deprotonated at the test pH and those that contained heterocyclic or aromatic nitrogen (Westerhoff et al., 2005). Snyder et al. (2003a) demonstrated that PAC adsorption could also remove 60-80% of nonylphenol, nonylphenol ethoxylates, triclosan, bisphenol A and octylphenol present in surface waters.

The effectiveness of oxidation processes such as chlorination, ozone and ozone/H_2O_2) has also been studied. With the use of chlorine, the reactive site of molecules can be predictable, where a single aromatic bond is broken from a double bond by chlorination (Shon et al., 2007). Chlorination has been found to remove up to 90% of compounds containing substituted phenols such as oxybenzone (96%) and triclosan (97%), while ketone compounds or other compounds lacking these groups (galaxolide: 39%, musk ketone: 25%) were generally less reactive with chlorine, (Westerhoff et al., 2005; Gallard and von Gunten, 2002). Bisphenol A, a phenolic compound was noted to be very reactive (>90%) with chlorine by Hu et al. (2002). The least reactive compounds such as DEET (16%) and TCEP (4%) typically have electron-withdrawing functional groups or lack conjugated carbon bonds (Westerhoff et al., 2005). Compounds easily oxidized (> 80%) by chlorine were always found to be oxidized at least as efficiently as ozone by Westerhoff et al. (2005) and that the addition of H_2O_2 prior to ozonation generally enhanced the extent of oxidation compared to ozone alone. Snyder et al. (2005b) noted that ozone/H_2O_2 treatment could remove 83% DEET, 89% galaxolide, 33% musk ketone, 96% oxybenzone, 15% TCEP and 82% triclosan. The low removal of TCEP was attributed to the lack of aromatic moieties in this compound (Westerhoff et al., 2005).

Removal of PCPs using membrane technologies including reverse osmosis (RO), microfiltration (MF), ultrafiltration (UF), nanofiltration (NF) and membrane bioreactors (MBR) have also been reported (Shon et al., 2007; Yoon et al., 2007; Yoon et al., 2006; Snyder et al., 2007). Yoon et al (2007) examined removal of PPCPs from one synthetic and 3 surface waters using UF and NF and reported removals of 0% and 58%, 83% and 97%, 32% and 82%, and 93% and 97% for DEET, oxybenzone, TCEP and triclosan, respectively. The NF membrane retained a larger fraction of compounds than the UF membrane implying that retention is affected by the pore size of the membrane. More polar, less volatile and less hydrophobic compounds exhibited relatively low retention, which suggested that retention by NF and UF is governed by hydrophobic adsorption. They observed a general separation trend as a function in terms of hydrophobic adsorption as a function of compound octanol-water partitioning coefficient between the hydrophobic compounds and porous hydrophobic membrane, where retention for the NF and UF membranes increased with increasing log K_{ow} (Yoon et al, 2007; Yoon et al. 2006). Berg et al. (1997) noted that the rejection of uncharged trace organic compounds by NF membranes was influenced by steric hindrance, while the rejection of polar trace organics could be explained by electrostatic interactions with the charged membranes. Snyder et al., noted that MBR produced marginal improvement in the treatment of organic contaminants compared to activated sludge (Snyder et al., 2007).

To date, studies involving the treatment of source waters, particularly surface waters, for drinking water purposes have shown that that conventional treatment processes including coagulation and chlorination would have low removal of many PCPs and that the addition of adsorption processes such as PAC and/or advanced oxidative processes such as ozone and ozone/H_2O_2 could substantially improve their removals. Some compounds such as TCEP have been found to exhibit low removals with all water treatment processes considered to date and removal processes capable of removing these types of compounds will require further investigation (Westerhoff et al., 2005; Shon et al., 2007).

4.4.2 Wastewater Treatment

The primary removal processes of PCPs in wastewater streams are biological degradation, under aerobic or anaerobic conditions (depending on redox conditions), chemical degradation via hydrolysis or photolysis processes, as well as sorption where some lipophilic substances, due to their physio-chemical properties, will bind to suspended solids. Hydrophilic PCPs including polar metabolites will generally remain in the aqueous phase be discharged to receiving environments with the treated wastewater effluent (Kenda et al., 2003). Advanced adsorptive and oxidative treatment processes, such as those described for water treatment, could also contribute significantly to the removal of these contaminants in wastewaters.

Ternes et al. (2004) stated that whether or not trace organic compounds can be removed in wastewater treatment processes is largely dependent on the effectiveness of the biological treatment stage. They reported that sorption onto suspended solids in the wastewater and subsequent removal by sedimentation as primary and secondary sludge was a significant elimination process. Sorption involves two processes: (1) absorption through hydrophobic interactions of the aliphatic and aromatic groups of a PCP compound with the lipophilic cell membrane of microorganisms and the sludge fat fraction; and (2) adsorption involving electrostatic interactions between positively charged groups and negatively charged surfaces of microorganisms. Due to the trace concentrations ($<10^{-4}$ g/L) of PCPs in wastewaters (Ternes, 1998; Heberer, 2002), biological transformation or degradation is believed to occur only if a primary substrate is available to sustain a relevant microbial population. Hence, it is likely that the biological processes rely on co-metabolism where the microorganisms would break down or partially convert the trace organic pollutant without using it as its primary carbon source. Biological decomposition of a number of compounds has been found to increase with sludge age suggesting the presence of greater microbial population diversity or the acclimation of slower growing microorganisms with time (Ternes et al., 2004).

Carballa et al. (2004, 2005a, 2005b) investigated the fate of 13 PPCPs including galaxolide and tonalide in a wastewater treatment facility of northwest Spain. They reported that a removal of 30-50% of the polycyclic musk fragrances was achieved during primary treatment coinciding with the removal of suspended solids which would indicate that sorption onto solid particles may have been a key removal mechanism. In addition, removals of 30-40% for galaxolide and 45-50% for tonalide were obtained during secondary treatment which was attributed to biodegradation during biological treatment and adsorption onto biomass (Carballa et al., 2004; Carballa et al., 2005b). In another study, Carballa et al. (2005a) demonstrated that galaxolide and tonalide could be significantly removed during coagulation and flocculation with efficiencies as high as 70%. This was attributed to the fact that lipophilic compounds like polycyclic musks with high sorption properties (high log K_d) are readily adsorbed onto the lipid fraction of the sludge.

Yu et al (2006) examined the biodegradation of 18 PPCPs including 6 antiseptic PCPs: p-chloro-m-cresol, biosol, p-chloro-m-xylenol, biphenylol; triclosan and clorophene in wastewater from a wastewater treatment facility in Baltimore (U.S.). They noted that, under aerobic conditions, 4 out of 6 antiseptics showed a biodegradation greater than 30% within a 4 day period, and that p-chloro-m-cresol, biphenylol, and clorophene exhibited greater than 80% biodegradation after 14 days, indicating relatively fast biodegradation rates of these 3 compounds. Slower biodegradation rates were observed for biosol and p-chloro-m-xylenol as less than 60% biodegradation was observed after 21 days. After 70 days, a minimum biodegradation of 80% was reported for each of the 6 antiseptics. Thomas and Foster (2005) also noted triclosan removals ranging between 51 and 99% during the secondary or biological treatment of wastewater.

Matamoros and Bayona (2006) investigated the removal efficiency 11 PPCPs including PCP fragrances methyl dihydrojasmonate, galaxolide and tonalide in a pilot-scale aerobic and anaerobic subsurface constructed wetlands treating wastewater from a community of 200 inhabitants in Barcelona (Spain). Biodegradation and sorption onto the substrate, solids or filter media were identified as the two main removal mechanisms. Shallower, more aerobic wetland conditions generally yielded higher removal efficiencies of all PPCPs due to the less negative redox potential. Methyl dihydrojasmonate exhibited a high removal efficiency (>80%) within the subsurface wetland, while galaxolide and tonalide were mostly removed and retained in the gravel bed via sorption onto organic matter. In another wetland study, Matamoros et al. (2007) examined the removal efficiencies 13 PPCPs including methyl dihydrojasmonate, hydrocinnamic acid, oxybenzone, galaxolide and tonalide in a pilot-scale vertical flow constructed wetland and sand filter treating wastewater at a wastewater treatment facility in Arhus (Denmark). Aerobic conditions were found to be more effective in the removal of PPCPs than anaerobic conditions. Removals of 99% and 98%, 99% and 99%, 97% and 95%, 90% and 92%, and 82% and 82% were reported for the vertical flow wetland and sand filter systems, for methyl dihydrojasmonate, hydrocinnamic acid, oxybenzone, galaxolide, and tonalide, respectively under aerobic unsaturated flow conditions; while removals of 78% and 76%, 82% and 69%, 88% and 64%, 88% and 88%, and 75% and 73%, were reported for these respective compounds in the vertical flow wetland and sand filter system operated under anaerobic saturated flow conditions. The sorption of galaxolide and tonalide onto the organic matter retained in the substrate was found to be a significant removal mechanism. Plant uptake and biotranformation was identified as potentially contributing to the removal of neutral compounds such as methyl dihydrojasmonate and oxybenzone, while it would not be considered to be significant for negatively charged or highly hydrophobic compounds. However, phytoremediation processes were not investigated in either of these studies.

Lishman et al. (2006) examined the fate of PPCPs including triclosan, celestolide, phantolide, traseolide, galaxolide and tonalide in 12 municipal wastewater treatment facilities discharging along the Thames River (Canada). Triclosan removals ranged from 74% to 98% for the 12 facilities with a median reduction efficiency of 93-95%. Singer et al. (2002) noted that the fate of triclosan in wastewater treatment facilities was 79% biological degradation, 15% sorption to sludge and 6% discharged to receiving water. Influent and effluent concentrations for the 5 polycyclic musks indicated that galaxolide and tonalide were present at the highest concentrations, and phantolide was predominantly present at non-quantifiable concentrations. Median reductions for the 5 musks ranged between 37% and 65% in conventional activated sludge systems, while conventional activated sludge systems with filtration were expected to exhibit higher reductions of musks bound to the solid fraction of the wastewaters, although this trend was not clearly observed. Higher removals of traseolide, galaxolide and tonalide appeared to coincide with longer solids retention times. In lagoon systems, musk reduction for galaxolide and tonalide were observed to be on the order of 98-99%.

The use of biofiltration systems for the removal of PPCPs including a number of PCPs has been proposed. Removal of PCPs within these systems would occur in a two stage process: (1) sorption of the organic compound, and (2) biological degradation of the organic compound sorbed onto the sorption media (Shon et al., 2007). Snyder et al. (2007) demonstrated that 35% TCEP and 44.4% DEET removal could be achieved in a biofiltration system using granular activated carbon as the adsorption medium. Magnetic ion exchange resins (MIEX) have also been used in the removal of trace contaminants in wastewaters. Bench-sale and pilot-scale studies have shown its capability in removing negatively charged compounds (Bourke et al., 1999). On average 70-80% of treated wastewater organic matter are weak organic acids present in ionized form at a pH range of 6 to 8. Smaller compounds (MW 1000-10000 Da) have been shown to be exchanged rapidly during both resin loading and regeneration, while larger compounds (MW > 10,000 Da) have been observed to have slower exchange kinetics and to form stronger ionic bonds with the resin (Zhang et al., 2007). Snyder et al. (2004) demonstrated that a few PCPs such as triclosan (84-94%) and oxybenzone (4-40%) could be treated using this process, but most PPCPs could not be removed effectively.

PCPs in wastewaters can also be removed by advanced oxidation processes. Nakada et al. (2007) investigated the removal efficiencies of 24 pharmaceutically active compounds including 2 phenolic antiseptics (triclosan and thymol) and 3 phenolic endocrine disrupting chemicals (nonylphenol, octylphenol and bisphenol A) during sand filtration and ozonation in an operating municipal wastewater treatment facility in Tokyo (Japan). Triclosan and thymol had high removal efficiencies during ozonation, 91% and 84% respectively, which was attributed to their chemical structure. It was noted that compounds with a C-C double bond or an aromatic structure with electron donors such as phenols were particularly susceptible to ozonation. In thymol, the aliphatic functional groups on the aromatic ring, in addition to its phenolic structure, were believed to be the contributing factor in achieving consistently high removals because of their ability to donate electrons. Comparatively, the removal of triclosan and thymol during sand filtration was generally inefficient, probably due to the low hydrophobicities of these compounds. Ozonation was not as effective at alkylphenol removal. Approximately 6-67% of nonylphenol and -65-33% of octylphenol in the water were removed by ozonation. The low nonylphenol removal efficiencies by ozonation was consistent with results obtained by Petrovic et al. (2003), and were partly attributed to the formation nonylphenol from the breakdown of ethoxylated nonylphonol compounds during ozonation. Sand filtration was more successful at removing nonylphenon, octylphenol and bisphenol A which was likely due to the removal of suspended particles to which these hydrophilic compounds were adsorbed. In generally, the combination of ozonation and sand filtration with activated sludge treatment yielded efficient removals (>80%) of all the PCPPs. Light oxidation processes (photocatalysis) such as ultraviolet/H_2O_2 and titanium dioxide (TiO_2) have also been employed in the treatment of PCPs in wastewaters. Removals higher than 98% were obtained by using TiO2 photocatalysis to treat bisphenols (Esplugas et al., 2007).

4.4.3 Solids Treatment and Disposal

Studies have indicated that the sorption to wastewater treatment process sludges is the main removal mechanism for a number of PCPs observed in wastewaters (Keller et al., 2003; Xia et al., 2005). As previously noted, sludges and biosolids have particles with large surface areas (0.8-1.7 m^2/g), negative surface charges and interstitial spaces which promote sorption, promote occlusion into the solid matrix, as well as strong bonding between charged compounds and the solids surfaces (Jones-Lepp and Stevens, 2007). Wang et al. (1993) proposed that the sorption of toxic organic compounds onto the solid matrix of sludges was a 2-stage process: initial adsorption onto the surface of the sludge, followed by partitioning into the interior of the sludge solid matrix. Throughout the wastewater treatment process, PCPs and their transformation products partition into the solid matrix depending on the hydrophobicity of the compound, which is a function of its octanol-water partitioning coefficient (K_{ow}), where the greater the K_{ow} of the compound, the more hydrophobic the compound (Rogers, 1996). The sorption mechanisms presented by Wang et al. (1993) for toxic organic compounds could also apply to the PCPs with similar log K_{ow} values in the solid matrix of sludges and biosolids. The following was proposed as a general guideline: compounds with log K_{ow} < 2.5 have low sorption potentials, compounds with 4.0 > log K_{ow} > 2.5 have medium sorption potentials, and compounds with log K_{ow} > 4.0 typically have high sorption potentials (Jones-Lepp and Stevens, 2007). As can be seen from Table 4.1, a number of PCPs have log K_{ow} values which would be indicative of medium to high sorption potentials onto the solid matrix of sludges and biosolids.

Sludge recycling can yield mean cell residence times of biological solids in a wastewater treatment facility ranging from a few days to 30 days. However, these residence times can be shorter than the half-lives of certain PCPs (Halling-Sorensen et al., 1998). In addition, PCPs can also move into microsites within the solid matrix of sludges and biosolids, where the biodegradability of organic compounds can be reduced significantly as these microsites can present physical barriers to microbial degradation (Hatzinger and Alexander, 1995; Kelsey et al., 1997; Alexander, 2000) as well as challenges in terms of aerobic versus anaerobic degradation. Nitro musks have been reported to represent significant ecotoxicological potential in receiving environment. Aminobenzene (reduced) transformation products of the nitro musks created under anaerobic conditions as can be found in sludges and biosolids can be highly toxic (Daughton and Ternes, 1999).

Carballa et al. (2007a) examined the removal efficiencies of PPCPs including the musks galaxolide and tonalide in mesophilic and thermophilic digesters using non-ozonated and ozonated municipal sludge. They attributed the reduction in PPCPs to biodegradation since the volatilization of the compounds or their transformation products was known to be negligible. All compounds were noted to be removed to some extent during the anaerobic digestion process. Musks, with the exception of tonalide in pre-ozonated sludge digested at 55°C, were eliminated to a significant degree (60-80%) in both digesters. As a sludge pretreatment, ozonation was not found to enhance the removal rates of any compounds with the exception of tonalide. This was attributed to the fact that most PPCPs including musk fragrances are sorbed onto the sludge and biosolids, and are thus not readily exposed to the ozone, or that they exhibit slow reaction rates with ozone. Reactions with OH radicals were noted to be of minor relevance, since this oxidant should be mainly consumed by the elevated total suspended solids content (Dodd et al., 2006). Similar results were achieved under the mesophilic and thermophilic temperature regimes, thus temperature was not found to significantly affect the removal of galaxolide and tonalide.

Composting has been proposed as a safe alternative to degrade a number of PCPs associated with sludges and biosolids prior to their land application (Xia et al., 2005). The diversity of microorganisms, range in aerobic and anaerobic microenvironments, high temperatures, changing pH and availability of substrate are conditions which can enhance composting and the degradation of organic contaminants (Buyuksonmez et al., 1999); Barker and Bryson, 2002). A controlled pilot-scale study demonstrated that nonylphenol could be effectively degraded during composting (Buyuksonmez et al., 1999). Biosolids initially containing 450 mg/kg of nonylphenol were mixed with wood shavings at different dry weight ratios, and in all cases a rapid degradation of nonylphenol was noted within 15 days with removals as high as 80%. Temperature was found to significantly affect nonylphenol degradation particularly in the first 15 days of incubation with higher temperatures yielding higher nonylphenol reductions over shorter periods of time. No significant difference was reported for nonylphenyl degradation between composts treated at 45°C and 65°C, where approximately 80% of nonylphenol was degraded after 15 days and approximately 92% was degraded after 43 days. Nonylphenol degradation was comparatively slower at 25°C, where between 50 and 60% nonylphenol degradation was noted after 15 days, and an additional 55 d was required to achieve > 90% nonylphenol degradation. Lower biosolids to wood shaving ratios appeared to favour higher nonylphenol degradation. The results of the study indicated that that composting could provide an effective approach for the removal of certain PPCPs in biosolids and sludges. However, more research is required to determine the benefits of composting on the degradation of other PPCPs in biosolids and to establish effective composing parameters.

4.4.4 Soil and Groundwater Systems

Few studies have been conducted that focus on the mobility and transport of PCPs in sludge and biosolids-amended soils. However, it has been suggested that land application of these sludges anc biosolids can be a major route through which some PCPs enter the environment. The potential for surface runoff and leaching can transport PCPs into surface and groundwater. Tolls (2001) suggested that the mechanisms that play a role in the sorption of compounds in soil matrices included hydrophobic portioning which plays a significant role in the sorption of some compounds, as well as ion exchange, ion bridging at clay surfaces, surface complexation, and hydrogen bonding. These interactions would be significantly affected by soil pH, soil mineral composition and soil solution chemistry. The major transport process for strongly sorbed PCPs would likely be leaching through soil macropores or preferential transport facilitated by dissolved soil colloids (Thiele-Bruhn, 2003).

Photolysis and biodegradation are thought to be the major removal mechanisms of PCPs in soils. Hesselsoe et al. (2001) demonstrated that the aggregate size of biosolids can affects the aerobic transformation of organic contaminants such as nonylphenol in biosolids-amended soils through limitations in oxygen availability, where homogenous mixtures achieved completed nonylphenol degradation within a shorter period of time than non-homogeneous mixtures. In a laboratory investigation, a 55% reduction of nonylphenol in land applied biosolids was reported within 30 days of light exposure by Xia and Jeong (2004). Matscheko et al. (2002) detected elevated concentrations of PBDEs in a number of biosolids-amended soils tested as well as a corresponding significant accumulation of PBDEs in earthworms of the sites. This study suggests that PBDEs are persistent in the soil matrix, primarily due to the high hydrophobicity of the compounds. The flame retardant decabromodiphynyl ether, was noted to be debrominated photolytically in a soil matrix to form more resistant and bioaccumulative lower brominated bromodiphenyl ether compounds in a study by Soderstrom et al. (2004). Fragrances were noted to be removed in biosolids-amended soils. In an investigation by Difrancesco et al. (2004), the removal of 22 fragrances was observed 3 months after the spiking of biosolids-amended soils. Complete removal was noted for most of the tested compound with the exception of musk ketone and tonalide was observed within 1 year in soils with a wide range of textures and organic matter contents.

Ternes et al. (2007) suggested that the use of treated wastewaters for the irrigation of soils could contribute to the removal and limit the movement of PPCPs in the environment. In their study investigating the fate of 52 PPCPs including including musk fragrances galaxolide and tonalide in a soil environment, the pollution a groundwater as a result of irrigation using wastewater treatment facility effluents with or without the addition of digested sludge was examined. None of the compounds of interest which were present in the treated wastewater were observed in the groundwater which was attributed to the fact that many polar compounds do not sorb to sludge and lipophilic compounds are generally not mobile in soil aquifers. They reported that the acidic musk fragrances were likely sorbed or transformed while passing the top soil layer which was supported by their soil sorption coefficients of 150 /kg and 180 /kg which for galaxolide and tonalide, respectively.

A limited number of studies have investigated to occurrence of PCPs in urban groundwater systems, where contamination could be related to landfill leachate contamination (Ellis, 2006). Previous studies have noted that sewer exfiltration could also contribute PCPs in the soil and groundwater environment. Ellis and Revitt (2002) reported evidence of limited exfiltrations and PPCP contamination of shallow depths immediately adjacent to and below the trunk sewer, generally limited to 25-30 cm depths.

Despite the sorption of a number of PCPs to solids in water and wastewater treatment processes, to date, no information is available which correlates concentrations of PCPs released to the environment to the land application of PCP containing sludges and biosolids. Xia et al. (2005) proposed that although the characteristics of environmental behaviors of organic compounds such as those found in PCPs may be unique due to their close association to biosolids, existing environmental behavior models developed for industrial compounds and agricultural pesticides could be used to evaluate the fate and transport of PCPs of interests.

4.5   Conclusion

A wide range of PCPs and PCP-derived chemical compounds are considered to be micropollutants. They include a broad spectrum of compounds with a variety of polarities, solubilities, partitioning coefficients, neutral, acidic or basic moieties, and molecular weights. These properties, coupled with occurrence at trace levels within a variety of environmental matrices create unique challenges for both analytical detection and removal processes, which has presented a challenge to the research community and regulatory bodies. Personal care products are can released to receiving environments directly into recreational waters or volatilized into the air or by "down-the-drain" activities" which often transports the PCPs through a wastewater treatments system prior to discharge of the liquid stream into receiving water bodies or application of sludges and biosolids onto land. Moreover, solid environmental matrices (sludges, biosolids, soils, sediments) can pose challenges in terms of the development of PCP analytical methods, including the presence of particles with large surface areas, negative surface charges an interstitial spaces, all of which promote sorption onto the solid matrix, foster occlusion into the solid matrix and strong bonding between the charged species and solid matrix surfaces.

To be effective, treatment approaches for the removal of PCPs will need to include a combination of conventional and advanced oxidative and adsorption processes. Collectively, these treatment processes will contribute to the reduction in the ecotoxicological risk associated exposure to these compounds. The majority of ecological studies have focused on the presence of PCPs in aquatic environments as these often act as sources and sinks for PCPs of concern. However, few ecotoxicological studies have investigated chronic exposure of aquatic organisms to PCPs at environmentally relevant concentrations. These could induce subtle effects at these sublethal concentrations, such as behavior, immune function, and fecundity.

References

Adolfsson-Erici, M.; Pettersson, M.; Parkkonen, J. and Sturve, J. (2002) Triclosan, a Commonly Used Bactericide Found in Human Milk and in the Aquatic Environment in Sweden. Chemosphere, 46, 1485-1489.

Alexander, M. (2000) Aging, Bioavailability, and Overestimation of Risk from Environmental Pollutants. Environmental Science and Technology, 34, 4259-4265.

Allchin, C.R.; Law, R.J. and Morris, S. (1999) Polybrominated Diphenyl Ethers in Sediments and Biota Downstream of Potential Sources in the UK. Environmental Pollution, 105, 197-207.

Azevedo, D.D.; Lacorte, S.; Bacelo, P.V. and Barcelo, D. (2002) Occurrence of Nonylphenol and Bisphenol A in Surface Waters from Portugal. Journal of the Brazilian Chemical Society, 12, 532-537.

Balmer, M. E.; Buser, H.-R.; Müller, M. D. and Poiger, T. (2005) Occurrence of Some Organic UV Filters in Wastewater, in Surface Waters, and in Fish from Swiss Lakes. Environmental Science and Technology, 39, 953-962

Barceló, D. (2003) Emerging Pollutants in Waster Analysis. Trends in Analytical Chemistry, 22, xiv-xvi.

Barceló, D. and Petrovic, M. (2007) Pharmaceuticals and Personal Care Products (PPCPs) in the Environment. Analytical and Bioanalytical Chemistry, 387, 1141-1142.

Barker, A.V. and Bryson, G.M. (2002) Bioremediation of Heavy Metals and Organic Toxicants by Composting. The Scientific World, 2, 407-420.

Behnisch, P.A.; Fujii, K.; Shiozaki, K; Kawakami, I. and Sakai, S. (2001) Estrogenic and Dioxin-Like Potency in Each Step of a Controlled Landfill Leachate Treatment Plant in Japan. Chemosphere, 43, 977-984.

Berg, P.; Hagmeyer, G. and Gimbel, R. (1997) Temoval of Pesticides and Other Micropollutants by Nanofiltration. Desalination, 113, 205-208.

Berset, J.-D.,; Bigler, P. and Herren, D. (2000) Analysis of Nitro Musk Compounds and Their Amino Metabolites in Liquid Sewage Sludges Using NMR and Mass Spectrometry. Analytical Chemistry, 72, 2124-2131.

Bester, K. (2003) Triclosan in a Sewage Treatment Process: Balances and Monitoring Data. Water Research, 37, 167-184.

Bourke, M.; Slunjski, M.; O'Leary, B. and Smith, P. (1999) Scale-Up of the MIEX DOC Process for Full-Scale Water Treatment Plants. 18th Federal Convention AWWA Proceedings, Adelaide.

Boyd, G.R.; Palmeri, J.M.; Zhang, S. and Grimm, D.A. (2004) Pharmaceuticals and Personal Care Products (PPCPs) and Endocrine Disrupting Chemicals (EDCs) in Stormwater Canals and Bayou St. John in New Orleans, Louisiana, USA. Science of the Total Environment, 333, 137-148.

Buyuksonmez, F.; Rynk, R.; Hess, T.F. and Bechinski, E. (1999) Occurrence, Degradation, and Fate of Pesticides During Composting. Part I: Composting, Pesticides, and Pesticide Degradation. Compost Science and Utility, 7, 66-82.

Canesi, L.; Ciacci, C.; Lorusso, L.C.; Betti, M.; Gallo, G.; Pojana, G. and Marcomini, A. (2007) Effects of Triclosan on Mytilus galloprovincialis Hemocyte Fuction and Destive Gland Enzyme Activities: Possible Modes of Action and Non-Target Organisms. Comparative Biochemistry and Physiology, Part C, 144, 464-472.

Carballa, M; Manterola, G.; Larrea, L.; Ternes, T.; Omil, F. and Lema, J.M. (2007a) Influence of Ozone Pre-Treatment on Sludge Anaerobic Digestion: Removal of Personal Care Products. Chemosphere, 67, 1444-1452.

Carballa, M.; Omil, F.; Ternes, T. and Lema, J.M. (2007b) Fate of Pharmaceuticals and Personal Care Products (PPCPs) During Anaerobic Digestion of Sewage Sludge. Water Research, 41, 2139-2150.

Carballa, M.; Omil, F. and Lema, J.M. (2005a) Removal of Cosmetic Ingredients and Pharmaceuticals in Sewage Primary Treatment. Water Research, 39, 4790-4796.

Carballa, M.; Omil, F.; Lema, J.M.; Llompart, M.; Garcia, C.; Rodriguez, I.; Gomez, M. and Ternes, T. (2005b) Behaviour of Pharmaceuticals and Personal Care Products in a Sewage Treatment Plant of Northwest Spain. Water Science and Technology, 52, 29-35.

Carballa, M.; Omil, F.; Lema, J.M.; Llompart, M.; Garcia-Jares, C.; Rodriguez, I.; Gomez, M. and Ternes, T. (2004) Behaviour of Pharmaceuticals, Cosmetics and Hormones in a Sewage Treatment Plant. Water Research, 38, 2918-2926.

Carlile, P.; Fielding, M.; Harding, L.; Hart, J.; Hutchinson, J. and Kanda, R. (1996) Effect of Water Treatment Processes on Oestrogenic Chemicals. UK WIR Report 96/DW/05/01, United Kingdom Water Industry Research Limited (UK WIR), London.

Chen, B.; Xuan, X.; Zhu, L.; Wang, J.; Gao, Y.; Yang, K.; Shen, X. and Lou, B. (2004) Distribution of Polycyclic Aromatic Hydrocarbons in Surface Waters, Sediments and Soils of Hangzhou City, China. Water Research, 38, 3558-3568.

Chuanchuen, R.; Beinlick, K.; Hoang, T.; Becher, A.; Karkhoff-Schweizer, R. and Schweizer, H. (2001) Cross-Resistance Between Triclosan and Antibiotics in Pseudomonas aeruginosa Is Mediated by Multidrug Efflux Pumps: Exposure of a Susceptible Mutant Strain to Triclosan Selects nfxB Mutants Overexpressing MexCD-OprJ. Antimicrobial Agents and Chemotherapy, 45, 428-432.

Covaci, A.; Voorspoels, S. and de Boer, J. (2003) Determination of Brominated Flame Retardants with Emphasis on Polybrominated Dipheny Ethers (PBDEs) in Environmental and Human Samples: A Review. Environment International, 29, 735-756.

Cuderman, P. and Heath, E. (2007) Determination of UV Filters and Antimicrobial Agents in Environmental Water Samples. Analytical and Bioanalytical Chemistry, 387, 1343-1350.

Cunningham, V.L.; Buzby, M.; Hutchinson, T.; Mastrocco, F.; Parke, N. and Roden, M. (2006) Effects of Pharmaceuticals on Aquatic Life: Next Steps. A-Pages Mag. Environmental Science and Technology, 40, 3456-3462.

Daughton, C.G. (2004) Non-Regulated Water Contaminants: Emerging Research. Environmental Impact Assessment Review, 24, 711-732.

Daughton, C.G. and Ternes, T.A. (1999) Pharmaceuticals and Personal Care Products in the Environment: Agents of Subtle Change? Environmental Health Perspectives, 107, 907-938.

DeBoer, J.; Weter, P.G.; Van der Horst, A. and Leonards, P.E.G. (2003) Polybrominated Diphenyl Ethers in Influents, Suspended Particle Matter, Sediments, Sewage Treatment Plant Effluents and Biota from the Netherlands. Environmental Pollution, 122, 63-74.

DeLorenzo, M.E. and Fleming, J. (2008) Individual and Mixture Effects of Selected Pharmaceuticals and Personal Care Products on the Marine Phytoplankton

Species Dunaliella tertiolecta. Archives of Environmental Contaminants and Toxicology, 54:203-210.

Difrancesco, A.M.; Chiu, P.C.; Standley, L.J.; Allen H.E. and Salvito, D.T. (2004) Dissipation of Fragrance Materials in Sludge-Amended Soils. Environmental Science and Technology, 38, 194-2001.

Dodd, M.C.; Buffle, M.-O. and von Gunten, U. (2006) Oxidation of Antibacterial Molecules by Aqueous Ozone: Moiety-Specific Reaction Kinetics and Application to Ozone-Based Wastewater Treatment, Environmental Science and Technology, 40, 1969-1977.

Dussault, E.B.; Balakrishnan, V.K.; Sverko, E.; Solomon, K.R. and Sibley, P.K. (2008) Toxicity of Human Pharmaceuticals and Personal Care Products to Benthic Invertebrates. Environmental Toxicology and Chemistry, 27, 425-432.

Ellis, J.B. (2006) Pharmaceutical and Personal Care Products (PPCPs) in Urban Receiving Waters. Environmental Pollution, 144, 184-189.

Ellis, J.B. and Revitt, D.M. (2002) Sewer Losses and Interactions with Groundwater Quality. Water Science and Technology, 45, 195-202.

Englande, A.J.; Jin, G. and Defrechou, C. (2003) Microbial Contamination in Lake Pontchartrain Basin and Best Management Practices on Microbial Contamination Reduction. Journal of Environmental Science and Health, 37, 1765-1779.

Esplugas, S.; Bila, D.M.; Krause, L.G. and Dezotti, M. (2007) Ozonation and Advanced Oxidation Technologies to Remove Endocrine Disrupting Chemicals (EDCs) and Pharmaceuticals and Personal Care Prodcuts (PPCPs) in Water Effluents. Journal of Hazardous Materials, 149, 631-642.

Fawell, J.K.; Sheahan, D.; James, H.A.; Hurst, M. and Scott, S. (2001) Oestrogens and Oestrogenic Activity in Raw and Treated Severn Trent Water. Water Research, 35, 1240-1244.

Fent, K.; Weston, A.A. and Caminada, D. (2006) Ecotoxicology of Human Pharmaceuticals. Aquatic Toxicology, 76, 122-159.

Ferrari, B.; Paxeus, N.; Giudice, R.L.; Pollio, A. and Garric, J. (2003) Ecotoxicological Impact of Pharmaceuticals Found in Treated Wastewaters: Study of Carbamezepine, Clofibric Acid and Diclofenac. Ecolotoxicological and Environmental Safety, 55, 359-370.

Ferrer, I. and Furlong, E.T. (2002) Accelerated Solvent Extraction Followed by On-line Solid-Phase Extraction Coupled to Ion Trap LC/MS/MS for Analysis of Benzalkonium Chlorides in Sediment Samples. Analytical Chemistry, 74, 1275-1280.

Field, J.A. and Reed, R.L. (1996) Nonylphenol Polyethoxy Carboxylate Metabolites of Nonionic Surfactants in U.S. Paper Mill Effluents, Municipal Sewage Treatment Plant Effluents, and River Waters. Environmental Science and Technology, 30, 3544-3550.

Fielding, M.; Harding, L.; James, C. and Mole, N. (1998) Removal of Nonylphenol Ethoxylates by Water Treatment Processes. UK WIR Report 98/TX/01/5. United Kingdom Water Industry Research Limited (UK WIR), London.

Fooken, C. (2004) Synthetic Musks in Suspended Particulate Matter (SPM), Sediment, and Sewage Sludge in The Handbook of Environmental Chemistry. Springer. pp. 29-47.

Fromme, H.; Otto, T. and Pilz, K. (2001) Polycyclic Musk Fragrances in Different Environmental Compartments in Berlin (Germany). Water Research, 35, 121-128.

Gallard, H. and von Gunten, U. (2002) Chlorination of Phenols: Kinetics and Formation of Chloroform. Environmental Science and Technology, 36, 884-890.

Gatermann, R.; Hellow, J.; Huhnerfuss, H.; Rimkus, G.G. and Zitko, V. (1999) Polycyclic and Nitro Musk in the Environment: A Comparison between Canadian and European Aquatic Biota. Chemosphere, 33, 17-28.

Giesy, J.P.; Pierens, S.L.; Snyder, E.M.; Miles-Richardson, S.; Kramer, S.; Snyder, V.J.; Nichols, K.M. and Villeneuve, D.A. (2000) Effects of 4-Nonylphenol on Fecundity and Biomarkers of Estrogenicity in Fathead Minnows (Pimephales promelas). Environmental Science and Toxicological Chemistry, 19, 1363-1377.

Golet, E.M.; Alder, A.C.; Hartmann, A.; Ternes, T.A. and Giger, W. (2001) Trace Determination of Fluoroquinolone Antibacterial Agents in Urban Wastewater by Solid Phase Extraction and Liquid Chromatography with Fluorescence Detection. Analytical Chemistry, 73, 3632-3638.

Guenther, K.; Heinke, V.; Thiele, B.; Kleist, E.; Prast, H. and Raecker, T. (2002) Endocrine Disrupting Nonylphenols are Ubiquitous in Food. Environmental Science and Technology, 36, 1676-1680.

Hale, R.C.; La Guardia, M.J.; Harvey, E.P.; Gaylor, M.O.; Mainor, T.M. and Duff, W.H. (2001) Persistent Pollutants in Land-Applied Sludges. Nature, 412, 140-141.

Halling-Sorensen, B.; Nielsen, S.N.; Lanzky, P.F.; Ingerslev, F.; Luzhoft, H.C.H. and Jorgensen, S.E. (1998) Occurrence, Fate and Effects of Pharmaceutical Substances in the Environment: A Review. Chemosphere, 36, 357-393.

Harris, C.A.; Santos, E.M.; Janbakhsh, A.; Pottinger, T.G.; Tyler, C.R. and Sumpter, J.P. (2001) Nonylphenol Affects Gonadotropin Levels in the Pituitary Gland and Plasma of Female Rainbow Trout. Environmental Science and Technology, 35, 2909-2916

Hatzinger, P.B. and Alexander, M. (1995) Effects of Aging of Chemicals in Soil on Their Biodegradability and Extractability. Environmental Science and Technology, 29, 537-545.

Heath, R.J and Rock, C.O. (2000) A Triclosan-Resistant Bacterial Enzyme. Nature, 406, 145.

Heemken, O.P.; Reincke, H.; Stachel, B. and Theobald, N. (2001) The Occurrence of Xenoestrogens in the Elbe River and the North Sea. Chemosphere, 45, 245-259.

Heberer, T. (2002) Occurrence, Fate, and Removal of Pharmaceutical Residues in the Aquatic Environment: A Review of Recent Research Data Toxicological Letters, 131, 5-17.

Heberer, T.; Gramer, S. and Stan, H.-J. (1999) Occurrence and Distribution of Organic Contaminants in the Aquatic System in Berlin. Part III: Determination of Synthetic Musks in Berlin Surface Water Applying Solid-Phase Microextraction (SPME) and Gas Chromatography-Mass Spectrometry (GC/MS). Acta Hydrochimica et Hydrobiologica, 27, 150-156.

Hesselsoe, M.; Jensen, D.; Skals, K.; Olesen, T.; Modrup, P. Roslev, P.; Mortensen, G.K. and Henriksen, K. (2001) Degradation of 4-nonylphenol in Homogeneous and Nonhomogeneous Mixtures of Soil and Sewage Sludge. Environmental Science and Technology, 35, 3695-3700.

Hites, R.A. (2004) Polybrominated Diphenyl Ethers in the Environment and in People: A Meta-Analysis of Concentrations. Environmental Science and Technology, 38, 945-956.

Hyotylainen, T. and Hartonen, K. (2002). Determination of Brominated Flame Retardants in Environmental Samples. Trends in Analytical Chemistry, 21, 13-29.

Jjemba, P.K. (2006) Excretion and Exotoxicity of Pharmaceutical and Personal Care Products in the Environment. Ecotoxicology and Environmental Safety, 63, 113-130.

Jjemba, P.K. (2004) Environmental Microbiology: Principles and Applications. Science Publisher, Inc. Enfield, New Hampshire.

Jjemba, P.K. (2002) The Potential Impact of Veterinary and Human Therapeutic Agents in Manure and Biosolids on Plants Grown on Arable Land: A Review. Agriculture, Ecosystems and Environment, 93,267-278

Jones-Lepp, T.L. and Stevens, R. (2007) Pharmaceuticals and Personal Care Products in Biosolids/Sewage Sludge: The Interface Between Analytical Chemistry and Regulation. Analytical and Bioanalytical Chemistry, 387:1173-1183.

Kannan, K.; Reiner, J.L.; Yun, S.-H.; Perrotta, E.E., Tao, L.; Johnson-Restopo, B. and Rodan, B.D. (2005) Polycyclic Musk Compounds in Higher Trophic Level Aquatic Organisms and Humans from the United States. Chemosphere, 61, 693-700.

Kasim, N.A.; Whitehouse, M.; Ramachandran, C.; Bermejo, M.; Lennernas, H.; Hussain, A.S.; Junginger, H.E.; Stavchansky, S.A.; Midha, K.K.; Shah, V.P. Amidon, G.L. (2004) Molecular Properties of WHO Essential Drugs and Provisional Biopharmaceutical Classification. Molecular Pharmacokinetics, 2, 389-402.

Keller, H.; Xia, K. and Bhandari, A. (2003) Occurrence and Degradation of Estrogenic Nonylphenol and Its Precursors in Northeast Kansas Wastewater Treatment Plants. Practice Periodicals of Hazardous, Toxic and Radioactive Waste Management, 7, 203-213.

Kelsey, J.W.; Kottler, B.D. and Alexander, M. (1997) Selective Chemical Extractants to Predict Bioavailability of Soil-Aged Organic Chemicals. Environmental Science and Technology, 31, 214-217.

Kenda, R.; Griffin, P.; James, H.A. and Fothergill, J. (2003) Pharmaceuticals and Personal Care Products in Sewage Treatment Works. Journal of Environmental Monitoring, 5, 823-830.

Khim, J.S.; Kannan, K.; Villeneuve, D.L.; Koh, C.H. and Giesy, J.P. (1999a) Characterization and Distribution of Trace Organic Contaminants in Sediment from Masan Bay, Korea. I: Instrumental Analyses. Environmental Science and Technology, 33, 4199-4205.

Khim, J.S.; Villeneuve, D.L.; Kannan, K.; Lee, K.T.; Snyder, S.A.; Koh, C.H. and Giesy, J.P. (1999b) Alkylphenols, Polycyclic Aromatic Hydrocarbons, and Organochlorines in Sediment from Lake Shihwa, Korea: Instrumental and Bioanalytical Characterization. Environmental Toxicology and Chemistry, 18, 2424-2432.

Kinney, C.A.; Furlong, E.T.; Zaugg, S.D.; Burkhardt, M.R.; Werner, S.L.; Cahill, J.D. and Jorgensen, G.J. (2006) Survey of Organic Wastewater Contaminants in Biosolids Destined for Land Application. Environmental Science and Technology, 40, 7207-7215.

Kolpin, D.W.; Furlong, E.T.; Meyer, M.T.; Thurman, E.M.; Zaugg, S.D.; Barber, L.B. and Buxton, H.T. (2002) Pharmaceuticals, Hormones, and Other Organic Wastewater Contaminants in US Streams, 1999-2000: A National Reconnaissance. Environmental Science and Technology, 36, 1202-1211.

Kummerer, K. (2004) Pharmaceuticals in the Environment. Springer-Verlag, New York, NY, USA.

Kuch, E. and Ballschmiter, K. (2001) Determination of Endocrine Disrupting Phenolic Compounds and Estrogens in Surface and In the Water Drinking Water by HRGC-(NCI)-MS in the Picogram per Liter range, Environment of the. Environmental Science and Technology, 35, 3201-3206.

Kupper, T.; Berset, J.D.; Etter-Holzer, R.; Furrer, R. and Tarradellas, J. (2004) Concentrations and specific loads of polycyclic musks in sewage sludge originating from a monitoring network in Switzerland. Chemosphere, 54, 1111-1120.

La Guardia, M.J.; Hale, R.C., Harvey, E. and Mainor T.M. (2001) Alkylphenol Ethoxylate Degradation Products in Land-Applied Sewage Sludge (Biosolids). Environmental Science and Technology, 35, 4798-4804.

Laguna, A.; Bacaloni, A.; De Leva, I; Faberi, A.; Fargo, G. and Marino, A. (2004) Analytical Methodologies for Determining the Occurrence of Endocrine Disrupting Chemicals in Sewage Treatment Plants and Natural Waters. Analytica Chimica Acta, 501, 79-88.

Levy, C.W.; Roujeinikova, A.; Sedelnikova, S.; Baker, P.J.; Stuitje, A.R.; Slabas, A.R.; Rice, D.W. and Rafferty, J.B. (1999) Molecular Basis of Triclosan Activity. Nature, 398, 383-384.

Liebig, M.; Moltmann, J.F. and Knacker, T. (2006) Evaluation of Measured and Predicted Environmental Concentrations of Selected Human Pharmaceuticals and Personal Care Products. Environmental Science and Pollution Research, 13, 110-119.

Lindstrom, A.; Buerge, I.J.; Poiger, T.; Bergqvist, P.-A.; Muller, M.D. and Buser, H.-R. Occurrence and Environmental Behavior of the Bactericide Triclosan and its Methyl Derivative in Surface Waters and Wastewater. Environmental Science and Technology. 35, 283-298.

Lishman, L.; Smyth, S.A; Sarafin, K.; Kleywegt, S.; Toito, J.; Peart, T.; Lee, B.; Servos, M.; Beland, M. and Seto, P. (2006) Occurrence and Reductions of Pharmaceuticals and Personal Care Products and Estrogens by Municipal Wastewater Treatment Plants in Ontario, Canada. Science of the Total Environment, 367, 544:558.

Loraine, G.A. and Pettigrove, M.E. (2006) Seasonal Variations in Concentrations of Pharmaceuticals and Personal Care Products in Drinking Water and Reclaimed Wastewater in Southern California. Environmental Science and Technology, 40, 687-695.

Luckenbach, T.; Corsi, I. and Epel, D. (2004) Fatal Attraction: Synthetic Musk Fragrances Compromise Multixenobiotic Defense Systems in Mussels. Marine Environmental Research, 58, 215-219.

MacLeod, S.L, McClure, E.L and Wong, C.S. (2007) Laboratory Calibration of Field Deployment of the Polar Organic Chemical Integrative Sampler for Pharmaceuticals and Personal Care Products in Wastewater and Surface Water. Environmental Toxicology and Chemistry, 26, 2517-2529.

Matamoros, V. and Bayona, J.M. (2006) Elimination of Pharmaceuticals and Personal Care Products in Subsurface Flow Constructed Wetlands. Environmental Science and Technology, 40, 5811-5816.

Matamoros, V.; Arias, C.; Brix, H. and Bayona, J. (2007) Removal of Pharmaceuticals and Personal Care Products (PPCPs) from Urban Wastewater in a Pilot Vertical Flow Constructed Wetland and a Sand Filter. Environmental Science and Technology, 41, 8171-8177.

Matscheko, N.; Tysklind, M.; DeWit, C.; Bergek, S.; Andersson, R. and Sellstrom, U. (2002) Application of Sewage Slduge to Arable Land – Soil Concentrations of Polybrominated Diphenyl Ethers and Polychlorinated Dibenzo-p-dioxins, Dibenzofurans, and Biphenyls, and Their Accumulation in Earthworms. Environmental and Toxicological Chemistry, 21, 2515-2525.

McMurry, L.M., Oethinger, M. and Levy, S.B. (1998) Triclosan Targets Lipid Synthesis. Nature, 394, 531-532.

Moldovan, Z. (2006) Occurrences of Pharmaceuticals and Personal Care Products as Micropollutants in Rivers of Romania. Chemosphere, 64, 1808-1817.

Morales-Munoz, S.; Luque-Garcia, J.L.; Ramos, M.J.; Fernandez-Alba, A. and de Castro, M.D.L. (2005) Sequential superheated liquid extraction of pesticides, pharmaceutical and personal care products with different polarity from marine sediments followed by gas chromatography mass spectrometry detection. Analytica Chimica Acta, 552, 50-59.

Muller, S.; Schmid, S. and Schlatter, C. (1996) Occurrence of Nitro and Non-Nitro Benzenoid Musk Compounds in Human Adipose Tissue. Chemosphere, 33, 17-28.

Nagtegall, M.; Ternes, T.A.; Baumann, W. and Nagel, R. (1997) Deterction of Suscreen Agents in Water and Fish of the Meerfelder Maar the Eifel, Germany. Umweltwiss. Schadst.-Forsch, 9, 79-86.

Nakada, N.; Shinohara, H.; Murato, A.; Kiri, K.; Managaki, S.; Sato, N. and Takada, H. (2007) Removal of Selected Pharmaceuticals and Personal Care Products (PPCPs) and Endocrine-Disrupting Chemicals (EDCs) During Sand Filtration and Ozonation at a Municipal Sewage Treatment Plant. Water Research, 41, 4373-4382.

Nasu, M.; Goto, M.; Kato, H.; Oshima, Y. and Tanaka, H. (2001) Study on Endocrine Disrupting Chemicals in Wastewater Treatment Plants. Water Science and Technology, 43, 101-108.

Osemwengie, L. (2006) Determination of Synthetic Musk Compounds in Sewage Biosolids by Gas Chromatography/Mass Spectrometry. Journal of Environmental Monitoring, 8, 897-903.

Oetken, M.; Nentwig, G.; Loffler, D.; Ternes, T. and Oehlmeann, J. (2005) Effects of Pharmaceuticals on Aquatic Invertebrates. Part I. The Antiepileptic Drug Carbamazepine. Archives of Environmental Contamination and Toxicology, 49, 353-361.

Okumura, T. and Nishikawa, Y. (1996) GC-MS Determination of Triclosan in Water Sediment and Fish Samples via Methylation with Diazomethane. Analytica Chimica Acta, 325, 175-184.

Parencevich, E.N.; Wong, M.T. and Harris, A.D. (2001) National and Regional Assessment of Antibactrial Soap Market: A step Toward Determining the Impact of Prevalent Antibacterial Soaps. American Journal of Infectious Control, 29, 281-283.

Patterson, D.B.; Brumsely, W.C.; Kelliher, V. and Ferguson, P.L. (2002) Application of U.S. EPA Methods to the Analysis of Pharmaceuticals and Personal Care Products in the Environment: Determination of Clofibric Acid in Sewage Effluent by GC-MS. American Laboratory (Shelton, Connecticut), 34, 20-28.

Pedersen, J.; Yeager, M.A. and Suffet, I.H. (2003) Xenobiotic Organic Compounds in Runoff from Fields Irrigated with Treated Wastewater. Journal of Agriculture and Food Chemistry, 51, 1360-1372.

Petrovic, M.; Diaz, A.; Ventura, F. and Barcelo, D. (2003) Occurrence and Removal of Estrogenic Short-Chain Ethoxy Nonylphenolic Compounds and Their Halogenated Derivatives During Drinking Water Production. Environmental Science and Technology, 37, 4442-4448.

Petrovic, M.; Eljarrat, E.; Lopez de Alda, M.J.; Barcelo, D. (2002) Recent Advances in the Mass Spectrometric Analysis Related to Endocrine Disrupting compounds in Aquatic Environmental Samples. Journal of Chromatography, A 974, 23-51.

Petrovic, M.; Eljarrat, E.; Lopez de Alda, M.J. and Barcelo, D. (2001) Analysis and Environmental Levels of Endocrine-Disrupting Compounds in Freshwater Sediments. Trends in Analytical Chemistry, 20, 637-648.

Rice, S.L. and Mitra, S. (2007) Microwave-Assisted Solvent Extraction of Solid Matrices and Subsequent Detection of Pharmaceuticals and Personal Care Products (PPCPs) Using Gas Chromatography-Mass Spectrometry. Analytica Chimica Acta, 589, 125-132.

Richarson, S.D. (2004) Environmental mass Spectrometry: Emerging Contaminants and Current Issues. Analytical Chemistry, 76, 3337-3364.

Richarson, S.D. (2003) Water Analysis: Emerging Contaminants and Current Issues. Analytical Chemistry, 75, 2831-2857.

Richardson, B.J.; Lam, P.K.S and Martin, M. (2005) Emerging Chemicals of Concern: Pharmaceuticals and Personal Care Products (PPCPs) in Asia, with Particular Reference to Southern China. Marine Pollution Bulletin, 50, 913-920.

Richter, G.E.; Jones, B.A.; Ezzell, J.L.; Porter, N.L.; Avdalovic, N. and Pohl, C. (1996) Accelerated Solvent Extraction: A Technique for Sample Preparation. Analytical Chemistry, 68, 1033-1039.

Rimkus, G.G. (1999) Polycyclic Musk Fragrances in the Aquatic Environment. Toxicology Letters, 111, 37-56.

Rimkus, G.G. and Wolf, M. (1999) Polycyclic Musk Frangrances in Human Adipose Tissue and Human Milk. Chemosphere, 33, 2033-2043.

Rimkus, G.; Gatermann, R. and Huhnerfuss, H. (1999) Musk Xylene and Musk Ketone Amino Metabolites in the Aquatic Environment. Toxicology Letters, 111, 5-15.

Rimkus, G.; Rimkus, B. and Wolf, M. (1994) Nitro Musks in Human Adipose Tissue and Breast Milk. Chemosphere, 28, 421-433.

Rogers, H.R. (1996) Sources, Behavior, and Fate of Organic Contaminants During Sewage Treatment and in Sewage Sludges. Science of the Total Environment, 185, 3-26.

Routledge, E.J.; Parker, J.; Odum, J.; Ashby, J. and Sumpter, J.P. (1998) Some Alkyl Hydroxy Benzoate Preservatives (Parabens) are Estrogenic. Toxicology and Applied Pharmacology, 153, 12-19.

Schlump, M.; Schmid, P.; Durrer, S.; Conscience, M.; Maerkel, K.; Henseler, M.; Gruetter, M.; Herzog, I.; Reolon, S.; Ceccatelli, R.; Faass, O.; Stutz, E.; Jarry, H.; Wuttke, W. and Lichtensteiger, W. (2004) Endocrine Activity and Developmental Toxicity of Cosmetic UV Filters: An Update. Toxicology, 205, 113-122.

Sellstrom, U.; Kierkegaard, A.; De Wit, C. and Jansson, B. (1998) Polybrominated Diphenyl Ethers and Hexabromocyclododecane in Sediment and Fish from a Swedish River. Environmental Toxicology and Chemistry, 17, 1065-1072.

Shon, H.K.; Vigneswaran, S. and Snyder, S.A. (2007) Effluent Organic Matter (EfOM) in Wastewater: Constituents, Effects, and Treatment. Critical Reviews in Environmental Science and Technology, 327-374.

Simonich, S.L.; Federle, T.W.; Eckhoff, W.S.; Rottiers, A.; Webb, S.; Sabaliunas, D. and De Wolf, W. (2002) Removal of Fragrance Materials During U.S. and European Wastewater Treatment. Environmental Science and Technology, 36, 2839-2847.

Simonich, S.L.; Begley, W.M.; Debaere, G. and Eckhoff, W.S. (2000) Trace Analysis of Fragrance Materials in Wastewater and Treated Wastewater. Environmental Science and Technology, 34, 959-965.

Singer, H.; Muller, S.; Tixier, C. and Pillonel, L. (2002) Triclosan: Occurrence and Fate of a Widely Used Biocide in the Aquatic Environment: Field Measurements in Wastewater Treatment Plants, Surface Waters, and Lake Sediments. Environmental Science and Technology, 36, 4998-5004.

Snyder, S.A.; Adham, S.; Redding, A.M.; Cannon, F.S.; DeCarolis, J.; Oppenheimer, J.; Wert, E. and Yoon, Y. (2007) Role of Membranes and Activated Carbon in the Removal of Endocrine Disruptors and Pharmaceuticals. Desalination, 202, 156-181.

Snyder, S.A.; Wert, E.; Edwards, J.; Budd, G.; Long, B. and Rexing, D. (2004) Magnetic Ion Exchange (MIEX) for the Removal of Endocrine Disruption Chemicals and Pharmaceyticals. AWWA Water Quality and Technology Conference, California.

Snyder, S.A; Vanderford, B.; Pearson, R.; Quinones, O. and Yoon, Y. (2003a) Analytical Methods Used to Measure Endocrine Disrupting Compounds in Water. Practice Periodicals of Hazardous, Toxic and Radioactive Waste Management, 7, 224-234.

Snyder, S.A; Westerhoff, P.; Yoon, Y. and Sedlak, D.C. (2003b) Pharmaceuticals, Personal Care Products and Endocrine Disrupting Compounds in Water:

Implications for Water Treating. Environmental Engineering Science, 20, 449-469.

Snyder, S.A; Kelly, K.; Grange, A.; Sovocool, G.W.; Snyder, E. and Giesy, J. (2001a) Pharmaceuticals and Personal Care Products in the Waters of Lake Mead, Nevada. Chapter 7 in Pharmaceuticals and Personal Care Products in the Environment; Scientific and Regulatory Issues. Dughton, C., Jones-Lepp, T., Eds. American Chemical Society. Washington, D.C.

Snyder, S.A.; Keith, T.L.; Pierens, S.L.; Snyder, E.M. and Giesy, J.P. (2001b) Bioconcentration of Nonylphenol in Fathead Minnows (Pimephales promelas). Chemosphere, 44, 1697-1702.

Snyder, S.A.; Keith, T.L.; Naylor, C.G.; Staples, C.A. and Giesy, J.P. (2001c) Identification and Quantification Method for Nonylphenol and Lower Oligomer Nonylphenol Ehtoxylates in Fish Tissues. Environmental Toxicology and Chemistry, 20, 1870-1873.

Snyder, S.A.; Snyder, E.; Villeneuve, D.; Kurunthachalam, K.; Villalobos, A.; Blankenship, A. and Giesy, J. (2000) Instrumental and Bioanalytical Measures of Endocrine Disruptors in Water. Analysis of Environmental Endocrine Disruptors, L.H. Keith, T.L. Jones-Lepp, L.L. Needham, Eds. American Chemical Society, Washington, D.C., 73-95.

Snyder, S.A.; Keith, T.L.; Verbrugge, D.A.; Snyder, E.M.; Gross, T.S.; Kannan, K. and Giesy, J.P. (1999) Analytical Methods for Detection of Selected Estrogenic Compounds in Aqueous Mixtures. Environmental Science and Technology, 33, 2814-2820.

Soderstrom, G.; Sellstrom, U.; DeWitt, C.A. and Tysklind, M. (2004) Photolytic Debromination of Decabromodiphenyl Ether (DBE 209). Environmental Science and Technology, 38, 127-132.

Sohoni, P.; Tyler, C.R.; Hurd, K.; Caunter, J.; Hetheridge, M.; Williams, T.; Woods, C.; Evans, M.; Toy, R.; Gargas, M. and Sumpter, J.P. (2001) Reproductive Effects of Long-Term Exposure to Bisphenol A in the Fathead Minnow (Pimephales promelas). Environmental Science and Technology, 35, 2917-2925.

Sole, M.; Castillo, M.; Lopez de Alda, M.J.; Porte, C.; Ladegaard-Pedersen, K. and Barcelo, D. (2000). Estrogenicity Determination in Sewage Treatment Plants and Surface Waters from the Catalonian Area (NE Spain). Environmental Science and Technology, 34, 5076-5083.

Stanford, B.D. and Weinberg, H.S. (2007) Isotope Dilution for Quantitation of Steroid Estrogens and Nonylphenols by Gas Chromatography with Tandem Mass Spectrometry in Septic, Soil and Groundwater Matrices. Journal of Chromatography A. 1176:26-36.

Steinberg, D. (2000) Encyclopedia of UV Filters Updated. Cosmetics and Toiletries (Allured), 115, 65-75.

Stevens, J.L.; Northcott, G.L.; Stern, G.A.; Tomy, G.T. and Jones, K.C. (2003) PAHs, PCBs, PCNs, Organochlorine Pesticides, Synthetic Musks, and Polychlorinated n-Alkanes in UK-Sewage Sludge: Survey Results and Implications. Environmental Science and Technology, 37, 462-467.

Takahashi, A.; Higashitani, T.; Yakou, Y.; Saitou, M.; Tamamoto, H. and Tanaka, H. (2003) Evaluating Bioaccumulation of Suspected Endocrine Disruptors into

Periphytons and Benthos in the Tama River. Water Science and Technology, 47, 71-76.

Ternes, T.A. (1998) Occurrence of Drugs in German Sewage Treatment Plants and Rivers. Water Research, 32, 3245-3260.

Ternes, T.A.; Bonerz, M.; Herrmann; N.; Teiser, B and Andersen, H.R. (2007) Irrigation of Treated Wastewater in Braunschweig, Germany: An Option to Remove Pharmaceuticals and Musk Fragrances. Chemosphere, 66, 894-904.

Ternes, T.A.; Joss, A. and Siegrist, H. (2004) Scrutinizing Pharmaceuticals and Personal Care Products in Wastewater Treatment. Environmental Science and Technology, 38, 393-399.

Ternes, T.A.; Meisenheimer, M.; McDowell, D.; Sacher, F.; Brauch, H.-J.; Haste-Gulde, B.; Preuss, G.; Wilme, U. and Zulei-Seibert, N. (2002) Removal of Pharmaceuticals During Drinking Water Treatment. Environmental Science and Technology, 36, 3855-3863.

Thiele-Bruhn, S. (2003) Pharmaceutical Antibiotic Compounds in Soils – A Review. Journal of Plant Nutrition and Soil Science, 166, 145-167.

Thomas, P.M. and Foster, G.D. (2005) Tracking Acidic Pharmaceuticals, Caffein, and Triclosan Through the Wastewater Treatment Process. Environmental Toxicology and Chemistry, 24, 25-30.

Tixier, C.; Singer, H.P.; Canonica, S. and Muller, S.R. (2002) Phototransformation of Triclosan in Surface Waters: A Relevant Elimination Process for This Widely Used Biocide-Laboratory Studies, Field Measurements, and Modeling. Environmental Science and Technology, 36, 3482-3489.

Tolls, J. (2001) Sorption of Veterinary Pharmaceuticals in Soils: A Review. Environmental Science and Technology, 35, 3397-3406.

Trenholm, R.A.; Vanderford, B.J.; Holady, J.C.; Rexing, D.J and Snyder, S.A. (2006) Broad Range Analysis of Endocrine Disruptors and Pharmaceuticals Using Gas Chromatography and Liquid Chromatography Tandem Mass Spectrometry. Chemosphere, 65, 1990-1998.

Tsuda, T.; Takino, A.; Muraki, K.; Harada, H. and Kijima, M. (2001) Evaluation of 4-Nonlyphenols and 4-tert-Octylphenol Contamination of Fish in Rivers by Laboratory Accumulation and Excretion Experiments. Water Research, 35, 1786-1792.

Voordeckers, J.W.; Fennell, D.E.; Jones, K. and Haggblom, M.M. (2002) Anaerobic Biotransformation of Tetrabromobisphenol A, Tetrachlorobisphenol A, and Bisphenol A in Esturine Sediments. Environmental Science and Technology, 36, 696-701.

Westerhoff, P.; Yoon, Y.; Snyder, S. and Wert, E. (2005) Fate of Endocrine-Disruptor, Pharmaceutical, and Personal Care Product Chemicals During Simulated Drinking Water Treatment Processes. Environmental Science and Technology, 39, 6649-6663.

Wang, L.; Govind, R. and Dobbs, R.A. (1993) Sorption of Toxic Organic Compounds on Wastewater Solids: Mechanisms and Modeling. Environmental Science and Technology. 27, 152-158.

Wilson, B.A.; Smith, V.H.; Denoyelles, F. and Larive, C.K. (2003) Effects of Three Pharmaceutical and Personal Care Products on Natural Algal Assemblages. Environmental Science and Technology, 37, 1713-1719.

Winkler, M.; Kopf, G; Hauptvogel, C. and Neu, T. (1998) Fate of Artificial Musk Fragrances Associated with Suspended Particulate Matter (SPM) from the River Elbe (Germany) in Comparison to Other Organic Contaminants. Chemosphere, 37, 1139-1156.

Wormuth, M.; Scheringer, M. and Hungerbühler, K. (2005) Linking the Use of Scented Consumer Products to Consumer Exposure to Polycyclic Musk Fragrances. Journal of Industrial Ecology, 9, 237-258.

Xia, K. and Jeong, C.Y. (2004) Photodegradation of Endocrine-Disrupting Chemical 4-nonylphenol in Biosolids Applied to Soil. Journal of Environmental Quality, 33, 1568-1574.

Xia, K. and Pillar, G.D. (2003) Anthropogenic Organic Chemicals in Biosolids from Selected Wastewater Treatment Plants in Georgia and South Carolina in Proceedings of the Georgia Water Research Conference. April 23-24. Athens, Georgia. University of Georgia Press.

Xia, K.; Bhandari, A.; Das, K. and Pillar, G. (2005) Occurrence and Fate of Pharmaceuticals and Personal Care Products (PPCPs) in Biosolids. Journal of Environmental Quality, 34, 91-104.

Yamagishi, T.; Miyazaki, T.; Horii, S. and Akiyama, K. (1983) Synthetic Musk Residues in Biota and Water from Tama River and Tokyo Bay (Japan). Archives of Environmental Contamination and Toxicology, 12, 83-89.

Yamagishi, T.; Miyazaki, T.; Horii, S. and Kaneko, S. (1981) Identification of Musk Xylene and Musk Ketone in Freshwater Fish Collected from the Tama River, Tokyo. Bulletin of Environmental Contamination and Toxicology, 26, 656-662.

Yoon, Y.; Westerhoff, P.; Snyder, S.A. and Wert, E.C. (2006) Nanofiltration and Ultrafiltration of Endocrine Disrupting Compounds, Pharmaceuticals and Personal Care Products. Journal of Membrane Science, 270, 88-100.

Yoon, Y.; Westerhoff, P.; Snyder, S.A.; Wert, E.C. and Yoon, J. (2007) Removal of Endocrine Disrupting Compounds and Pharmaceuticals by Nanofiltration and Ultrafiltration Membranes. Desalination, 202, 16-23.

Yu, J.T.; Bouwer, E.J. and Colehan, M. (2006) Occurrence and Biodegradability Studies of Selected Pharmaceuticals and Personal Care Products in Sewage Effluent. Agricultural Water Management, 86, 72-80.

Zehringer, M. and Herrman, A. (2001) Analysis of Polychlorinated Biphenyls, Pyrethroid Insecticides and Fragrances in Human Milk Using a Laminar Cup Liner in the GC Injector. European Food Research and Technology, 212, 247-251.

Zhang, H. and Huang, C.-H. (2003) Oxidative Transformation of Triclosan and Chlorophene by Manganese Oxides. Environmental Science and Technology, 37, 2421-2424.

Zhang, R.; Vigneswaran, S.; Ngo, H.H. and Nguyen, H. (2007) Magnetic Ion Exchanger (MIEX) as a Pre-Treatment to a Submerged Membrane System in the Treatment of Biologically Treated Wastewater. Desalination, 192, 296-302.

Zhang, S.; Zhang, Q.; Darisaw, S.; Ehie, O. and Wang, G. (2007) Simultaneous Quantification of Polycyclic Aromatic Hydrocarbons (PAHs), Polychlorinated Byphenyls (PCBs), and Pharmaceuticals and Personal Care Products (PPCPs) in Mississippi River Water, in New Orleans, Louisiana, USA. Chemosphere, 66, 1057-1069.

CHAPTER 5

Antimicrobials and Antibiotics

Say Kee Ong, Warisara Lertpaitoonpan, Alok Bhandari, Tawan Limpiyakorn

5.1   Introduction

A recent national study conducted by the United States Geologic Survey in thirty states indicated that contaminants of emerging environmental concern such as pharmaceuticals, personal care products and antibiotics were detected in 80% of the 139 streams sampled near urban and livestock production areas (Kolpin et al., 2002). Many researchers have identified discharges from wastewater treatment plants as major point sources of these contaminants of concern especially pharmaceuticals and personal care products (Heberer, 2002; Ternes, 1998; and Jørgensen and Halling-Sørensen, 2000) while concentrated animal feeding operations (CAFOs) and aquaculture (fish farming) are generally identified as both point and non-point sources of anti-microbials and antibiotics which are used for the maintenance of the health and well-being of livestock.   Runoff and seepage of contaminated water from land-applied biosolids, animal feces, spills from livestock feed, and from leakage of wastes from waste storage lagoons have been found to contaminate surface waters and groundwater (DEFRA, 2002; Daughton and Ternes, 1999; and Jørgensen and Halling-Sørensen, 2000).

Natural and synthetic antibiotics were introduced in the late 1930s and their usage for human and animal production has since increased (Cromwell, 2000). It has been shown that the presence of antibiotics in the environment correlates well with their production and use for both human and for livestock consumptions (Daughton and Ternes, 1999; Lindberg et al., 2005). There are more than 10 different classes of antibiotics which are characterized according to their structural and chemical properties. Table 5.1 summarizes the common classes and examples of antibiotics within each class.

Table 5.1 Common classes of antibiotics

Chemical Class	Examples of antibiotics	Effect
Aminoglycosides	Neomycin, streptomycin, tobramycin, paromomycin	Aerobic bacteria and gram-negative bacteria antimicrobial properties
Carbapenems	Ertapenem, meropenem, doripenem	Both gram-positive and gram-negative antimicrobial properties
Cephalosporins	Cefalotin, cefamandole, cefoxitin, cefeprime	Gram-negative antimicrobial properties
Glycopeptides	Teicoplanin, vancomycin	Gram-positive microbial properties
Marcolides	Azithromycin, clarithromycin, erythromycin, telithromycin, spectinimycin	For streptococcal infections, respiratory infections
β-lactams (Penicillin)	Amoxicillin, ampicillin, carbenicillin, cloxacillin, penicillin, meticillin, cephalexin, penicillin, cefprozil, cefuroxime, loracarbef	For wide range of infections, streptococcal infections
Polypeptides	Bacitracin, colistin, polymyxin B	For eye, ear and bladder infections
Fluoroquinolones	Ciprofloxacin, enoxacin, ofloxacin, levofloxacin	For urinary tract infections
Sulfonamides	Sulfamethizole, sulfanilamide, sulfamethazine, trimethoprim, sulfamethoxazole	For urinary tract infections
Tetracyclines	Tetracyclines, chlortetracycline, oxytetracyclines	For infections of respiratory tract, urinary tract
Others	Lincomycin, Mupirocin	

The total worldwide market for antibiotics and antifungals was estimated to be US$25 billion in 2005 (Spectra Intelligence, 2006). Data on the amount of antibiotics for human usage is not easily available. The more popular antibiotics used by humans are β-lactams (amoxicillin and penicillin), macrolides (azithromycin, clarithromycin, and erythromycin), fluoroquinolones (ciprofloxacin and levofloxacin), aminoglycosides (neomycin and tobramycin), sulfonamide (sulfamethoxazole), and tetracycline (tetracycline). For human therapy, penicillin or β-lactam antibiotics is the most used antibiotic, followed by macrolide, sulfonamide, and fluoroquinolone antibiotics (Huang et al., 2001).

The total amount of veterinary antibiotics sold in the U.S. in 2001 was approximately 21.8 million lbs (Vansickle, 2002). Table 5.2 provides the amount of different classes of antibiotics sold to farms and for different livestock in the U.S. in 2001 as reported by the Animal Health Institute (Vansickle, 2002). Data provided by Prescott (1977) indicated that the swine industry used the most antibiotics in comparison to the other livestock industries. A national survey by USDA (2002) showed that about 92% of swine CAFOs used antibiotics. A summary of different types of antibiotics used in different countries for veterinary purposes can be found in Sarmah et al. (2006).

The most frequently administered antibiotics for nursery pigs were chlortetracycline, tylosin and carbadox which represented about 30.1%, 23.1% and 22.8% of the frequency of use as reported in a survey by the National Animal Health Monitoring System in 1995 (USDA, 2002). This survey constitutes about 94% of the U.S. swine population. For antibiotics used in grower/finisher farms, the most common antibiotics used were tylosin (56.3% of use), oxytetracycline (48%), and procaine penicillin G (40%) (USDA, 2002).

Table 5.2 Sales of veterinary antibiotics in U.S. in 2001 as reported by Animal Health Institute (adapted from Vansickle, 2002; Kolz, 2004).

Chemical Class	Antibiotics	Use	Sales (lbs)
Ionophores/Arsenicals[a]	Monensin	Ruminant Growth Promoter	7,758,490
Tetracyclines	Tetracycline, Chlor- and Oxy-tetracycline	Therapeutic	7,144,520
Macrolides & other minor classes	Tylosin, Bacitracin, Carbadox Erythromycin, Lincomycin	Growth Promoter, Therapeutic	4,268,660
Sulfonamides	Sulfamethazine	Therapeutic	592,000
β-Lactams	Penicillin-G	Therapeutic	1,814,070
Aminoglycosides	Streptomycin, Neomycin	Therapeutic	257,250
Fluoroquinolones	Enrofloxacin	Therapeutic	36,200
		Total	**21,871,190**

[a] Ionophores/arsenicals are for animal production and not related to traditional antibiotics

5.2   Properties of Common Antibiotics

5.2.1 Macrolides

Macrolides have a macrocyclic lactone ring, usually 14, 15 or 16-membered, to which two or more sugar moieties are attached. Antibiotic macrolides are used to treat respiratory tract and soft tissue infections. They are primarily bacteriostatic and are specifically effective against gram-positive bacteria and *Mycoplasma* and have good activity against anaerobic bacteria (Prescott and Baggot, 1993). It functions by binding to the 50S subunit of the ribosome, inhibiting the bacterial protein synthesis. Common macrolide antibiotics include erythromycin, tylosin, spiramycin, and roxithromycin.

5.2.1.1 Tylosin

Tylosin is a macrolide antibiotic. Figure 5.1 provides the structure of various forms typically referred to as tylosin A, B, C and D. The empirical formula for tylosin is $C_{46}H_{80}N_2O_{13}$ with a molecular weight of 869.15. Tylosin A is the predominant form administered to animals. pK_a values of tylosin A range between 7.1 and 7.7 (Loke et al., 2002; O'Neil et al., 2001; Tolls, 2001) with the protonated form present below its pK_a and the non-ionized form present above the pK_a (Loke et al 2002). Tylosin has an estimated octanol-water partition coefficient (log K_{ow}) between 2.5 - 3.5 (Loke et al., 2002, and Wollenberger et al., 2000). Its melting point is between 128 and 132° C. Solubility of tylosin in water at 25° C is 5 mg/mL. The amount of tylosin and other macrolide antibiotics sold in 2001 was between 4 to 5 million pounds (Vansickle, 2002). Tylosin is more active against mycoplasma but is less active against bacteria (Prescott and Baggot, 1993).

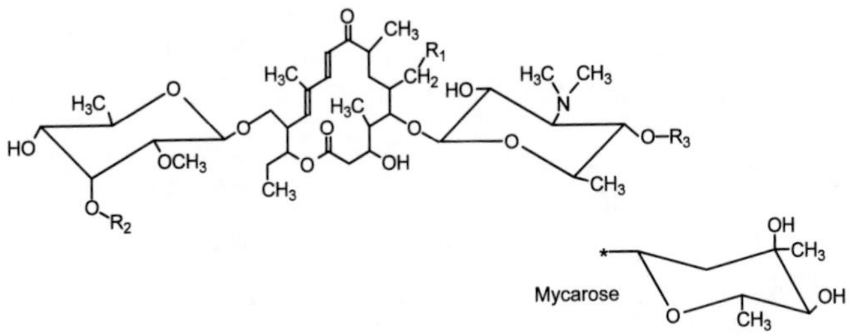

	Tylosin A	Tylosin B	Tylosin C	Tylosin D	Dihydrodesmycosin
R_1	-CHO	-CHO	-CHO	-CH$_2$OH	-CH$_2$OH
R_2	-CH$_3$	-CH$_3$	-H	-CH$_3$	-CH$_3$
R_3	-Mycarose	-H	-Mycarose	-Mycarose	-H

Figure 5.1. Chemical structures of various tylosin forms

5.2.1.2 Erythromycin

Erythromycin consists of two sugars, desosamine and cladinose, attached to erythronolide, a macrocyclic lactone (Figure 5.2). It is a white or slightly yellow, crystalline powder and is odorless or practically odorless. Erythromycin is produced during fermentation by a strain of *Streptomyces erythreus*. Three erythromycins, designated as A, B and C, are produced during fermentation. Erythromycin A is the major and most important component of the erythromycins produced. Erythromycin A has the same sugar moieties as erythromycin B but differ in position 12 of the aglycone (erythronolide) where A has a hydroxyl substituent. In the case of erythromycin C, the sugar cladinose is replaced by the sugar mycarose. The empirical formula is $C_{37}H_{67}NO_{13}$ with a molecular weight of 733.94. Its base has a pK_a of 8.8 (Merck, 1989). The solubility of erythromycin in soft blended water at $22 \pm 2^\circ$ C is 2.29 mg/mL. The chloroform/water partition coefficient is $12,587 \times 10^4$ at pH 7.4, indicating a high affinity for lipid, non-polar solvents (Burton and Schanker, 1974).

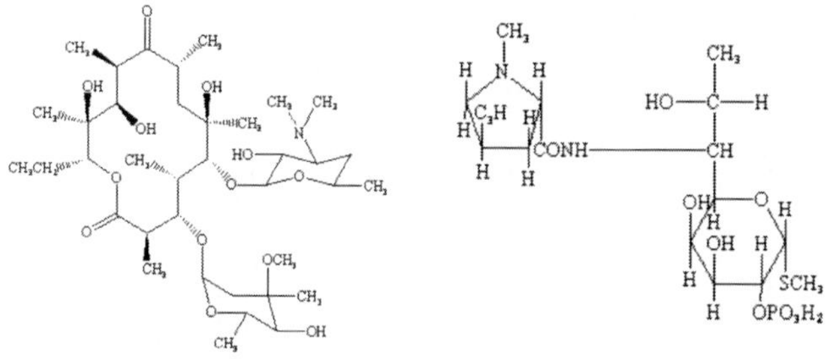

Erythromycin   Lincomycin

Figure 5.2 Chemical structures of erythromycin and lincomycin

5.2.1.3 Lincomycin

Lincomycin comes under the class of lincosamide antibiotic (Figure 5.2) and is produced from *Stretomyces lincolnensis*. It is similar in structure and mechanism of action as macrolides. Lincomycin inhibits the synthesis of bacterial protein by binding onto the 50S subunit and disrupting the process of peptide chain initiation. Lincomycin has a wider antibacterial spectrum than macrolides and is effective on other species such as *Actinomycetes, Staphylococcal, Streptococcal* and *Mycoplasma*. Due to its adverse effects, it is only used for patients that are allergic to penicillin or bacteria that has developed resistance to antibiotics. The empirical formula for lincomycin is $C_{18}H_{34}N_2O_6S$ and its molecular weight of 406.54. Its log K_{ow} is 0.56 and its water solubility is 927 mg/L at 25° C. The pK_a for lincomycin is 12.9 (Sarmah et al., 2006).

5.2.2 Tetracycline

Tetracyclines are broad-spectrum polyketide antibiotics produced by the *Streptomyces* bacterium. Tetarcyclines are primarily used to treat infections of the respiratory tract due to *Hemophilus influenzae, Streptococcus pneumoniae*, or *Mycoplasma pneumoniae*, sinuses, middle ear, urinary tract, intestines, and gonorrhea. It is now used in humans mostly to treat acne and other skin disorder such as rosacea. It is also used to treat Rocky Mountain spotted fever, typhus, cancroids, cholera, brucellosis, anthrax, and syphilis. When used with other medications, it is effective in treating *Helicobacter pylori*, the bacteria causing ulcers. They function by binding to the 30S subunit of the ribosome resulting in inhibition of bacterial protein synthesis. Some of the common tetracycline antibiotics include tetracycline, chlortetracycline and oxytetracycline. The chemical structures of various forms of tetracycline are shown in Figure 5.3.

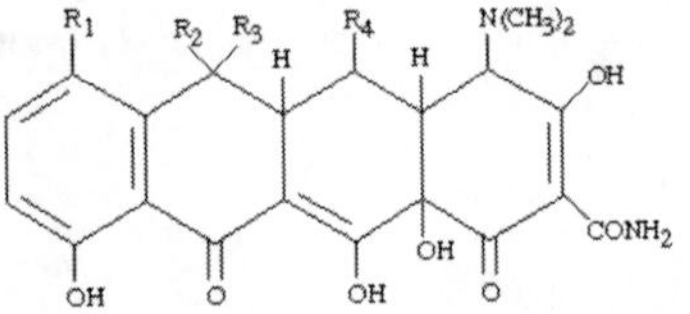

Compound	R1	R2	R3	R4
Tetracycline	H	CH$_3$	OH	H
Chlortetracycline	Cl	CH$_3$	OH	H
Oxytetracycline	H	CH$_3$	OH	H
Minocycline	N(CH$_3$)$_2$	H	H	H
Doxycycline	H	CH$_3$	H	OH

Figure 5.3 Chemical structures of various forms of tetracycline

5.2.2.1 Tetracycline

Tetracycline has an empirical formula of $C_{22}H_{24}N_2O_3$ and a molecular weight of 444.44 (Figure 5.3). The pK$_a$s for tetracycline are 3.32, 7.78 and 9.58. It is slightly soluble in water at pH 7.0 and is available as the hydrochlorides. Its melting point is between 170° and 173° C. The chloroform-water partition coefficient at pH 7.4 and log K_{ow} at pH 7.5 for tetracycline are 0.09 and -1.44, respectively (Barza et al., 1975). It is prescribed for many different infections particularly respiratory tract infections. However, it is used for both treatment and growth promotion in cattle, swine and poultry.

5.2.2.2 Chlortetracycline

The empirical formula for chlortetracycline is $C_{22}H_{23}ClN_2O_8$ with a molecular weight of 478.89 and a log K_{ow} of - 0.62. Its pK$_a$s are 3.33, 7.55 and 9.33. Its solubility in water is between 630 mg/L at 25° C but becomes very soluble in aqueous solution with pH above 8.5. In comparison to other antibiotics in its class, chlortetracycline is the most commonly used antibiotics for confined animal operations.

5.2.2.3 Oxytetracycline

Oxytetracycline has a molecular weight of 460.43 and its empirical formula is $C_{22}H_{24}N_2O_9$. The pK_as for oxytetracycline are 3.22, 7.46 and 8.94. The chloroform-water partition coefficient at pH 7.4 and log K_{ow} at pH 7.5 for oxytetracycline are 0.007 and -0.9, respectively (Barza et al., 1975). Solubility of oxytetracycline in water is approximately 313 mg/L at 25° C (Kay et al., 2005). Oxytetracycline is used for growth promotion and is used as a prophylaxis for cattle. It is very slightly soluble in water and has a half-life in humans of about 10.6 hours.

5.2.3 Sulfonamides

Sulfonamides, also known as sulfa drugs, are synthetic antimicrobials, used to treat bacterial and some fungal infections. The sulfonamides act by competitively inhibiting the conversion of p-aminobenzoic to dihydropteroate which is needed by the bacteria for folic acid synthesis. Sulfanilic amide has been known since 1930's as bacterial infection fighter and although over 5000 compounds in the sulfonamide class have been synthesized, only about 100 compounds have been commercialized (Holm at al., 1995). This class of antibiotics is the second largest group used in France, Germany, and U.K., consisting of between 11 and 23% of total antibiotics (Thiele-Bruhn, 2003).

Compounds within this class of antibiotics include sulfamethoxazole, sulfanilamide, sulfametahzine, sulfamethizole, sulfadoxine, and sulfadiazine. The general structure of the sulfonamide group is presented in Figure 5.4. Sulfonamides typically have two pK_a values with the protonation of the amino group occurring at pK_1 and deprotonation of RSO_2NHR occurring at pK_2 (Ingerslev and Halling-Sørensen, 2001).

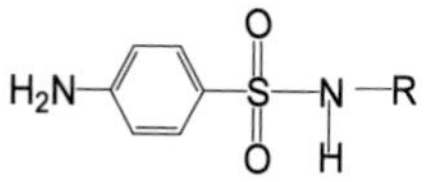

Figure 5.4 General structure of sulfonamide group

5.2.3.1 Sulfamethazine

Sulfamethazine, also known as sulfadimidine, is one of the major antibiotics in the sulfonamide group used in CAFOs (Figure 5.5). Sulfamethazine has an empirical formula of $C_{12}H_{14}N_4O_2S$ and a molecular weight of 278.34. Its log K_{ow} is 0.89 with a solubility of 1,500 mg/L in water at 29° C (NLM, 2008). The pK_as for sulfamethazine are 2.65 and 7.4. Sulfamethazine is widely used for the treatment of food-producing animals such as pig, cattle, sheep, chicken, and turkey. In pigs, it is normally used in combination with chlortetracycline and penicillin for maintenance of weight gain. Following ingestion, sulfamethazine can be converted by enzymes from the liver to various metabolites. One of the metabolites is N⁴-acetylsulfamethazine, which forms the majority of the excreted drug in the urine.

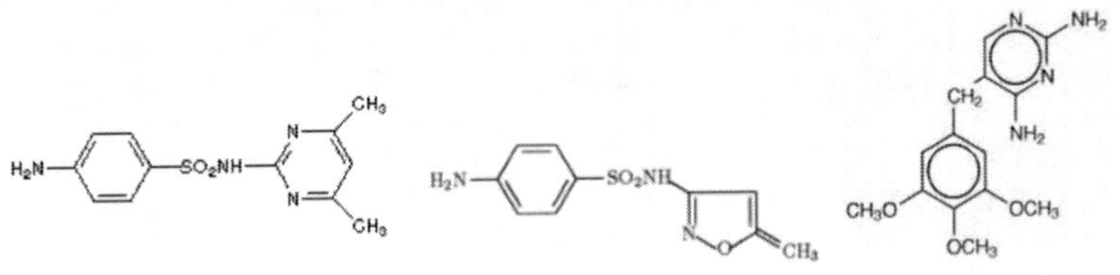

| Sulfamethazine | Sulfamethoxazole | Trimethoprim |

Figure 5.5 Chemical structures of sulfamethazine, sulfamethoxazole and trimethoprim

5.2.3.2 Sulfamethoxazole

Sulfamethoxazole is a bacteriostatic antibiotic and is often used in combination with trimethoprim. Its empirical formula is $C_{10}H_{11}N_3O_3S$ with a molecular weight of 253.28 (Figure 5.5). Its log K_{ow} is 0.89 with a solubility of 610 mg/L in water at 37° C. The pK$_a$s for sulfamethoxazole are 1.6 and 5.7. It is generally used to treat urinary tract infections and is effective against *Streptococcus*, *Staphylococcus aureus*, *Escherichia coli*, *Haemophilus influenzae*, and oral anaerobes.

5.2.3.3 Trimethoprim

The chemical formula for trimethroprim is $C_{14}H_{18}N_4O_3$ and it has a molecular weight of 290.3 (Figure 5.5). The pK$_a$s for trimethroprim are 3.23 and 6.76. Its log K_{ow} is 0.91 with a solubility of 400 mg/L in water at 25° C. It is typically used as a prophylaxis and for the treatment of ear infections or bladder and urinary tract infections.

5.2.4 β-Lactams

β-Lactams are antibiotics with a structural nucleus of a β-lactam ring. All β-lactams operate by binding to and inactivating the enzymes required for bacterial cell wall synthesis resulting in defective cell walls formation. The activity of transpeptidase and peptidoglycan-active enzymes which catalyze cross-linkage of glycopeptides polymer units of cell wall are inhibited by the antibiotic. This will eventually lead to lysis of the cells and death of the organism. Examples of β-lactam include penicillin-G, amoxicillin and the cephalosporins, carbapenems and penems family of antibiotics.

5.2.4.1 Penicillin-G

Of the different β-lactam antibiotics, penicillin-G is the most widely for medicinal purposes. Penicillin-G is typically white in color, crystalline, and odorless. Its empirical formula is $C_{16}H_{18}N_2O_4S$ with a molecular weight of 334.4 (Figure 5.6). It has a pKa of 2.7 and a log K_{ow} of 1.83 (Veith, et al, 1980). Its solubility in water is 210 mg/L at 25° C (NLM, 2008).

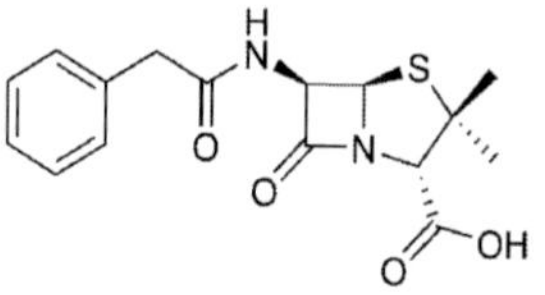

Penicillin-G 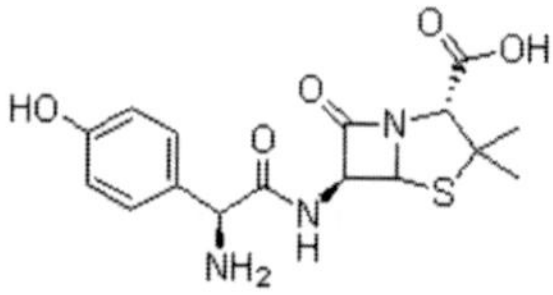   Amoxicillin

Figure 5.6 Chemical structures of penicillin-G and amoxicillin

5.2.4.2 Amoxicillin

The empirical formula for amoxicilin is $C_{16}H_{19}N_3O_5S$ with a molecular weight of 365.4 (Figure 5.6). Its water solubility is 3,430 mg/L at 25° C and a log K_{ow} of 0.87 (NLM, 2008). Amoxicillin is used for the treatment of many different types of bacterial infections including ear infections, bladder infections, pneumonia, gonorrhea, and *E. coli* or *Salmonella* infection. It is used more than the other drugs within the β-lactam class because it is easily sorbed after ingestion. Metabolism in humans is less than 30% and it is biotransformed in liver with a half life of 61.3 minutes.

5.2.5 Fluoroquinolones

The main structural characteristic of fluoroquinolones consists of a 4-pyridone-3-carboxylic acid moiety with a bi-cyclic hetero-aromatic ring with a nitrogen at position 1 (Figure 5.7). It is generally used for gram-negative but has also been effective on gram-positive bacteria. It exhibits bactericidal activity by inhibiting the activity of DNA gyrase and topoisomerase. These are essential enzymes for the replication of bacterial DNA. Fluoroquinolones are very useful and widely used in large amounts for humans and animal medical care. Examples of fluoroquinolones include ciprofloxacin, levofloxacin, and enrofloxacin.

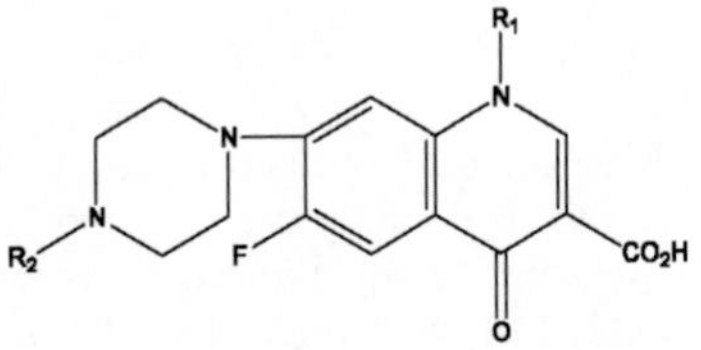

Figure 5.7 General structure of fluoroquinolones

5.2.5.1 Ciprofloxacin

Ciprofloxacin has an empirical formula of $C_{17}H_{18}FN_3O_3$ and a molecular weight of 331.35. Its pK_as are 6.20 and 8.59 (Figure 5.8). The log K_{ow} is 0.28 while the water solubility is 30,000 mg/L at 20° C (NLM, 2008). Ciprofloxacin is widely used for the treatment of human diseases rather than animal disease - typically for the treatment of lower respiratory infections such as pneumonia, urinary tract infections, and septicemia.

5.2.5.2 Levofloxacin

Levofloxacin has an empirical formula of $C_{18}H_{20}FN_3O_4 \cdot \frac{1}{2}H_2O$ and a molecular weight of 361.4 (Figure 5.8). Its pK_as are 6.26 and 7.81 (Escribano et al., 1997). Water solubility is maximum at 272 mg/mL at pH 6.7. Levofloxacin is a second-generation fluoroquinolone with excellent activity against gram-positive, gram-negative and anaerobic bacteria and is used against a wide range of infections such as pneumonia, and urinary tract infections before the specific organism is identified. After administration, it is metabolized to demethyl-levofloxacin and levofloxacin-N-oxide, which are excreted via urine (Dumka, 2007).

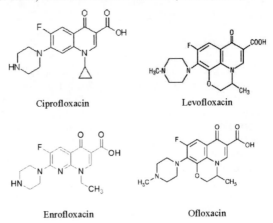

Ciprofloxacin   Levofloxacin

Enrofloxacin   Ofloxacin

Figure 5.8 Chemical structures of some common fluoroquinolones

5.2.5.3 Enrofloxacin

Enrofloxacin was originally developed as a veterinary medicine for livestock and is currently used for the treatment of individual pets and livestock. The empirical formula is $C_{19}H_{22}FN_3O_3$ (Figure 5.8). Its log K_{ow} is 0.70 and its water solubility is 3,400 mg/L at $25°$ C (NLM, 2008). Its activity is concentration dependent and bacterial cells are killed within 20-30 minutes of exposure. It is effective for both gram-negative and gram-positive bacteria and the mechanism is thought to be the inhibition of bacterial DNA-gyrase which prevents DNA supercoiling and synthesis.

5.2.5.4 Ofloxacin

Ofloxacin has a molecular weight of 361.4 g/mol (Figure 5.8) and an empirical formula of $C_{18}H_{20}FN_3O_4$ (Isidori et al., 2005). Solubility of ofloxacin in water is 28,300 mg/L at $25°$ C. Its log K_{ow} is -0.39 (NLM, 2008). The pharmakocinetical excretion rate is 70%. Ofloxacin is effective against most gram-negative and gram-positive bacteria and is used for treating acute and chronic urinary tract infections. Its mechanism of action is to inhibit the supercoiling activity of bacterial DNA gyrase -- halting DNA replication.

5.2.6 Ionophores

Ionophores are antimicrobial compounds used as growth promoters in animal feed and for improving feed efficiency. These are specifically used to target the ruminal bacterial population and for altering the ecology of the intestinal microbial consortium. Ionophores acts by transporting ions across cell membranes and disrupting the transmembrane ion concentration gradients of microorganisms and consequently killing the bacteria. Common ionophores used are monensin, ionomycin, lasalocid, laidlomycin, salinomycin, valinomycin, and calixarene.

5.2.6.1 Monesin

Monensin is a naturally found polyether ionophore antibiotics. Monesin has a chemical formula of $C_{36}H_{62}O_{11}$ (Figure 5.9) with a molecular weight of 692 and a pK_a of 6.65. The log K_{ow} range from 2.75–3.89 and it has a water solubility of 3,000 mg/L. It operates by blocking intracellular protein transport and by interrupting glycoprotein secretions affecting the microorganism ability to transport monovalent and divalent metal ions through cell membrane. It is soluble in methanol, ethanol, and chloroform. It is broadly used as growth promoter in beef and diary cattle.

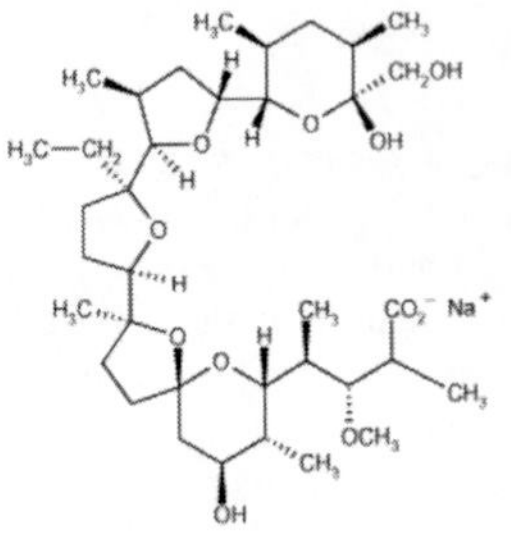

Monesin 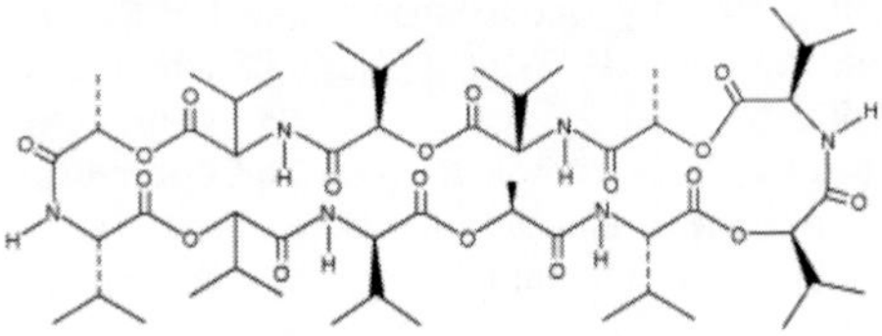   Valinomycin

Figure 5.9 Chemical structures of monesin and valinomycin

5.2.6.2 Valinomycin

Valinomycin is a cyclododecadepsipeptide potassium-selective ionophore which is isolated from *Streptomyces*. The structure consists of twelve alternating amino acids and esters to form a macrocyclic molecule (Figure 5.9). The empirical formula for valinomycin is $C_{54}H_{90}N_6O_{18}$ and has a molecular weight 1,111.3. Its melting point is between $172°$ C -177 °C. Its log K_{ow} is 4.49. It is soluble in methanol, ethanol, ethyl acetate, petrol-ether, and dichloromethane but is considered to be insoluble in water. It plays a role as an ion-exchange agent by disrupting the potassium ion movements through lipid membranes, thus disrupting transmembrane ion concentration gradients that are required for the proper functioning and survival of microorganisms.

5.3   Incomplete Metabolism of Antibiotics

Antibiotics administered to humans and livestock are not completely metabolized in vivo with the release of the parent compound and pharmaceutically active metabolites in the feces or manure (Boxall et al., 2001, Sieck et al., 1978). In humans, between 30 to 90% of the administered amount may be excreted from the body without metabolism (Rang et al., 1999; Costanzo et al., 2005). In the case of livestock, Table 5.3 summarizes the percent of administered antibiotics in animal wastes. As such, the main pathway of release of antibiotics into the environment is through the release of manure from animal feedlots (Halling-Sørensen et al., 2002). This has been confirmed from the detection of antibiotics and resistant bacteria near CAFOs (Boxall et al., 2001; Campagnolo et al., 2002; Chee-Sanford et al., 2001; Kolpin et al., 2002; and Thurman and Hostetler, 2000).

Animal wastes are typically placed in anaerobic lagoons or in aerated lagoons with mechanical aeration to reduce and stabilize the carbonaceous components, nutrients, odor and solids volume (Zhang, 2001). Digested and stabilized solids are typically applied to agricultural fields as a source of nutrients. Runoff from land-applied biosolids including direct liquid discharges from the storage of the manure in lagoons may result in the transport and dispersion of antibiotics in the environment.

Manure management is important in reducing antibiotic discharge but there are not many data available that correlates antibiotic presence and different manure management techniques (Boxall et al., 2001, and Daughton and Ternes, 1999). To understand the fate and release of antibiotics and their metabolites into watersheds, it is important to understand the factors controlling the biodegradation, sorption, hydrolysis, photolysis of antibiotics within municipal and industrial wastewater treatment plants, manure waste lagoons and after the treated effluent are discharged into receiving streams and solids applied to the fields. However, there are many challenges in assessing the risk of antibiotics. These challenges include developing the appropriate analytical methods, measurement of antibiotics at low levels and near the detection limits, and interferences from other compounds in the analysis.

Table 5.3 Percentage of some administered antibiotics in animal waste (adapted from Kolz, 2004).

Antibiotic	Percent (%) in waste	References
Tylosin	Up to 40% as tylosin A (5%) and active metabolites (35%)	Sieck et al., 1978
	28-76% as tylosin	Feinman and Matheson, 1978 as cited by Jjemba, 2002
Chlortetracycline	>80% as chlortetracycline	Hirsch et al., 1999
	17-75% as chlortetracycline	Montforts et al., 1999
Oxytetracycline	40-90%	Boxall et al., 2001
Penicillin-G	50-70% as penicillin-G	Hirsch et al., 1999
Streptomycin	50-60% as dihydrostreptomycin	Huber, 1971a
Sulfonamides	30-95% as sulfonamides	Boxall et al., 2002
Valnemulin	2% as valnemulin	Boxall et al., 2002

5.4 Occurrence of Antibiotics in Natural Systems

5.4.1 Macrolides

The concentrations of various macrolides in various matrices are presented in Table 5.4. Concentrations of erthyromycin and tylosin in surface waters including rivers and streams downstream of wastewater treatment plants discharges are typically less than 1 µg/L while concentrations of trimethoprim may be as high as 7 μg/L. Other antibiotics in this class have concentrations in surface waters in the low nanogram/L (ng/L) range. It is important to note that the concentrations reported are very much dependent on the analytical methods used and their limits of detection. Concentrations of macrolides in manures are dependent on their usage and can be as high as 10 mg/kg. In wastewater of manure lagoons, the concentration of macrolides can be as high as 100 μg/L.

Table 5.4 Macrolides concentrations in soil, manure and surface water.

Compounds	Concentration	Conditions	References
General Macrolides	94 μg/L	Manure lagoons, 8 sites in Iowa and Ohio	Campagnolo et al., 2002
Lincomycin	0.02 μg/L	Surface water grab samples	Kolpin et al., 2002
Erythromycin	1.7 μg/L	Surface water grab samples	Kolpin et al., 2002
	0.175 μg/L	Surface water downstream of WWTP*	Haggard et al., 2006
	BLD** to 71 ng/L	Surface water downstream of WWTP	Gros et al., 2007
	1.8 – 4.8 ng/L	Surface water downstream of WWTP in Korea	Kim et al. 2007
Tylosin	0.3 – 0.5 mg/kg	Manure-amended soil	Halling-Sørensen et al., 2002
	10.3 mg/kg	Manure	Royal Danish School of Pharmacy, 2002
	~ 0.2 μg/kg	Soil	Boxall et al., 2001
	0.13 – 0.42 μg/L	Groundwater	Boxall et al., 2001
	0.012 μg/L	Surface water downstream of WWTP	Haggard et al., 2006
Trimethoprim	0.19 μg/L	Surface water downstream of WWTP	Haggard et al., 2006
	10 to 69 ng/L	Surface water downstream of WWTP	Gros et al., 2007
	0.090 - 6.0 μg/L	Surface water downstream of WWTP	Batt et al., 2006
	<0.1 mg/kg (wet)	Manure lagoon grab samples	Haller et al., 2002
	0.013 μg/L	Surface water grab samples	Kolpin et al., 2002
	17.4 ng/L	Surface water in Italy	Zuccato et al., 2000
Clindamycin	0.043 – 0.076 μg/L	Surface water downstream of WWTP	Batt et al., 2006
Azithromycin	9 - 68 ng/L	Surface water downstream of WWTP	Gros et al., 2007

* - WWTP – wastewater treatment plant ** - BLD – below level of detection

Table 5.5 presents the partition coefficients of tylosin for different environmental matrices. Linear sorption coefficient (K_d) values of tylosin ranged between 0.9 - 240 L/kg. Most studies reported isotherms that were slightly non-linear. Gupta et al. (2002) found that K_{oc} values for tylosin in a sandy loam (organic carbon (OC) = 2.2%) and a clay loam (OC = 4.4%) were similar. On the contrary, the K_{oc} values for soils (553 L/kg – 7,990 L/kg) were found to be higher than in manure (109.6 L/kg). This generally implies that sorption to organic carbon may not be a major factor but other factors such as cation exchange, cation bridging at clay surfaces, surface complexation, and hydrogen or dipole bonding may influence sorption to soils and manure (Loke et al., 2002; Tolls, 2001). Sorption studies in agricultural soils showed that tylosin partitioning has also been found to be positively correlated to the clay content, matrix pH and cation exchange capacity (Oliveira et al., 2002, and Rabølle and Spliid, 2000). The K_{oc} value for lincomycin was found to be 59 by Boxall et al. (2005) while the K_{oc} values for trimethoprim ranged from 1680 – 3900 L/kg.

Table 5.5 Tylosin sorption in various adsorbents (adapted from Kolz, 2004).

Adsorbent	Typical f_{oc} (%)	K_d (L/kg)	K_{oc} (L/kg)	K_f	n	Reference
Sandy soils and sandy loam soils	1.1%-1.6%	8.3 -128	553-7990	2.3-7.0	0.8-0.9	Rabølle and Spliid, 2000
Soil: 88% Sand 5% Clay 5% silt	1.4% w/w	Not given	Not given	2.03±0.9	0.67± 0.07	Ingerslev and Halling-Sørensen, 2001
Soil: 76% Sand 11% Clay 11% silt	1.6% w/w	Not given	Not given	32.1±13.2	0.67± 0.06	Ingerslev and Halling-Sørensen, 2001
Webster clay loam: 34% clay	4.4%	92	2090	Not given	Not given	Gupta et al., 2002
Hubbard sandy loam: 10.4 % clay	2.2%	66	3000	Not given	Not given	Gupta et al., 2002
Fresh (solid) swine manure	42%	240	571	Not given	Not given	Loke et al., 2002
Manure slurries	13 – 16%	91 - 106	564 - 831	39 - 100	1.02 – 1.32	Kolz et al., 2005a

Half-lives of tylosin averaged between 4 and 8 days in swine, calf and chicken manure and between 2 and 8 days in aqueous manure and mixtures of soils and manure (De Liguoro et al., 2003; Ingerslev and Halling-Sørensen, 2001; Loke et al., 2000; Teeter and Meyerhoff, 2003). Kolz et al. (2005b) found that 90% of tylosin disappeared between 12 and 26 hours in aerated swine manure slurries. Under anaerobic conditions, 90% of tylosin disappeared between 30 and 130 hours indicating that besides biodegradation, sorption may also be involved in the disappearance of tylosin. Tylosin disappearance followed a biphasic pattern, where initial loss was rapid followed by a slow phase. In surface waters, the half-lives were between 10 and 40 days (Ingerslev et al., 2001; Loke et al., 2000; Teeter and

Meyerhoff, 2003). Metabolites found in biodegradation studies included dihydrodesmycosin and an unknown degradate product with molecular mass of 933.5 amu (Kolz et al., 2005b). After 8 months of incubation in anaerobic swine manure slurries, tylosin remained in the slurries indicating that tylosin degradation in lagoons was incomplete (Kolz et al., 2005b).

Because the hydrolysis of erythromycin is very fast, anhydro-erythromycin is more frequently detected in the environment rather than the parent compound (Kümmerer, 2004). Persistence of erythromycin in the environment is normally longer than one year (Zuccato et al., 2000). Sorption isotherms of erythromycin A on sediments typically follow Langmuir-type isotherm and are impacted by clay and organic matter contents of the sediments (Kim et al., 2004). The Langmuir distribution coefficient and the exponential constant for sediments with 2.9% OC, 93.4% sand, 4.8 % silt, 1.8% clay were 815 ± 48 and 0.52 ± 0.022, respectively (Kim et al., 2004). For a sediment with 90.7% sand, 7.8 % silt, 1.5% clay and OC of 1.6%, the Langmuir distribution coefficient and the exponential constant were 481 ± 23 and 0.43 ± 0.016, respectively.

The half-life of erythromycin was estimated to be about 20 days in soil by Schlusener and Bester (2006). Using the OECD Closed Bottle Test (OECD, 1992), Alexy et al. (2004) found that there were no apparent degradation of erythromycin after 28 days of incubation. In a follow-up study, similar results were obtained (Gartiser et al., 2004). Gavalchin and Katz (1994) found that the half-life of erythromycin in sandy loam soil and cattle faeces was about 30 days. Smolenski et al. (2002) reported that the half-life of ^{14}C-lincomycin ranged from 2 to 9 days in aerobically incubated soils and the metaobolites found were lincomycin sulfoxide, lincomycin sulfone and 14 other unknown degradates.

Using the OECD Closed Bottle Test (OECD, 1992), trimethoprim naphthoate were found not to be biodegradable in 28 days (Alexy et al., 2004) which was confirmed by a similar study conducted over 28 days by Gartiser at al. (2004). In another study using a microcosm containing aquatic communities, the half life of trimethoprim was estimated to be 5.7 days, however, photodegradation was found to be more important than biodegradation (Lam et al., 2004).

5.4.2 Tetracyclines

Table 5.6 shows the concentrations of tetracycline compounds in surface water, groundwater, sediments, soils and manure. The concentrations in different matrices are dependent on several factors including the type of medium, pH of the solution and ion content of the matrix (Bakal and Stoskopf, 2001; Tolls, 2001). For example, tetracyclines may still leach into watersheds even though the low solubility of tetracyclines in water indicates that they may be highly immobile. In a study by Boxall et al. (2001), oxytetracycline traveled through clay loam soil into nearby surface waters within thirty-six hours of application to the soil. In a study by Thurman and Hostetler (2000), both tetracyclines and sulfonamides were found to be present in groundwater at µg/L levels near hog lots. A peak oxytetracycline concentration of 71.7 µg/L resulted when plots of soils amended with pig slurry with 18.85 mg/L of oxytetracycline were irrigated after 24 hours, indicating the fairly transportable portion of the oxytetracycline (Kay et al., 2005).

Table 5.6 Tetracyclines concentrations in soil, manure and water.

Compound	Concentrations	Matrix	References
Tetracycline	4.0 mg/kg	Manure applied soil (30-50 m³/ha-year)	Hamscher et al, 2002
	198.7 µg/kg	Manure applied soil (10-20 cm depth)	Hamscher et al., 2002
	1 – 39.6 µg/kg	Soil	Boxall et al., 2001
	0.11 – 0.27µg/L	Groundwater	Boxall et al., 2001
	~1 µg/L	River water	Halling-Sørensen et al., 1998
	1.34 µg/L	Surface water	Lindsey et al., 2001
	0.11 µg/L	Surface water	Kolpin et al., 2002
Oxytetracycline	171 µg/L	Manure lagoons, 8 sites in Iowa and Ohio	Campagnolo et al., 2002
	413.8 mg/kg	Manure	Halling-Sørensen et al., 2002
	2.1 – 19 mg/kg	Manure	De Liguoro at al., 2003
	0.82 mg/kg	Matured manure after 5 months	De Liguoro at al., 2003
	0.1 – 11 mg/kg	Sediment	Halling-Sørensen et al., 1998
	0.1 – 285 mg/kg	Sediment	Boxall et al., 2001
	0.1 – 4.9 mg/kg	Anoxic, fish farm pond sediment	Halling-Sørensen et al., 2002
	0.9 – 8.6 µg/kg	Soil	Boxall et al., 2001
	1.4 – 1.5 mg/kg	Swine manure amended soil	Halling-Sørensen et al., 2002
	1.14 mg/kg	Swine manure-amended soil (0.87 kg/ha)	Boxall et al., 2001
	0.85 – 2.73 µg/L	Soil water after manure applied (0.87 kg/ha)	Boxall et al., 2001
	1250 µg/kg	Swine manure-amended soil (0 – 2 ins. soil depth)	Kulshrestha et al., 2005
	11 µg/kg	Swine-manure-amended soil (12 – 14 ins. soil depth)	Kulshrestha et al., 2005
	<5 µg/kg	Cattle manure-amended soil (60 cm depth)	De Liguoro et al. (2003)
	0.34 µg/L	Surface water	Kolpin et al., 2002
	0.15 – 0.19 µg/L	Groundwater	Boxall et al., 2001

Concentrations of oxytetracycline and chlortetracycline in swine manure were found to be as high as 414 mg/kg and 7.7 mg/L, respectively (Halling-Sorensen et al., 2002; Kumar et al., 2004). Kumar et al. (2005) estimated that if the swine manure with a chlortetracycline concentration of 7.7 mg/L was applied at a typical rate of 50,000 L/ha, the amount of chlortetracycline may be as high as 387 g/ha (Kumar et al., 2005). Concentrations of tetracycline in manure-amended soils are typically less than 10 mg/kg while concentrations in surface and groundwater were typically less than 1 μg/L.

Table 5.6 Tetracyclines concentrations in soil, manure and water (cont'd).

Compound	Concentrations	Matrix	References
Chlortetracycline	7.73 mg/ L	Swine manure	Kumar et al., 2004
	0.1 mg/kg	Manure applied soil (30-50 m^3/ha-year)	Hamscher et al, 2002
	4.6 – 7.3 mg/kg	Manure applied sandy soil (depth 0- 30 cm.)	Hamscher et al, 2002
	0.7 – 41.8 μg/kg	Soil	Boxall et al., 2001
	0.17 – 0.22 μg/L	Groundwater	Boxall et al., 2001
	0.5 μg/L	Surface Water	Boxall et al., 2001
	0.69 μg/L	Surface water	Kolpin et al., 2002

Rabole and Spliid (2000) reported K_{oc} values of 27,792 - 93,317 L/kg for sorption of oxytetracycline to sandy loam soils with OC between 1.2 to 1.6%. The K_d values were between 417 to 1,026 L/kg. Loke et al. (2002) found that the sorption of oxytetracycline to manure had a log K_d of 1.89, giving a log K_{oc} of 2.29, with the sorption coefficient much higher than the estimated values based on the log K_{ow} of the compound. Loke et al. (2002) speculated that sorption of oxytetracycline to manure was influenced by ionic binding to divalent metal ions as such Mg^{2+} and Ca^{2+} as well as other charged compounds in the manure. They found that binding of oxytetracycline to soil was stronger than the binding to manure which was most likely due to the strong mineral-related metal complexes formed between soil, metal ion and oxytetracycline (Loke et al., 2002). Samuelsen et al., (1992) speculated that tetracyclines can form complexes with double-charged cations, such as calcium, which occur in high concentrations in soils.

The half-life of oxytetracycline in calf manure was found to be approximately 30 days with detectable concentrations at low mg/kg even after 5 months (De Liguoro et al., 2003). The half-life of oxytetracycline in marine sediment was found to be longer than in manure or soil at approximately 151 days (Hektoen et al., 1995). In shake flask studies, the half lives of oxytetracycline under aerobic conditions ranged from 42 – 46 days and were not affected by the initial concentrations of the antibiotics (Ingerslev et al., 2001). However, the rate of removal increased when the solution was amended with 1 g/L of sediment or 3 g/L of activated sludge.

Tetracycline in liquid manure was found to degrade by 50% in 5 months (Winckler and Grafe, 2000). Chlortetracycline in soil fertilized with manure was found to decrease by approximately 80% to 90% after 5 months of application, and the degradation corresponded to first order kinetics (Kulshrestha et al., 2005). Aga and O'Conner (2003) found that the half-life of chlortetracycline was 2.4 days in manure-amended soils with isochlortetracycline as a degradation product.

5.4.3 Sulfonamides

Table 5.7 shows the concentration of various sulfonamides in soils, manure, groundwater and surface waters. Concentrations of sulfonamides in manure were typically less than 20 mg/kg while the concentrations in the aqueous phase of manure in lagoons were in the ng/L range. Soils fertilized with manure typically have sulfonamides of less than 5 mg/kg. In Germany, up to 15 µg/ kg of sulfamethazine in field soil was found after seven months of manure application (Christiana et al., 2003). Sulfonamides were detected in 7 out of 144 water samples collected from groundwater and surface waters throughout the U.S. which implies the mobility of sulfonamides (Lindsey et al., 2001). Kolpin et al. (2002) monitored 95 organic contaminants from 104 stream samples in the U.S. from 1999 to 2000, and found that the maximum concentrations of sulfadimethoxine, sulfamethazine, sulfamethizole, and sulfamethoxazole ranged from 0.06 to 1.9 µg/L while sulfachloropyridazine, sulfamerazine, and sulfathiazole were not detected. Other studies as listed in Table 5.7 indicate that concentrations in surface waters were typically less than 1 µg/L.

Sulfonamides are amphoteric compounds and are ionized at certain pHs (Langhammer, 1989; Thiele, 2000; Boxall et al., 2002; Thiele-Bruhn et al., 2004) and are impacted by the pH of soils. Since the pH of soils and the aqueous phase are higher than the pK_1 of sulfonamides (pK_1 between 2 to 3), the major species of sulfonamides in the environment are neutral or negatively charged with a very small fraction as positively charged compounds. The higher the pH of the soil, the more dissociated are the sulfonamide species and therefore resulting in a decrease in sorption (Thiele-Bruhn et al., 2004). The study conducted by Boxall et al. (2002) showed that sorption of sulfachloropyridazine onto soils at high pH was less than sorption at low pH. Sulfonamides are normally weakly sorbed to soils with the potential to leach from soil and are practically mobile in soils, polluting groundwater and surface waters (Thiele, 2000; Tolls, 2001; Boxall et al., 2002).

Table 5.7 Sulfonamides concentrations in manure, soils and water.

Compound	Concentrations	Matrix	References
General Sulfonamides	~ 20 µg/L	Manure lagoons, measured average of 8 sites in Iowa and Ohio	Campagnolo et al., 2002
	Up to 20 mg/kg	Manure pits in Switzerland	Haller at al., 2002
Sulfamethazine	0.08 – 0.16 µg/L	Groundwater	Boxall et al., 2001
	0.13 – 8.7 mg/kg	Manure lagoon grab samples	Haller et al., 2002
	0.22 µg/L	Surface water	Lindsey et al., 2001; Kolpin et al., 2002
	11 µg/kg soil	Soil fertilized with manure	Höper at al., 2002
	Up to 0.05 µg/L	River	Thiele-Bruhn & Aust, 2004
[4] N -acetyl-sulfamethazine	<0.1 – 2.6 mg/kg	Manure lagoon grab samples	Haller et al., 2002
Sulfadimethoxine	0.06 - 15 µg/L	Surface water	Lindsey et al., 2001; Kolpin et al., 2002
Sulfadimethoxine	0.004	Downstream of WWTP	Haggard et al., 2006
Sulfamethizole	0.13 µg/L	Surface water	Kolpin et al., 2002
Sulfathiazole	<0.1 – 12.4 mg/kg	Manure lagoon grab samples	Haller et al., 2002
	0.08 µg/L	Surface water	Lindsey et al., 2001
Sulfachloro-pyridazine	1.55 mg/kg	Soil after 1.18 kg/ha applied	Boxall et al., 2001
Sulfamethoxazole	~1 µg/L	River water	Halling-Sørensen et al., 1998
	~1 - 2 µg/L	River water/surface water	Halling-Sørensen et al., 1998; Lindsey et al., 2001; Kolpin et al., 2002
	0.22 µg/L	Groundwater	Lindsey et al., 2001
	Up to 0.48 µg/L	Surface water	Hirsch et el., 1999
	Up to 0.47 µg/L	Ground water	Hirsch et el., 1999
	0.5	Downstream of WWTP	Haggard et al., 2006
	300 ng/L	In Rio Grande River	Brown et al. (2006)
	5 to 169 ng/L	In Ebro River Basin in Spain	Gros et al., 2007
	1.7 – 36 ng/L	Downstream of WWTP (Korea)	Kim et al. 2007

The K_d values as presented in Table 5.8 for the various sulfonamides typically range from 1 - 3 L/kg. K_{oc} values ranged from 30 to 200 L/kg with several studies giving K_{oc} values in the range of 300 L/kg. Proposed mechanisms of sorption for sulfonamides include hydrophobic partitioning, cation exchange, cation bridging, surface complexes, hydrogen bonding, and electrostatic interactions (Holten Lützhøft et al., 2000; Tolls, 2001). Sorption of sulfonamides onto soils is impacted by soil pH, soil organic matter, soil texture, soil minerals, and cation exchange capacity. Since the sulfonamides are polar compounds, they can be sorbed to both soil organic matter and soil minerals even though sorption to soil minerals is less than to organic matter (Kaiser and Zech, 1998). An increase in sorption of sulfonamides with increasing organic carbon content (Langhammer and Büning-Pfaue 1989) may be explained as partitioning. Ong et al. (2006) found that the sorption of sulfamethazine onto two soils increased with the organic carbon content of the soil but decreased for pH greater than 7. Linear K_d's for a soil with 1.42% OC were 1.17, and 0.49 L/kg at pH of 6 and 9, respectively. The K_d's for a soil with 4.0% OC were 3.00 and 1.04 L/kg at pH 6 and 9, respectively.

The impact of pig manure on the sorption of sulfonamides to soils has been investigated and sorption was found to decrease for the soils with manure added (Thiele-Bruhn and Aust, 2004). Impact of manure can be explained by the competition of dissolved organic matter in manure for the soil sorption sites (Tolls, 2001; Thiele-Bruhn and Aust, 2004). Thiele (2000) found that sorption of sulfapyridine was influenced by the quantity, composition, and the structure of soil colloids and, as such, the mobility of sulfonamides may be enhanced via colloidal-facilitated transport.

Degradation of sulfonamides has been found to occur during storage in manure lagoons. However, 40% and 60% of the initial concentrations of sulfamethazine and sulfathiazole, respectively, were found to remain in the manure slurry after 5 weeks of storage (Langhammer, 1989). Using the OECD Closed Bottle Test (OECD, 1992), about 4% of sulfamethoxazole were found to be degraded in 28 days (Alexy et al., 2004). In another study using a microcosm containing aquatic communities, the half life of sulfamethoxazole was estimated to be about 19 days (Lam et al., 2004). The half-lives of sulfadimethoxine in cattle were found to range from 1.36 to 2.56 days depending on the initial concentrations (Wang et al., 2007).

5.4.4 Penicillin

There is not much information available on sorption and degradation of penicillin in the environment. Haggard et al. (2006) speculated that these compounds hydrolyze rapidly and are, therefore, are not commonly detected in surface waters. The β-lactam ring is of poor stability and can be opened by β-lactamase, a widespread enzyme in bacteria, or by chemical hydrolysis. As such, intact penicillin compounds do not occur frequently in the environment (Myllyniemi et al., 2000). Alexy et al. (2004) found that 27% of benzylpenicillon were degraded within 28 days using the Closed bottle Test (OECD 301 D). Using the OECD 302 B test, Gartiser et al. (2004) found that 78 – 87% of benzylpenicillin was degraded over 28 days under aerobic conditions. In another study by Gartiser and co-workers, ultimate degradation of benzylpenicillin was observed after 60 days in an anaerobic digestion test according to ISO standard 11734 (Gartiser et al., 2007).

Table 5.8 Sorption coefficients of various sulfonamides onto different sorbents.

Compound	K_f[1]	K_d[2] (L/kg)	K_{oc} (L/kg)	Conc. (μg/g)	pH	Matrix	References
Sulfachloro - pyridazine	- -	- -	41-82	- -		Not available	Kay at al., 2005
	- -	1.8	- -	0.05-20	6.5	Clay loam	Boxall et al., 2002
	- -	0.9	- -	0.05-20	6.8	Sandy loam	Boxall et al., 2002
	- -	4	129	- -	6.2	Clay loam, 3.1% OC	Tolls et al., 2002
Sulfa-nilamide	1.65	0.57	35.4	0.1-40	7.5	Unfertilized silt loam, 1.6% OC	Thiele-Bruhn and Aust, 2004
	- -	0.59	36.6	0.1-40	7.4	Soil:slurry 1:50 (w/w), silt loam, 1.6 % OC	Thiele-Bruhn and Aust, 2004
	- -	1.7	- -	0-10	7.0	Unfertilized silt loam, 1.6% OC	Thiele-Bruhn et al., 2004
Sulfa-methazine	2.72	0.79	49.1	0.1-40	7.5	Unfertilized soil, silt loam, 1.6% OC	Thiele-Bruhn and Aust, 2004
	- -	0.74	45.9	0.1-40	7.4	Soil:slurry 1:50 (w/w), silt loam, 1.6 % OC	Thiele-Bruhn and Aust, 2004
	- -	1.2	174	0.2-25	5.2	Sand, 0.9% OC	Langhammer, 1989
	- -	3.1	125	0.2-25	5.6	Loamy sand, 2.3% OC	Langhammer, 1989
	- -	2.0	208	0.2-25	6.3	Sandy loam, 1.2% OC	Langhammer, 1989
	- -	1.0	82	0.2-25	6.9	Clay silt, 1.1% OC	Langhammer, 1989
	- -	2.4	- -	- -	7	soil 1.6 % OC,	Thiele et al., 2002
	- -	3	97	- -	6.2	Clay loam, 3.1% OC	Tolls et al., 2002
	- -	2.4	- -	0-10	7.0	Unfertilized silt loam, 1.6% OC,	Thiele-Bruhn et al., 2004

[1] Freundlich sorption coefficient K_f [2] Linear sorption coefficient

Table 5.8 Sorption coefficients of various sulfonamides onto different sorbents (cont'd).

Compound	K_f[1]	K_d[2]	K_{oc}	Conc.	pH	Matrix	References
		(L/kg)	(L/kg)	(μg/g)			
Sulfa-diazine	3.27	2.0	124	0.1-40	7.5	Unfertilized soil, silt loam, 1.6% OC	Thiele-Bruhn and Aust, 2004
	- -	1.18	73.2	0.1-40	7.4	Soil:slurry 1:50 (w/w), silt loam, 1.6 % OC	Thiele-Bruhn and Aust, 2004
	- -	2.0	- -	0-10	7.0	Unfertilized silt loam, 1.6% OC	Thiele-Bruhn et al., 2004
	- -	2.5	81	- -	6.2	Clay loam, 3.1% OC	Tolls et al., 2002
Sulfadi-methoxine	4.41	0.73	45.3	0.1-40	7.5	Unfertilized soil, silt loam, 1.6% OC, pH	Thiele-Bruhn and Aust, 2004
	- -	0.62	38.4	0.1-40	7.4	Soil:slurry 1:50 (w/w), silt loam, 1.6 % OC	Thiele-Bruhn and Aust, 2004
	- -	2.3	- -	0-10	7.0	unfertilized silt loam, 1.6% OC, pH	Thiele-Bruhn et al., 2004
	- -	10	323	- -	6.2	Clay loam, 3.1% OC	Tolls et al., 2002
Sulfa-pyridine	4.30	1.02	63.4	0.1-40	7.5	Unfertilized soil, silt loam, 1.6% OC	Thiele-Bruhn and Aust, 2004
	- -	1.22	75.7	0.1-40	7.4	Soil:slurry 1:50 (w/w), silt loam, 1.6 % OC	Thiele-Bruhn and Aust, 2004
	- -	3.5	- -	0-10	7.0	unfertilized silt loam, 1.6% OC	Thiele-Bruhn et al., 2004
	- -	1.6	101	0.1-500	7.0	Silt loam, 1.6% OC	Thiele, 2000
	- -	7.4	308	0.1-500	6.9	Silt loam, 2.4% OC	Thiele, 2000
Sulfa-thiazole	- -	3	97	- -	6.2	Clay loam, 3.1% OC	Tolls et al., 2002

[1] Freundlich sorption coefficient K_f [2] Linear sorption coefficient

5.4.5 Fluoroquinolones

Table 5.9 presents concentrations of some of the fluoroquinolones in the environment. Concentrations in surface waters are typically in the ng/L range.

Table 5.9 Concentrations of fluoroquinolones in the environment

Class	Compound	Concentration	Conditions	Source
Fluoroquinolones	Enrofloxacin	0.02 µg/L	River water	Kolpin et al., 2002
	Ciprofloxacin	0.039	Downstream of WWTP	Haggard et al., 2006
		0.043 to 0.076 µg/L	Downstream of WWTP	Batt et al., 2006
	Ofloxacin	0.109	Downstream of WWTP	Haggard et al., 2006
		BLD* to 146 ng/L	In Ebro River Basin in Spain	Gros et al., 2007

*BLD – Below limit of detection

Sukul and Spiteller (2007) indicated that fluoroquinolones are bound strongly to topsoil and therefore have less potential to migrate by surface waters and groundwater. Its strong binding characteristic, however, discourages biodegradation in soil and sediment and may be persistent in the environment. Sorption of ciprofloxacin was reported as log K_{oc} = 4.8 L/kg for soils, and log K_{oc} = 4.3 - 4.9 L/kg for particulate matters (Nowara et al., 1997; Cardoza et al., 2005). In experiments conducted by Belden et al. (2007), log K_d's for ciprofloxacin were found to be 4.54 L/kg and 2.92 L/kg for the sorption onto fine particulate organic matter and coarse particulate organic matter, respectively. Zhang and Huang (2007) studied the sorption of several fluoroquinolones on goethite and found that adsorption can be described by the Langmuir isotherm. The Langmuir constant (K_L) and the maximum concentration, C_{max} for ciprofloxacin at pH 5 for two different geothites (Fe-OOH) were found to be 79.96 l/mmol and 0.15 mmol/gm and 4.19 l/mmol and 0.10 mmol/gm, respectively. In addition, they also found the sorption was accompanied by slow oxidation by goethite forming a range of hydroxylated and dealkylated products.

In aquatic systems, fluoroquinolones are prone to photodegradation and biodegradation is not a significant pathway of disappearance for ciprofloxacin (Kümmerer et al., 2000). The half-life of ciprofloxacin in natural water is, somewhat short, at about 2 hours (Cardoza et al., 2005) as it is rapidly photodegraded (Lam et al., 2003; Cardoza et al., 2005).

Using the OECD Closed Bottle Test (OECD, 1992), ofloxacin was not degraded while about 5% of amoxicillin was found to be degraded in 28 days (Alexy et al., 2004). A study by Gartiser et al. (2004) showed that between 11 – 63% of amoxicillin was biodegraded in 28 days using the OECD 302 D test while ofloxacin was not degraded.

5.4.6 Ionophores

Not much information is available for ionophores. The most common ionophore, monensin was found in the concentration range of 0.8 – 1.08 mg/kg in soils (Boxall et al., 2001). Donoho (1984) found that about 60 to 70% of monensin was degraded in manure under aerobic conditions with a half-life of 70 days. Using the OECD Closed Bottle Test (OECD, 1992), about 4% of monensin sodium salt was found to be degraded in 28 days (Alexy et al., 2004) while Gartiser et al. (2004) found that monensin sodium salt was not degraded after 28 days using the OECD 302 B test.

5.5 Fate of Antibiotics in Engineered Systems

5.5.1 Typical Concentrations of Antibiotics in Wastewater

Municipal wastewater is one of the major sources of antibiotics in the environment. Extensive work has been done in many countries to quantify the extent of antibiotics contamination in municipal wastewaters and the removal and fate of antibiotics in conventional and advanced wastewater treatment systems. The common antibiotics found in municipal wastewaters include macrolides (erythromycin), tetracycline (oxytetracycline and tetracycline), sulfonamides (sulfamethoxazole), and fluoroquinolones (ciprofloxacin and ofloxacin) with concentrations typically ranging from ng/L to low µg/L.

Various studies from different regions of the US have been conducted to estimate the extent of antibiotics occurrence in municipal wastewater and wastewaters from different facilities. For example, sulfamethoxazole, trimethoprim, ciprofloxacin, and ofloxacin were detected in the influent of six wastewater treatment plants in New Mexico with concentrations ranging from 390 to 1,000 ng/L (Brown et al., 2006). For one of the treatment plants studied, the effluent concentrations of sulfamethoxazole, trimethoprim, and ofloxacin ranged from 110 to 310 ng/L. In a study in Wisconsin where samples were collected from seven wastewater treatment plants ranging from activated sludge plants, oxidation ditches and aerated lagoons, the most frequently detected antibiotics were tetracycline and trimethoprim (80% of the samples), followed by sulfamethoxazole (70%), erythromycin-H_2O (45%), ciprofloxacin (40%), and sulfamethazine (10%) (Karthikeyan and Meyer, 2006). Influent and effluent average concentrations of each of the antibiotics were as follows: tetracycline (0.52 µg/L and 0.17 µg/L), tr imethoprim (0.33 µg/L and 0.17 µg/L), sulfamethoxazole (0.30 µg/L and 0.2 µg/L), erythromycin-H_2O (0.34 µg/L and 0.27 µg/L), ciprofloxacin (0.15 µg/L and 0.06 µg/L) and sulfamethazine (0.16 µg/L and non detect (ND)). A study by Batt et al. (2007) at four wastewater treatment plants in New York reported the following concentrations for antibiotics (influent and effluent): ciprofloxacin (0.6 to 1.4 µg/L and 0.22 to 0.45 µg/L, respectively) , sulfamethoxazole (0.72 to 2.8 µg/L and 0.22 to 0.68 µg/L), tetracycline (0.32 to 1.1 µg/L and 0.061 to 0.29 µg/L), and trimethoprim (2.1 to 7.9 µg/L and 0.21 to 0.54 µg/L).

A study covering 11 wastewater treatment plants in 10 states in the US showed a maximum concentration of 0.589 µg/L of sulfamethoxazole, 0.353 µg/L of trimethoprim and 1 µg/L of triclosan in the effluent of the wastewater treatment plants (Glassmeyer et al., 2005). The presence for fluoroquinolone antibiotics in wastewater effluents and surface river/lake waters in the US and Canada was investigated by Nakata et al. (2005) and ofloxacin and ciprofloxacin were detected in secondary and final wastewater treatment plant effluents in East Lansing, Michigan, at concentrations of 204 and < 19 ng/l, respectively.

Miao et al. (2004) surveyed the treated effluents of eight WWTPs in five Canadian cities for 31 antimicrobials. Ciprofloxacin, clarithromycin, erythromycin-H2O, ofloxacin, sulfamethoxazole, sulfapyridine, and tetracycline were frequently detected in the effluents with maximum concentrations of 0.4, 0.536, 0.838, 0.506, 0.871, 0.228 and 0.977 µg/L, respectively.

In Europe, Lindberg et al., (2005) investigated the fate of 12 antibiotics at five municipal wastewater treatment plants in Sweden and found that the most frequently detected antibiotics were norfloxacin, ofloxacin, ciprofloxacin, trimethoprim, sulfamethoxazole, and doxycycline. The influent and effluent concentrations measured were norfloxacin (66 – 174 ng/L (influent) and < 7 – 34 ng/L (effluent)), ofloxacin (< 6 – 287 ng/L and < 6 – 52 ng/L), ciprofloxacin (90 – 300 ng/L and < 6 – 60 ng/L), trimethoprim (99 – 1300 ng/L and 66 – 1340 ng/L), sulfamethoxazole (< 80 – 674 ng/L and < 80 – 304 ng/L), and doxycycline (< 64 – 2480 ng/L and < 64 – 915 ng/L). They estimated the environmental loads for each of the antibiotics as 0.8 mg/person/week for norfloxacin and as high as 1.3 mg/person/week for ciprofloxacin. In another study by Lindberg and his co-workers in Sweden, the mean influent concentrations in municipal wastewater for norfloxacin, ciprofloxacin and trimethoprim were 293, 220, and 1373 ng/L, respectively (Lindberg et al., 2006). Also in Sweden, the presence of antibiotics in the influent and effluent of a small wastewater treatment plant was investigated by Brendz et al., (2005). The influent and effluent concentrations for trimetoprim were 0.08 µg/L and 0.04 µg/L while for sulfamethoxazole, the effluent concentration at 0.07 µg/L was found to be higher than the influent of 0.02 µg/L.

A study by Gobel et al. (2005) at two wastewater treatment plants in Switzerland gave maximum influent and effluent concentrations of sulfamethazine as 570 ng/L and 860 ng/L, trimethoprim as 440 ng/L and 310 ng/L, and erythromycin–H2O as 190 ng/L and 110 ng/L. Castiglioni et al. (2006) assessed six sewage treatment plants in Italy and found that the most common antibiotics present in the municipal wastewaters were ciprofloxacin, ofloxacin, and sulfamethoxazole. The estimated influent loads were 50-500 mg/day/1000 inhabitants and after removal in STPs, the loads discharging into the rivers were 25 - 280 mg/day/1000 inhabitants. Hirsch et al. (1999) examined sewage treatment plant effluents, surface water, and ground water in Germany for 18 antibiotics. The target substances in sulfonamide class were sulfamethoxazole and sulfamethazine. Only sulfamethoxazole was detected in both effluents and surface water while both sulfa drugs were detected in ground water (2 out of 59 samples).

Studies conducted in 7 wastewater treatment plants along the Ebro River basin of Spain showed that the concentrations of erythromycin, azithromycin, sulfamethoxazole, trimethoprim, and ofloxacin were in the low μg/L to high ng/L range but specific concentrations data were not available (Gros et al., 2007). However, Gros and co-workers estimated that the total antibiotic loads to be in the range of 0.007 to 0.07 g/d/1000 inhabitants in the influent wastewater and from ND to 0.06 g/d/1000 inhabitants in the effluent.

In Finland, Vieno et al. (2007) assessed three antibiotics (ciprofloxacin, norfloxacin, and ofloxacin) in the raw and treated sewage of 12 municipal waste water treatment plants and found that the average concentrations were nd to 4230 ng/L (influent) and nd to 130 (effluent) for ciprofloxacin, nd to 960 ng/L (influent) and nd to 110 ng/L (effluent) for norfloxacin, nd to 350 ng/L (influent) and ND to 30 ng/L (effluent) for ofloxacin.

In Asia, Yasojima et al. (2006) measured the occurrence of 3 antibiotics, levofloxacin, clarithromycin and azithromycin in each process of six activated sludge wastewater treatment plants in Japan. Levofloxacin, clarithromycin and azithromycin were detected in the influent at concentrations of 552, 647 and 260 ng/L, respectively. Effluents from 7 wastewater treatment plants in South Korea discharging into the Youngsan River basin were analyzed using liquid chromatography/tandem mass spectrometry with electrospray ionization and atmospheric pressure chemical ionization (Kim et al. 2007). Erythromycin and sulfamethoxazole were the two antibiotics tested and were found in to be in the range of 8.9 - 294 ng/L and 3.8 - 407 ng/L, respectively. For triclosan, the effluent concentration was 1.3 - 3.2 ng/L. Also in Korea, Choi et al. (2007) used solid-phase extraction with liquid chromatography mass spectrometry to analyze for sulfonamide antibiotics and tetracycline antibiotics in an activated sludge municipal wastewater treatment plant. The seven sulfonamide antibiotics tested were sulfamonomethoxine, sulfadimethoxine, sulfamethoxazole, sulfathiazole, sulfachloropyridazine, sulfamerazine, and sulfamethazine while the seven tetracycline antibiotics tested were oxytetracycline-HCl, minocycline-HCl, doxycycline-cyclate, meclocycline-sulfosalicylate, chlortetracycline-HCl, democlocycline- HCl, and tetracycline. The sulfonamide and tetracycline antibiotic concentrations ranged of 0.46–4.01 μg/L and 0.11 - 0.97 μg/L, respectively in the influent of the treatment plant and ND to 0.18 μg/L and ND to 0.18 μg/L, respectively, in the effluent.

Another source of antibiotics is wastewater from hospitals, retirement facilities and nursing homes. Brown et al. (2006) found that four out of five hospital effluents (4 hour composites and grab samples) contained at least one antibiotic while three out of five have four or more antibiotics. The antibiotics detected were sulfamethoxazole, trimethoprim, ciprofloxacin, ofloxacin, lincomycin, and penicillin-G ranging from 300 to 35,500 ng/L), the highest being ofloxacin with concentrations in the range of 20,000 to 30,000 ng/L for three hospital effluents. Concentrations of ofloxacin as high as 23,500 ng/L were found in the effluent of an assisted living and retirement facility in New Mexico (Brown et al., 2006).

In comparison to studies in Europe and Canada, Karthikeyan amd Meyer (2006) concluded that the concentrations detected in the US were within an order of magnitude of that in Europe but were within a factor of two in comparison to those reported for Canada. A comparison of the occurrence of ofloxacin, norfloxacin and ciprofloxacin in wastewater effluents in the US (Michigan, in particular) and in various countries in Europe was conducted by Nakata et al. (2005). They found that the concentrations of ofloxacin and ciprofloxacin in France, Italy and Greece were 2 to 3 times higher than in the US while ofloxacin was found at a much lower concentration (< 5 ng/L) in Switzerland than in the US but ciprofloxacin was about 3 - 5 times higher than in the US. The study by Miao et al. (2004) seemed to suggest that there are differences in the relative concentrations for fluoroquinolone and sulfonamide compounds detected in the final effluents of wastewater treatment plants in Canada relative to concentrations reported in northern Europe.

5.5.2 Removal of Antibiotics in Wastewater Treatment Plants

Antibiotics are removed to varying degrees in municipal wastewater treatment plants depending on the antibiotics type and the treatment processes. Currently, there is no consensus or trend on which biological waste treatment types are best in the removal of antibiotics. There is agreement that sorption of antibiotics to sludge plays an important role in the final fate of antibiotics and the longer the solids retention time (SRT) the higher is the removal (Kim et al., 2005; Batt et al., 2007). A summary of different studies conducted in various countries are discussed.

Batt at al. (2007) compared the removal of antibiotics for 4 different wastewater treatment schemes (conventional two-stage activated sludge with nitrification, extended aeration activated sludge with ferrous chloride, rotating biological contactor and a pure oxygen activated sludge plant) and found that extended aeration activated sludge with ferrous chloride gave the highest removal for 2 antibiotics, sulfamethoxazole ($\approx$75%), and trimethoprim ($\approx$ 97%) while the two-stage activated sludge gave the highest removal for tetracycline ($\approx$ 84%) and the rotating biological contactor had the highest removal rate for ciprofloxacin ($\approx$ 76 %). Batt et al. concluded that the SRT of the treatment plant is an important process operation parameter in removal of antibiotics from wastewater. Brown et al. (2006) reported percent removals of 20%, 70%, and 77% for sulfamethoxazole, trimethoprim, and ofloxacin, respectively, for a conventional activated sludge treatment plant.

In Europe, Lindberg et al. (2005) found that the mean percent removal of norfloxacin, ofloxacin, and ciprofloxacin from 5 wastewater treatment plants in Sweden were 87%, 86%, and 87%, respectively. Similar results were reported by Golet et al. (2002). For sulfamethoxazole, trimethoprim, and doxycycline, the average percent removals were 42%, 3%, and 70%, respectively. Lindberg et al. (2005) concluded that large amounts of the antibiotics were in the sludge of the wastewater treatment plants.

In another study by Lindberg and co-workers (Lindberg et al., 2006), the total removal efficiencies of norfloxacin and ciprofloxacin from the aqueous phase in a wastewater treatment plant in Sweden were 80% and 78%, respectively. In the same study, they found that trimethiprim was not removed from the wastewater treatment plant. Using mass balances, they found that more than 70% of the total amount of these compounds was in the digested sludge (Lindberg et al., (2006). Brendz et al., (2005) found 49% removal of trimetoprim for a small wastewater treatment plant in Sweden but the sulfamethoxazole effluent concentration was found to be higher than the influent concentration.

Work done by Gros et al., 2007 for 6 activated sludge plants in Spain showed that about 30% to 100% of fluoroquinolone antibiotics were removed. However, in several wastewater treatment plants the effluent concentrations of the sulfonamide and macrolide antibiotics were higher than the concentrations in the influent. This may be due to the reconversion of the conjugates to the parent compounds during the treatment process (Langhammer, 1989). Also in Spain, a study by Carballa et al., 2004 indicated that 57% of sulfamethoxazole was removed from a conventional activated sludge plant operating with a hydraulic retention time of 24 hours.

In a study in Italy, the percent removals of antibiotics in municipal wastewater treatment plants were found to be affected seasonally with percent removals for amoxicillin (winter 49-100%, summer 100%), ciprofloxacin (winter 45-78%, summer 53-69%), clarithromycin (winter 0-24%, summer 0%), erythromycin (winter 0%, summer 0%), lincomycin (winter 0%, summer 0%) and sulfamethoxazole (winter 0-84%, summer 71%) (Castiglioni et al., 2006). A study in Finland for 12 municipal wastewater treatment plants showed percent removals of ciprofloxacin and ofloxacin were 84 and 82%, respectively. For different treatment processes, conventional activated sludge was found to remove 86% and 83%, denitrifying treatment plants removed 79% and 88%, and oxidation ditches removed 96% and 75% of ciprofloxacin and ofloxacin, respectively (Vieno et al., 2007).

In Japan, Yasojima et al. (2006) found that average removal efficiencies of levofloxacin, clarithromycin and azithromycin in 6 activated sludge wastewater treatment plants were 42%, 43% and 49%, respectively. In their study, they found clarithromycin and azithromycin concentrations to decrease as the treatment progressed but for levofloxacin, the concentration increased in activated sludge reactors in some cases. Since the removal efficiency were similar despite the differences in octanol-water partition coefficients of the three antibiotics, Yasojima and his team concluded that removal cannot be explained by simple adsorption by the activated sludge. A study in Korea by Choi et al. (2007) indicated that an activated sludge treatment removed between 64–100% of the incoming tetracycline antibiotics and 96–100% of incoming sulfonamides.

Bench-scale experiments have been conducted to assess the fate of antibiotics under well-controlled conditions. A study conducted by Kim et al. (2005) using bench-scale SBRs showed that removal efficiency of tetracycline ranged from 78.4% to 86.4% with decreased tetracycline removal for a reduction in SRT from 10 days to 3 days. However, they concluded that the principal removal mechanism of tetracycline was by sorption to activated sludge as there was no evidence of biodegradation. The batch adsorption coefficient (K_{ads}) was estimated to be 8400 $\pm$ 500 mL/g and a desorption coefficient (K_{des}) of 22,600 $\pm$ 2200 mL/g.

Using a laboratory-scale sewage treatment plant as described by the Organization for Economic Cooperation and Development (OECD) Guideline No. 303A and ^{14}C-labeled benzylpenicillin, ceftriaxone, and trimethoprim at μg/L concentrations, Junker et al. (2006) found that approximately 25% of benzylpenicillin was mineralized, whereas ceftriaxone and trimethoprim were not mineralized at all.

Kim et al. (2007) showed that erythromycin was hardly removed in a bench-scale MBR but more than 99% of erythromycin was removed using membrane methods such as reverse osmosis and nanofitration. In the case of sulfamethoxazole, between 60 -70% removal was found in the bench-scale MBR and more than 99% was removed using reverse osmosis and nanofiltration. In the same study, approximately 66 – 72% of triclosan was removed with the bench-scale MBR and more than 99% was removed using reverse osmosis and nanofiltration.

Carballa et al. (2007) found that greater than 95% of roxithromycin and more than 99% of sulfamethoxazole were removed in a mesolphilic anaerobic reactor with a SRT of 6 to 30 days. They also found that slightly better removal was obtained under thermophilic conditions than mesophilic conditions. Using municipal wastewater as the feedwater for a bench-scale SBR, approximately 69% of the lincomycin (initial concentration 2 mg/L) was removed (Carucci et al., 2006).

5.6   Antibiotic-Resistant Bacteria

Antibiotic resistance is the ability of microorganisms to withstand the effects of antibiotics. The development and proliferation of antibiotic resistance in bacteria is of public health concern because a patient can develop an antibiotic resistant infection by contacting a resistant organism, or by having a resistant microbe emerge in the body as treatment with antibiotic begins (Lewis 1995). In 1970, non-medical uses of antibiotics were questioned and antimicrobial agents were described as potential environmental contaminants and a threat to public health (Huber, 1971b). Since that time, several studies have reported the occurrence of antibiotic resistant organisms in environmental samples and advocated a global public health concern due to these bacteria (Pillai et al., 1997; Ash et al., 2002).

The important mechanisms by which microorganisms exhibit resistance to antibiotics include drug inactivation or modification, alteration of the target site, alteration in the metabolic pathway, and reduced drug accumulation (Katzung, 2004). These mechanisms are described below and illustrated in Figure 5.10.

Drug inactivation or modification: Resistant bacteria synthesize and secret enzymes which affect the antimicrobial activity of the antibiotics. For example β-lactamases synthesized by antibiotic resistant bacteria hydrolyze the β-lactone ring of penicillin thereby inactivating the antibiotic.

Alteration of target site: Penicillin acts on bacteria by attaching to penicillin binding proteins (PBP), which are essential components for synthesis of bacterial cell wall. Bacteria develop resistance to penicillin either by the overproduction of PBPs or by synthesis of PBPs, which have low affinity to penicillin.

Alteration of metabolic pathway: Bacteria are able to modify their metabolic pathways in order to evade the action of antibiotics. For example, sulfonamides inhibit the synthesis of folic acid, and sulfonamide resistant bacteria develop alternate routes for synthesis of folic acid or repress its synthesis.

Reduced drug accumulation: Bacteria developing resistance to antibiotics are able to reduce the uptake of the antibiotic by either altering the permeability of the drug or by enhancing active efflux of the drug.

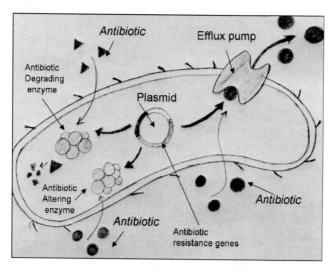

Figure 5.10 Mechanisms exploited by bacteria for antibiotic resistance (adapted from Yim, 2007).

Previously it was believed that resistance in bacteria was acquired by spontaneous mutation, which is called as primary resistance. The wide spread development of multiple antibiotic resistance in many species of bacteria led researchers to believe that another mechanism beyond spontaneous mutation was responsible for the acquisition of antibiotic resistance. The mechanism responsible for the development of resistance was through lateral or horizontal gene transfer. Horizontal gene transfer (HGT) has three possible mechanisms: transduction, transformation and conjugation as shown in Figure 5.11.

Transduction occurs when bacteria-specific viruses or bacteriophages transfer DNA between two closely related bacteria. Transformation is a process where parts of DNA are taken up by the bacteria from the external environment. This DNA present in the external environment is due to death of another bacterium. Conjugation occurs when there is direct cell-cell contact between two bacteria and transfer of small pieces of DNA called plasmids takes place (Yim, 2007).

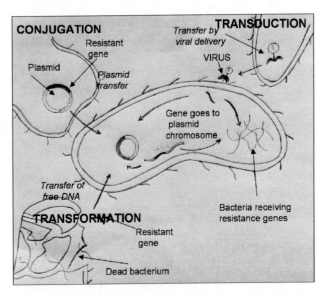

Figure 5.11 Mechanisms of horizontal gene transfer in bacteria (Yim, 2007).

5.6.1 Minimum Inhibitory Concentration

Antibiotics, just like other toxic agents, exhibit a dose-response relationship in bacterial cultures. The response typically is mortality. Figure 5.12 illustrates a hypothetical dose response curve. Exposure to an antibiotic results in a reduction in observed bacterial concentration or population. The lowest concentration in a dose-response assay at which no bacteria are affected is called the 'no observed adverse effect level' or NOAEL. The lowest antibiotic concentration at which a significant decrease in the bacteria concentration is noted is considered to be the 'lowest observable adverse effect level' or LOAEL. As the antibiotic concentration is increased, the concentration resulting in a 50% kill of the bacterial population is described as the lethal concentration for 50% kill or LC_{50}. NOAEL, LOAEL and LC_{50} depend on the type of bacteria, type of antibiotic and environmental conditions, including matrix chemistry.

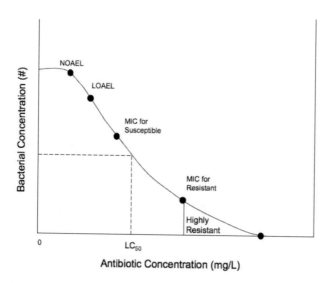

Figure 5.12 Dose-response curve for antibiotic toxicity and resistance in bacteria (Nagulapally, 2007). MIC = Minimum inhibitory concentration, NOAEL = Non-observable adverse effect level, LOAEL = Low-observable adverse effect level.

Bacteria may be considered 'susceptible' to antibiotic resistance at a given concentration if a significant fraction survives exposure to the antibiotic at that concentration. These clinical concentrations for various antibiotics were described by Wu (1995) and are summarized in Table 5.10. The experimental protocol for determination minimum inhibitory concentration (MIC) is described in Andrews (2001). Bacteria susceptible to antibiotic resistance require a high dose of antibiotic for deactivation. Bacteria that grow at exposures lower than the susceptible level of an antibiotic are considered potentially antibiotic resistant. The minimum inhibitory concentration (MIC) is defined as the lowest concentration of antibiotic that inhibits the visible growth of bacteria (Andrews, 2001).

Bacteria that are observed to grow at MIC level exposure are defined as antibiotic-resistant organisms while those that grow at even higher antibiotic exposures are considered highly resistant to the antibiotic. Bacteria that grow at exposures higher than the 'susceptible concentration' but lower than MIC are considered to have intermediate resistance to the antibiotic. MIC is important in the field of medicine to confirm the resistance of microorganisms to an antibiotic and to monitor the activity of newly developed antibiotics. MICs for some of the target antibiotics are tabulated in Table 5.10.

Table 5.10 Equivalent minimum inhibitory concentration (MIC) breakpoints (Wu, 1995).

Antibiotic	Susceptible Level (mg/L)	Resistant Level (mg/L)
Ciprofloxacin	< 1	> 4
Sulfamethoxazole/Trimethoprim	< 2/38	> 8/152
Vancomycin	< 8	> 32

5.6.2 Antibiotic-Resistant Bacteria in Municipal Wastewater

Gallert et al. (2005) observed multi-resistant antibiotic fecal coliforms and enterococci in influent and effluent wastewater from treatment plants. Such multiple anitibiotic resistant organisms have been observed in wastewater treatment plants across the world. Reinthaler et al. (2003) tested 767 *E. coli* isolates, obtained from sewage treatment plants in Australia, for their resistance to 24 different antibiotics. Among the antimicrobial agents evaluated, highest resistances in bacteria were observed for ampicilin (18%), piperacillin (12%) cefalothin (35%), cefuroxime-axetil (11%) sulfamethoxazole/trimethoprim (13%) and tetracycline (57%). In another study conducted in Finland, more than 20% of fecal coliforms were observed to be resistant to ampicillin, chloramphenicol, sulfanomide, tetracycline and streptomycin in one of the treatment plant effluents (Niemi et al., 1983). Still other studies conducted across the world revealed that fecal coliforms and *E.coli* in raw sewage were resistant to ampillicin, gentamycin, kanamycin, neomycin and streptomycin (Qureshi and Qureshi, 1991). Boczek et al. (2007) recently reported isolation of *E. coli* belonging to clonal group A (CGA) which causes drug-resistant urinary tract infections in humans, in four of seven wastewater treatment plant effluents collected from across the US. Although CGA isolates constituted 1% of the total *E. coli* cultured from these samples, all CGA isolates were resistant to sulfamethoxazole/trimethoprim (SMT). Approximately 5.2% of the total *E. coli* isolates were resistant to SMT.

Enterococci resistant to ciprofloxacin, sulfamethoxazole/trimethoprim and vancomycin at minimum inhibitory concentrations have also been observed in raw and treated effluent wastewater (Gallert et al., 2005). Hassani et al. (1992) reported *E. coli* were resistant to a wide range of antibiotics in raw and treated sewage in Morocco. Nagulapally (2007) recently reported the occurrence of fecal coliforms, *E. coli* and enterococci that were resistant or highly resistant to one or more target antibiotics in the influent and secondary clarifier effluent of a municipal wastewater treatment plant.

The past few years have witnessed an increasing interest in the study of antibiotic resistant *Enterococci*. *Enterococci* are the second or third most important bacterial genus responsible for hospital infections (Klare et al., 2003). *Enterococci* are intrinsically resistant to a wide range of antibiotics (Gilmore, 2002). Hence, this has always limited the choice of antibiotics available for use against these organisms. Nosocomial infections are caused by *Enterococci* and, therefore, antibiotics have been used in greater frequency in hospitals. Murray (1990) observed in his research that 12 species of *Enterococci* were pathogenic for humans including most common human isolates *Enterococcus faecalis* and *Enterococcus faecium*. *Enterococcus faecalis* causes 80% to 90% of human enterococcal infections, while *E. faecium* accounts for a majority of the remainder (Moellering, 1992; Murray, 1990; Schnell, 1992). Resistance in *Enterococci* was developed by acquiring resistance genes on plasmids or transposons from other organisms or by spontaneous mutations (Gilmore, 2002). Resistant *Enterococci* were entering the environment through wastewater effluents and hospital wastewater.

Recent studies have identified fluoroquinolone resistant *E. coli* isolates in leukemia patients (Kern et al., 1994). Antibiotic resistance provides a survival benefit to microorganisms and makes it difficult to eliminate the infections caused by them. Infections caused by antibiotic resistant bacteria are hard to treat. Hence, physicians have to prescribe higher dosage of alternative antibiotics to cure the infections. High doses have side effects and the potential to produce more antibiotic-resistant strains of bacteria. Hence, there is a need to study antibiotic resistance patterns in wastewater bacteria.

5.7   Conclusion

The studies to determine the concentrations and fate of antibiotics in water, wastewaters, and soils are the first logical steps in mapping and understanding the extent of antibiotics in the environment. Armed with this data, the health risks to humans and the ecological impact of the environment can be assessed accordingly. Our current level of knowledge of the environmental fate of these emerging contaminants of concern has grown over the last decade and there is sufficient evidence to speculate the fate of these emerging contaminants. For the more common antibiotics, the ability to measure these compounds at very low concentrations has been overcome with the latest analytical instruments. However, the analytical procedures are still tedious and time consuming. There is a need to improve measurements of the metabolites or degradation products of these emerging contaminants in various matrices and also the measurement of less common antibiotics. Many antibiotics are strongly sorbed to soils and sludge of wastewater treatment plants. Further information are needed to understand the ultimate fate of these compounds that are sorbed to soils and sludge. i.e., whether they are biodegraded, chemically altered or degraded, photodegraded or continue to persist in these matrices. Additionally, there is a need to understand the conditions and circumstances these compounds are degraded in soils, waste treatment plants, and livestock lagoons. An understanding of the fate and degradation of the emerging contaminants byproducts/metabolites in the different matrices will further improve the understanding of the risks posed by these compounds and their metabolites since the metabolites may be 'reactivated' through microbial processes to their original antibiotic potencies. The information available on resistant microorganisms in the environment is very much in its infancy state and much work is needed to have a complete assessment and understanding of conditions and factors that will result in resistant microorganisms.

References

Aga, D.S.; O'Connor, S.K. (2003) "Biodegradation of chlortetracycline in manure fertilized soil and identification of degradates by electrospray ion-trap mass spectrometry." 24th Meeting of the Society of Environmental Toxicology and Chemistry, 09 – 13, November, Austin, TX

Alexy, R.; Kumpel, T.; Kummerer, K. (2004) "Assessment of degradation of 18 antibiotics in the closed bottle test." Chemosphere 57:05–512

Andrews, J. M. (2001) "Determination of minimum inhibitory concentrations." Journal of Antimicrobial Chemotherapy, 48:5-16.

Ash, R. J., Mauck, B., and Morgan, M. (2002) "Antibiotic resistance of gram-negative bacteria in rivers, United States." Emerging Infectious Disease, 8:713-716.

Bakal, R. S.; and Stoskopf, M.K. (2001) "In vitro studies of the fate of sulfadimethoxine and ormetoprim in the aquatic environment." Aquaculture,195(1-2):95-102.

Barza, M.; Brown, R.B.; Shanks, C.; Gamble, C.; Weinstein, L. (1975) "Relation between lipophilicity and pharmacological behavior of minocycline, doxycycline, tetracycline, and oxytetracycline in dogs." Antimicrobial Agents Chemotheraphy, 8(6):713–720.

Batt, A.L.; Kim, S.; Aga, D.S. (2007) "Comparison of the occurrence of antibiotics in four full-scale wastewater treatment plants with varying designs and operations." Chemosphere, 68(3):428-435.

Belden, J.B.; Maul, J.D.; Lydy, M.J. (2007) "Partitioning and photodegradation of ciprofloxacin in aqueous systems in the presence of organic matter." Chemosphere, 66 :1390–1395

Bendz, D.; Paxeus, N.A.; Ginn, T.R.; Loge, F.J. (2005) "Occurrence and fate of pharmaceutically active compounds in the environment, a case study: Hoje River in Sweden." J. Hazardous Materials, 122(3):195-204.

Boczek, L.A.; Rice, E.W.; Johnston, B.; Johnson, J.R. (2007). "Occurrence of antibiotic-resistant uropathogenic Escherichia coli clonal group A in wastewater effluents." Applied and Environ. Microbiology, 73:4180-4184.

Boxall, A. B. A.; Fogg, L.; Blackwell, P.A.; Kay, P.; Pemberton, E. (2001) Review of Veterinary Medicines in the Environment. Environment Agency Report P6-012/8TR. U.K. Environment Agency, Bristol, UK.

Boxall, A. B. A.; Blackwell, P. A.; Cavallo, R.; Kay, P.; and Tolls, J. (2002) "The sorption and transport of a sulphonamide antibiotic in soil systems." Toxicology Letters, 131(1-2):19-28.

Boxall, A.B.A.; Fogg, L.A.; Baird, D.J.; Lewis, C.; Telfer, T.C.; Kolpin, D.; Gravell, A. (2005) Targeted monitoring study for veterinary medicines in the UK environment. Final Report, UK Environment Agency, London, UK.

Brown, K. D.; Kulis, J.; Thomson, B.; Chapman, T.H.; Mawhinney, D.B. (2006) "Occurrence of antibiotics in hospital, residential, and dairy effluent, municipal wastewater, and the Rio Grande in New Mexico." Science of the Total Environment, 366(2-3):772-783.

Burton, J.A.; Schanker, L.S. (1974) "Absorption of antibiotics from the rat lung." The Society for Experimental Biol. and Med. 145:752-756.

Campagnolo, E. R.; Johnson, K. R.; Karpati, A.; Rubin, C. S.; Kolpin, D. W.; Meyer, M. T.; Esteban, J. E.; Currier, R. W.; Smith, K.; Thu, K. M.; McGeehin, M. (2002) "Antimicrobial residues in animal waste and water resources proximal to large-scale swine and poultry feeding operations." Science of the Total Environment, 299(1-3):89-95.

Carballa, M.; Omil, F.; Lema, J.M.; Llompart, M.; Garcia-Jares, C.; Rodriguez, I.; Gomez, M.; Ternes, T. (2004) "Behavior of pharmaceuticals, cosmetics and hormones in a sewage treatment plant." Water Research, 38(12):2918-2926.

Carballa, M.; Omil, F.; Ternes, T.; Lema, J.M.; (2007) "Fate of pharmaceutical and personal care products (PPCPs) during anaerobic digestion of sewage sludge." Water Research, 41(10):2139-2150

Cardoza, L.A.; Knapp, C.W.; Larive, C.K.; Belden, J.B.; Lydy, M.; Graham, D.W. (2005) "Factors affecting the fate of ciprofloxacin in aquatic field systems." Water, Air and Soil Pollution, 161(1-4):383-398.

Carucci, A.; Cappai, G.; Piredda, M. (2006) "Biodegradability and toxicity of pharmaceuticals in biological wastewater treatment plants." J. Environ. Sci. and Health, Part A, 41(9):1831-1842.

Castiglioni, S.; Bagnati, R.; Fanelli, R.; Pomati, F.; Calamari, D.; Zuccato, E. (2006) "Removal of pharmaceuticals in sewage treatment plants in Italy." Environ. Sci. Technol. 40(1):357-363.

Chee-Sanford, J.C.; Aminov, R. I.; Krapac, I. J.; Garrigues-Jeanjean, N.; and Mackie, R. I. (2001) "Occurrence and diversity of tetracycline resistance genes in lagoons and groundwater underlying two swine production facilities." Applied and Environmental Microbiology, 67(4):1494-1502.

Choi, K.; Kim, S.; Kim, C.; Kim, S. (2007) "Determination of antibiotic compounds in water by on-line SPE-LC/MSD." Chemosphere, 66(6): 977-984.

Christiana, T.; Schneidera, R.J.; Färberb, H.A.; Skutlarekb, D.; Meyerc, M.T.; Goldbacha, H.E. (2003) "Determination of antibiotic residues in manure, soil, and surface waters." Acta hydrochim. hydrobiol. 31(1):36–44

Costanzo, S.D.; Murby, J.; Bates, J. (2005) "Ecosystem response to antibiotics entering the aquatic environment." Marine Pollution Bulletin, 51(1-4):218-223.

Cromwell, G. L. (2000) Why and how antibiotics are used in swine production. Presentation at Animal Health Institute (no city of publication given). http://www.ahi.org/Slide%20Show/cromwell.ppt (accessed 2 February 2003)

Daughton, C. G.; Ternes, T. A. (1999) "Pharmaceuticals and personal care products in the environment: Agents of subtle change?" Environmental Health Perspectives, 107(Supplement 6):907-938.

DEFRA (2002) A Diffuse Pollution Review: The Government's Strategic Review of Diffuse Water Pollution From Agriculture in England. Dept. for Environment, Food and Rural Affairs Agriculture and Water, (no city of publication given), UK.

De Liguoro, M.; Cibin, V.; Capolongo, F.; Halling-Sørensen, B.; and Montesissa, C. (2003) "Use of oxytetracycline and tylosin in intensive calf farming: Evaluation of transfer to manure and soil." Chemosphere, 52(1):203-212.

Donoho, A.L. (1984) "Biochemical studies on the fate of monensin in animals and in the environment." J. Anim. Sci. 58:1528-1539.

Dumka, V.K. (2007) "Disposition kinetics and dosage regimen of levofloxacin on concomitant administration with paracetamol in crossbred calves." J. Vet. Sci. 8(4):357-360.

Escribano, E.; Calpena, A.C.; Garrigues, T.M.; Freixas, J.; Domenech, J.; Moreno, J. (2000) "Structure-absorption relationships of a series of 6-fluoroquinolones." Antimicrobial Agents and Chemotherapy, 41(9):1996-2000.

Gallert, C.; Fund, K.; and Winter, J. (2005). "Antibiotic resistance of bacteria in raw and biologically treated sewage and in groundwater below leaking sewers." Applied Microbiology and Biotechnology, 69:106-112.

Gartiser, S.; Urich, E.; Alexy, R.; Kummerer, K. (2004) "Ultimate biodegradation and elimination of antibiotics in inherent tests." Chemosphere, 67:604-613.

Gartiser, S.; Urich, E.; Alexy, R.; Kummerer, K. (2007) "Anaerobic inhibition and biodegradation of antibiotics in ISO test schemes." Chemosphere, 66:1830-1848.

Gavalchin, J.; Katz, S.E. (1994) "The persistence of fecal-borne antibiotics in soil." J. AOAC Int. 177:481-485.

Gilmore, S. M. (2002). In The Enterococci: Pathogenesis, Molecular Biology and Antibiotic Resistance, ASM Press, Washington, DC.

Glassmeyer, S.T.; Furlong, E.T.; Kolpin, D.W.; Cahill, J.D.; Zaugg, S.D.; Werner, S.L.; Meyer, M.T.; Kryak, D.D.; (2005) "Transport of chemical and microbial compounds from known wastewater discharges: potential for use as indicators of human fecal contamination." Environ. Sci. Technol. 39:5157–5169.

Gobel, A.; Thomsen, A.; Mcardell, C.S.; Joss, A.; Giger, W. (2005) "Occurrence and sorption behavior of sulfonamides, macrolides, and trimethoprim in activated sludge treatment." Environ. Sci. Technol. 39:3981–3989.

Gros, M.; Petrovicacute, M.; Barcelo, D. (2007) "Wastewater treatment plants as a pathway for aquatic contamination by pharmaceuticals in the Ebro River Basin (Northeast Spain)." Environmental Toxicology and Chemistry, 26(8):1553-1562.

Gupta, S.; Singh, A.; Kumar, K.; Thompson, A.; and Thoma, D. (2002) "Antibiotic losses in runoff and drainage from manure-applied fields." (no city of publication given).   http://water.usgs.gov/wrri/01grants/national/prog-compl-reports/2001MN1041G.pdf (accessed 21 June 2004).

Haggard, B. E.; Galloway, J.M.; Green, W.R.; Meyer, M.T. (2006) "Pharmaceuticals and other organic chemicals in selected North-Central and Northwestern Arkansas streams." J. Environ. Quality. 35(4):1078-1087.

Haller, M. Y.; Müller, S. R.; McArdell, C. S.; Alder, A. C.; Suter, M. J.-F. (2002) "Quantification of veterinary antibiotics (sulfonamides and trimethoprim) in animal manure by liquid chromatography-mass spectrometry." J. of Chromatography A, 952(1-2):111-120.

Halling-Sørensen, B.; Nielsen, S. N.; Lanzky, P. F.; Ingerslev, F.; Holten Lützhøft, H. C.; and Jørgensen, S. E. (1998) "Occurrence, fate and effects of pharmaceutical substances in the environment - A review." Chemosphere, 36(2):357-393.

Halling-Sørensen, B.; Nielsen, S. N.; and Jensen, J. (2002) "Environmental fate and occurrence of veterinary medicinal products." In Environmental Assessment of Veterinary Medicinal Products in Denmark, Environmental Project No. 659. Danish Environmental Protection Agency (no city of publication given). http://www.mst.dk/homepage/default.asp?Sub=http://www.mst.dk/udgiv/publicati ons/2002/87-7944-971-9/html/ (accessed 19 November 2002).

Hamscher, G.; Sczesny, S.; Höper, H.; and Nau, H. (2002) "Determination of persistent tetracycline residues in soil fertilized with liquid manure by high-performance liquid chromatography with electrospray ionization tandem mass spectrometry." Analytical Chemistry, 74(7):1509-1518.

Hassani, L.; Imziln, B.; Gauthier, M. J. (1992) "Antibiotic-resistant Escherichia coli from wastewater before and after treatment in stabilization ponds in the arid region of Marrakech, Morocco." Letters in Applied Microbiology, 15:228-231.

Heberer, T. (2002) "Occurrence, fate, and removal of pharmaceutical residues in the aquatic environment: a review of recent research data." Toxicology Letters, 131(1-2):5-17.

Hektoen, H.; Berge, J.A.; Hormazabal, V.; Yndestad, M. (1995) "Persistence of antimicrobial agents in marine sediments." Aquaculture, 133:175-184.

Hirsch, R.; Ternes, T.; Haberer, K.; Kratz, K.L. (1999) "Occurrence of antibiotics in the aquatic environment." Science of the Total Environment, 225(1-2):109-118.

Holm, J.V.; Rugge, K.; Bjerg, P.L.; Christensen, T.H. (1995) "Occurrence and distribution of pharmaceutical organic compounds in the groundwater downgradient of a landfill (Grindsted, Denmark)." Environ Sci. Technol. 29:1415-20.

Holten Lützhøft, H.C.; Vaes, W.H.J.; Freidig, A.P.; Halling-Sorensen, B.; Hermens, J.L.M. (2000) "1-octanol/water distribution coefficient of oxolinic acid: influence of pH and its relation to the interaction with dissolved organic carbon." Chemosphere, 40:711-714.

Huang C.H.; Renew J.E.; Smeby K.L.; Pinkerston K.; and Sedlak D.L., B.E. (2001) "Assessment of potential antibiotic contaminants in water and preliminary occurrence analysis." Water Resource, 20:30-40.

Huber, W.G. (1971a) "The impact of antibiotic drugs and their residues." Advances in
Veterinary Science and Comparative Medicine, 15(1):101-132.

Huber, W. G. (1971b) "Antibacterial drugs as environmental contaminants." In Advances in Environmental Science and Technology, Pitts, J.N., Jr and Metcalf, R. L. (Eds.), Vol. 2, New York: Wiley-Interscience, 289-320.

Ingerslev, F.; Halling-Sørensen, B. (2001) "Biodegradability of metronidazole, olaquindox, and tylosin and formation of tylosin degradation products in aerobic soil-manure slurries." Ecotoxicology and Environmental Safety, 48(4):311-320.

Ingerslev, F.; Toräng, L.; Loke, M.-L.; Halling-Sørensen, B.; and Nyholm, N. (2001) "Primary biodegradation of veterinary antibiotics in aerobic and anaerobic surface water simulation systems." Chemosphere, 44(4):865-872.

Isidori, M.; Lavorgna, M.; Nardelli, A.; Pascarella, L.; Parrella, A. (2005) "Toxic and genotoxic evaluation of six antibiotics on non-target organisms." Science of the Total Environment, 346:87-98.

Jjemba, P. K. (2002) "The potential impact of veterinary and human therapeutic agents in manure and biosolids on plants grown on arable land: A review." Agriculture, Ecosystems & Environment, 93(1-3):267-278.

Jørgensen, S. E.; Halling-Sørensen, B. (2000) "Drugs in the environment." Chemosphere, 40(7):691-699.

Junker, T.; Alexy, R.; Knacker, T.; Kuemmerer, K. (2006) "Biodegradability of ^{14}C-labeled antibiotics in a modified laboratory scale sewage treatment plant at environmentally relevant concentrations." Environ. Sci. Technol. 40(1):318-324.

Karthikeyan, K.G.; Meyer, M.T. (2006) "Occurrence of antibiotics in wastewater treatment facilities in Wisconsin, USA." Science of the Total Environment, 361(1-3):196-207.

Katzung, B. G. (2004). In Basic and Clinical Pharmacology, Lange Medical Books, McGraw-Hill, 9th edition, New York.

Kay, P.; Blackwell P.A.; Boxall, A.B.A. (2005) "Transport of veterinary antibiotics in overland flow following the application of slurry to arable land." Chemosphere 59:951-959.

Kern, W. V.; Andriof, E.; Oethinger, M.; Kern, P.; Hacker, J.; and Marre, R. (1994) "Emergence of fluroquinolone-resistant E. coli at a cancer center." Antimicrobial Agents and Chemotherapy Journal, 38:681-687.

Klare, I.; Konstabel, C.; Badstubner, D.; Werner, G.; and Witte, W. (2003) "Occurrence and spread of antibiotic resistances in Enterococcus faecium." International J. of Food Microbiology, 88:269-290.

Kim, Y.H.; Heinze, T.M.; Kim, S.J.; Cerniglia, C.E. (2004) "Adsorption and clay catalyzed degradation of Erythromycin A on homoionic clays." J. Environ. Qual. 33: 257-264.

Kim, S.; Eichhorn, P.; Jensen, J.N.; Weber, A.S.; Aga, D.S. (2005) "Removal of antibiotics in wastewater: effect of hydraulic and solid retention times on the fate of tetracycline in the activated sludge process." Environ. Sci. Technol. 39:5816-5823.

Kim, S. D.; Cho, J.; Kim, I.S.; Vanderford, B.J.; Snyder, S.A.; (2007) "Occurrence and removal of pharmaceuticals and endocrine disruptors in South Korean surface, drinking, and waste waters." Water Research, 41(5):1013-1021.

Kolpin, D. W.; Furlong, E. T.; Meyer, M. T.; Thurman, E. M.; Zaugg, S. D.; Barber, L. B.; and Buxton, H. T. (2002) "Pharmaceuticals, hormones, and other organic wastewater contaminants in U.S. streams, 1999-2000: A National Reconnaissance." Environ. Sci. and Technol. 36(6):1202-1211.

Kolz, A.C. (2004) Degradation and sorption of tylosin in swine manure lagoons, Master's thesis, Iowa State University, Ames, Iowa.

Kolz, A.C.; Ong, S.K.; Moorman, T.B. (2005a) "Sorption of tylosin onto swine manure." Chemosphere, 60(2):284-289.

Kolz, A.C.; Moorman, T.B; Ong, S.K.; Scoggin, K.D.; Douglas, E.A. (2005b) "Degradation and metabolite production of tylosin in anaerobic and aerobic swine manure slurries." Water Environment Research, 77(1): 49-56.

Kulshrestha, P.; Giese, R.F.; Wood, T.D. (2005) "Detection of residues of tetracycline antibiotics in soil fertilized with manure and wastewater using enzyme linked immunosorbent assay." Physical Chemistry Posters, 37th Middle Atlantic Regional Meeting, May 22-25, New Brunswick, NJ.

Kumar, K.; Thompson, A.; Singh, A.K.; Chander, Y.; Gupta, S.C. (2004) "Enzyme-linked immunosorbent assay for ultratrace determination of antibiotics in aqueous samples." J. Environ. Qual. 33:250-256.

Kumar, K.; Gupta, S.C.; Baidoo, S.K.; Chander, Y.; Rosen, C.J. (2005) "Antibiotic uptake by plants from soil fertilized with animal manure." J Environ Qual. 34:2082-2085.

Kummerer, K.; Al Ahmad, A.; Mersch-Sundermann, V. (2000) "Biodegradability of some antibiotics, elimination of the genotoxicity and affection of wastewater bacteria in a simple test." Chemosphere, 40:701-710.

Kummerer, K. (2004) Pharmaceuticals in the Environment: Sources, Fate, Effects and Risks. 2nd edition, Springer, NY.

Lam, M. W.; Young, C.J.; Brain, R.A.; Johnson, D.J.; Hanson, M.A.; Wilson, C.J.; Richards, S.M.; Solomon, K.R.; Mabury, S.A. (2004) "Aquatic persistence of eight pharmaceuticals in a microcosm study." Env. Toxicology and Chemistry, 23(6):1431-1440.

Langhammer, J.P. (1989) Untersuchungen zum Verbleib antimikrobiell wirksamer Arzneistoffe als Rückstände in Gülle und im landwirtschaflichen Umfeld. PhD thesis, Rheinische Friedrich-Wilhelms-Universität, Bonn, Germany.

Langhammer, J.-P., Büning-Pfaue, H. (1989) "Bewertung von Arzneistoff-Ruckstanden aus der Gulle in Boden." Lebensmit-telchem, Gerichtl. Chem. 43:103-113.

Lewis, R. (1995). "The rise of antibiotic-resistant infections." FDA Consumer Magazine, Vol. 29 (September).

Lindberg, R.H.; Wennberg, P.; Johansson, M.I.; Tysklind, M.; Andersson, B.A.V. (2005) "Screening of human antibiotic substances and determination of weekly mass flows in five sewage treatment plants in Sweden." Environ. Sci. Technol. 39(10):3421-3429.

Lindberg, R.H.; Olofsson, U.l.; Rendahl, P.; Johansson, M.I.; Tysklind, M.; Andersson, B.A.V. (2006) "Behavior of fluoroquinolones and trimethoprim during mechanical, chemical, and active sludge treatment of sewage water and digestion of sludge." Environ. Sci. Technol. 40(3):1042-1048.

Lindsey, M.E.; Meyer, M.; Thurman, E.M. (2001) "Analysis of trace levels of sulfonamide and tetracycline antimicrobials in groundwater and surface water using solid-phase extraction and liquid chromatography/mass spectrometry." Anal. Chem. 73:4640-4646.

Loke, M.-L.; Ingerslev, F.; Halling-Sørensen, B.; and Tjørnelund, J. (2000) "Stability of tylosin A in manure containing test systems determined by high performance liquid chromatography." Chemosphere, 40(7):759-765.

Loke, M.-L.; Tjørnelund, J.; Halling-Sørensen, B. (2002) "Determination of the distribution coefficient (log K_d) of oxytetracycline, tylosin A, olaquindox and metronidazole in manure." Chemosphere, 48(3):351-361.

Merck Index (1989) An Encyclopedia of Chemicals and Drugs, Eleventh Edition. S. Budavari, M. O'Neil, A. Smith, P. Heckelman, (Eds.) Merck & Co., Inc., Rahway, N. J.

Miao, X.-S.; Bishay, F.; Chen, M.; Metcalfe, C.D. (2004) "Occurrence of antimicrobials in the final effluents of wastewater treatment plants in Canada." Environ. Sci. Technol. 38(13):3533-3541.

Moellering, R. C., Jr. (1992) "Emergence of Enterococcus as a significant pathogen." Clinical Infectious Diseases, 14:1173-1178.

Montforts, M. H.; Kalf, D. F.; van Vlaardingen, P. L. A.; Linders, J. B. (1999) "The exposure assessment for veterinary medicinal products." Science of the Total Environment, 225(1-2):119-133.

Murray, B. E. (1990) "The life and times of the enterococcus." Clinical Microbiology Reviews, 3:46-65.

Myllyniemi, A.-L.; Rannikko, R.; Lindfors, E.; Niemi, A. (2000) "Microbiological and chemical detection of incurred penicillin G, oxytetracycline, enrofloxacin and ciprofloxacin residues in bovine and porcine tissues." Food Addit. Contam. 17: 991-1000.

Nagulapally, S. (2007) Antibiotic resistance patterns in municipal wastewater bacteria. MS Thesis. Kansas State University, Manhattan, KS.

Nakata, H.; Kannan, K.; Jones, P.D.; Giesy, J.P.; (2005) "Determination of fluoroquinolone antibiotics in wastewater effluents by liquid chromatography-mass spectrometry and fluorescence detection." Chemosphere, 58(6):759-766.

National Library of Medicine (NLM) (2008) Toxicology Data Network, http://toxnet.nlm.nih.gov/ (accessed on Feb. 28, 2008)

Niemi, M.; Sibakov, M.; and Niemela, S. (1983) "Antibiotic resistance among different species of fecal coliforms isolated from water samples." Applied and Environmental Microbiology, 45:79-83.

Norwara, A.; Burhenne, J.; Spiteller, M. (1997) "Binding of fluoroquinolone carboxylic acid derivatives to clay minerals." J. Agric. Food Chem. 45:1459-1462.

OECD (1992) Guideline for Testing of Chemicals (301D). Closed Bottle Test. Organization of Economic Cooperation and Development, Paris, France.

Oliveira, M. F.; Sarmah, A. K.; Lee, L. S.; and Rao, P. S. C. (2002) "Fate of tylosin in aqueous manure-soil systems." Presented at the Soil Science Society of America National Meeting, November 10-14, Indianapolis, Indiana.

O'Neil, M.J.; Smith, A.; Heckelman, P. E. (Eds.). (2001) The Merck Index, 13th Ed., Merck, Whitehouse Station, NJ.

Ong, S. K.; Lertpaitoonpan, W.; Moorman, T. (2006) "Sorption of sulfamethazine to soils: Effect of organic carbon and pH." Int'l Conf. on Waste Management for a Sustainable Future, Jan. 10 – 12, Nat'l Research Center for Environmental and Hazardous Waste Management, Bangkok, Thailand.

Pillai, S. D.; Widmer, K.W.; Maciorowski, K.G.; Ricke, S.C.; (1997) "Antibiotic resistance profiles of Escherichia coli isolated from rural and urban environment." J. of Environmental Science and Health, 32:1665-1675.

Prescott, J.F. (1997) "Antibiotics: Miracle drugs or pig food." Canadian Veterinary Journal, 38(12):763.

Prescott, J.F.; Baggot J.D., (1993) Antimicrobial Therapy in Veterinary Medicine. 2nd ed. Iowa State University Press, Ames, IA.

Qureshi, A.A.; Qureshi, M. A. (1991) "Multiple antibiotic resistant fecal coliforms in raw sewage." Water, Soil and Air Pollution, 61:47-56.

Rabølle, M.; Spliid, N.H. (2000) "Sorption and mobility of metronidazole, olaquindox, oxytetracycline and tylosin in soil." Chemosphere, 40(7):715-722.

Rang, H.P.; Dale, M.M.; Ritter, J.M. (1999) Pharmacology. Churchill Livingstone, Edinburgh, Scotland.

Reinthaler, F.F.; Poshch, J.; Feierl, G.; Wust, G.; Haas, D.; Ruckenbauer, G.; Mascher, F.; Marth, E. (2003) "Antibiotic resistance of E. coli in sewage and sludge." Water Research, 37:1685-1690.

Royal Danish School of Pharmacy (2002) Biodegradability of medical compounds in environmental matrices – summary (no city of publication given). http://www.dfh.dk/insta/pdf/flemming_ingerslev_uk.pdf (accessed 19 November 2002).

Samuelsen, O. B.; Torsvik, V.; Ervik, A. (1992) "Long-range changes in oxytetracyline concentration and bacterial resistance towards oxytetracycline in a fish farm sediment after medication." Sci. Total Environ. 114:25-36.

Sarmah, A.K.; Meyer, M.T.; Boxall, A.B.A. (2006) "A global perspective on the use, sales, exposure pathways, occurrence, fate and effects of veterinary antibiotics (VAs)." Chemosphere, 65(5):725-759.

Schnell, N.; Engelke, G.; Augustin, J.; Rosentein, R.; Ungermann, V.; Gotz, F.; Entian, K.D. (1992) "Analysis of genes involved in biosynthesis of the antibiotic epidermin." European Journal of Biochemistry, 204:57-68.

Schlusener, M.P.; Bester, K. (2006) "Persistence of antibiotics such as macrolides, tiamulin and salinomycin in soil." Environmental Pollution, 143(3):565-571.

Sieck, R.F.; Graper, L.K; Giera, D.D.; Herberg, R.J.; Hamill, R.L. (1978) [14]C Tylosin Tissue Residue Study in Swine. Unpublished report dated November 1978 from Agricultural Biochemistry, Lilly Research Laboratories, Greenfield, Indiana. Submitted to WHO by Lilly Research Centre Ltd., Windlesham, Surrey, England (no city of publication given). http://www.inchem.org/documents/jecfa/jecmono/v29je08.htm (accessed 22 March 2004).

Smolenski, W.J.; Hummel, B.D.; Lesman, S.P. (2002) "Fate of lincomycin, a veterinary antimicrobial used in livestock production in aerobically incubated soil." 7th Congress of the International Pig and Veterinary Society, Iowa State University, Ames, Iowa.

Spectra Intelligence (2006) The Global Market for Antimicrobials and Antifingals; Challenging Resistance to 2015, Spectra Intelligence, http://www.researchandmarkets.com/reportinfo.asp?report_id=362456&t=d&cat_ id= (Accessed 28 Feb., 2008)

Sukul, P.; Spiteller, M. (2007) "Fluoroquinolone antibiotics in the environment." Rev. Environ. Contam. Toxicol. 191:131-62.

Teeter, J. S.; Meyerhoff, R. D. (2003) "Aerobic degradation of tylosin in cattle, chicken, and swine excreta." Environmental Research, 93(1):45-51.

Ternes, T.A. (1998) "Occurrence of drugs in German sewage treatment plants and rivers." Water Res. 32(11):3245-3260.

Thiele, S. (2000) "Adsorption of the antibiotic pharmaceutical compound sulfapyridine by a long-term differently fertilized loess Chernozem." J. Plant Nutr. Soil Sci. 163:589–594.

Thiele, S.; Seibicke, T.; Leinweber, P. (2002) "Sorption of sulfonamide antibiotic pharmaceuticals in soil particle size fractions." SETAC Europe 12th Annual Meeting, May 12-16, Vienna, Austria.

Thiele-Bruhn, S. (2003) "Pharmaceutical antibiotic compounds in soils—a review." J. Plant Nutr. Soil Sci. 166:145–167.

Thiele-Bruhn, S.; Aust, M.-O. (2004) "Effects of pig slurry on the sorption of sulfonamide antibiotics in soil." Arch. Environ. Contam. Toxicol. 47:31-39.

Thiele-Bruhn, S.; Seibicke, T.; Schulten, H.-R.; Leinweber, P. (2004) "Sorption of sulfonamide pharmaceutical antibiotics on whole soils and particle-size fractions." J. Environ. Qual. 33:1331–1342.

Thurman, E.M.; Hostetler, K.A. (2000) "Analysis of tetracycline and sulfamethazine antibiotics in ground water and animal-feedlot wastewater by high-performance liquid chromatography/mass spectrometry using positive-ion electrospray." In F. D. Wilde, L. J. Britton, C. V. Miller, and D. W. Kolpin (Eds.), Effects of Animal Feeding Operations on Water Resources and the Environment--Proceedings of the Technical Meeting, Fort Collins, Colorado, August 30-September 1, 1999: U.S. Geological Survey Open-File Report 00-204. U. S. Geological Survey, Fort Collins, CO.

Tolls, J. (2001) "Sorption of veterinary pharmaceuticals in soils: A review." Environmental Science and Technology, 35(17):3397-3406.

Tolls, J.; Gebbink, W.; Cavallo, R. (2002) "ph-dependence of sulfonamide antibiotic sorption: data and model evaluation." SETAC Europe 12th Annual Meeting, May 12-16, Vienna, Austria.

USDA (2002) Preventative Practices in Swine: Administration of Iron and Antibiotics. USDA Animal Plant Health Inspection Service, Fort Collins, CO.

Vansickle, J. (2002) "Drug use drops again." National Hog Farmer, 47(Nov. 15):11-16.

Veith, G.D.; Macek, K.J.; Petrocelli, S.R.; Carroll, J. (1980) "An evaluation of using partition coefficients and water solubility to estimate bioconcentration factors for organic chemicals in fish." In Eaton, J.G., Parish, P.R., and Hendricks, A.C.

(eds.). Aquatic Toxicology, ASTM STP 707. American Society for Testing and Materials, Washington, DC.

Vieno, N.; Tuhkanen, T.; Kronberg, L. (2007) "Elimination of pharmaceuticals in sewage treatment plants in Finland." Water Research, 41(5):1001-1012

Wang, Q.Q.; Bradford, S.A.; Zheng, W.; Yates, S.R. (2006) "Sulfadimethoxine degradation kinetics in manure as affected by initial concentration, moisture and temperature." J. Environ. Quality, 35:2162-2169.

Winckler ,C.; Grafe, A. (2000) "Abschätzung des Stoffeintrags in Böden durch Tierarzneimittel und pharmakolo-gisch wirksame Futterzusatzstoffe unter besonderer Berücksichtigung der Tetracycline Herausgeber: Umweltbundesamt." Forschungsbericht 297(33):911.

Wollenberger, L.; Halling-Soerensen, B.; Kusk, K.O. (2000) "Acute and chronic toxicity of veterinary antibiotics to Daphnia magna." Chemosphere, 40(7):723-730.

Wu, W. G. (1995). In Medical Microbiology: A Laboratory Study, Star Publishing Company, 3rd Edition, CA.

Yasojima, M.; Nakada, N.; Komori, K.; Suzuki, Y.; Tanaka, H. (2006) "Occurrence of levofloxacin, clarithromycin and azithromycin in wastewater treatment plant in Japan." Water and Wastewater Management for Sustainable Development of Chemical Industries, 53(11):227-233.

Yim, G. (2007). "Attack of the superbugs: Antibiotic resistance." The Science Creative Quarterly, Issue 3, http://www.scq.ubc.ca/attack-of-the-superbugs-antibiotic-resistance/ (accessed Mar 2008)

Zhang, R. (2001) Biology and Engineering of Animal Wastewater Lagoons (no city of publication given). http://ucce.ucdavis.edu/files/filelibrary/5049/678.pdf (accessed 13 July 2004).

Zhang, H.; Huang, C.H. (2007) "Adsorption and oxidation of fluoroquinolone antibacterial agents and structurally related amines with goethite." Chemosphere, 66:1502-1512.

Zuccato, E.; Calamari, D.; Natangelo, M.; Fanelli, R. (2000) "Presence of therapeutic drugs in the environment." Lancet, 355:1789-1790.

CHAPTER 6

Hormones

Tawan Limpiyakorn, Supreeda Homklin and Say Kee Ong

6.1   Introduction

Hormones are the body's chemical messengers that travel in the bloodstream to tissues or organs, carrying a signal from one or a group of cells to another. Hormones are produced by endocrine glands but are also produced by nearly every organ system and tissue type in a multicellular organism.   Endocrine hormones are molecules released directly into the bloodstream while exocrine hormones are released directly into a duct which then either enter the bloodstream or flow from cell to cell by diffusion. The major endocrine glands are the pituitary, pineal, thymus, thyroid, adrenal glands and pancreas. Men also produce hormones in their testes and women produce them in their ovaries.   They affect many different processes, including growth and development, metabolism, sexual function, reproduction and moods. A small amount of hormone is needed to cause changes in cells or to the whole body.

Based on their chemical structure, hormones can be divided into 3 broad classes: peptide hormones, amine hormones and steroid hormones. Peptide hormones consist of polypeptide hormones and oligopeptide hormones. Peptide hormones are from the anterior pituitary gland, placenta, stomach, duodenum, pancreas and liver. Oligopeptide hormones are from the posterior pituitary gland and the hypothalamus. Peptide hormones consist of chains of amino acids and examples would include insulin and growth hormones. Amine hormones are derived from the amino acids, tyrosine and tryptophan.   Examples of major amine hormones are dopamine, norepinephrine, and epinephrine. Steroid hormones are synthesized by chemical modification of cholesterol and are derived from the adrenal cortex, gonads and placenta.   Steroid hormones can be subcategorized into corticosteriods and sex steroids. Sex steroids can be divided into androgen, estrogens and prosgestagens. Examples of androgens are 20-hydroxyecdysone, androgen, androstenedione, androsterone,   dehydroepiandrosterone,   dihydrotestosterone,   epitestosterone, methyltestosterone, superdrol, testosterone, and trestolone. Common compounds in the estrogen group include equilin, equol, estradiol, estriol, estrone, estropipate, premarin, and xenoestrogen. Progestagens include drospirenone, dydrogesterone, 17-hydroxyprogesterone, progestagen, progesterone, progestin, prometrium.

Of concern in the environment are the peptide hormones and steroid hormones. In the mid-1990s, several researchers reported the presence of steroidal estrogen hormones in surface waters, municipal wastewaters and wastes from concentrated animal feeding operations (Ternes et al., 1999a; 1999b; Matsui et al., 2000; Kolpin et al., 2002; Hanselman et al., 2003). Estrogens have endocrine-disrupting properties and are known to disrupt the endocrine system of humans, aquatic and animal species at ng/L (Tyler et al., 2005; Purdom et al., 1994). Human and animal steroidal hormones have extremely high estrogenic potency in comparison to synthetic chemicals such as organochlorine aromatic compounds. For example, the estrogenic potency of endogenous steroidal estrogens was reported to be 10,000-100,000 times higher than exogenous endocrine disrupting compounds (Hanselman et al., 2003). Studies by several researchers have found that aquatic species such as turtles, trout, and minnows were sexually inhibited or reversed by the presence of natural estrogens at concentrations as low as few tens of ng/L (Jobling et al., 1998; Irwin et al., 2001; Lange et al., 2001). At a concentration of 5 ng/L of estradiol, production of female specific proteins was found to be induced in male Japanese medaka (Tabata et al., 2001).

In recent years, there is an increase in use of estrogenic compounds in hormone therapy of humans and in livestock farming and since estrogenic compounds are naturally produced, it is expected that the presence of these compounds in wastewaters and in the environment will continue to increase. The fate and behavior of these compounds in the environment are not well understood. Consequently, the risks of these chemicals to humans, aquatic organisms and animals are also not well understood. Many estrogenic compounds are not routinely analyzed or the analysis methods are too tedious or are not available in medium such as biosolids and sediments. This chapter will focus on the fate and behavior of certain androgens (methyltestosterone and testosterone) and estrogens (estradiol (17β-E2 and 17α-E2), estriol (E3), estrone (E1)) in municipal wastewaters and livestock farming where data are available.

6.2   Properties and Structure of Common Hormones

The basic structure of steroidal estrogens consists of a tetracyclic molecular framework of four rings: a phenol, two cyclohexanes, and a cyclopentane (see Figure 6.1). Steroidal estrogens are also known as the C18 steroidal compounds with differences in the various functional groups in the configuration of the D-ring at positions C16 and C17. Estrone has a carbonyl group while estradiol has a hydroxyl group on the C17 position. Estriol has an alcohol group on C16 and C17. The C17 hydroxyl group of the estradiol can either point downward in relation to the molecular plane, forming the 17α-E2 compound, or upward, forming the 17β-E2 compound. Ethinyl estradiol (EE2) is the first orally active synthetic steroidal estrogen and was synthesized in 1938. It is an active ingredient in oral contraceptive pills and in hormone therapy. The physical and chemical properties of the free estrogens, also known as unconjugated estrogens are presented in Table 6.1. The conjugated forms of estrogens are due to the esterification of C3 and/or C17 position(s) of free estrogens by glucuronide (GLU) and/or sulfate (SUL) groups.

The estrogenic potentials of estrogens depend significantly on the forms of estrogens. Matsui et al. (2000) compared the estrogenic activity of various natural estrogens in terms of EC50 as measured by yeast estrogen screening assay (YES) and found that 17β-E2 had the most estrogenic activity among the natural estrogens tested, while E1 and E3 showed a relative activity of 0.21 and 0.0013 with respect to 17β-E2. Estrogenic activity of estrogen conjugates such as sulfate and glucoronide were more than four orders of magnitude less than that of 17β-E2 with relative activity of 5.3×10^{-5}, 3.1×10^{-5}, and 5.9×10^{-7} with respect to 17β-E2 for βestradiol 3-sulfate (E2-3S), βestradiol 3-(βD-glucuronide) (E2-3G), and βestradiol 17-(βD-glucuronide) (E2-17G), respectively (Matsui et al., 2000). In another study, in vitro assays showed that the estrogenic potency of 17α-E2 was only 1% of that of 17β-E2 (Hoogenboom et al., 2001). In comparison, the estrogenic activity of nonylphenol and bisphenol A were 1.0×10^{-3} and 2.7×10^{-4} times that of 17β-E2.

The chemical structures of testosterone and methyltestosterone (MT) are presented in Figure 6.1 and their physical chemical properties presented in Table 6.1. Testosterone is secreted in the testes of males and the ovaries of females and with small amounts secreted by the adrenal glands. MT is a synthetic derivative of testosterone with a methyl group added to the C-17α position of the molecule. MT was developed synthetically in the search of an androgen that could be given orally without loss of bioavailability. Testosterone itself is ineffective when taken orally since the majority of the compound is metabolized and destroyed by the liver during the "first pass" with only about 5 - 10% of the compound entering the blood stream. MT is not broken down and deactivated as fast as oral testosterone by the liver. It is used to treat men with a testosterone deficiency and also used in women to treat breast cancer, breast pain, swelling due to pregnancy and, with the addition of estrogen, treat symptoms of menopause.

MT is biologically active and may impact the metabolic activity of living cells. MT is a questionable human carcinogen, producing nonmalignant tumors in the liver (Soe et al., 1992) and causes developmental abnormalities in the urogenital system (Lewis, 1991). Exposure of the fish, *Oryzias latipes* (medaka) to 27.75 ng/L showed male secondary sex characters in which no fish with ovary could be discerned (Masanori et al., 2004). This suggests that MT at low concentration can affect aquatic organisms.

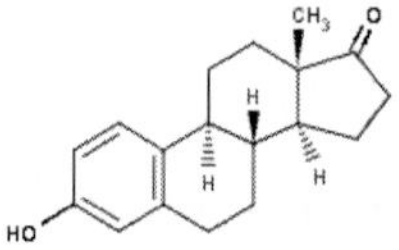

Estrone 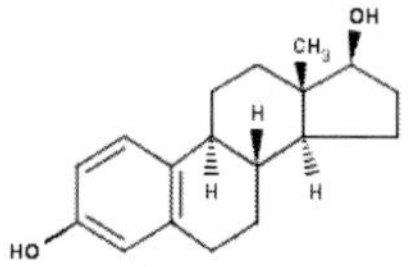   17β - Estradiol 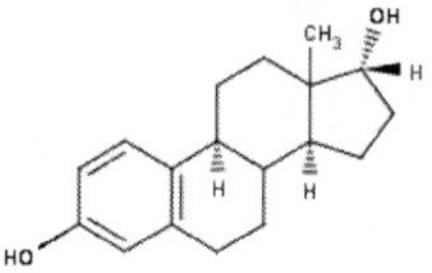   17α - Estradiol

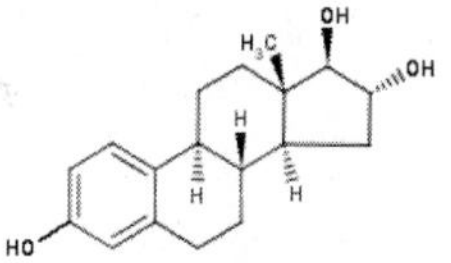

Estriol 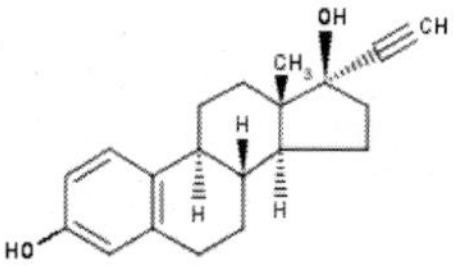   Ethinyl estradiol

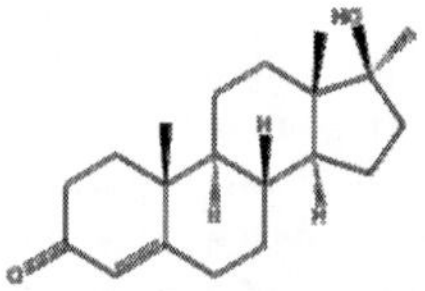

Methyltestosterone 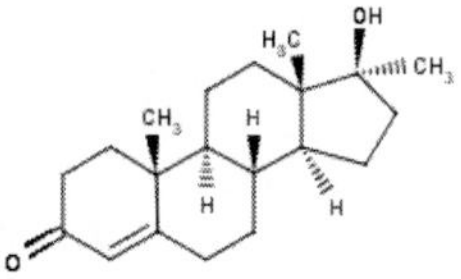   Testosterone

Figure 6.1 Chemical structures of steroidal estrogens

Table 6.1 Physical-chemical properties of steroidal estrogens (ChemIDplus Lite, 2008)

Hormone	Acronym	Structure	MW*	S* mg/L	log K_{ow} *	pK_a	Half-life (hrs)	VP (mm Hg)	K_H (atm-m^3/mole)
Estrone	E1	$C_{18}H_{22}O_2$	270.4	30	3.13	10.3-10.8	19	$1.42\ 10^{-7}$	3.8×10^{-10}
17β-estradiol	17β-E2 or E2	$C_{18}H_{24}O_2$	272.4	3.6	4.01	10.5-10.7	13	1.26×10^{-8}	3.64×10^{-11}
17α-estradiol	17α-E2	$C_{18}H_{24}O_2$	272.4	3.9	3.94	NA	NA	NA	NA
Estriol	E3	$C_{18}H_{24}O_3$	288.4	441	2.45	10.4	NA	1.07×10^{-10}	1.33×10^{-12}
Ethinyl estradiol	EE2	$C_{20}H_{24}O_2$	296.4	11.3	3.67	NA	36	2.67×10^{-9}	7.94×10^{-12}
Testosterone	--	$C_{19}H_{28}O_2$	288.4	23.4	3.32	NA	2-4	2.23×10^{-9}	3.53×10^{-9}
Methyl-testosterone	MT	$C_{20}H_{30}O_2$	302.45	3.4	3.36	NA	NA	1.85×10^{-8}	4.68×10^{-9}

*MW – molecular weight, S- solubility in water, log K_{ow} – octanol-water partition coefficient, VP – vapor pressure, NA – not available

6.3   Sources of Androgens and Estrogens

6.3.1   MT and Testosterones

Sources of testosterones are generally from humans and livestock farming (Kolodziej and Sedlak, 2007). Measurements of peripheral and spermatic vein plasma samples in humans indicate an average concentration of 95 ± 6.9 ng/L of testosterone (Kelch et al., 1972). Testosterone secretion by the adult human testis is pulsatile and, in the gonadal vein, testosterone levels ranged from 1 to 1,540 µg/L with a frequency of 4.0 ± 0.3 pulses per 4 hours (Winters and Troen, 1986). In urine, testosterone was detected ranging from 25 to 86 µg/L in males and 6.5 to 13.5 µg/L in females (Stopforth et al., 2007). Using a different approach in specifying the amount of testosterone produced, the mean excretion rates of testosterone was reported as 51.7 µg per 24 hours for males aged between 21 - 63 years and 6.5 µg per 24 hours for females aged 20 - 55 years (Ismail and Harkness, 1966).

Rongone   and   Segaloff   (1962)   showed   that   radioactive   17α-methyltestosterone administered to a male was extensively metabolized with two thirds of the radioactivity present in the feces and one third in the urine. The main metabolites excreted in the urine were found to be the glucuronides of 17α-methyl-5α-androstan-3α,17βdiol and 17α methyl-5βandrostan-3α,17βdiol (Quincey and Gray, 1967). In a recent study by Shinohara et al. (2000) where 500 mg of 17α-methyltestosterone was administered to four healthy adult males, the main metabolites found were 17α-methyl -5α-androstan-3α, 17βdiol and 17α-methyl-5β androstan-3α, 17βdiol which were similar to the work of Quincey and Gray (1967).

Blokland et al, (2004) studied the metabolism and excretion of 17β methyltestosterone and its metabolites in a heifer treated by intra-muscular injection and found that the main metabolic pathway was the hydroxylation of the 17β methyltestosterone with the formation of 17α-methyl-5βandrostane-3α, 17βdiol, 17α-methyl-5βandrostane-3β 17βdiol and17α-methyl-5βandrostane-3β 17ß-diol. In a study by Dumasia (2003), two castrated thoroughbred horses were injected with 17β methyltestosterone and the metabolites found in the urine of the horses were glucuronide and sulfate conjugates. The pathways of the in vivo metabolism were assumed to be through the reduction of the A-ring (di- and tetrahydro), epimerization of C-17 and oxidation at the C-6 and C-16.

A large amount of MT is used in the livestock industry especially in the aquaculture industry. MT is used for the masculinization of fish by using MT-impregnated foods or by single immersion treatment of fish fry after hatching in water containing MT. For example, farmed raised nile tilapia are fed with food containing 60 mg of MT/kg of food or by immersing the fry in water containing 1,800 µg/L of MT (Wattanodorn et al., 2007; Barry et al., 2007). It is highly probable that the remaining and unmetabolized MT-impregnated foods may accumulate in the sediments of the masculinization ponds and release into the receiving waters.

Other sources of androgenic compounds include pulp and paper mills effluent where the presence of androgens have resulted in the masculinization of fish exposed to the effluent (Larsson et al., 2000; Parks et al., 2001).

6.3.2. Estrogens - Livestock

Livestock wastes such as feces and urine especially from concentrated animal feeding operations (CAFOs) are among the major sources of estrogens in the environment. Research has focused on natural estrogens (E1, 17β-E2, 17α-E2, and E3) which are naturally produced in the bodies of animals and synthetic estrogens used as growth promoters such as benzoate and palmitate esters of E2, which can be hydrolyzed into E2 in animal bodies (Schiffer et al., 2001; Casey et al., 2003).

Several researchers have provided evidence to show that livestock can be a major source of estrogens and may be more significant than urban/human sources (Raman et al., 2004; Johnson et al., 2006). In some countries, the population of livestock is considerably larger than that of humans. In addition, certain animals such as cattle excrete two times more estrogens than humans while swine excrete ten times more than humans (Johnson et al., 2006).

Table 6.2 lists the few studies providing data on the estimated annual excretion of estrogens from different animals in the European Union (EU), United States (US) and United Kingdom (UK). The annual amounts of estrogens excreted from livestock were 33 and 49 tons/yr in EU and US, respectively (Lange et al., 2002). In UK, livestock generated approximately four times more estrogens at 0.79 tons/yr of 17β-E2 equivalent than humans at 0.22 tons/yr of 17β-E2 equivalent (Johnson et al., 2006). In terms of tons of estrogen excreted per animal, cattle excrete far more estrogens than the other livestock while swine was the second major contributor. Johnson et al. (2006) suggested that even though the combined cattle and swine populations in some countries may be less than humans, livestock may produce the vast majority of estrogens.

Different animals excreted estrogens through different routes. In cattle, about 58% of estrogens were found in the feces whereas in swine (96%) and poultry (69%) of the estrogens were mostly excreted with the urine (Ivie et al., 1986; Ainsworth et al., 1962; Palme et al., 1996; Hanselman et al., 2003). Nevertheless, these ratios change accordingly for pregnant female animals (Hoffmann et al., 1997; Hanselman et al., 2003).

Both 17α-E2 and 17β-E2 were found in cattle wastes (*Bos taurus*) (Ainsworth et al., 1962; Common et al., 1969; Moore et al., 1982; Hanselman et al., 2003) while only 17β-E2 was presented in swine (*Sus scrofa*) and poultry (*Gallus domesticus*) wastes (Ainsworth et al., 1962; Common et al., 1969; Moore et al., 1982; Hanselman et al., 2003). More than 90% of estrogens excreted by cattle were 17α-E2, 17β-E2 and E3 in free and conjugated forms with 17α-E2 more prevalent than 17β-E2 (Hanselman et al., 2003). On the contrary, swine and poultry excreted E1, 17β-E2, and E3 with free and conjugated forms. The presence of the two E2 epimers (α and β) in receiving water may be used to deduce the species of animals contaminating the water.

Table 6.2 Estimated annual excretion of estrogens from livestock

Location	Species	Head million	Total ton/yr	E1 ton/yr	E2* ton/yr	E1+E2 ton/yr	E2 equivalent ton/yr	Reference
EU	Cattle	82	26	- -	- -	- -	- -	Lange et al., 2002
	Swine	122	3.0					
	Chickens	1002	2.8					
	Sheep	112	1.3					
USA	Cattle	98	45	- -	- -	- -	- -	Lange et al., 2002
	Swine	59	0.83					
	Chickens	1816	2.7					
	Sheep	7.7	0.092					
UK	Human	59	- -	0.219	0.146	0.365	0.219	Johnson et al., 2006
	Cattle	2.2		0.693	0.365	1.058	0.596	
	Swine	5		0.367	0.19	0.386	0.141	
	Chickens	141.2		0.015	0.034	0.306	0.039	
	Sheep	9.1		0.0206	0.0064	0.027	0.013	

* E2 refers to both 17β-E2 and 17α-E2

Most estrogens excreted via urine are in the conjugated form whereas the free form of estrogens is mainly presented in the feces (Palme et al., 1996; Hoffmann et al., 1997; Hanselman et al., 2003). In the urine of cattle, E1-sulfate and 17α-E2-glucuronide were reported to be the dominant estrogens (48 - 92% and 5 - 36%, respectively) followed by E1-glucuronide and 17β-E2-glucuronide (Hoffmann et al., 1997).

The amounts of estrogens excreted from animals depend on factors such as animal species, gender, age, mass, diet, circadian variation, health conditions, and reproductive stage (Lucas and Jones, 2006; Schwarzenberger et al., 1996; Lange et al., 2002; Sarmah et al., 2006). Table 6.3 (modified from Hanselman et al., 2003) shows the amounts of estrogens excreted from different animals. As expected, the amounts of estrogens excreted increase during pregnancy, typically the later days of pregnancy. Non-pregnant diary cattle excrete in the range of 400 - 600 μg/d of total estrogens per 1,000 kg of body weight in feces and about 500 μg/d/1000 kg in urine. Amounts excreted pregnant dairy cattle can be 30 times more in the feces and 320 times more in urine. For a non-pregnant sow, estrogens range from 100 - 900 μg/d/1000 kg in feces and 400 - 600 μg/d/1000 kg in urine whereas a pregnant sow may excrete up to 1,600 μg/d/1000 kg in feces and up to 108,000 μg/d/1000 kg in urine. The amount of estrogens excreted by poultry is in the same range as cattle and swine in terms of μg/d/1000 kg but the amounts excreted for egg-laying hens are about six times more than a non-egg laying hens.

Table 6.3 Fecal and urinary excretion of estrogens from different animal species (Modified from Hanselman et al., 2003)

Animal species/ excretion	Reproductive stage	Excretion rate/1000 kg live animal mass (μg/d)	Measured estrogens	Method[+]	Reference
Dairy cattle/feces	Non-pregnant	600±200	17α-E2	RIA	Möstl et al., 1984
	Non-pregnant	400 ±10	E1, 17α-E2, 17β-E2[a]	RIA	Desaulniers et al., 1989
	0-80 d pregnant	300 ±nd	E1, 17α-E2, 17β-E2	RIA	Hoffman et al., 1997
	0-84 d pregnant	400 ±20	E1, 17α-E2, 17β-E2[a]	RIA	Desaulniers et al., 1989
	80-210 d pregnant	1500 ±nd	E1, 17α-E2, 17β-E2	RIA	Hoffman et al., 1997
	140-200 d pregnant	11400 ±1200	E1, 17α-E2[a], 17β-E2	RIA	Desaulniers et al., 1989
	210-240 d pregnant	5400± nd	E1, 17α-E2, 17β-E2	RIA	Hoffman et al., 1997
Dairy cattle/urine	Non-pregnant	500± 40	E1, 17α-E2[b], 17β-E2	RIA	Monk et al., 1975
	55-81d pregnant	700 ±60	E1, 17α-E2[b], 17 β -E2	RIA	Monk et al., 1975
	101-123 d pregnant	14400± nd	E1, 17α-E2, 17β-E2, E3	FL	Erb et al., 1968 a
	111 d pregnant	34300 ±nd	E1, 17α-E2, 17β-E2, E3	FL	Erb et al., 1968 b
	107-145 d pregnant	3400± 1200	E1, 17α-E2[b], 17β-E2	RIA	Monk et al., 1975
	165-175 d pregnant	28800 ±nd	E1, 17α-E2, 17β-E2, E3	FL	Erb et al., 1968 a
	205-209 d pregnant	22300 ±2500	E1, 17α-E2[b], 17β-E2	RIA	Monk et al., 1975
	250-254 d pregnant	86800 ±28000	E1, 17α-E2, 17β-E2, E3	FL	Erb et al., 1968 a
	271-285 d pregnant	163000 ±20000	E1, 17α-E2, 17β-E2, E3	FL	Erb et al., 1968 a
Sow/feces	Non-pregnant	800 ±nd	E1, 17β -E2, E3	RIA	Vos, 1996
	Non-pregnant	100 ±70	E1	EIA	Vos et al., 1999
	Non-pregnant	600 ±250	E1[c], 17α-E2, 17β-E2, E3	RIA	Choi et al., 1987
	Non-pregnant	900 ± nd	Not specified	RIA	Szenci et al., 1993
	14-34 d pregnant	1500 ± nd	E1, 17β-E2, E3	RIA	Vos ,1996
	25-33 d pregnant	1000 ± 680	E1	EIA	Vos et al., 1999
	0-35 d pregnant	1600 ± nd	E1[c], 17α-E2, 17β-E2, E3	RIA	Choi et al., 1987
Sow/urine	Non-pregnant	600 ± 350	E1	FL	Lunaas, 1965
	Non-pregnant	500 ± 600	E1	FL	Raeside, 1963
	Non-pregnant	400 ± 300	E1	FL	Lunaas, 1962
	0-42 d prengant	4400 ± 6200	E1	CL	Raeside,1963
	42-77 d pregnant	5000 ± 6200	E1	CL	Raeside, 1963
	77-105 d pregnant	108000 ± 106000	E1	CL	Raeside, 1963
Hen chicken/ Urine	Non-laying	600± 30	E1	CL	Common et al., 1965
	Non-laying	500± nd	E1, E3	CL	Mathur et al., 1966
	Non-laying	400± 20	E1	CL	Common et al., 1965
	Non-laying	1400± 550	E1, 17β-E2	CL	Mathur et al., 1969
	Non-laying	900± nd	E1, E3	CL	Mathur et al., 1966
	Laying	1600± nd	E1, 17β-E2	FL	Chan et al., 1972
	Laying	2100± 80	E1	CL	Common et al., 1965
	Laying	2700± 130	E1, E3	CL	Mathur et al., 1966
	Laying	1400± 50	E1	CL	Common et al., 1965
	Laying	3500± 430	E1, 17β-E2	CL	Mathur et al., 1969
	Laying	1600± nd	E1, E3	CL	Mathur et al., 1966

nd - no data; [a] 11% 17α-estradiol (17α-E2) cross reactivity; [b] 32% 17α-estradiol (17α-E2) cross reactivity [c] 122% estrone (E1), 30% 17α-estradiol (17α-E2), 100% 17β-estradiol (17β-E2), and 64% estriol (E3) cross reactivity; [+] RIA (radioimmunoassay), EIA (enzyme immunoassay), FL (fluorimetry), CL (colorimetry).

It should be noted that different analytical methods were used to determine the concentrations presented in Table 6.3 which can bias the amounts of estrogens measured. Colorimetry (CL) was a method used during the 1950s to 1970s era. Due to its low sensitivity and selectivity for estrogens, radioimmunoassay (RIA) and enzyme immunoassay (EIA) replaced the use of CL (Cohen, 1969; Hanselman et al, 2003). EIA and RIA are capable of detecting estrogens in the μg/L range. However, these methods can be affected by cross-reactivity and other interferences such as protein binding and complex compounds such as humic substances, and endogenous enzymes (Wood, 1991; Maxey et al, 1992; Nunes et al, 1998; Huang and Sedlak, 2001; Taieb et al, 2002; Hanselman et al, 2003). Interferences in environmental samples can cause an overestimation of the estrogen present. The use of EIA and RIA can be confirmed by using selective analytical instruments such as gas chromatography with mass spectrometry (GC-MS or GC-MSMS) and liquid chromatography with mass spectrometry (LC-MS or LC-MSMS). In general, the GC-MS has a larger library of mass spectrum than LC-MS. However, GC-MS requires labor intensive sample preparation due to the low vapor pressure of estrogens. Silylation reagent is widely used in the derivatization of estrogen into a compatible form for analysis with GC-MS (Jeannot et al., 2002; Braun et al., 2003; Ding and Chiang, 2003; Gomes et al., 2004; Hernando et al., 2004; Zuo and Zhang, 2005; Shareef et al., 2006). Another approach is to present the amounts of estrogen excreted per animal as presented by Johnson et al., 2006 (see Table 6.4).

Table 6.4 Estrogens excreted per animal for cattle and swine as estimated by Johnson et al. (2006)

Species	Route	Reproductive stage	Estrone (E1) (ug/head/d)	17β-estradiol (17β-E2) (ug/head/d)	estradiol - E2 equivalent (ug/head/d)
Dairy	Feces	- -	51	148	165
	Urine	- -	787	236	498
Swine	Feces	Non-pregnant	12	9	13
		Pregnant	33	28	39
	Urine	Non-pregnant	82	BDL*	27
		Pregnant	1400	n.d.**	467

* BDL – below detection limit
** n.d. – no data

6.3.3 Estrogens - Humans

Humans, in general, excrete estrogens in the conjugated forms rather than free forms (D'Ascenzo et al., 2003). The liver is responsible for the transformation of steroid estrogens to the conjugated forms of glucuronic acid and sulfate (Lauritzen, 1988; Mutschler, 1996). Normally, conjugated estrogens are more soluble in water and have lower estrogenic potentials than those in free form. Conjugated estrogens including E1-3G, E1-3S, E2-3G, E2-17G, E2-3S, E3-3G, E3-16G, and E3-3S are usually found in urine of females (Table 6.5). Males also excrete estrogens; however, detailed information is not available (Johnson et al., 2000).

D'Ascenzo et al (2003) investigated the amounts of conjugated estrogens in 50 cycling women, 22 non-cycling women and 1 pregnant woman. Only conjugated estrogens were found, while those in free forms were not observed. The amounts of conjugated estrogens excreted depended on the women's conditions. Pregnant women excreted conjugated estrogens at two to three orders of magnitude higher than that of cycling and non-cycling women, and cycling women excreted higher amounts of conjugated estrogens than non-cycling women. Glucoronide estrogens were found as the most predominant forms. In addition, Caccamo et al. (1998) and Adlercreutz et al. (1982) found that conjugated estrogens were excreted higher in the luteal (or secretory) phase which is the phase of estrous cycle of women than in follicular (or proluferative) phase. On average, humans excrete about 10.5 μg/d/person of E1 and 6.6 μg/d/person of 17β-E2 while EE2 is normalized at 1 μg/d/person (Johnson et al., 2004).

6.4   Fate and Behavior in Natural Systems

6.4.1   *Fate of MT and Testosterone in Natural Environments*

In a study by Kolpin et al. (2002), testosterone, progesterone, and cis-androseterone at a concentration of 5 ng/L were found to be present in 3 to 14% of the samples taken from 139 streams throughout the US. A study conducted by Shore et al. (2004) found that testosterone concentrations, as high as 6 ng/L, were found in the runoff from cattle pasture and fish pond effluent. Kolodziej et al. (2004) found that the maximum concentrations of testosterone and medroxyprogesterone in nearby rivers of dairy animal feeding operations in California were 0.6 ng/L and < 0.4 ng/L, respectively while irrigation canals contained concentrations of testosterone and medroxyprogesterone of 1.9 ng/L and 1.0 ng/L, respectively. Testosterone and androstenedione in the effluent of several fish hatcheries in California were found to be between 0.5 and 2 ng/L (Kolodziej et al., 2004). Kolodziej and Sedlak (2007) found that the maximum concentrations of testosterone in surface waters of a grazing range, dairy farms and effluent of a municipal wastewater treatment plant were 4.3 ng/L, 1.9 ng/L, and 8.0 ng/L, respectively. Androstenedione was also found at a maximum concentration of 44 ng/L in surface waters of grazing rangeland. A recent study showed that MT concentrations in soils were between 2.8 and 2.9 ng/g demonstrating the persistence of MT in soil for nearly three months after cessation of treatment (Mcelwee et al., 2000).

Table 6.5 Excretion of conjugated estrogens from female urine ($\mu g/d$)

Category	E1-3G	E1-3S	E2-3G	E2-17G	E2-3S	E3-3G	E3-16G	E3-3S	References
Menstruating female[a]	39.6*	--				45.1*	20.9*	--	Adlercreutz et al, 1982
Menstruating female[a]	--	--				10.1*	8.9*	--	Caccamo et al, 1998
Menstruating female[b]	78.1*	--				86.9*	30.8*	--	Adlercreutz et al, 1982
Menstruating female	16 ± 8	5.4 ± 1.9	5.5 ± 3.5	2.5 ± 1.4	3.5 ± 3.2	10 ± 5	7.2 ± 5.4	3.7 ± 1.7	D'Ascenzo et al, 2003
Menopausal female	9.5 ± 6.1	3.2 ± 1.7	4.2 ± 4	1.5 ± 1	1.3 ± 0.9	3.6 ± 2.3	2.1 ± 1.2	0.6 ± 0.3	D'Ascenzo et al, 2003
Pregnant women	--	641	--	--	66	--	--	890	Fotsis et al., 1980
Pregnant women	490	450	104	90	64	2025	3300	1170	D'Ascenzo et al, 2003

[a] follicular phase; [b] luteal phase; *$\mu g/l$ multiply with 1.1 liter of urine per day

Casey et al. (2003, 2004) showed that the Freundlich sorption coefficients, K_f, of testosterone ranged from 0.05 to 1.20 mL/g. In a study by Stumpe amd Marschner (2007), the Freundlich sorption coefficients and log K_{oc} values of testosterone ranged from 13.23 to 21.27 mL/g and 2.91 to 3.29, respectively for 4 different types of soils. A study by Lee et al. (2003) showed that the average log K_{oc} of testosterone for 5 different soils was 3.34. Sorption of MT to different soils by Chotisukarn et al. (2007) showed that the estimated K_d values ranged from 12.3 mL/g to 168.8 mL/g and was a function of the organic content of the soil.

Dissipation half lives of testosterone were estimated to range from 0.3 and 7.3 days depending on the soil types in soil-water slurries (Lee et al., 2003). Evolution of $^{14}CO_2$ using ^{14}C-testosterone showed that approximately 50 – 56% was mineralized in 4 different soils over a 23-day incubation period (Stumpe and Marschner, 2007). Degradation rate coefficients of testosterone in column studies were found to be between 0.404 and 0.600 per hour (Casey et al., 2004). Aerobic degradation of testosterone has been studied in detail (Shi and Whitlock, 1968; Horinouchi et al., 2004); testosterone degrades via the 9, 10-seco pathway, where the cleavage is via the C-19 steroids nucleus.

MT is biodegradable and its degradation rate constants in sediments were found to range from 0.001 to 0.011 per hour (Homklin et al., 2007). Analysis of 16S rRNA gene sequences in sediments suggest that strain *Pimelobacter simplex* strain S151 and *Nocardioidaceae* were responsible for the degradation (Wattanodorn et al., 2007, Homklin et al., 2007).

6.4.2   Release of Estrogens from Livestock

Estrogens from livestock can be released to environments by the application of animal wastes or treated wastes for the fertilization of soils. Estrogens are nonvolatile and of low water solubility and, therefore, tend to be sorbed onto sediments and soils or onto the waste sludge. Estrogens are strongly sorbed onto soils by binding with organic matter and are not easily desorbed with sorption coefficients (K_d) ranging from 86 to 6,670 L/kg (Shore et al., 1993; Casey et al., 2003). Even though estrogens are strongly sorbed onto soil, they can be dissipated to water environments via desorption in runoff waters or via facilitated transport in colloidal matter. Hanselman et al. (2003) found an increase in the transportation of estrogens in soil to groundwater and receiving water in the presence of surfactants.

6.4.2.1 Fate of estrogens in soil environment

17β-E2 was reported by Colucci et al. (2001) to rapidly oxidize to E1 in loam soil, sandy loam and silty loam soils. In loamy soil, E1 was mineralized whereas in sandy loam and silty loam soil, E1 was present for 3 months. In addition, they found that temperature and the moisture content in the soil affected the mineralization of estrogens.

6.4.2.2 Fate of estrogens in water environment

Nichols et al. (1997) showed that surface runoff was contaminated with 17β-E2 when surface runoff flow through land which was amended with poultry litter at the rate of 4 tons per acre containing 133 µg/kg of 17β-E2. They found that the estrogen concentrations increased with the application rate of manure and the highest concentration measured in the first flush runoff was 1.28 µg/L. 17β-E2 concentration in the second runoff decreased to 66% of the concentration of the first runoff. Similar results were found in the study by Finlay-Moore et al. (2000) for broiler litter applied to land for 7 months. In their study, the concentrations of 17β-E2 in runoffs increased from 50 - 150 ng/L (background level) to 20 – 2,530 ng/L when broiler litter was applied at loading rates of 2,550 – 8,260 kg/ha. The concentrations of 17β-E2 in soils also increased from 55 to 675 ng/kg after application with broiler litter and 17β-E2 persisted in the soils for several weeks. It was suggested that the concentrations of 17β-E2 in runoffs and in the soil depended on the rate and time of application of the broiler litter. Nichols et al (1998) found that grass filter strips of 6.1, 12.2, and 18.3 m wide reduced the transportation of estrogens in runoffs from land application by 58%, 81%, and 94%, respectively.

Data from the study by Kolpin et al. (2002) for 139 streams in 30 states in the US impacted by both urban and livestock production showed maximum concentrations of 112, 74, 93, 51, and 831 ng/L and median concentrations of 27, 30, 9, 19 and 73 ng/L, respectively for E1, 17α-E2, 17β-E2, E3 and EE2, respectively. Shore et al. (1995) investigated the concentrations of E1 plus 17β-E2 in surface water (small stream), irrigation pond and well water (farm well) after the application of poultry litter. They found that the concentrations of E1 plus 17β-E2 increased from < 0.5 to 5 ng/L in surface water while the concentration decreased from 23 to 5 ng/L in irrigation pond and was < 0.1 ng/l in well water. Peterson et al. (2000) found 17β-E2 concentrations in the range of 6 - 66 ng/L in five springs from the mantled karst aquifer system of northwest Arkansas. This area supports high production of poultry and cattle and the concentrations found suggest that estrogens from livestock may have contaminated the groundwater. A study by Ying and Kookana (2003) found that 17β-E2 and EE2 were recalcitrant to degradation under anaerobic conditions.

6.5   Fate of Livestock Waste Estrogens in Engineered Systems

6.5.1   Fate of Estrogens in Livestock Wastes

Since the amounts of estrogens excreted are dependent on the animal species, the concentrations of estrogens in livestock wastes will vary accordingly. The concentrations of estrogens in different animal wastes are presented in Table 6.6. Concentrations in cattle slurries of holding ponds ranged from 41 – 3,057 µg/kg for E1, 19 – 1,028 µg/kg for 17α-E2 and 29 – 1,229 µg/kg for 17β-E2. Concentrations in cattle manure solids or semi-cakes were less than 500 µg/kg. It is interesting to note that higher total estrogen concentrations were found on the surface of cow manure pile (1,000 µg/kg) than inside the manure pile (27 µg/kg) (Möstl et al., 1997). Concentrations in swine slurries were as high as 4,728 µg/kg for E1, 890 µg/kg for 17α-E2, 1,251 µg/kg for 17β-E2 and 402 µg/kg for E3. E1 and17α-E2 were typically

Table 6.6 Concentrations of estrogens in livestock wastes

Animal	Location	Waste type	Estrone (E1) (µg/kg)*	17α-estradiol (17α-E2) (µg/kg)*	17β-estradiol (17β-E2) (µg/kg)*	Estriol (E3) (µg/kg)*	Total estrogen (µg/kg)*	Method	References
Cattle	Germany	Diary (slurry)	254-592	--	167-1229	--	--	--	Wenzel et al. (1998)
		Bulls (slurry)	84	--	15-65	--	--	--	Wenzel et al. (1998)
	USA	Dairy (press cake solids)	426±78	139±7	BDL	--	--	GC-MS	Raman et al. (2001)
		Dairy, dry-stack (solid)	203±176	289±207	113±67	--	--	GC-MS	Williams (2002)
		Holding ponds	543±269	370±59	239±30	--	--	GC-MS	
		Dairy, dry-stack (semisolid)	13-80	--	13-27	--	--	GC-MS/ELISA	Raman et al. (2004)
		Dairy, dry-stack (solid)	12-37	--	5.8-25	--	--	GC-MS/ELISA	
		Dairy, holding pond	2.5-5.6	--	0.8-1.9	--	--	GC-MS/ELISA	
	Not specified	Dairy manure pile (surface)	--	--	--	--	1000	--	Möstl et al, 1997
		Dairy manure pile (inside)	--	--	--	--	27	--	
	New Zealand	Dairy, oxidation pond	41-3057**	19-1028**	29-289**	--	--	GC-MS	Sarmah et al. (2006)
Swine	Germany	Piggery (slurry)	<2	--	<2	--	--	--	Williams (2002)
	USA	Piggery (finishing lagoon)	1507	BDL	BDL	--	--	GC-MS	
		Piggery (finishing hoops)	217	BDL	160	--	--	GC-MS	
		Piggery (farrowing lagoon)	1295	BDL	BDL	--	--	GC-MS	
		Piggery (farrowing pit)	4728±427	890±120	1251±275	--	--	GC-MS	
	Not specified	Swine manure pit	2006.54**	--	153.95**	401.91**	--	--	Shappell et al, 2007
	USA	Swine, finishing hoop structure	30-78	--	32-49	--	--	GC-MS/ELISA	Raman et al. (2004)
		Swine, farrowing pit	31-150	--	12-29	--	--	GC-MS/ELISA	

Table 6.6 Concentrations of estrogens in livestock wastes (Cont'd)

Animal	Location	Waste type	Estrone (E1) (μg/kg)*	17α-estradiol (17α-E2) (μg/kg)*	17β-estradiol (17β-E2) (μg/kg)*	Estriol (E3) (μg/kg)*	Total estrogen (μg/kg)*	Method	References
Poultry	Not specified	Broiler litter	nd	nd	33±12	--	--	EIA	Finlay-Moore et al. (2000)
		Broiler litter	nd	nd	133 ±6	--	--	EIA	Nicols et al. (1997)
		Broiler litter (at treat)	nd	nd	101± 2	--	--	EIA	
		Broiler litter	nd	nd	904	--	--	EIA	Nicols et al. (1998)
		Broiler litter (female)	nd	nd	65± 7	--	--	EIA	Shore et al (1993)
		Broiler litter (male)	nd	nd	14± 4	--	--	EIA	
		Layer litter	nd	nd	533± 40	--	--	EIA	
		Rooster litter	nd	nd	93 ±13	--	--	EIA	
Goat	New Zealand	Effluent (slurry)	157**	172**	47**	--	--	GC-MS	Sarmah et al. (2006)

**ng/l
nd, not detect
BDL, below detection limit
EIA, enzyme immunoassay

non-detect and can be as high as 533 μg/kg of 17β-E2 in poultry litter and 904 μg/kg in broiler litter.

Schlenker et al., 1998 found that 80% of estrogens were degraded when they were incubated with cattle feces for 12 weeks. Burnison et al. (2003) found that high concentrations (about 6.9 to 16.6 ppm) of estrogens (E1 and 17β-E2) were present in pig manure tank in farm at Southwestern Ontario (Canada) after storage for more than 7 months. However, Bamberg et al (1986) found that the concentration of estrogens in equine and bovine feces did not change during one week storage.

Schlenker et al. (1999) observed no change in estrogen concentrations in cattle manure at 5° C for a 12-week experiment. However, when temperature was raised to 30° C, estrogen was found to reduce to non-detect levels within 3 weeks. Raman et al. (2001) found that the degradation rate coefficients increased from 0.03 to 0.12 per day when the temperature was raised from 5 to 50 °C. Shore et al. (1993) found that at pH 5 and 7, estrogen concentrations in broiler litter were reduced whereas, at pH 8 and 12, estrogen concentrations remained the same. At pH 2, transformation of estrogens in cattle manure was reduced and only 31% was lost in spite of incubating the samples at 30° C (Raman et al., 2001).

6.5.2   Fate of Estrogens in Livestock Waste Treatment Systems

Information on the fate of estrogens in livestock waste treatment systems is very limited. This is probably due to the difficulty in determining estrogen in a complex matrix (Sumpter and Jobling, 1995; Raman et al., 2001; Raman et al., 2004). The concentrations of estrogens in the influent and effluent of different waste treatment systems are summarized in Table 6.7.

6.5.2.1 Lagoon systems and oxidation ponds

Lagoons are used widely as a reservoir and treatment facility for wastes and wastewaters from livestock. The few studies available for lagoon systems showed E1 and 17β-E2 concentrations in the aqueous phase were in the range of 1.1 - 14 and 1.8 - 3.3 μg/L, respectively, in a swine finishing lagoon and 5.9 and 3.9 μg/L, respectively, in a swine farrowing lagoon (Raman et al., 2004). The order of estrogens present in swine lagoon from the highest to lowest concentrations was E1 > E3 > 17β-E2 (Fine et al., 2003; Shappell et al., 2007). Lagoons appeared to have high removal efficiencies of estrogens at more than 90% with 96%, 98%, and 98% removal of E1, 17β-E2, and E3, respectively (Shappell et al., 2007). Sarmah et al. (2006) found the concentrations of E1, 17α-E2, and 17β-E2 were 27, 11, 8 ng/L, respectively, in the effluent of an oxidation pond treating dairy cattle, swine and goat wastes.

Table 6.7 Estrogens in influents and effluents of waste and wastewater treatment systems

System	Animal Species	Influent (ng/L)						Effluent (ng/L)						References
		E1	17α-E2	17β-E2	E3	EE2	EEQ	E1	17α-E2	17β-E2	E3	EE2	EEQ	
Finishing lagoon	Swine	1100-14000	--	1800-3300	--	--	--	--	--	--	--	--	--	Raman et al., 2004
Farrowing lagoon	Swine	5900	--	3900	--	--	--	--	--	--	--	--	--	Raman et al., 2004
Lagoon	Swine	2006.54± 1160.22	153.95±117.17		401.91± 334.60	--	330.52± 150.14	81.20± 83.11	3.00±2.18		8.72 ± 8.17	--	16.35 ± 14.99	Shappell et al., 2007
Oxidation pond	Swine	27.3	10.9	8.0	nd	nd	33.4± 4.51	--	--	--	--	--	--	Sarmah et al., 2006
	Dairy cattle	40.9- 3123	18.8- 1028	28.8-331	nd	nd	<5-521	--	--	--	--	--	--	Sarmah et al., 2006
	Goat	157	172	47.1	nd	nd	60.9± 10.6	--	--	--	--	--	--	Sarmah et al., 2006
Holding pond	Dairy	2500- 5600	--	800-1900	--	--	--	--	--	--	--	--	--	Raman et al., 2004
Marsh constructed wetland	Swine	74.9± 92.37	1.6±2.48		7.08± 8.77	--	11.17± 8.99	11.44± 4.90	1.6±1.91		3.54 ± 2.12	--	1.36 ± 0.84	Shappell et al., 2007
Trickling filter	Swine	3600 - 3800	590 - 660	410 -4600	1500 - 2700	<6.3- 9.5	--	4.5-6.9	<0.3-72	<0.3-24	<0.2-6.9	<1.4-18	--	Furuichi et al., 2006
UASB	Swine	5200 - 5400	650 - 680	1000 - 1500	2200 - 3000	<6.3- 9.5	--	3600 - 3800	590 -660	410 - 4600	1500 -2700	<6.3-9.5	--	Furuichi et al., 2006

nd - not detect

6.5.2.2 Constructed wetlands

Constructed wetlands are artificial marshes or swamps which act as biofilters removing pollutants such as heavy metals and organic compounds from wastewater. Shappell et al. (2007) found that about 85% of E1 and 50% of E3 were removed in a constructed wetland with no removal of 17β-E2 with influent and effluent concentrations of 1.6 ng/L. The reduction of estrogenic potential in constructed wetland system ranged from 83 - 93% resulting in an effluent estrogenic potentials of between 1.63 - 2.18 ng/L of E2 equivalent which were lower than the lowest observable effect concentration (LOEC) of estradiol (10 ng/L) and higher than the predicted no-effect concentration (PNEC) (1 ng/L).

6.5.2.3 Tricking filters

Fate of estrogens was investigated in a pilot-scale trickling filter for the treatment of swine wastewaters by Furuichi et al. (2006). In their study, the influent contained between 3.6 - 3.8, 5.9 - 6.6, 4.1 - 4.6, and 1.5 - 2.7 μg/L of E1, 17α-E2, 17β-E2, and E3, respectively and between 6.3 - 9.5 ng/L of EE2. Estrogen removals were between 44% and 99% resulting in concentrations of less than 70 ng/L of each estrogen in the effluent (see Table 6.7).

6.5.2.4 Up-flow anaerobic sludge blanket system

Furuichi et al. (2006) studied the removal of estrogens from swine wastewater in a pilot-scale up-flow anaerobic sludge blanket (UASB) system in Japan. The influent E1, 17α-E2, 17β-E2, and E3 concentrations were in the range of between 1 - 7 μg/L while the concentrations of EE2 were between 6.3 - 9.5 ng/L (see Table 7). Removal rates in the UASB ranged from 30% to 50% with EE2 passing through the UASB system without any treatment. This suggests that estrogens may not be easily degraded under anaerobic condition as compared to aerobic conditions.

6.6   Fate of Estrogens in Septic Tanks and Sewer Lines

Humans, in general, excrete estrogens in the final conjugation forms of glucuronic acid and sulfate. These conjugated estrogens are transported in septic tanks or and into the sewers before entering the wastewater treatment plants. In a septic tank, the percent of sulfate estrogens of the total conjugated estrogens were found to increase from 22% in the influent to 55% in the effluent indicating that glucoronide estrogens rather than sulfate estrogens were cleaved in the septic tanks (D'Ascenzo et al., 2003). It is believed that microorganisms such as E. coli which are capable of producing β-glucuronidase enzyme are responsible for the deconjugation or cleavage of the glucuronide estrogens. These microorganisms are known to produce less arylsulfatase enzyme than β-glucuronidase enzyme (Fürhacker et al., 1999; Baronti et al., 2000; D'Ascenzo et al., 2003) and were confirmed by batch studies with bacteria from a septic tank where glucuronide estrogens were found to rapidly deconjugate within 24 hours (D'Ascenzo et al., 2003). In addition, D'Ascenzo et al. found that glucuronide estrogens with glucuronide acid at A-ring position 3 cleaved more easily than the glucoronide acid at D-ring position 16 or 17. E1-3S and E2-3S were found to persist as long as 5 days whereas E3-3S persisted as long as 8 days.

In the sewers, the ratio of free to conjugated estrogens increased from 0.91 at the entrance to 2 at the exit of sewers with conjugated estrogen removal rates of about 7%, 57%, 63%, 100%, 42%, 70%, 51%, 100% and 24 % for E1-3S, E1-3G, E2-3S, E2-17G, E2-3G, E3-3S, E3-16G and E3-3G, respectively (D'Ascenzo et al, 2003). The percentage of sulfate estrogens increased from 55% at the entrance to 60% at the exit of the sewers indicating that the mechanism of degradation was probably similar to that in the septic tank with most of the deconjugation on glucoronide estrogens rather than on sulfate estrogens (D'Ascenzo et al., 2003).

6.7   Fate of Estrogens in Municipal Sewage Treatment Systems

Municipal wastewater treatment plants have the capability of removing estrogens from wastewater thus attenuating the widespread dispersion of estrogens in natural environments. Data collected from various wastewater treatment plants in 13 countries are summarized in Table 6.8. Information listed includes the type of wastewater, system configuration, operational conditions such as hydraulic retention time (HRT) and solids retention time (SRT), concentrations of estrogens (E1, 17β-E2, E3, and EE2) and estrogenic potentials in the influents and effluents of the treatment plants and their removal efficiencies. Since the concentrations of estrogens measured in the wastewaters are impacted by the analytical method, Table 6.8 also presents the analytical method. For 17α-E2, they are not too many studies reporting the presence of this compound (see Table 6.9). Table 6.10 reports the estrogenic potentials and their removals in wastewater treatment plants.

6.7.1 Concentrations of Estrogens in Influents of Municipal Wastewater Treatment Systems

The amounts of estrogens as well as estrogenic potentials in influents of municipal wastewater treatment plants varied significantly, ranging from non-detect (nd) to several hundred ng/L. Concentrations of E1, 17β-E2, E3 and EE2 and their standard deviations, if available, are presented in Table 6.8. Concentrations of E1, 17β-E2, E3 and EE2 in the influent of wastewaters ranged between < 2.5 - 130.0, 4.5 - 35.0, 11.4 - 118, and 0.4 - 59 ng/L with average concentrations of 47.9 ± 32.6, 14.3 ± 8.1, 51.6 ± 38.2, and 11.6 ± 17.6 ng/L, respectively. In many cases, E3 seemed to be the most dominant estrogen followed by E1 with 17β-E2 and EE2 at much lower concentrations. The above concentrations were similar to the concentrations found in septic tanks. The estrogen concentrations in the wastewater of different countries did not vary much. In Australia, the average influent E1, 17β-E2, and EE2 concentrations for an inland treatment plant were 54.8, 22, and < 0.5 ng/L, respectively whereas the concentrations of a coastal treatment plant were 58, 14, and < 0.5 ng/L, respectively (Braga et al., 2005). In Italy, Lagana et al. (2004) reported that the average influent E1, 17β-E2, E3, and EE2 concentrations were 35, 25, 31 ng/L and less than the detection limit, respectively. Influent of a German municipal wastewater treatment plant contained 67.5, 15.8, and 8.2 ng/L of E1, 17β-E2, and EE2, respectively (Andersen et al, 2003). Ternes et al. (1999b) reported that the influent of a Brazilian wastewater treatment plant contained 40 and 21 ng/L of E1 and 17β-E2, respectively.

Table 6.8 Influent and effluent concentrations of various estrogens in municipal wastewater treatment systems

Ref.	Place	System*,+	Operation		Anal. Meth.[i]	E1			17β-E2			E3			EE2		
			HRT (hr)	SRT (d)		Infl	Eff	% Rem	Infl	Eff	% Rem	Infl	Eff	% Rem	Infl	Eff	% Rem
Kim et al., 2007	South Korea	Pr/AS/RBC[1]	--	--	LM	--	2.2-36	--	--	<1.0	--	--	8.9-25	--	--	1.3	--
Hintemann et al., 2006	Germany	Pr	--	--	E	--	--	--	--	3.1-3.8	--	--	--	--	--	1.6-3.3	--
		Pr/AS	--	--	E	--	--	--	--	9.5	--	--	--	--	--	0.9	--
		Pr/AS	--	--	E	--	--	--	--	12	--	--	--	--	--	0.4-2.9	--
		Pr/Lag (Aer)	--	--	E	--	--	--	--	15-51	--	--	--	--	--	1.8-3.1	--
Servos et al., 2005	Canada	Pr/AS/N/PoPPt(Al)/SF/UV	8.5	5.5	G, Y	--	--	66.7	--	--	82.9	--	--	--	--	--	--
		Pr/AS/N/PoPPt(Fe)/Cl	6.6	9.6	G, Y	--	--	72.7	--	--	96.8	--	--	--	--	--	--
		Pr/AS/PoPPt(Fe)/Cl	6.7	2.7	G, Y	--	--	0	--	--	39.5	--	--	--	--	--	--
		Pr/AS/N/PoPPt(al)/UV	2.8	0.9	G, Y	--	--	76.7	--	--	92.7	--	--	--	--	--	--
		Pr/AS/N/UV	8	4.1	G, Y	--	--	85.4	--	--	98.3	--	--	--	--	--	--
		Pr/AS/N/PoPPt(Al)/UV	6.5	4.7	G, Y	--	--	0	--	--	75.9	--	--	--	--	--	--
		AS(Ex)/N/PoPPt(Al)/SF/UV	43	13.6	G, Y	--	--	96.5	--	--	93.3	--	--	--	--	--	--
		Pr/AS(Ex)/N/PoPPt(Fe)/GA/UV	16.8	53	G, Y	--	--	97.8	--	--	98.8	--	--	--	--	--	--
		Pr/AS/N/PoPPt(Fe)/Cl	12.3	35.5	G, Y	--	--	95.1	--	--	98.2	--	--	--	--	--	--
		Lag(Aer/Fac)/PoPPt(Al)	>150	>150	G, Y	--	--	93.3	--	--	98.4	--	--	--	--	--	--
		Lag(Fac/Aer)/PoPPt(Al)/SSF	>150	>150	G, Y	--	--	46.4	--	--	80.5	--	--	--	--	--	--
		Flag(Fac/Aer)/PoPPt(Al)	>150	>150	G, Y	--	--	95.3	--	--	95.9	--	--	--	--	--	--
		Lag(Ana)/PoPPt(Al)	>150	>150	G, Y	--	--	96.1	--	--	98.1	--	--	--	--	--	--

+ Treating municipal wastewater unless indicated, [1] 85% domestic; 18% Industrial; 2% livestock [2] 70% Domestic + 30% industrial [3] 90% Domestic + 10% industrial [4] industrial

* Pr – Primary, AS- Activated Sludge, AS(Ex) – Extended Aeration Activated Sludge, Lag – Lagoons, Lag(Aer) – Lagoons (aerated), Lag(Fac) – Lagoons (Facultative), Lag(Ana) – Lagoons (Anaerobic), SBR – Sequencing batch reactor, ASBR – Anaerobic Sequencing Batch Reactor, HRO – High Rate Oxygen, N – nitrogen removal, P – Phosphorus removal, PePPt (Fe or Al) – pre-precipitation of phosphorus by iron or alum, PoPPt (Fe or Al) – Post-precipitation of phosphorus by iron or alum, MBR – membrane bioreactor, TF – Trickling filter, RBC – rotating biological contactors, BF – Biological Filters, CS – contact stabilization, PeF – Percolating filters or sand filters, UV – ultraviolet, Oz – ozonation, Cl – chlorination, AC – activated carbon, CMF – continuous microfiltration, RO – reverse osmosis

[i]-Analytical methods: L – LC MS, LM – LC-MSMS, E- ELISA, G – GC MS, GM – GC-MSMS, ER – ERBA, ES – E-Screen, Y – YES, LE – LC- ESI-MSMS, H – HPLC, HL - HPLC-LCMS, HR – HRGC-MS, T – TLC, GL- GLC

Table 6.8 Influent and effluent concentrations of various estrogens in municipal wastewater treatment systems (Cont'd)

Ref.	Place	System[*]	Operation HRT (hr)	Operation SRT (d)	Anal. Meth.[/]	E1 Infl	E1 Effl	E1 % Rem	17β-E2 Infl	17β-E2 Effl	17β-E2 % Rem	E3 Infl	E3 Effl	E3 % Rem	EE2 Infl	EE2 Effl	EE2 % Rem
Braga et al., 2005	Australia	Pr/SBR/CMF/ROCl	4	16	G	54.8±14	<1	>98	22±15.9	<1	>95	--	--	--	<5	<1	--
		Pr	0.75	--	G	58±15	54 ± 13	7	14±11	14±10	0	--	--	--	<5	<5	--
Servos et al., 2005	Canada	Pr/AS/N/PoPPt(Al)/UV	11.1	12.6	G, Y	--	--	82.1	--	--	94.7	--	--	--	--	--	--
		Pr/HRO/AS	4.1	2.7	G, Y	--	--	80.6	--	--	96.1	--	--	--	--	--	--
		Pr/HRO/AS	2.7	2.2	G, Y	--	--	95.1	--	--	97.1	--	--	--	--	--	--
		Pr/TF/SC/Cl	1.0	1.9	G, Y	--	--	0	--	--	0	--	--	--	--	--	--
		Pr/PoPPt(Al)	--	na	G, Y	--	--	0	--	--	0	--	--	--	--	--	--
Cargouët et al., 2004	France	Pr/AS/N[2]	--	--	G	17.6±0.5	7.2±0.8	59	11.1±1.7	4.5±1	60	14.9±1.1	7.3 ± 1.4	51	5.4±0.6	3.1±0.6	43
		Pr/AS/N	10	--	G	15.2±1.8	6.5±1.2	57	17.4±1.7	7.2±0.8	59	15.2±1.4	5.0 ± 0.8	67	7.1±0.9	4.4±1.2	38
		Pr/UpBF/N[3]	2.5- 4	--	G	9.6±1.5	4.3±0.6	55	11.6±0.6	6.6±1.4	43	12.3±1.9	5.7 ± 1.6	54	4.9±1.0	2.7±0.8	45
		Pr/AS	2-3	--	G	11.2±2.3	6.2±0.8	44	17.1±0.6	8.6±0.9	49	11.4±1.4	6.8 ± 0.6	40	6.8±1.4	4.5±0.8	34
Lagana et al., 2004	Italy	Pr/AS	--	--	G	15-60	5-30	54	10-31	3-8	76	23-48	nd-1	--	nd	nd	--
Pawlowski et al., 2004	Germany	AS/N/P	--	--	GM, Y	--	1.2-19	--	--	1-5.6	--	--	--	--	--	<1-1.5	--
Xie et al., 2004	USA	Nr/Nr/P	144	--	Y	--	--	--	--	--	--	27	0.7	--	--	--	--
		Nr/N/Ter/Cl	--	--	Y	--	--	--	--	--	--	103	--	--	--	--	--
Andersen et al., 2003	Germany	Pr/AS/N/PoPPt(Fe)	--	--	G	65.7	<1	>98	15.8	<1	>94	--	--	--	8.2	<1	>87.8
Isobe et al., 2003	Japan	Pr/AS	--	--	LE	--	34	--	--	2.5	--	--	--	--	--	--	--
	Japan	Pr/AS	--	--	LE	--	2.5	--	--	0.3	--	--	--	--	--	--	--

+ Treating municipal wastewater unless indicated. [1] Industrial; [2] 85% domestic; 18% domestic; 2% livestock [3] 70% Domestic + 30% industrial [4] 90% Domestic + 10% industrial [5] industrial

* Pr – Primary, AS- Activated Sludge, AS(Ex) – Extended Aeration Activated Sludge, Lag – Lagoons, Lag(Aer) – Lagoons (aerated), Lag(Fac) – Lagoons (Facultative), Lag(Ana) – Lagoons (Anaerobic), SBR – Sequencing batch reactor, ASBR – Anaerobic Sequencing Batch Reactor, HRO – High Rate Oxygen, N – nitrogen removal, P – Phosphorus removal, PePPt (Fe or Al) – pre-precipitation of phosphorus by iron or alum, PoPPt (Fe or Al) – Psot-precipitation of phosphorus by iron or alum, MBR – membrane bioreactor, TF – Trickling filter, RBC – rotating biological contactors, BF – Biological Filters, CS – contact stabilization, PeF – Percolating filters or sand filters, UV– ultraviolet, Oz – ozonation, Cl – chlorination, AC – activated carbon, CMF – continuous microfiltration, RO – reverse osmosis

/ –Analytical methods: L – LC MS, LM– LC-MSMS, E- ELISA, G – GC MS, GM – GC-MSMS, ER – ERBA, ES – E-Screen, Y – YES, LE – LC- ESI-MSMS, H – HPLC, HL – HPLC-LCMS, HR – HRGC-MS, T – TLC, GL - GLC

Table 6.8 Influent and effluent concentrations of various estrogens in municipal wastewater treatment systems (Cont'd)

Ref.	Place	System++*	Operation HRT (hr)	Operation SRT (d)	Anal. Meth./	E1 Infl	E1 Effl	E1 %Rem	17β-E2 Infl	17β-E2 Effl	17β-E2 %Rem	E3 Infl	E3 Effl	E3 %Rem	EE2 Infl	EE2 Effl	EE2 %Rem
Williams et al., 2003	UK	Pr/AS/PPPt	15-18	17	GM	--	0.8-11.2	--	--	<0.1-2	--	--	--	--	--	<1.0-1.9	--
		AS	13.5	13	GM	--	<0.4-2.2	--	--	<0.4-1.7	--	--	--	--	--	<0.4-1.1	--
		PeF	10.5	--	GM	--	3.5-12.2	--	--	<0.4-4.3	--	--	--	--	--	<0.5-3.4	--
Bruchet et al., 2002	France	AS	--	--	GM	20	8	60	10	3	70	--	--	--	2.4	1.4	42
Petrovic et al., 2002	Spain	NR	--	--	HL	<2.5-115	<2.5-8.1	--	11-30.4	<5	--	--	--	--	<5	--	--
		NR	--	--	HL	<2.5-4.6	<2.5-2.7	--	<5	<5-7.6	--	--	--	--	--	--	--
		NR	--	--	HL	<2.5-13.1	<2.5	--	<5	<5	--	--	--	--	--	--	--
		NR	--	--	HL	<2.5-56.5	<2.5-7.2	--	<5-14.5	<5	--	--	--	--	--	--	--
Schullerer et al., 2002	NR	AS/TF/N/P	--	--	G	130	11	92	32	2.7	92	--	--	--	55	7	87
		Pr/AS/N/P/AC	--	--	G	120	nd	>99	35	nd	>97	--	--	--	20	1.8	91
		Pr/AS/N/P/AC	--	--	G	49	nd	>98	31	nd	>97	--	--	--	59	2.2	96
Adler, 2001	Germany		--	--	G	<0.5-20	<0.1-18	87	<0.5-4	<0.05-0.6	70	--	--	--	1.0-14	<0.05-0.6	98
Kuch & Ballschmiter, 2001	Germany	Pr/AS	--	--	HR	--	0.35-18	--	--	0.15-5.2	--	--	--	--	--	0.1-8.9	--
Baronti et al., 2000	Italy	AS	12	--	LE	132	13	90	22	2.9	87	70	3.3	95	2.9	1	66
						69	17	75	8.1	2.2	73	56	7.3	87	0.45	nd	--
						59	6.9	88	25	0.74	97	61	5.7	91	13	0.49	96
						42	5.8	86	9.4	0.55	94	44	1.3	97	1.6	nd	--
				--		53	5.4	90	16	1	94	188	1.1	99	1.7	0.44	74

+ Treating municipal wastewater unless indicated, [1] 85% domestic; 18% Industrial; 2% livestock [2] 70% Domestic + 30% industrial [3] 90% Domestic + 10% industrial [Industrial]

* Pr – Primary, AS– Activated Sludge, AS(Ex) – Extended Aeration Activated Sludge, Lag – Lagoons, Lag(Aer) – Lagoons (aerated), Lag(Fac) – Lagoons (Facultative), Lag(Ana) – Lagoons (Anaerobic), SBR – Sequencing batch reactor, ASBR – Anaerobic Sequencing Batch Reactor, HRO – High Rate Oxygen, N – nitrogen removal, P – Phosphorus removal, PePh (Fe or Al) – pre-precipitation of phosphorus by iron or alum, PoPh (Fe or Al) – Post-precipitation of phosphorus by iron or alum, MBR – membrane bioreactor, TF – Trickling filter, RBC – rotating biological contactors, BF – Biological Filters, CS – contact stabilization, PeF – Percolating filters or sand filters, UV – ultraviolet, Oz – ozonation, Cl – chlorination, AC – activated carbon, CMF – continuous microfiltration, RO – reverse osmosis

/ – Analytical methods: L – LC MS, LM – LC-MSMS, E- ELISA, G – GC MS, GM – GC-MSMS, ER – ERBA, ES – E-Screen, Y – YES, LE – LC-ESI-MSMS, H – HPLC, HL - HPLC-LCMS, HR – HRGC-MS, T – TLC, GL - GLC

Table 6.8 Influent and effluent concentrations of various estrogens in municipal wastewater treatment systems (Cont'd)

Ref.	Place	System[*][+]	HRT (hr)	SRT (d)	Anal. Meth.[']	E1 Infl	E1 Effl	E1 % Rem	17β-E2 Infl	17β-E2 Effl	17β-E2 % Rem	E3 Infl	E3 Effl	E3 % Rem	EE2 Infl	EE2 Effl	EE2 % Rem
Baronti et al., 2000	Italy	AS	14		LE	53	9.7	82	8.6	0.82	90	46	0.64	99	1.8	0.36	80
						59	8	86	6.3	0.72	89	43	0.43	99	0.4	nd	--
						38	3.7	90	11	0.62	94	60	0.83	99	4.6	0.35	92
						68	6.9	90	9.6	0.81	92	33	0.84	97	2.9	0.73	75
	Italy	AS	12		LE	34	10	71	11	0.8	93	146	1.4	99	1.7	0.31	82
						87	6.5	93	6.6	2.1	68	52	1.6	97	3.6	1.7	53
						58	2.5	96	6.4	0.56	91	78	2.2	97	0.44	nd	--
						75	3.7	95	16	0.35	98	71	0.57	99	6.8	0.3	96
						74	4.3	94	4	0.41	90	26	0.44	98	3.9	0.31	92
		AS	14		LE	41	3.3	92	13	1.2	91	127	0.94	99	2.2	0.42	81
						49	10	80	11	0.93	92	69	1.1	98	2.8	0.33	88
						30	6.4	79	6.3	0.44	93	75	0.72	99	0.46	nd	--
						33	6.4	81	14	0.91	94	70	1.7	98	2.4	0.56	77
						30	6.6	78	12	0.74	94	35	1	97	6.8	0.53	92
		AS	12		LE	42	40	5	14	1.9	86	148	8.4	94	2.3	0.5	78
						42	51	--	10	3.1	69	53	11	79	2.9	1.2	59
						28	30	--	4.7	1.9	60	42	6.7	84	0.43	nd	--
						25	22	12	8.3	1.6	81	26	5.8	78	3.3	0.51	85
						48	8.7	82	10	0.53	95	24	1.8	93	6.1	0.52	91
						33	40		10	2.3	77	127	18	86	2	0.4	80

+ Treating municipal wastewater unless indicated, [1] 85% domestic; 18% Industrial; 2% livestock [2] 70% Domestic + 30% industrial [3] 90% Domestic + 10% industrial [4] industrial

* Pr – Primary, AS- Activated Sludge, AS(Ex) – Extended Aeration Activated Sludge, Lag – Lagoons, Lag(Aer) – Lagoons (aerated), Lag(Fac) – Lagoons (Facultative), Lag(Ana) – Lagoons (Anaerobic), SBR – Sequencing batch reactor,ASBR – Anaerobic Sequencing Batch Reactor, HRO – High Rate Oxygen, N – nitrogen removal, P – Phosphorus removal, PePPt (Fe or Al) – pre-precipitation of phosphorus by iron or alum, PoPPt (Fe or Al) – Psot-precipitation of phosphorus by iron or alum, MBR – membrane bioreactor, TF – Trickling filter, RBC – rotating biological contactors, BF – Biological Filters, CS – contact stabilization, PeF – Percolating filters or sand filters, UV – ultraviolet, Oz – ozonation, Cl – chlorination, AC – activated carbon, CMF – continuous microfiltration, RO – reverse osmosis

' -Analytical methods: L – LC MS, LM– LC-MSMS, E- ELISA, G – GC MS, GM – GC-MSMS, ER – ERBA, ES – E-Screen, Y – YES, LE – LC- ESI-MSMS, H – HPLC, HL – HPLC-LCMS, HR – HRGC-MS, T – TLC, GL – GLC

Table 6.8 Influent and effluent concentrations of various estrogens in municipal wastewater treatment systems (Cont'd)

Ref.	Place	System**	Operation		Anal. Meth.'	E1			17β-E2			E3			EE2		
			HRT (hr)	SRT (d)		Infl	Effl	% Rem	Infl	Effl	% Rem	Infl	Effl	% Rem	Infl	Effl	% Rem
Baronti et al., 2000	Italy	AS	14	--	LE	67	82.1	--	18	3.3	82	130	1.4	99	2.4	1.1	54
						33	13	61	5.9	0.72	88	137	0.63	100	0.52	nd	--
						54	46	15	19	3	84	135	1.5	99	3.6	0.47	87
						44	35	20	8.5	1.7	80	54	0.74	99	4.8	0.8	83
						55	47	15	22	3.5	84	187	1.1	99	1.1	nd	--
Johnson et al., 2000	Italy	Not reported	--	--	L	13	<0.5	96.15	10	<0.5	95.00	2	<0.5	75	10	<0.5	95.00
						75	--	--	7	--	--	35	18	48.57	4.1	--	--
	Italy	Not reported	--	--	L	<0.5	31	--	<0.5	3	--	120	--	--	6	0.6	90.00
						11	54	--	20	6	70	70	4	94.29	<0.5	<0.5	0
						32	--	--	5.3	--	--	33	5	84.85	2.2	--	--
	Italy	Not reported	--	--	L	3	2	33.33	6	4	33.33	60	--	--	<0.5	<0.5	0
						50	3	94	17	7	58.82	76	11	85.53	3	2.2	26.67
						71	--	--	16	--	--	43	28	34.88	5.4	--	--
	Italy	Not reported	--	--	L	33	11	66.67	10	3	70.00	102	--	--	<0.5	<0.5	0
						14	19	--	4	2	50	52	7	86.54	<0.5	<0.5	0
						38	--	--	9	--	--	57	20	64.91	3.2	--	--
	Italy	Not reported	--	--	L	18	20	--	8	3	62.50	22	--	--	<0.5	<0.5	0
						13	52	--	<0.5	4	--	--	--	--	<0.5	<0.5	0
						32	--	--	4	--	--	--	--	--	4.8	--	--

+ Treating municipal wastewater unless indicated, [1] 85% domestic; 18% Industrial; 2% livestock [2] 70% Domestic + 30% industrial [3] 90% Domestic + 10% industrial [4] industrial

* Pr – Primary, AS– Activated Sludge, AS(Ex) – Extended Aeration Activated Sludge, Lag – Lagoons, Lag(Aer) – Lagoons (aerated), Lag(Fac) – Lagoons (Facultative), Lag(Ana) – Lagoons (Anaerobic), SBR – Sequencing batch reactor, ASBR – Anaerobic Sequencing Batch Reactor, HRO – High Rate Oxygen, N – nitrogen removal, P – Phosphorus removal, PePPt (Fe or Al) – pre-precipitation of phosphorus by iron or alum, PoPPt (Fe or Al) – Psot-precipitation of phosphorus by iron or alum, MBR – membrane bioreactor, TF – Trickling filter, RBC – rotating biological contactors, BF – Biological Filters, CS – contact stabilization, PeF – Percolating filters or sand filters, UV – ultraviolet, Oz – ozonation, Cl – chlorination, AC – activated carbon, CMF – continuous microfiltration, RO – reverse osmosis

' -Analytical methods: L – LC MS, LM – LC-MSMS, E– ELISA, G – GC MS, GM – GC-MSMS, ER – ERBA, ES – E-Screen, Y – YES, LE – LC- ESI-MSMS, H – HPLC, HL – HPLC-LCMS, HR – HRGC-MS, T – TLC, GL - GLC

Table 6.8 Influent and effluent concentrations of various estrogens in municipal wastewater treatment systems (Cont'd)

Ref.	Place	System**	Operation HRT (hr)	Operation SRT (d)	Anal. Meth.	E1 Infl	E1 Effl	E1 % Rem	17β-E2 Infl	17β-E2 Effl	17β-E2 % Rem	E3 Infl	E3 Effl	E3 % Rem	EE2 Infl	EE2 Effl	EE2 % Rem
Johnson et al., 2000	Italy	Not reported	--	--	L	11	2.7	75.45	11	--	--	--	--	--	<0.5	<0.5	0
	Italy	Not reported	--	--	L	42	15	64.29	14	1.1	92.14	--	--	--	<1.4	<1.4	0
			--	--		18	<0.4	97.78	--	--	--	--	--	--	<0.2	<0.2	0
	Italy	Not reported	--	--	L	100	6.3	93.7	31	0.7	97.74	--	--	--	<1.4	<1.8	--
			--	--		87	2.1	97.59	9	<0.6	93.33	--	--	--	8.8	<0.2	97.73
			--	--		140	47	66.43	48	12	75	--	--	--	1.3	<0.3	76.92
Johnson et al., 2000	Sweden	Not reported	--	--	L	--	6	--	15	1	--	--	--	--	--	4.5	--
	Denmark	Not reported	--	--	L	27	25	7.41	15	5.4	64	--	--	--	2.4	2.4	0
	Brazil	Not reported	--	--	L	40	6.8	83	21	<0.4	98.10	--	--	--	4.2	0.9	78.57
	UK	Not reported	--	--	L	--	2-4	--	--	3-6	--	--	--	--	--	nd	--
	UK	Not reported	--	--	L	--	2-13	--	--	4-12	--	--	--	--	--	nd	--
Belfroid et al., 1999	Holland	AS			H, GM	--	2.7	--	--	na	--	--	--	--	--	<1.4	--
						--	15	--	--	1.1	--	--	--	--	--	<0.2	--
		AS			H, GM	--	<0.4	--	--	na	--	--	--	--	--	<1.8	--
						--	6.3	--	--	0.7	--	--	--	--	--	<0.2	--
		AS			H, GM	--	2.1	--	--	<0.6	--	--	--	--	--	<0.3	--
						--	47	--	--	1.2	--	--	--	--	--	7.5	--
		AS[4]				--	11	--	--	<0.6	--	--	--	--	--	<1.8	--
		AS[4]			H, GM	--	0.7	--	--	1.8	--	--	--	--	--	2.6	--

+ Treating municipal wastewater unless indicated, [1] 85% domestic; 18% Industrial; 2% livestock; [2] 70% Domestic + 30% industrial; [3] 90% Domestic + 10% industrial; [4] industrial

* Pr – Primary, AS- Activated Sludge, AS(Ex) – Extended Aeration Activated Sludge, Lag – Lagoons, Lag(Aer) – Lagoons (aerated), Lag(Fac) – Lagoons (Facultattive), Lag(Ana) – Lagoons (Anaerobic), SBR – Sequencing batch reactor,ASBR – Anaerobic Sequencing Batch Reactor, HRO – High Rate Oxygen, N – nitrogen removal, P – Phosphorus removal, PePPt (Fe or Al) – pre-precipitation of phosphorus by iron or alum, PoPPt (Fe or Al) – Psot-precipitation of phosphorus by iron or alum , MBR – membrane bioreactor, TF – Trickling filter, RbC – rotating biological contactors, BF – Biological Filters, CS – contact stabilization, PeF – Percolating filters or sand filters, UV – ultraviolet, Oz – ozonation, Cl – chlorination, AC – activated carbon, CMF – continuous microfiltration, RO – reverse osmosis

**-Analytical methods: L – LC MS, LM– LC-MSMS, E- ELISA, G – GC MS, GM – GC-MSMS, ER – ERBA, ES – E-Screen, Y – YES, LE – LC- ESI-MSMS, H – HPLC, HL – HPLC-LCMS, HR – HRGC-MS, T – TLC, GL - GLC

Table 6.8 Influent and effluent concentrations of various estrogens in municipal wastewater treatment systems (Cont'd)

Ref.	Place	System*,+	Operation HRT (hr)	Operation SRT (d)	Anal. Meth.[	E1 Infl	E1 Effl	E1 % Rem	17β-E2 Infl	17β-E2 Effl	17β-E2 % Rem	E3 Infl	E3 Effl	E3 % Rem	EE2 Infl	EE2 Effl	EE2 % Rem
Belfroid et al., 1999	Holland	AS[4]	--	--	H, GM	--	<0.4	--	--	<0.7	--	--	--	--	--	<0.3	--
Larsson et al., 1999	Sweden	Pr/Bio	--	--	G, E	--	<0.1	--	--	<0.4	--	--	--	--	--	<0.2	--
Ternes et al., 1999a	Brazil	Pr/AS/TF	--	--	GM	--	5.8	--	--	1.1	--	--	--	--	--	--	78
	German y	Pr/AS/PePPt(Fe	--	--	GM	40	9-70	83	21	<0.001-3	99.9	--	--	--	--	--	--
	Canada	Pr/AS/PePPt(Al)	--	--	GM	27	10-48	67	15	14-64	>99.9	--	--	--	--	4-15	--
Deshrow et al., 1998	England	Pr	--	--	G	--	32-48	--	--	29-48	--	--	--	--	--	29-42	--
		PeF	--	--	G	--	5.2-8.9	--	--	3.7-7.1	--	--	--	--	--	nd-7	--
		AS	--	--	G	--	1.8-3.6	--	--	2.7-6.3	--	--	--	--	--	nd	--
		AS	--	--	G	--	2-13	--	--	4.3-12	--	--	--	--	--	nd	--
		Pr/BF	--	--	G	--	15-76	--	--	6.5-10	--	--	--	--	--	0.6-4.3	--
		BF	--	--	G	--	6.1-12	--	--	4-5.7	--	--	--	--	--	0.2-0.8	--
		AS (Ex)	--	--	G	--	1.4-9.9	--	--	6.1-7.4	--	--	--	--	--	nd	--
Lee & Peart, 1998	Canada	AS	--	--	G	26±1	6±0.5	77±2	7±0.6	<5	>29-6	--	--	--	--	--	--
		AS	--	--	G	53±4	8±1	85±2	14±2	--	>64±5	--	--	--	--	--	--
		AS	--	--	G	69±1	9±1	87±1	7±0.7	--	>29-7	--	--	--	--	--	--
		AS	--	--	G	109±5	72±2	34±4	<5	--	nr	--	--	--	--	--	--
		AS	--	--	G	66.5±12	17±1	74±5	<5	--	nr	--	--	--	--	--	--
		AS	--	--	G	41±4	14±1	66	15±2	--	67±4	--	--	--	--	--	--

+ Treating municipal wastewater unless indicated. [1] 85% domestic; 18% Industrial; 2% livestock [2] 70% Domestic + 30% industrial [3] 90% Domestic + 10% industrial [4] industrial

* Pr – Primary, AS– Activated Sludge, AS(Ex) – Extended Aeration Activated Sludge, Lag – Lagoons, Lag(Aer) – Lagoons (aerated), Lag(Fac) – Lagoons (Facultative), Lag(Ana) – Lagoons (Anaerobic), SBR – Sequencing batch reactor,ASBR – Anaerobic Sequencing Batch Reactor, HRO – High Rate Oxygen, N – nitrogen removal, P – Phosphorus removal, PePt (Fe or Al) – pre-precipitation of phosphorus by iron or alum, PoPPt (Fe or Al) – Psot-precipitation of phosphorus by iron or alum , MBR – membrane bioreactor, TF – Trickling filter, RBC – rotating biological contactors, BF – Biological Filters, CS – contact stabilization, PeF – Percolating filters or sand filters, UV – ultraviolet, Oz – ozonation, Cl – chlorination, AC – activated carbon, CMF – continuous microfiltration, RO – reverse osmosis

[Analytical methods: L – LC MS, LM – LC-MSMS, E- ELISA, G – GC MS, GM – GC-MSMS, ER – ERBA, ES – E-Screen, Y – YES, LE – LC- ESI-MSMS, H – HPLC, HL – HPLC-LCMS, HR – HRGC-MS, T – TLC, GL - GLC

Table 6.8 Influent and effluent concentrations of various estrogens in municipal wastewater treatment systems (Cont'd)

Ref.	Place	System[*]	Operation HRT (hr)	SRT (d)	Anal. Meth.[†]	E1 Infl	E1 Effl	E1 % Rem	17β-E2 Infl	17β-E2 Effl	17β-E2 % Rem	E3 Infl	E3 Effl	E3 % Rem	EE2 Infl	EE2 Effl	EE2 % Rem
Tabak et al., 1981	USA	TF	--	--	T, GL	20±20	10±7	50±61	10	<10	--	--	--	--	1070±10	670±590	37±73
		AS	--	--	T, GL	10±7	10	0±70	10	<10	--	--	--	--	890±630	530±430	40±64
		NR	--	--	T, GL	20±14	10	50±35	10	<10	--	--	--	--	880±520	540±370	39±56
		TF	--	--	T, GL	40±20	20±14	50±43	10±7	10	--	--	--	--	1330±610	910±630	32±57
		CS	--	--	T, GL	40±20	20±7	50±31	20±7	10	50±18	--	--	--	1290±710	920±580	29±60
		TF	--	--	T, GL	30±20	20±14	33±64	10±7	10	--	--	--	--	1000±640	660±470	34±63
		NR	--	--	T, GL	30±20	10±7	67±32	10±7	10	--	--	--	--	990±520	630±400	36±52
		Pr	--	--	T, GL	50±30	30±14	40±46	20±7	10±7	50±39	--	--	--	1770±710	1320±640	25± 47
		CS	--	--	T, GL	40±14	10±14	75±36	10±7	10	--	--	--	--	1480±640	1040±530	30± 47
		Pr	--	--	T, GL	50±20	30±14	40±37	20±7	10	50±18	--	--	--	1590±750	1160±660	27± 54
		AS	--	--	T, GL	20±14	10±7	50±50	10	<10	--	--	--	--	1270±400	780±320	39± 32
		TF	--	--	T, GL	30±20	20±7	33±50	10±7	10	--	--	--	--	1000±690	600±490	40± 64

+ Treating municipal wastewater unless indicated, [1] 85% domestic; 18% Industrial; 2% livestock [2] 70% Domestic + 30% industrial [3] 90% Domestic + 10% industrial [4] industrial

* Pr – Primary, AS– Activated Sludge, AS(Ex) – Extended Aeration Activated Sludge, Lag – Lagoons, Lag(Aer) – Lagoons (aerated), Lag(Fac) – Lagoons (Facultative), Lag(Ana) – Lagoons (Anaerobic), SBR – Sequencing batch reactor,ASBR – Anaerobic Sequencing Batch Reactor, HRO – High Rate Oxygen, N – nitrogen removal, P – Phosphorus removal, PePPt (Fe or Al) – pre-precipitation of phosphorus by iron or alum, PoPPt (Fe or Al) – Psot-precipitation of phosphorus by iron or alum, MBR – membrane bioreactor, TF – Trickling filter, RBC – rotating biological contactors, BF – Biological Filters, CS – contact stabilization, PeF – Percolating filters or sand filters, UV – ultraviolet, Oz – ozonation, Cl – chlorination, AC – activated carbon, CMF – continuous microfiltration, RO – reverse osmosis

'-Analytical methods: L – LC MS, LM– LC-MSMS, E- ELISA, G – GC MS, GM – GC-MSMS, H – HPLC, HL – HPLC-LCMS, HR – HRGC-MS, T – TLC, GL - GLC

Estrogenic potentials in raw municipal wastewaters ranged from 1.1 to 150 ng/L with an average value of 28.9 ± 29.6 ng E2 equivalent/L and were similar for different countries. For example, the influent estrogenic potentials of a municipal wastewater in Australia (Leusch et al, 2005) ranged from 20 - 54 ng/L E2 equivalent while influents of four municipal wastewaters in France contained 46 - 63 ng/L E2 equivalent (Cargouët et al., 2004).

Table 6.9 Effluent concentrations of 17α-E2 in wastewater treatment systems

Ref.	Place	System[+]	Operation		Anal. Method [*]	17α-E2
			HRT (hr)	SRT (d)		Effl
Kuch & Ballschmiter, 2001	Germany	Pr/AS	- -	- -	HR	0.15-4.5
Belfroid et al., 1999	Holland	AS	- -	- -	H, GM	<0.1 - 1.3
		AS	- -	- -	H, GM	1.2 -1.7
		AS	- -	- -	H, GM	<0.1 - 5
		AS[4]	- -	- -	H, GM	<0.5 – 2.1
		AS[4]	- -	- -	H, GM	<0.1

+ Pr – primary treatment, AS – Activated sludge
*GM – GC-MSMS, H – HPLC, HR – HRGC-MS,

Table 6.10 Estrogenic potentials and removals in wastewater treatment systems

Ref.	Place	System[+][*]	Operation HRT (hr)	Operation SRT (d)	Anal. Meth.[/]	EEQ Infl	EEQ Effl	EEQ % Rem
Kim et al., 2007	South Korea	Pr/AS/RBC[1]	--	--	LM	--	--	--
Leusch et al., 2005	Australia	Pr/ASBR/N/SF/Oz/A C/UV	--	--	ER, ES	20-54	<0.75	>95
Cargouët et al., 2004	France	Pr/AS/N[2]	--	--	G	57 ± 9	2 ± 1	97
		Pr/AS/N	10	--	G	56 ± 4	13 ± 3	77
		Pr/UpBF/N[3]	2.5-4	--	G	46 ± 7	9 ± 3	81
		Pr/AS	2-3	--	G	63 ± 11	24 ± 2	62
Lagana et al., 2004	Italy	Pr/AS	--	--	G	--	--	--
Pawlowski et al., 2004	Germany	AS/N/P	--	--	GM, Y	--	34.1-65.96	--
Xie et al., 2004	USA	Nr/Nr/P	144	--	Y	2.57	<1	61
		Nr/N/Ter/Cl	--	--	Y	--	<1	--
Svenson et al., 2003	Sweden	Pr/PoPPt(Al)	--	--	Y	11-12.8	11.4-13.4	0
		Pr/PoPPt(Al)	--	--	Y	10.7-10.9	11.5-13.9	0
		Pr/PoPPt(Fe)	--	--	Y	5.3-5.6	5.9±0.3	0
		Pr/PoPPt(Lime)	--	--	Y	3.6-4.7	1.1	73
		Pr/AS/PoPPt(Fe)	2-8	--	Y	29.8	12.1-12.5	59
		Pr/AS/PoPPt(Al)	--	--	Y	4.6-5.4	0.1-0.5	94
		Pr/AS/PPPt(Al)	--	--	Y	110.1-10.3	4.2-4.4	58
		Pr/AS/PPPt(Fe)	--	--	Y	4.5-5.5	1.4-1.8	69
		Pr/AS/BioSorp/ PoPPt(Al)	--	--	Y	5.9-6.2	1.2	81
		Pr/TF/PoPPt(Al)	--	--	Y	20.6-24.1	13.3-16.	33
		Pr/TF/PoPPt(Al)	--	--	Y	2.9-3.2	10.1-11.4	0
		Pr/RBC/PoPPt(Fe)	--	--	Y	1.1-2.1	5.1-5.4	0
		Pr/AS/PePoPPt(Al)	--	--	Y	7.7-8.3	2.3-2.8	68
		Pr/TF/PePPt(Al)	--	--	Y	6.6-6.9	1.7	75
		Pr/AS/TF/ PePoPPt(Al)	12	--	Y	6.1-7.8	<0.1-0.1	99
		Pr/AS/PPoPt(Fe)	--	--	Y	12.1-12.9	1.4-1.5	89

+ Treating municipal wastewater unless indicated, [1] 85% domestic; 18% Industrial; 2% livestock [2] 70% Domestic + 30% industrial [3] 90% Domestic + 10% industrial [4] industrial

* Pr – Primary, AS- Activated Sludge, AS(Ex) – Extended Aeration Activated Sludge, Lag – Lagoons, Lag(Aer) – Lagoons (aerated), Lag(Fac) – Lagoons (Facultattive), Lag(Ana) – Lagoons (Anaerobic), SBR – Sequencing batch reactor,ASBR – Anaerobic Sequencing Batch Reactor, HRO – High Rate Oxygen, N – nitrogen removal, P – Phosphorus removal, PePPt (Fe or Al) – pre-precipitation of phosphorus by iron or alum, PoPPt (Fe or Al) – Psot-precipitation of phosphorus by iron or alum , MBR – membrane bioreactor, TF – Trickling filter, RBC – rotating biological contactors, BF – Biological Filters, CS – contact stabilization, PeF – Percolating filters or sand filters, UV – ultraviolet, Oz – ozonation, Cl – chlorination, AC – activated carbon, CMF – continuous microfiltration, RO – reverse osmosis

[/] -Analytical methods: L – LC MS, LM – LC-MSMS, E- ELISA, G – GC MS, GM – GC-MSMS, ER – ERBA, ES – E-Screen, Y – YES, LE – LC- ESI-MSMS, H – HPLC, HL - HPLC-LCMS, HR – HRGC-MS, T – TLC, GL - GLC

Table 6.10 Estrogenic potentials and removals in wastewater treatment systems (Cont'd)

Ref.	Place	System++	Operation		Anal. Meth.¹	EEQ		
			HRT (hr)	SRT (d)		Infl	Effl	% Rem
Svenson et al., 2003	Sweden	Pr/AS/N/PoPPt(Al)	20	--	Y	2.8-4.9	<0.1	>97
		Pr/AS/N/Wet/PePPt(Fe)	7d	--	Y	17.9-21.1	<0.1	>99
		Pr/AS/TF/N/PePoPPt(Al)	--	--	Y	na	5.0-7.7	--
		Pr/AS/TF/N/PoPPt(Al)	--	--	Y	na	1.9-3.8	--
Holbrook et al., 2002	USA	Pr/MBR/PePPt(Al)	8.5	20-25	Y	18.5	5	--
		Pr/AS/PePPt(Fe)	10	20-25	Y	17.8	7.2	--
Kirk et al., 2002	UK	Pr/AS	14	8-12	Y	23.9	10.6	--
		Pr	2-6	--	Y	43	40	7
						17	14	18
		Bio/UV	4	--	Y	75	8	89
						22	7	68
		Pr/BF	13.5	--	Y	38	nd	>97
						10	2	80
		Pr	13	--	Y	33	nd	>97
						27	1	96
		Pr/AS	13	--	Y	77	nd	>98
						40	7	83
Kirk et al., 2002	UK	Bio	--	--	NA	--	nd-13	--
		Pr	--	--	NA	--	40	--
Bolz et al., 2002	Germany	Pr/AS/N/P/AC	--	--	--	13	0.19	99
		Pr/AS/N/P/AC	--	--	ES	82	0.74	99
		Pr/AS/TF/N/P	--	--	ES	22	1.6	93
		AS/N/P	--	--	ES	16	1.3	92
		Pr/AS/N/P	--	--	ES	38	1.9	95
		TF/N/P	--	--	ES	46	2.6	94
Körner et al., 2000	Germany	Pr/AS/N/P	--	--	ES	51-66	6.4	89

+ Treating municipal wastewater unless indicated, [1] 85% domestic; 18% Industrial; 2% livestock [2] 70% Domestic + 30% industrial [3] 90% Domestic + 10% industrial [4] industrial

* Pr – Primary, AS- Activated Sludge, AS(Ex) – Extended Aeration Activated Sludge, Lag – Lagoons, Lag(Aer) – Lagoons (aerated), Lag(Fac) – Lagoons (Facultattive), Lag(Ana) – Lagoons (Anaerobic), SBR – Sequencing batch reactor,ASBR – Anaerobic Sequencing Batch Reactor, HRO – High Rate Oxygen, N – nitrogen removal, P – Phosphorus removal, PePPt (Fe or Al) – pre-precipitation of phosphorus by iron or alum, PoPPt (Fe or Al) – Psot-precipitation of phosphorus by iron or alum , MBR – membrane bioreactor, TF – Trickling filter, RBC – rotating biological contactors, BF – Biological Filters, CS – contact stabilization, PeF – Percolating filters or sand filters, UV – ultraviolet, Oz – ozonation, Cl – chlorination, AC – activated carbon, CMF – continuous microfiltration, RO – reverse osmosis

[1]-Analytical methods: L – LC MS, LM– LC-MSMS, E- ELISA, G – GC MS, GM – GC-MSMS, ER – ERBA, ES – E-Screen, Y – YES, LE – LC- ESI-MSMS, H – HPLC, HL - HPLC-LCMS, HR – HRGC-MS, T – TLC, GL – GLC

Table 6.10 Estrogenic potentials and removals in wastewater treatment systems (Cont'd)

Ref.	Place	System[++]	Operation HRT (hr)	Operation SRT (d)	Anal. Meth.[/]	EEQ Infl	EEQ Effl	EEQ % Rem
Körner et al., 2000	Germany	Pr/AS/N/P	--	--	ES	51-66	5.4-7) 6.3	89.23
		Pr/AS/N/PePPt(Fe)	--	--	ES	--	2.5-4.1	--
		Pr/AS/N/PePPt(Al)	--	--	ES	--	2.8	--
		Pr/AS/N/PePPt(Fe)	--	--	ES	--	22±1.2	--
			--	--		--	25±8.7	--
			--	--		--	7.8±1.5	--
		Pr/AS/N/PePPt(Fe)	--	--	ES	--	12.5±2.7	--
			--	--		--	21.1±4.3	--
Matsui et al., 2000	Japan	Pr/N/Cl	--	--	Y	150	10	93
			--	--	E	50	10	80

+ Treating municipal wastewater unless indicated, [1] 85% domestic; 18% Industrial; 2% livestock [2] 70% Domestic + 30% industrial [3] 90% Domestic + 10% industrial [4] industrial
* Pr – Primary, AS- Activated Sludge, AS(Ex) – Extended Aeration Activated Sludge, Lag – Lagoons, Lag(Aer) – Lagoons (aerated), Lag(Fac) – Lagoons (Facultattive), Lag(Ana) – Lagoons (Anaerobic), SBR – Sequencing batch reactor,ASBR – Anaerobic Sequencing Batch Reactor, HRO – High Rate Oxygen, N – nitrogen removal, P – Phosphorus removal, PePPt (Fe or Al) – pre-precipitation of phosphorus by iron or alum, PoPPt (Fe or Al) – Psot-precipitation of phosphorus by iron or alum , MBR – membrane bioreactor, TF – Trickling filter, RBC – rotating biological contactors, BF – Biological Filters, CS – contact stabilization, PeF – Percolating filters or sand filters, UV – ultraviolet, Oz – ozonation, Cl – chlorination, AC – activated carbon, CMF – continuous microfiltration, RO – reverse osmosis
[/]-Analytical methods: L – LC MS, LM– LC-MSMS, E- ELISA, G – GC MS, GM – GC-MSMS, ER – ERBA, ES – E-Screen, Y – YES, LE – LC- ESI-MSMS, H – HPLC, HL - HPLC-LCMS, HR – HRGC-MS, T – TLC, GL – GLC

6.7.2 Percent Removal of Estrogens in Municipal Wastewater Treatment Systems

Overall, the percent removals of estrogens in municipal wastewater treatment systems were highly variable depending on the biological processes of the treatment plant system. Removal percentages ranged from 0 to > 99% for E1, 0 to > 99% for 17β-E2, 40 - 100% for E3, and 38 - 96% for EE2. Based on the data presented in Table 6.8, the average removal percentages were 69 ± 29% for E1, 77 ± 26% for 17β-E2, 78 ± 23% for E3, and 70 ± 21% for EE2. Similarly, the estrogenic potentials removed by municipal wastewater treatment systems ranged from 0 to > 99% (Table 6.10) with an average percentage of 69 ± 34%. Similarly, removal of estrogenic potentials varied greatly from country to country. For example, Leusch et al. (2005) reported more than 95% removal of estrogenic potential in an Australian municipal treatment plant while Servos et al. (2005) reported a reduction in estrogenic potential of 47% in a Swedish municipal treatment plant. An average removal of 79.3% was reported by Cargouët et al. (2004) for four domestic wastewater treatment plants in France. Murk et al. (2002) reported 90 - 95% reduction of estrogenic potentials in four municipal wastewater plants in Netherlands. Several mechanisms, including deconjugation, sorption, biodegradation, precipitation are involved in estrogen removals in municipal wastewater treatment systems.

Estrogens in the effluent of wastewater treatment plants contained between nd - 82 ng/l of E1, nd - 64 ng/l of 17β-E2, nd - 18 ng/l of E3, and nd - 42 ng/l of EE2 with average concentrations of 12.9 ± 15.3, 6.2 ± 8.2, 5.0 ± 4.4, and 3.8 ± 6.1 ng/L, respectively (Table 6.8). One observation is that the average concentration of E1 in the effluent was about two to three times higher than other estrogens. A probable reason is that E1 is a product of the biodegradation of 17β-E2 by microorganisms in municipal wastewater treatment systems (Ternes et al., 1999 a). Servos et al. (2005) suggested that although removal of E1 was as high as 98%, in several plants in Canadian wastewater treatment plants, the concentrations of E1 in the final effluents were elevated and were above the influent concentrations. A study by Baronti et al., 2000 using six Rome municipal wastewater treatment plants indicated that 4 out of 30 test events during the 5-month study have effluent E1 concentrations higher than the influents. In some cases the levels of estrogens in the effluents were above known effective levels for aquatic organisms.

The estrogenic potentials of the effluents were reported to be between nd and 40 ng/l EEQ with an average value of 6.0 ± 6.7 ng/L (Table 6.10). For example, Leusch et al. (2005) reported that a municipal wastewater treatment plant in Australia reduced the estrogenic potentials to a level less than < 0.75 ng/L EEQ. Average level in the effluents of four French domestic wastewater treatment plants was 12 ng/L EEQ (Cargouët et al., 2004) while Kirk et al. (2002) reported estrogenic potentials between nd - 40 ng/l EEQ in the effluents of several wastewater treatment plants in UK.

6.7.2 Deconjugation of Estrogens in Municipal Wastewaters

Andersen et al. (2003) found that both glucuronide and sulfate estrogens in a municipal wastewater were cleaved in the first denitrification tank of an anoxic-aerobic (AO) system comprising of two denitrification tanks and a nitrification tank. Work done by D'Ascenzo et al. (2003) suggested that an activated sludge wastewater treatment system can completely remove all conjugated estrogens except for E1-3G, E1-3S, and E3-3S (see Table 11). A study by Isobe et al. (2003) in Japan found E1-3S and E2-3S in the effluent of wastewater treatment systems at concentrations of 0.3 - 2.2 and 1.0 ng/L, respectively. Work done by Belfroid et al. (1999) showed that no glucuronide estrogens were found above the limit of detection in most effluents of the wastewater treatment plants tested. In batch studies, conjugated estrogen (E2-17G) was cleaved by activated sludge from a sewage treatment system under aerobic conditions (Ternes et al., 1999 a).

6.7.4 Impact of Treatment Processes

Based on Table 6.8, several observations can be made with respect to the impact of different biological processes and configurations on estrogen removal. Treatment plants with secondary (biological) treatment systems tend to remove a larger amount of estrogens that those without secondary treatment. For example, Braga et al. (2005) reported that removal of 7%, 0%, and 0 % of E1, 17β-E2, and EE2, respectively, for an enhanced primary waste treatment plant whereas an advanced wastewater treatment plant with secondary treatment resulted in < 0.1 ng/L for all estrogens in the effluent. In a study of 20 Swedish municipal STP by Svenson et al. (2003), higher removals of estrogens were found in plants with secondary biological treatment systems than in the plants without secondary biological treatment systems. Kirk et al. (2002) showed that most estrogenicities in 5 wastewater treatment plants in UK were removed in secondary treatment systems with additional removal associated with tertiary treatment systems. Leusch et al. (2005) studied the estrogenic profile along each unit of an advanced biological nutrient removal plant in Australia and found that activated sludge followed by nitrification/denitrification removed > 95% of the estrogenic activity (to < 0.75 - 2.6 ng/L).

Table 6.11 Occurrence of conjugated estrogens in septic tanks and in wastewater treatment systems (adapted from D'Ascenzo et al., 2003).

Estrogen	Influent of wastewater treatment system (ng/L)	Effluent of wastewater treatment system (ng/L)
E1-3G	4.3	0.7
E1-3S	25	9
E2-3G	5.2	nd
E2-17G	nd	nd
E2-3S	3.3	nd
E3-3G	n.d	nd
E3-16G	19	nd
E3-3S	14	2.2
E1	44	2.3
17β-E2	11	1.6
E3	72	17

n.d. - not detect; LOQ - limit of quantity

Suspended growth seemed to have better removal than attached growth systems as reported by Servos et al. (2005) who suggested that activated sludge systems tended to be more effective in removing estrogenic potentials (58 to >99% with an average of 81%) than attached growth systems (0 - 75% with 28% average) such as trickling filters and rotating biological contactors. Korner et al. (2000) showed that the effluent from trickling filters had higher estrogen concentrations than from activated sludge plants.

When removal percentages of aerated lagoons were compared with activated sludge systems, contradictory results were obtained. Servos et al. (2005) reported that aerated lagoons were effective in removing 17β-E2 (80-98%) whereas Hintemann et al. (2006) reported relatively high 17β-E2 concentrations (15 - 51 ng/L) and 1.8 - 3.1. ng/L in the effluent. Possible reasons are that the aerated lagoons were non-optimized operations and may not have nitrification (Hintemann et al., 2006).

Wastewater treatment plants with long HRTs and SRTs have relatively high removal of E1, 17β-E2 and estrogenic potentials, while treatment plants with short HRTs and SRTs tended to have more variability and lower removal. However, there were no significant statistical correlations between the treatment plant HRTs or SRTs and estrogens or estrogenic potentials removal (Svenson et al., 2003).

Wastewater treatment plants that nitrify tend to remove a higher amount of estrogens as well as estrogenic potentials than those without nitrification (Servos et al., 2005, Svenson et al., 2003, Layton et al., 2000). Andersen et al. (2003) found that EE2 was removed only in the nitrifying tank of a German municipal wastewater treatment plant when the treatment plant was upgraded to include nitrification with an SRT of 11 - 13 days. Reasons for higher estrogen removal in processes with nitrification are: longer SRTs for the nitrification process resulting in a more diverse heterotrophic bacteria community, availability of bacteria that can degrade estrogens and the ability of nitrifying bacteria, ammonia-oxidizing bacteria (AOB) and nitrite-oxidizing bacteria (NOB) to degrade organic compounds via cometabolism during uptake of ammonia for energy (Sermwaraphan et al., 2007).

EE2 was found to be degraded by nitrifying activated sludge during ammonia oxidation but no degradation of EE2 was detected using sludge without nitrifying capacity (Vader et al., 2000). By using allythiourea, an inhibitor for ammonia monooxygenase, Shi et al. (2004) confirmed that AOB in activated sludge, pre-cultivated with mineral-salt medium containing 1,400 mg N/L, played an important role in degrading EE2. Shi et al. (2004) also found that *Nitrosomonas europaea* were capable of cometabolizing EE2 during ammonia oxidation with no further degradation of intermediates.

Chemical precipitation using iron and aluminum salts without biological treatment showed little removal efficiency (18%) and no significant reduction of the estrogenicity (0%) while treatment with lime showed more effective removal in estrogenicity (73%) (Svenson et al., 2003).

6.7.5 Mechanisms of Estrogen Removal in Wastewater Treatment Systems

6.7.5.1 Sorption

The two likely removal mechanisms for estrogens in biological wastewater systems are sorption onto biomass and biodegradation. The log octanol-water partition coefficients (log K_{ow}) of estrogens are between 2.45 and 4.15 (Lee et al., 2003; Lai et al., 2000) suggesting that estrogens are moderately to strongly sorbed to the solid phases (Roger et al., 1996). K_d values and organic carbon normalized sorption coefficient (K_{oc}) values are listed in Table 6.12.

Lai et al. (2000) found that sorption of estrogens increased with an increase in organic content of the sorbent. In their study they found that sorption of estrogens was well correlated (R = 0.86 - 0.94) with the TOC content. Suzuki et al. (2006) found that E1 and 17β-E2 were rapidly sorbed onto sludge floc in the first hour and then were biodegraded within the next 4 hours and sorption of estrogen were reduced at lower temperatures. Using autoclaved sterilized sludge, the sorption of E1 was reduced whereas the sorption of 17β-E2 was the same as with nonsterilized activated sludge. However, Clara et al (2004) found no significant differences in the sorption of 17β-E2 and EE2 onto sterilized and nonsterilized sludge. In addition, Clara et al. (2004) found that at pH of 9, 17β-E2 and EE2 (pKa between 10.4 - 10.7 for 17β-E2 and EE2) were desorbed from the solid phase due to the deprotonated form of the estrogen. This result suggests that 17β-E2 and EE2 can be desorbed from sludge during the dewatering process when the sludge is conditioned with lime. To estimate the sorption of estrogens onto the sludge floc, Matter-Müller et al. (1980) suggested the the following equation

$$K_d = 0.39 + 0.67 * K_{ow}$$

Several studies suggested that sorption onto the sludge floc was not an important mechanism in the removal of estrogens in activated sludge systems (Andersen et al, 2005; Ternes et al, 1999 a, Vader et al, 2000; Lee and Liu, 2002). Braga et al. (2005) using an estrogen mass balances for a sequencing batch reactor showed that only 25% of the E1+17β-E2 mass load was removed or accumulated in the mixed liquor suspended solids. Andersen et al (2005) estimated that the percent of sorbed estrogens removed was only 1.5 - 1.8% of the total estrogens loading in the excess wasted sludge while Holbrook et al. (2002) estimated about 30% of 17β-E2 was removed with excess wasted sludge. Layton et al., 2000 estimated that only 10 - 20% were due to sorption onto biosolids even though the total removal of 17β-E2 from the aqueous phase was over 90%.

Table 6.12 Sorption of estrogens onto sludge

Parameter	E1	17β-E2	EE2	References
log K_{oc}	- -	3.3	3.31	Clara et al., 2004
	- -		2.93	Ternes et al., 2004
	3.16	3.24	3.32	Andersen et al., 2005
log K_d	2.49	3	3.21	Karickhoff et al., 1981
	2.58	2.74	2.79	Karickhoff et al., 1981
	2.69	3.03	3.17	Matter-Muller et al., 1980
	2.6	2.68	2.77	Andersen et al., 2005
	- -	2.84	2.84	Clara et al., 2004
	- -	- -	2.54	Ternes et al., 2004
	- -	2.41	- -	Layton et al., 2000

6.7.5.2 Biodegradation of estrogens

A study by Andersen et al. (2003) on fate of estrogens in each compartment of a German municipal wastewater treatment plant, showed that E1 and 17β-E2 were largely degraded biologically in the denitrifying and nitrifying tanks, while EE2 was degraded only in the nitrifying tank. Layton et al. (2000) showed that in wastewater treatment plant, about 70 - 80% of 17β-E2 was degraded within 24 hours.

Several batch experiments showed that natural estrogens (E1 and 17β-E2) were degradable in contact with activated sludge, whereas synthetic estrogen (EE2) was principally persistent (Weber et al., 2005; Ternes et al., 1999 a). 17β-E2 was oxidized to E1, which was further eliminated without the detection of further degradation products (Ternes et al., 1999 a). 16α-hydroxy-E1, a possible human metabolite of estrone, was rapidly eliminated and the degradation products of this compound could not be identified (Ternes et al., 1999 a).

Information on estrogen degradation under anaerobic conditions is not available for activated sludge. However, Czajka and Londry, 2006 investigated biodegradation of 17β-E2 and EE2 using lake water and sediments under anaerobic conditions (nitrate-reducing, iron-reducing, sulfate-reducing and methanogenic conditions). Degradation of EE2 was not observed over three years in all four anaerobic conditions. In contrast, 17β-E2 was transformed to E1 under all conditions. Other than E1, 17α-E2 was also found as one of the metabolites under iron-reducing, sulfate-reducing, and methanogenic conditions. The extent of conversion of 17β-E2 to E1 differs for each electron acceptor, and the correlation between the extent of transformation and the use of electron acceptors was unclear. Under all four anoxic conditions, the transformation of 17β-E2 would only result in partial reduction of estrogenicity.

6.7.5.3 Microorganisms responsible for degradation of estrogens

Several researchers have isolated microorganisms which are known to degrade 17β-E2. Some of the microorganisms isolated from activated sludge include *Novosphingobium* species ARI-1 (Fujii et al., 2002), *Rhodococcus zopfii* (strain Y 50158) (Yoshimoto et al., 2004), and *Rhodococcus equi* (strains Y 50155, Y 50156, and Y50157) (Yoshimoto et al., 2004). *Novosphingobium* species ARI-1 was found to be capable of degrading E1 and E3 while the *Rhodococcus* strains were found to degrade E1, E3, and EE2. The products of 17β-E2 degradation were not E1, but compounds without estrogenicity. Chao et al. (2004) isolated *Sphingomonas* strain D12 from soils which was also found to be capable of degrading E1. Weber et al. (2005) isolated *Achromobacter xylosoxidans* and *Ralstonia sp.* which can use E1 and 17β-E2 as growth substrates and transform E3 and 16α-hydroxy-E1.

Yu et al., 2007 isolated fourteen phylogenetically diverse 17β-E2-degrading bacteria from activated sludge of a municipal wastewater treatment plant. The fourteen isolates can be classified into 8 different genera: two strains of *Aminobacter*, one strain of *Brevundimonas, Escherichia, Flavobacterium, Microbacterium, Nocardioides,* and *Rhodococcus,* and five strains of *Sphingomonas*. Although all fourteen strains could convert 17β-E2 to E1, three different degradation patterns were elucidated using these strains: stoichiometric conversion of 17β-E2 to E1 without further degradation of E1; conversion of 17β-E2 to E1 at a slower degradation rate than the earlier pattern but further elimination of E1; and with one out of 14 isolates showing rapid degradation of E1 with no residual estrogenic activity.

Although EE2 was found to be persistent in activated sludge (Weber et al., 2005; Ternes et al., 1999a), EE2-degrading microorganisms have been isolated. *Fusarium proliferatum* strain HNS-1, an EE2-degrading fungus, was first isolated from hundreds of sediment, soil, and manure samples (Shi et al., 2002). Haiyan et al. (2007) isolated an EE2-degrading bacterium, *Sphingobacterium sp.* JCR5 from an activated sludge treatment plant treating wastewater from oral contraceptives factory. The bacterium was found to be capable of cultivating on E1, 17β-E2, E3, and methy-EE2. They found that EE2 was first oxidized to E1 followed by 2-hydroxy-2, 4-dienevaleric acid and 2-hydroxy-2, 4-diene-1,6-dioic acid as the main catabolic intermediates.

6.7.5.4 Estrogen in sludge disposal

Sludge from a wastewater treatment plant may act as a sink for estrogens by sorbing the estrogens although as discussed earlier the amount sorbed is in dispute. If the sludge is land disposed and estrogens are not destroyed during sludge stabilization, sludge from wastewater treatment plants can be a source of estrogens. Andersen et al. (2003) found that the estrogen concentrations in digested sludge were approximately 25.2 and 5.1 ng/gm for E1 and 17β-E2, respectively. Similarly, Braga et al (2005) found that the concentrations of E1, 17β-E2 and EE2 in activated sludge were about 11.8, 0.31 and 0.42 ng/g, respectively and about 14.3, 0.57 and 0.61 ng/g in dewatered sludge, respectively. Joss et al (2004) found that the concentrations of E1, 17β-E2 and EE2 sorbed onto sludge were about < 2 ng/g, < 2 - 2.1 ng/g and < 2 - 2.4 ng/g, respectively.

Sorbed estrogens in digested sludge were found to desorb into the dewatering water at a concentration of about 67.1 and 5.4 ng/L which were similar to the concentrations found in the influent of the wastewater (Clara et al, 2004). In a study by Matsui et al (2000), similar high concentrations of estrogen were found in the supernatant from dewatered sewage sludge. Based on the concentrations in digested sludge, Clara et al (2004) suggested that estrogens were probably persistent under methanogenic conditions in the anaerobic digesters.

6.8   Conclusion

Hormones, especially estrogens and androgens, are considered contaminants of emerging environmental concern. They are widely present as they are produced by humans and are synthetically synthesized to treat various health conditions of humans and livestock. Research has been conducted to understand the fate and behavior of estrogens in the environment and in engineered systems. Much progress has been made to improve our understanding. However, there are still many lingering questions on the fate and impact of these compounds on humans, aquatic organisms and animals. Of importance is a need to further standardize the analytical method to measure the estrogens and the estrogenic activities of these compounds. Of concern are the measurements of estrogens in complex matrices such as manure and soils where other large molecule compounds may interfere with the analysis of estrogens.

The biodegradation pathways of estrogens need further elucidation and work is needed to investigate whether estrogens are used as sole carbon and energy source or the mechanism of degradation is due to cometabolism. The debate on whether secondary processes or long SRTs and HRTs and the role of nitrification (ammonia oxidizing bacteria) will improve estrogen removals need to be carefully studied and confirmed. In addition, the role of sorption in the removal of estrogen needs to be better investigated. Not much is known about the fate of estrogens in the biosolids and in the sludge treatment systems. The final fate of estrogens in digested sludge when land applied needs further investigation. The fate of conjugated estrogens also needs further investigation as most of the work are focused on free estrogens. Besides the common potent estrogens discussed above, there are many other estrogens, androgens and synthetic estrogens which needs to be investigated. Finally, the long term effects of these compounds at the nanogram level concentrations on humans and animals are still unknown and should be tested.

References

Adler, P. (2001) "Distribution of natural and synthetic estrogenic steroid hormones in water samples from Southern and Middle Germany." Acta Hydrochi. Hydrobiol. 29(4): 227–241.

Adlercreutz, H.; Brown, J.; Collins, W.; Goebelsmann, U.; Kellie, A.; Campbell, H.; Spieler, J.; Braissand G. (1982) "The measurement of urinary steroid glucuronides as indices of the fertile period in women." J. Steroid Biochem. 17:695-702.

Ainsworth, L.; Common, R.H.; Carter, A.L. (1962) "A chromatographic study of some conversion products of estrone-16-C-14 in the urine and feces of the laying hen." Can. J. Biochem Physiol. 40:23-135.

Andersen, H.; Siegrist, H.; Halling-Sorensen, B.; Ternes, T.A. (2003) "Fate of estrogens in municipal sewage treatment plant." Environ. Sci. Technol. 37(18):4021-4026.

Andersen, H.R.; Hansen, M.; KjØlholt, J.; Sture-Lauridsen, F.; Ternes, T.; Halling-SØrensen, B. (2005) "Assessment of the importance of sorption for steroid estrogens removal during activated sludge treatment." Chemosphere, 61:139-146.

Bamberg, E.; Choi, H.S.; Möstl, E. (1986) "Estrogen determination in feces for the pregnancy diagnosis in the horse, cow, pig, sheep and goats." Tierarztl Umsch. 41: 406–408.

Baronti, C.; Curini, R.; D'Ascenzo, G.; Di Corcia, A.; Gentili, A.; Samperi, R. (2000) "Monitoring natural and synthetic estrogens at activated sludge sewage treatment plants and in receiving river water." Environ. Sci. Technol. 34(24):5059-5066.

Barry, T.P.; Marwah, A.; Marwah, P. (2007) "Stability of 17α-methyltestosterone in fish feed." Aquaculture, 271(1-4):523-529.

Belfroid, A.C.; Van der Horst, A.; Vethaak, A.D.; Shafer, A.J.; Rijs, G.B.J.; Wegener, J.; Cofino, W.P. (1999) "Analysis and occurrence of estrogenic hormones and their glucuronides in surface water and waste water in The Netherlands." Science of the Total Environment, 225:101-108.

Blokland, M.H.; van Rossum, H.J.; Herbold, H.A.; Sterk, S.S.; Stephany, R.W.; van Ginkel, L.A. (2005) "Metabolism of methyltestosterone, norethandrolone and methylboldenone in a heifer." Analytica Chimica Acta, 529:317-323.

Bolz, U.; Kuch, B.; Metzeger, J.W.; Korner, W. (2002) "Input/output balance of total estrogenic activity of sewage treatment plants with different technical equipment applying bioassay." Vom Waser, 98:81–91.

Braga, O.; Smythe, G.A.; Schafer, A.I.; Feitz, A.J. (2005) "Fate of steroid estrogens in Australian inland and coastal wastewater treatment plants." Environ. Sci. Technol. 39:3351-3358.

Braun, P.; Moeder, M.; Schrader., S.; Popp, P.; Kuschk, P.; Engewald, W. (2003) "Trace analysis of technical nonylphenol, bisphenol A and 17α-ethinylestradiol in wastewater using solid-phase microextraction and gas chromatography–mass spectrometry." J. Chromatography A, 988:41-51.

Bruchet, A.; Prompsy, C.; Filippi, G.; Souaki, A. (2002) "A broad spectrum analytical scheme for the screening of endocrine disruptors, pharmaceuticals and personal care products in wastewaters and natural waters." Water Sci. Technol. 46(3):97–104.

Burnison, B.K.; Hartmann, A.; Lister, A.; Servos, M.R.; Ternes, T.; Van Der Kraak, G. (2003) "A toxicity identification evaluation approach to studying estrogenic substances in hog manure and agricultural runoff." Environ Toxicol Chem. 22:2243-2250.

Caccamo, F.; Carfagnini, G.; Di Corcia, A.; Sampert, R. (1998) "Measurement of urinary estriol glucuronides during the menstrual cycle by high-performance liquid chromatography." J. Chromatography, 434:61-70.

Cargouët, M.; Perdiz, D. ; Mouatassim-Souali, A. ; Tamisier-Karolak, S.; Levi, Y. (2004) "Assessment of river contamination by estrogenic compounds in Paris area (France)." Science of the Total Environment, 324:55-66.

Czajka, C.P.; Londry, K.L. (2006) "Anaerobic biotransformation of estrogens." Science of the Total Environment, 367:932-941.

Casey, F.X.M.; Larsen, G.L.; Hakk, H.; Simunek, J. (2003) "Fate and transport of 17beta-estradiol in soil-water systems." Environ. Sci. Technol., 37:2400-2409.

Casey, F.X.M.; Hakk, H.; Simunek, J.; Larsen, G.L. (2004) "Fate and transport of testosterone in agricultural soils." Environ. Sci. Technol. 38:790-798.

Chao, Y.; Kurisu, F.; Saitoh, S.; Yagi, O. (2004) "Degradation of 17β-estradiol by Sphingomonas sp. strain D12 isolated from soil." J. Environmental Biotechnology, 3(2):89-94

Chan, A.H.H.; Common, R.H. (1972) "Four-hour excretion of estrone and estradiol-17β in the urine of the hen." Poultry science, 51:1772-1777.

ChemlDplus Lite, (2008) US National Library of Medicine, 8600 Rockville Pikem bethesda, MD. http://chem.sis.nlm.nih.gov/chemidplus/chemidlite.jsp Accessed on May 15, 2008.

Choi, H. S.; Kiesenhofer, E.; Gantner, H.; Hois, J.; Bamberg, E. (1987) "Pregnancy diagnosis in sows by estimation of oestrogens in blood, urine or faeces." Animal Reprod. Sci. 15:209-216.

Chotisukarn, P.; Ong, S.K.; Limpiyakorn, T. (2008) "Sorption of methyltestosterone to sediments and soils." Pure and Applied Chemistry International Conference (PACCON 2008), Jan 30 - Feb. 1, Bangkok, Thailand.

Clara, M.; Strenn, B.; Saracevic, E.; Kreuzinger, N. (2004) "Adsorption of bisphenol-A, 17β-estradiole and 17α-ethinylestradiole to sewage sludge." Chemosphere, 56:843-851.

Cohen, S.L. (1969) "Removal of substances interfering with the rapid assay of estrogen in pregnancy urine." J. Clin Endocrinol Metab. 29:47-54.

Colucci, M.S.; Bork, H.; Topp, E. (2001) "Persistence of estrogenic hormones in agricultural soils: 17β-estradiol and estrone." J. Environ Qual. 30:2070-2076.

Common, R.H.; Ainsworth, L; Hertelendy, F.; Mathur, R.S. (1965) "The estrone content of hen's urine." Can. J. Biochem. 43:539-547.

Common, R.H.; Mathur, R.S.; Mulay, S.; Henneberry, G.O. (1969) "Distribution patterns of in vivo conversion products of injected estradiol-17-beta-4-14C and estrone-4-14C in the urines of the nonlaying and laying hen." Can. J. Biochem. 47(5):539-545.

D'Ascenzo, G.; Di Corcia, A.; Gentili, A.; Mancini, R.; Mastropasqua, R.; Nazzari, M.; Samperi, R. (2003) "Fate of natural estrogen conjugates in municipal sewage transport and treatment facilities." Science of the Total Environment, 302(1-3):199-209.

Desaulniers, D. M.; Goff, A. K.; Betteridge, K. J.; Rowell, J.; Flood, P. F. (1989) "Reproduction hormone concentrations in faeces during the oestrous cycle andpregnancy in cattle (Bos taturus) andmuskoxen (Ovibos moschatus)." Canadian J. Zoology, 67:1148-1154.

Desbrow, C.; Routledge, E.J.; Brighty, G.C.; Sumpter, J.P.; Waldock, M. (1998) "Identification of estrogenic chemicals in STW effluent 1: Chemical fractionation and in vitro biological screening." Environ. Sci. Technol. 32(11):1549-1558.

Ding, W.H.; Chiang, C.C. (2003) "Derivatization procedures for the detection of estrogenic chemicals by gas chromatography/mass spectrometry." Rapid Communications in Mass Spectrometry, 17:56-63.

Dumasia, M.C. (2003) "In vivo biotransformation of 17 alpha-methyltestosterone in the horse revisited: Identification of 17-hydroxymethyl metabolites in equine urine by capillary gas chromatography/mass spectrometry." Rapid Commun. Mass Spectrometry, 17(4):320-329.

Erb, R.E.; Randel, R.D.; Mellin, T.N.; Estergreen, V.L. Jr. (1968a) "Urinary estrogen excretion rates during pregnancy in the bovine." J. Dairy Sci. 51:416-419.

Erb, R.E.; Gomes, W.R.; Randel, R.D.; Estergreen, V.L. Jr., Frost, O.L. (1968b) "Effect of ovariectomy on concentration of progesterone in blood plasma and urinary estrogen excretion rate in the pregnant bovine." J. Dairy Sci. 51:420-427.

Fine, D.D.; Breidenbach, G.P.; Price, T.L.; Hutchins, S.R. (2003) "Quantification of estrogens in ground water and swine lagoon samples using solid-phase extraction, pentafluorobenzyl/trimethysilyl derivatization and gas chromatography-negative ion chemical ionization tandem mass spectrometry." J. Chromatography A, 1017:167-185.

Finlay-Moore, O.; Hartel, P.G.; Cabrera, M.L. (2000) "17β-estradiol and testosterone in soil and runoff from grasslands amended with broiler litter." J. Environ Qual. 29:1604-1611.

Fotsis, T.; Jarvenpaa, P.; Adlercreutz, H. (1980) "Purification of urine for quantification of the complete steriod profile." J. Steroid Biochem. 12:503-508.

Fujii, K.; Kikuchi, S.; Satomi, M.; Ushio-Sata, N.; Morita, N. (2002) "Degradation of 17β-estradiol by a gram-negative bacterium isolated from activated sludge in a sewage treatment plant in Tokyo, Japan." Applied and Environ. Microbiology, 68(4): 2057-2060

Furuichi, T.; Kannan, K.; Suzuki, K.; Tanaka, S.; Giesy, J.P.; Masunaga, S. (2006) "Occurrence of estrogenic compounds in and removal by a swine farm waste treatment plant." Environ. Sci. Technol. 40:7896-7902.

Fürhacker, M.; Breithofer, A.; Jungbauer, A. (1999) "17β-estradiol: Behavior during waste water analyses." Chemosphere, 39:1903-1909.

Gomes, R.L.; Avcioglu, E.; Scrimshaw, M.D.; Lester, J.N. (2004) "Steroid estrogen determination in sediment and sewage sludge: a critique of sample preparation and chromatographic/mass spectrometry considerations, incorporating a case study in method development." Trends in Analytical Chemistry, 23:737-744.

Haiyan, R.; Shulan, J.; ud din Ahmed, N.; Dao, W.; Chengwu, C. (2007) "Degradation characteristics and metabolic pathway of 17α-ethynylestradiol by Sphingobacterium sp. JCR5. " Chemosphere, 66:340-346.

Hanselman, T. A.; Graetz, D. A.; Wilkie, A.C. (2003) Manure-borne estrogens as potential environmental contaminants." Environ. Sci. Technol. 37:5471-5478.

Hernando, M.D.; Mezcua, M.; Gómez, M.J.; Malato, O.; Agüera, A.; Fernández-Alba, A.R. (2004) "Comparative study of analytical methods involving gas chromatography–mass spectrometry after derivatization and gas chromatography–tandem mass spectrometry for the determination of selected endocrine disrupting compounds in wastewaters." J. Chromatography A, 1047:129–135.

Hintemann, T.; Schneider, C.; Scholer, H.F.; Schneider, R.J. (2006) "Field study using two immunoassays for the determination of estradiol and ethinylestradiol in the aquatic environment." Water Research, 40:2287-2294.

Holbrook, R.D.; Novak, J.T.; Grizzard, T.J.; Love, N.G. (2002) "Estrogen receptor agonist fate during wastewater and biosolids treatment processes: A mass balance analysis." Environ. Sci. Technol. 36(21):4533-4539.

Hoffmann, B.; de Pinho, T.G.; Schuler, G. (1997) "Determination of free and conjugated oestrogens in peripheral blood plasma, feces and urine of cattle throughout pregnancy." Experimental and Clinical Endocrinology & Diabetes. 105:296–303.

Homklin, S.; Wattanodorn, T.; Ong, S.K.; Limpiyakorn, T. (2007) "Biodegradation of 17 alpha-methyltestosterone and isolation of 17 alpha-methyltestosterone-degrading bacteria from sediment of masculinizing pond of Nile tilapia fry." 2nd IWA-ASPIRE Conference, Oct. 28 - Nov. 1, Perth, Australia.

Hoogenboom, L.A.P.; De Haan, L.; Hooijerink, D.; Bor, G.; Murk, A.J.; Brouwer; A. (2001) "Estrogenic activity of estradiol and its metabolites in the ER-CALUX assay with human breast cells." APMIS. 109:101-107.

Horinouchi, M.; Kurita, T.; Yamamoto, T.; Hatori, E.; Hayashi, T.; Kudo, T. (2004) "Steroid degradation gene cluster of Comamonas testosteroni consisting of 18 putative genes from meta-cleavage enzyme gene tesB to regular gene tesR." Biochemical and Biophysical Research Communication, 324:597-604.

Huang, C.H.; Sedlak D.L. (2001) "Analysis of estrogenic hormones in municipal wastewater effluent and surface water using ELISA and GC/MS/MS." Environ. Toxicology and Chemistry, 20:133-139.

Irwin, L.K.; Gray, S.; Oberdorster, E. (2001) "Vitellogenin induction in painted turtle, Chrysemys picta, as a biomarker of exposure to environmental levels of estradiol." Aquat. Toxicol. 55:49-60.

Ismail, A.A.A.; Harkness, R.A. (1966) "A method for the estimation of urinary testosterone." Biochem. J. 99:717.

Isobe, T.; Shiraishi, H.; Yasuda, M.; Shinoda, A.; Suzuki, H.; Morita, M. (2003) "Determination of estrogens and their conjugates in water using solid-phase extraction followed by liquid chromatography-tanden mass spectrometry." J. Chromatography A, 984:195-202.

Ivie, C.W.,; Christopher, R.J.; Munger, C.E.; Coppock, C.E. (1986) "Fate and residues of [4-^{14}C] estradiol-17β after intramuscular injection into Holstein steer calves." J. Animal Sci. 62:681-690.

Jeannot, R.; Sabik, H.; Sauvard, E.; Dagnac, T.; Dohrendorf, K. (2002) "Determination of endocrine-disrupting compounds in environmental samples using gas and liquid chromatography with mass spectrometry." J. Chromatography A. 974:143–159.

Jobling, S.; Nolan, M.; Tyler, C.R.; Brighty, G.; Sumpter, J.P. (1998) "Widespread sexual disruption in wild fish." Environ. Sci. Technol. 32:2498-2506.

Johnson, A.C.; Belfroid, A.; Di Corcia, A. (2000) "Estimating steroid oestrogen inputs into activated sludge treatment works and observations on their removal from the effluent." Science of the Total Environment, 256:163-173.

Johnson, A.C.; Williams, R.J. (2004) "A model to estimate influent and effluent concentrations of estradiol, estrone and ethinylestradiol at sewage treatment works." Environ. Sci. Technol. 38:3694-3658.

Johnson, A.C.; Williams, R.J.; Matthiessen, P. (2006) "The potential steroid hormone contribution of farm animals to freshwaters, the United Kingdom as a case study." Science of the Total Environment, 362:166-178.

Joss, A.; Andersen, H.; Ternes, T.; Richle, P.R.; Siegrist, H. (2004) "Removal of estrogens in municipal wastewater treatment under aerobic and anaerobic conditions: consequences for plant optimization." Environ. Sci. Technol. 38:3047-3055.

Karickhoff, S.W. (1981) Chemosphere 10: 833-846.

Kim, S.D.; Cho, J.; Kim, I.S.; Vanderford, B.J.; Snyder S.A. (2007) "Occurrence and removal of pharmaceuticals and endocrine disruptors in South Korean surface, drinking , and waste waters." Water Research, 41:1013-1021.

Kirk, L.A.; Tyler, C.R.; Lye, C.M.; Sumpter J.P. (2002) "Changes in estrogenic and androgenic activities at different stages of treatment in wastewater treatment works." Environ. Toxicol. Chem. 21(5):972–979.

Korner, W.; Bolz, U.; Sußmuth, W.; Hiller, G.; Schuller, W.; Hanf, V.; Hagenmaier, H. (2000) "Input/output balance of estrogenic active compounds in a major municipal sewage plant in Germany." Chemosphere. 40:1131-1142.

Kuch, H.M.; Ballschmiter, K. (2001) "Determination of endocrine-disrupting phenolic compounds and estrogens in surface and drinking water by HRGC-(NCI)-MS in the pictogram per liter range." Environ. Sci. Technol. 35:3201-3206.

Kelch, R. P.; Jenner, M. R.; Weinstein, R.; Kaplan, S. L.; Grumbach, M. M. (1972) "Estradiol and testosterone secretion by human, simian, and canine testes in males with hypogonadism and in male pseudohermaphrodites with the feminizing testes syndrome." J. Clin. Invest. 51(4):824–830.

Kolpin, D.W.; Furlong, E.T.; Meyer, M.T.; Thurman, E.M.; Zaugg, S.D.; Barber, L.B.; Buxton, H.T. (2002) "Pharmaceuticals, hormones and other organic wastewater contaminants in US streams, 1999-2000: A national reconnaissance." Environ. Sci. Technol. 36:1202-1211.

Kolodziej, E.P.; Harter, T.; Sedlak, D.L. (2004) "Dairy wastewater, aquaculture, and spawning fish as sources of steroid hormones in the aquatic environment." Environ. Sci. Technol. 38(23):6377-6384.

Kolodziej, E.P.; Sedlak, D.L. (2007) "Rangeland grazing as a source of steroid hormones to surface waters." Environ. Sci. Technol. 41(10):3514-3520.

Lagana, A.; Bacaloni, A.; De Leva, I.; Faberi, A.; Fago, G.; Marino, A. (2004) "Analytical methodologies for determining the occurrence of endocrine disrupting chemicals in sewage treatment plants and natural waters." Analytica Chimika Acta, 501:79-88.

Lai, K.M.; Johnson, K.L.; Scrimshaw, M.D.; Lester, J.N. (2000) "Binding of waterborne steroid estrogens to solid phases in river and estuarine systems." Environ. Sci. Technol. 34:3890-3894.

Länge, R.; Hutchinson, T. H.; Croudace, C. P.; Siegmund, F.; Schweinfurth, H.; Hampe, P.; Panter, G. H.; Sumpter, J. P. (2001) "Effects of the synthetic estrogen 17α-ethinylestradiol on the life-cycle of the Fathead Minnow (Pimephales promelas)." Environ. Toxicol. Chem. 20:1216-1227.

Lange, I.G.; Daxenberger, A.; Schiffer, B.; Witters, H.; Ibarreta, D.; Meyer, H.H.D. (2002) "Sex hormones originating from different livestock production systems: fate and potential disrupting activity in the environment." Analytica Chimica Acta. 473:27-37.

Larsson, D.G.J.; Adolfsson-Erici, M.; Parkkonen, J.; Pettersson, M.; Berg, A.H.; Olsson, P.E.; Forlin, L. (1999) "Ethinyloestradiol-an undesired fish contraceptive?" Aquatic Toxicology, 45:91-97.

Larsson, D. G. J.; Hallman, H.; Forlin, L. (2000) "More male fish embryos near a pulp mill." Environ. Toxicol. Chem. 19:2911-2917.

Lauritzen, C. (1988) "Natuerliche und synthetische Sexualhormone - biologische Grundlage und Behandlungsprinzipien. Natural and synthetic sexual hormones - biological basis and medical treatment principles." In: Schneider HPG, Lauritzen C, Nieschlag E, (Editors). Grundlagen und Klinik der menschlichen Fortpflanzung. Walter de Gruyter, Berlin, Germany.

Layton, A.C.; Gregory, B.W.; Seward, J.R.; Schultz, T.W.; Sayler, G.S. (2000) "Minieralization of steroidal hormones by biosolids in wastewater treatment systems in Tennessee U.S.A." Environ. Sci. Technol. 34:3925-3931.

Lee, L.S.; Strock, T.J.; Sarmah, A.J.; Rao, P.S.C. (2003) "Sorption and dissipation of testosterone, estrogens, and their primary transformation products in soils and sediment," Environ. Sci. Technol. 37(18):4098-4102.

Lee, H.B.; Liu, D. (2002) "Degradation of 17β-estrodiol and its metabolites by sewage bacteria." Water Air and Soil Pollution, 134:353-368.

Lee, H.B.; Peart, T.E. (1998) "Determination of 17b-estradiol and its metabolites in sewage effluent by solid-phase extraction and gas chromatography/mass spectrometry." J. AOC Int. 81(6):1209–1216.

Leusch, F.D.L.; Chapman, H.F.; Korner, W.; Gooneratne, S.R.; Tremblay, L.A. (2005) "Efficacy of an advanced sewage treatment plant in Southeast Queensland, Australia to remove estrogenic chemicals." Environ. Sci. Technol. 39:5781-5786.

Lewis, R.J. (1991) Carcinogenically Active Chemicals. Van Nostrand Reinhold, New York, U.S.A.

Lucas, S.D. ; Jones, D.L. (2006) "Biodegradation of estrone and 17β-estradiol in grassland soils amended with animal wastes." Soil Biology and Biochemistry, 38:2803-2315.

Lunaas, T. (1962) "Urinary estrogen levels in. the sow during estrous cycle and early. pregnancy." J. Reprod. Fert. 4:13-20.

Lunaas, T. (1965) "Urinary excretion of oestrone and. oestradiol and of Zimmermann chromogens in the sow during oestrus." Acta Vet. Scand. 6:16-29.

Masanori, S.; Hirofumi,Y.; Haruki, M.; Masanobu, M.; Hiroshi, T.; Kunio, K. (2004) "Fish full life-cycle testing for androgen methyltestosterone on Medaka (Oryzias latipes)." Environ. Toxicol. Chem. 23:774-781.

Matter-Müller, C.; Gujer, W.; Giger, W.; Stumm, W. (1980) "Non-biological elimination mechanisms in biological sewage treatment plant. " Prog. Water Technol. 12:299-314.

Mathur, R.S.; Anastass, P.A.; Common, R.H. (1966) "Urinary excretion of estrone and of 16-epiestriol plus 17-epiestriol by the hen." Poultry Science. 45:946-952.

Mathur, R. S.; Common, R. H. (1969) "A note on the daily urinary excretion of estradiol-17β and estrone by the hen." Poultry Science. 48:100-104.

Matsui, S.; Takigami, H.; Matsuda, T.; Tniguchi, N.; Adachi, J.; Kawami, H.; Shimizu, Y. (2000) "Estrogen and estrogen mimics contamination in water and the role of sewage treatment." Water Sci. Technol. 42(12):173-179.

Matsui, S.; Takigami, H.; Matsuda, T.; Tniguchi, N.; Adachi, J.; Kawami, H.; Shimizu, Y. (2000) "Estrogen and estrogen mimics contamination in water and the role of sewage treatment." Water Sci. Technol. 42(12):173-179.

Maxey, K.M.; Maddipati, K.R.; Birkmeier, J. (1992) "Interference in Enzyme Immunoassays." J. Clin. Immunoassay. 15:116-120.

Mcelwee, K.; Burke, D.; Niles, M.; Cummings, X.; Egna, H. (2000) "Pond dynamics/aquaculture CRSP." Seventeenth Annual Technical Report, Oregon State University, Corvallis, Oregon, pp. 109-112.

Moore, A. B.; Bottoms, G. D.; Coppoc, G. L.; Pohland, R. C.; Roesel, O. F. (1982) "Metabolism of estrogens in the gastrointestinal tract of swine." J. Anim. Sci. 55:124-134.

Monk, E. L.; Erb, R. E.; Mollett, T. A. (1975) "Relations between immnnoreactive estrone and estradiol in milk, blood and urine of dairy cows." J. Dairy Sci. 58:34-40.

Möstl, E.; Choi, H. S.; Wurm, W.; Ismail, N.; Bamberg, E. (1984) "Pregnancy diagnosis in cows and heifers by determination of oestradiol-17α in faeces." British Veterinary J. 140:287-291.

Möstl, E.; Dobretsberger, A.; Palme, R. (1997) "Östrogenkonzentration im stallmist trächtiger rinder." Wien Tieärztl Mschr. 84:140-143.

Murk, A.J.; Legler, J.; Van Lipzig, M.M.H.; Meerman, J.H.N.; Belfroid, A.C.; Spenkelink, A.; Van Der Burg, B.; Rijs, G.B.J.; Vethaak, D. (2002) "Detection of estrogenic potency in wastewater and surface water with three in vitro bioassays." Environ. Toxicol. Chem. 21(1):16-23.

Mutschler, E. (1996) Arzneimittelwirkungen, Lehrbuch der Pharmakologie und Toxikologie. Textbook of pharmacology and toxicology. Stuttgart: Wissenschaftliche Verlagsgesellschaft mbH.

Nichols, D.J.; Daniel, T.C.; Edwards, D.R.; Moore, P.A.; Pote, D.H. (1998) "Use of grass filter strips to reduce 17β-estradiol in runoff from fescue-applied poultry litter." J. soil Water Conserve. 53:74-77.

Nichols, D.J.; Daniel, T.C.; Moore, P.A.; Edwards, D.R.; Pote, D.H. (1997) "Runoff of estrogen hormone 17β-estradiol from poultry litter applied to pasture." J. Environ. Qual. 26:1002-1006.

Nunes, G.S.; Toscano, I.A.; Barcelo D. (1998) "Analysis of pesticides in food and environmental samples by enzyme-linked immuno sorbent assays." TrAC Trend Anal. Chem. 17:79-87.

Palme, R.; Fischer, P.; Schildorfer, H.; Ismail, M. N. (1996) "Excretion of infused ^{14}C steroid hormones via faeces and urine in domestic livestock." Anim. Reprod. Sci. 43: 43-63.

Parks, L.G.; Lambright, C.S.; Orlando, E.F.; Guillette, L.J.; Ankley, G.T.; Gray, L.E. (2001) "Masculinization of female mosquito fish in kraft mill effluent-contaminated Fenholloway River water is associated with androgen receptor agonist activity." Toxicol. Sci. 62:257-267.

Pawlowski, S.; Ternes, T.A.; Bonerz, M.; Rastall, A.C.; Erdinger, L.; Braunbeck, T. (2004) "Estrogenicity of solid phase-extracted water samples from two municipal sewage treatment plant effluents and river Rhine water using the yeast estrogen screen." Toxicology In Vitro, 18:129-138.

Peterson, E.W.; Davis, R.K.; Orndorff, H.A. (2000) "17β-Estradiol as an indicator of animal waste contamination in mantled karst aquifers." J. Environ. Quality, 29:826-834.

Petrovic, M.; Eljarrat, E. ; López de Alda, M.J. ; Barceló, D. (2002) "Recent advances in the mass spectrometric analysis related to endocrine disrupting compounds in aquatic environmental samples." J. Chromatography A, 974:23-51.

Purdom, C.E.; Hardiman, P.A.; Bye, V.J.; Eno, N.C.; Tyler, C.R.; Sumpter, J.P. (1994) "Estrogenic effects of effluents from sewage treatment works." Chem. Ecol. 8:275-285.

Quincey, R.V.; Gray, C.H. (1967) "The metabolism of [1,2-^{3}H]17α-methyltestosterone in human subjects." J. Endocrin. 37:37–55

Raeside, J.I. (1963) "Urinary oestrogen secretion in pigs at oestrus and during the oestrous cycle." J. Reprod. Fert. 6:421-426.

Raman D.R.; Layton, A.C.; Moody, L.B.; Easter, J.P.; Sayler, G.S.; Burns, R.T.; Mullen, M.D. (2001) "Degradation of estrogens in dairy waste solids: Effects of acidification and temperature." Trans ASAE. 44:1881-1888.

Raman, D.R.; Williams, E.L.; Layton, A.C.; Burns, R.T.; Easter, J.P.; Daugherty, A.S.; Mullen, M.D.; Sayler, G.S. (2004) "Estrogen content of dairy and swine wastes." Environ. Sci. Technol. 38:3567-3573.

Rongone, E.L.; Segaloff, A. (1962) "Isolation of urinary metabolites of 17α-methyltestosterone." J. Biol. Chem. 237:1066–1067.

Rogers, H.R. (1996) "Sources, behaviour and fate of organic contaminants during sewage treatment and sewage sludges." Science of the Total Environment, 185:3–26.

Sarmah, A.K.; Northcott, G.L.; Leusch, F.D.L.; Tremblay, L.A. (2006) "A survey of endocrine disrupting chemicals (EDCs) in municipal sewage and animal waste effluents in the Waikato region of New Zealand." Science of the Total Environment. 335:135-144.

Schiffer, B.; Daxenberger, A.; Meyer, K.; Meyer, H.H.D. (2001) "The fate of trenbolone acetate and melengestrol acetate after application as growth promoters in cattle: environmental studies." Environ Health Perspect. 109:1145-1151.

Schwarzenberger, F.; Mostl, E.; Palme, R.; Bamberg, E. (1996) "Faecal steroid analysis for non-invasive monitoring of reproductive status in farm, wild and zoo animals." Anim. Reprod. Sci. 42:515-526.

Schlenker, G.; Muller, W.; Glatzel, P. (1998) "Analysis for the stability of sexual steroids in feces of cows over 12 weeks." Berl. Muench Tieraerztl Wochenschr, 111: 248-252.

Schlenker, G.; Muller, W.; Birkelbach, C.; Glatzel, P. (1999) "Experimental investigations into influence of Escherichia coli and Clostridium perfringens on the steroid estrone." Berliner und Muenchener Tieraerztliche Wochenschrift, 112(1):14-17.

Schullerer S.; Spengler, P.; Metzger, J.W. (2002) "Influence of different water treatment steps in sewage plant technology on the concentration of estrogens in sewage." Vom Waser, 98: 65–80.

Sermwaraphan P.; Kurisu F.; Limpiyakorn, T. (2007) "Degradation of 17 alpha-ethynylestradiol (EE2) by nitrifying activated sludge containing different ammonia-oxidizing bacterial communities." 2nd IWA – ASPIRE Asia-Pacific Regional Group Conference & Exhibition, Oct. 28 – Nov. 1, Perth, Australia.

Servos, M.R.; Bennie, D.T.; Burnison, B.K.; Jurkovie, A.; McInnis, R.; Neheli, T.; Schnell, A.; Seto, P.; Smyth, S.A.; Ternes, T.A. (2005) "Distribution of estrogens, 17β-estradiol and estrone in Canadian municipal wastewater treatment plants." Science of the Total Environment, 336:155-170.

Shappell, N.W.; Billey, L.O.; Forbes, D.; Matheny, T.A.; Poach, M.E.; Reddy, G.B.; Hunt, P.G. (2007) "Estrognic activity and steroid hormones in swine wastewater through a lagoon constructed-wetland system." Environ. Sci. Technol. 41:444-450.

Shareef, A.; Angove, M.J.; Wells, J.D. (2006) "Optimization of silylation using N-methyl-N-(trimethylsilyl)-trifluoroacetamide, N,O-bis-(trimethylsilyl)-trifluoroacetamide and N-(tert-bu tyldimethylsilyl)-N-methyltrifluoroacetamide for the determination of the estrogens estrone and 17_-ethinylestradiol by gas chromatography–mass spectrometry." J. Chromatography A, 1108:121-128.

Shi, C.J.; Whitlock, H.W. (1968) "Biochemistry of steroids." Annual Review of Biochemistry, 37:661-694.

Shi, J.H.; Suzuki, Y.; Lee, B-D.; Nakai, S.; Hosomi, M. (2002) "Isolation and characterization of the ethynylestradiol-biodegrading microorganism. Fusarium proliferatum strain HNS." Water Sci. Technol. 45:175-179.

Shi, J.; Fujisawa, S.; Nakai, S.; Hosomi, M. (2004) "Biodegradation of natural and synthetic estrogens by nitrifying activated sludge and ammonia-oxidizing bacterium Nitrosomonas europaea." Water Research, 38:2323-2330

Shore, L. S.; Reichmann, O.; Shemesh, M.; Wenzel, A.; Litaor, M.I. (2004) "Washout of accumulated testosterone in a watershed." Science of the Total Environment, 332:193–202

Shore, L.S.; Correll, D.L.; Chakraborty, P.K. (1995) "Relationship of fertilization with chicken manure and concentrations of estrogens in small streams." In: Steele K, (editor), Animal waste and the land-water interface. Boca Raton: CRC Press. 155-162.

Shore, L.S.; Gurevitz, M.; Shemesh, M. (1993) "Estrogen as an environmental pollutant." Bulletin of Environmental Contamination and Toxicology, 51(3):361-366.

Soe, KL.; Soe, M.; Gluud, C. (1992) "Liver pathology associated with the use of anabolicandrogenic steroids." Liver International, 12:73-9.

Stopforth, A.; Grobbelaar, C.J.; Crouch, A.M.; Sandra, P. (2007) "Quantification of testosterone and epitestosterone in human urine samples by stir bar sorptive extraction - thermal desorption - gas chromatography/mass spectrometry: Application to HIV-positive urine samples." J. Separation Sci., 30(2):257-265.

Stumpe, B.; Marschner, B. (2007) "Long-term sewage sludge application and wastewater irrigation on the mineralization and sorption of 17β-estradiol and testosterone in soils." Science of the Total Environment, 374:282–291.

Szenci, O.; Taverne, M.A.; Palme, R.; Bertoti, B.; Merics, I. (1993) "Evaluation of ultrasonography and the determination of unconjugated oestrogen in faeces for the diagnosis of pregnancy in pigs." Vet Rec. 132:510-512.

Sumpter, J.P.; Jobling, S. (1995) "Vitellogenesis as a biomarker for estrogenic contamination of the aquatic environment." Environmental Health Perspectives Supplements, 103:173-178.

Suzuki, T.; Maruyama, T. (2006) "Fate of natural estrogens in batch mixing experiments using municipal sewage and activated sludge." Water Research, 40:1061-1069.

Svenson, A.; Allard, A-S.; Ek, M. (2003) "Removal of estrogenicity in Swedish municipal sewage treatment plants." Water Research, 37:4433-4443.

Tabak, H.H.; Bloomhuff, R.N.; Bunch, R.L. (1981) "Steroid hormones as water pollutants II: Studies on the persistence and stability of natural urinary and synthetic ovulation inhibiting hormones in untreated and treated wastewaters." Dev. Ind. Microbiol. 22:497-519.

Tabata, A.; Kashiwada, S.; Ohnishi, Y.; Ishikawa, H.; Miyamoto, N.; Itoh, M.; Magara, Y. (2001) "Estrogenic influences of estradiol-17b, p-nonylphenol and bis-phenol-A on Japanese Medaka (Oryzias latipes) at detected environmental concentrations." Water Sci. Technol., 43:109-116.

Taieb, J.; Benattar, C.; Birr, A.S.; Lindenbaum, A. (2002) "Limitations of steroid determination by direct immunoassay." Clinical Chemistry, 48(3):583-585.

.Ternes, T.A.; Kreckel, P.; Mueller, J. (1999a) "Behavior and occurrence of estrogens in municipal sewage treatment plants II:Aerobic batch experiments with activated sludge." Science of the Total Environment, 225:91-99.

Ternes T.A.; Stumpf, M.; Mueller, J.; Haberer, K.; Wilken, R-D.; Servos, M. (1999b) "Behavior and occurrence of estrogens in municipal sewage treatment plants I: Investigations in Germany, Canada and Brazil." Science of the Total Environment, 225:81-90.

Ternes, T.A.; Herrmann, N.; Bonerz, M.; Knacker, T.; Siegrist, H.; Joss, A. (2004) "A rapid method to measure the solid-water distribution coefficient (K_d) for pharmaceuticals and musk fragrances in sewage sludge." Water Research, 38:4075–4084.

Tyler, C. R.; Spary, C.; Gibson, R.; Santos, E. M.; Shears, J.; Hill, E. M. (2005) "Accounting for differences in estrogenic responses in rainbow trout and rouch exposed to effluents from wastewater treatment works." Environ. Sci. Technol., 39:2599-2607.

Vader, J.S.; van Ginkel, C.G.; Sperling, F.M.G.M.; de Jong, J.; de Boer, W.; de Graaf, J.S.; van der Most, M.; Stokman, P.G.W. (2000) "Degradation of ethinyl estradiol by nitrifying activated sludge." Chemosphere, 41:1239-1243

Vos, E.A. (1996) "Direct ELISA for estrone measurement in the feces of sows: prospects for rapid, sow-side pregnancy diagnosis." Theriogenology, 46:211-231.

Vos, E.A.; van Oord, R.; Taverne, M.A.M.; Kruip, T.A.M. (1999) "Pregnancy diagnosis in sows: direct ELISA for estrone in feces and its prospects for an on-farm test, in comparison to ultrasonography." Theriogenology, 51:829-840.

Wattanodorn, T., Homklin, S.; Ong, S.K.; Limpiyakorn, T. (2007) "Biodegradation of 17βmethyltestosterone by bacteria from wastewater treatment systems and sediments under aerobic conditions." PSU- UNS International Conference on Engineering and Environment (ICEE 2007), May 10 - 11, 2007, Phuket Graceland Resort & Spa, Thailand.

Weber, S.; Leushner, P.; Kampfer, P.; Dott, W.; Hollender, J. (2005) "Degradation of estradiol and ethynylestradiol by activated sludge and by a defined mixed culture." Applied Microbial and Cell Physiology, 67:106-112.

Wenzel, A.; Kuechler, T.; Mueller, J. (1998) Konzentrationen oestrgogen wirksamer substanzen in umweltmedien. Report, Project sponsored by the German Environmental protection Agency (UBA), Project No 216 02 011/11.

Williams, E. (2002) Estrogen concentrations in animal waste holding facilities: implications for manure management practices. MS Thesis, Univ. of Tennessee, Knoxville. TN, USA.

Williams, R.J.; Johnson, A.C.; Smith, J.J.L.; Kanda, R. (2003) "Steroid estrogens profiles along river stretches arising from sewage treatment works discharges." Environ. Sci. Technol. 37:1744-1750.

Winters, S. J.;Troen, P. (1986) "Testosterone and estradiol are co-secreted episodically by the human testis." J. Clin. Invest. 78(4):870–873.

Wood, W.G. (1991) "Matrix effects in immonoassays." Scandinavian Journal of Clinical Laboratory Investigation. 51:105-112.

Xie, L.; Sapozhinkova, Y.; Bawardi, O.; Schlenk, D. (2004) "Evaluation of wetland and tertiary wastewater treatments for estrogenicity using in vivo and in vitro assays." Archives of Environmental Contamination and Toxicology, 48:81-86.

Yoshimoto, T.; Nagai, F.; Fujimoto, J.; Watanabe, K.; Mizukoshi, H.; Makino, T.; Kimura, K.; Saino, H.; Sawada, H.; Omura, H. (2004) "Degradation of estrogens by Rhodococcus zopfii and Rhodococcus equi isolates from activated sludge in wastewater treatment plants." Applied and Environmental Microbiology, 70(9):5283-5289.

Ying, G.G.; Kookana, R.S. (2003) "Degradation of selected endocrine disrupting chemicals in seawater and marine sediment." Environ. Sci. Technol. 37:1256-1260.

Yu, C-P.; Roh, H.; Chu, K-H. (2007) "17β-estradiol-degrading bacteria isolated from activated sludge." Environ. Sci. Technol. 41:486-492.

Zuo, Y.; Zhang, K. (2005) "Discussion: Suitability of N,O-bis(trimethylsilyl)trifluoroacetamide as derivatization reagent for the determination of the estrogens estrone and 17 α-ethinylestradiol by gas chromatography–mass spectrometry." J. Chromatography A, 1095:201-202.

CHAPTER 7

Phthalate Plasticizers and Degradation Products

S. Barnabé, I. Beauchesne, S.K. Brar, Y. Song, R.D. Tyagi, and R.Y.
Surampalli

7.1   Introduction

Plastic products have always been of major concern in terms of toxicity and persistence in the environment. They contain a myriad of additives including the plasticizers, which can make up to 40% of plastic formulations. Plasticizers are low molecular weight organic compounds that are essential for effective processing and tailoring of plastic formulations. They allow the production of flexible plastics with multiple applications ranging from automotive industry to medical and commodity products. They are manufactured in hundred millions of tons annually and represent an overwhelmingly large bracket of the plastic industry.

It has been established that plasticizers are toxic to a certain extent and some can exhibit endocrine disrupting properties. In fact, these compounds may also leach out from the plastics as they are not chemically bound to the plastic polymers. Hence, leaching is a major process for contamination of the environment by the ubiquitous plasticizers. Meanwhile, it has been estimated that the average ingestion rate of plasticizers could be about 8 mg per person per day. The abundance, potential endocrine disrupting effects and the growing interest in their fate, transport and treatment makes them emerging contaminants of major concern. It is also corroborated by the fact that their biodegradation in the environment can release breakdown products which can be potentially more toxic than the parent compounds. Hence, there is a need to understand the fate and behavior of plasticizers and their degradation products in natural and engineered environments, including industrial and municipal effluents, sewage sludges, and landfill leachates.

This chapter presents the current state of knowledge of the fate and treatment of plasticizers with a special focus on the important class of plasticizers called phthalate esters (PAE), and related compounds such as di-(2-ethylhexyl) phthalate (DEHP), di-butyl phthalate (DBP), and 2-ethylhexanol.

7.2   Background

7.2.1   Functions and Uses

Plasticizers are a significant part of the polymer industry from technical and economical perspectives. Plasticizers are the most common and least expensive plastic additives among flame retardants, processing aids, heat stabilizers, lubricants, antioxidants, organic peroxides and light stabilizers. The principal function of plasticizers is to lower the second transition temperature for improving flexibility and processing of polymers. They form secondary bonds with polymer chains and spread them apart. Plasticizers are available in two varieties: internal and external. External plasticizers are more sensitive to loss through evaporation, extraction and migration, while internal plasticizers are inherently part of the product. A plasticizer can present itself as a sole compound or a mixture of compounds where some compounds improve the plasticizing properties or compensate cost of others (Rahman and Brazel, 2004; Wypych, 2004).

Plasticizers are used in building materials (insulation of cables and wires, tubes, flooring, out-door walls and roof covering, sealing pastes and isolation mass), home furnishings, automobile materials (car under-coating and car seats made of imitation leather), clothing (footwear, outwear and rainwear), dental materials (teething rings), personal care products (anti-wrinkle cream, hair conditioning and nail varnish), shower curtains, adhesives, sealants paperboard, printer inks, food containers, animal glue, surface lubricants, insect repellents, pesticide granules and agricultural seed coating, aircraft materials, electric cables and, to a limited extent, in food packaging and medical products (Gomez-H. and Aguilar-C., 2003; Wypych, 2004). Poly (vinyl chloride) or PVC, poly (vinyl butyral), poly (vinyl acetate), acrylics, cellulose modelling compounds, nylon and polyamides are the most frequently plasticized polymers. PVC as a polymer accounts for more than 80% of all plasticizers consumed around the world (Rahman and Brazel, 2004; Wypych, 2004).

7.2.2   Types of Plasticizers

There are many types of plasticizers with a variety of characteristics and end uses including phthalate, phosphates, adipate, azelate and sebecate esters, epoxidized fatty acid esters or vegetable oils, benzoates, polymeric plasticizers (e.g. poly(ethylene) glycol), low molecular weight polymers (e.g. polybutadiene di-methylacrylate), trimellitates, sulfonic acid esters, sulfamides, monocarboxylic acid esters, chlorinated hydrocarbons, citrates, polymerizable plasticizers (e.g. allyl phthalate), elastomers (e.g. vinyl acetate copolymers), etc. (Rahman and Brazel, 2004; Wypych, 2004). The general criteria for choosing plasticizers for formulations include: (i) compatibility with the polymer; (ii) processing characteristics; (iii) target thermal, electrical and mechanical properties of the end product; (iv) resistance to abiotic and biotic degradation; (v) effect on polymer rheology; (vi) toxicity; and (vii) volume-cost analysis (Rahman and Brazel, 2004).

As plastic products are consumed in greater quantities, the global demand for plasticizers continues to mushroom. Rhaman and Brazel (2004) mentioned that the global demand for plasticizers in 1999 was 10.1 millions pounds, representing a market of US $7 billion, with a production growth rate of about 2.8% per year. PAEs were the most commonly used plasticizers and accounted for 92% of the compound produced worldwide (Wypych, 2004). PAEs are primarily used as plasticizers to impart flexibility to PVC. Their massive use in plastic products is due to many desirable properties such as: (i) minimal interaction with resins at room temperature; (ii) good fusion properties; (iii) satisfactory insulation for cables; (iv) highly elastic compounds with reasonable cold strength; (v) relatively non volatile at ambient conditions and; (vi) low cost (Rahman and Brazel, 2004).

DEHP is the most commonly used PAE plasticizer. At a global scale, more than 18 billion pounds of PAEs are used each year and above 2 million tons of DEHP alone are produced annually. Other important PAEs, in terms of production and application, are di-ethylphthalate (DEP), DBP, di-iso- and di-n-butylphthalate (DiBP, DnBP), butyl-benzyl phthalate (BBP), di-isononyl phthalate (DiNP) or di-n-octylphthalate (DnOP) (Latini, 2005). The United States Environmental Protection Agency (USEPA) and its international counterparts have classified most of the PAEs as priority organic pollutants and endocrine disrupting compounds (Birkett and Lester, 2003; Chang et al., 2007). In fact, DEP, BBP, DBP and DEHP occupy the top positions in the list, along with di-methyl phthalate (DMP).

7.2.3 Leaching and Migration of Plasticizers

As mentioned previously, PAEs have a tendency to leach from the plastic products or migrate to the surrounding media. Leaching and migration of plasticizers are critical issues that determine not only the material's useful life, but also pose health and environmental risks when environmental receptors are exposed to these compounds. Evaporation of plasticizers is also another mode of exposure to these compounds.

Plasticizers generally have a high mobility due to their relatively low molecular weight and are able to easily diffuse into the surrounding media (e.g., water and food). Temperature, plasticizer concentration, and characteristics (solubility, diffusion coefficients) are common factors that influence leaching and migration of plasticizers (Rahman and Brazel, 2004; Wypych, 2004; Goulas et al., 2007).

Plasticizers have been shown to be toxic in animal models and their mobility is an important issue for human health. Plasticizers first gained notoriety when they were suspected to favor premature puberty of young girls. High levels of these compounds were later found in baby toys. Since then, the plasticizer content of these products has become more regulated. However, plasticizers continue to find increasing use in food and medical products or materials. The exposure to these compounds through food packaging material is frequently investigated (Koo and Lee,

2004; Lopez-Espinosa et al., 2007). Goulas et al. (2007) reported many studies focusing on the migration of plasticizers from flexible PVC films into various foods. The amount of plasticizers migrating in packaged foods depends on many factors including, the fat content of the food, the type and the concentration of plasticizers, the packaging material, the storage time and temperature, and the contact time (Goulas et al., 2007). Hence, there are reasons to raise queries on exposure to plasticizers and their threat to human health. The general health issues regarding plasticizers are discussed in the next sub-section.

7.3   General Health Issues

There are numerous health issues related to plasticizers and the materials containing these compounds. Plasticizers are actually being scrutinized for environmental problems by government authorities and the scientific community. These compounds have been reported to show adjuvant and carcinogenic effects in animal models, and teratogenic and feminizing effects at low levels in rats and fishes (Larsen et al., 2002; Nuti et al., 2005). Plasticizers have known toxic effects in developing and adult animals in multiple organs such liver, reproductive tract, kidneys, heart, lungs.

7.3.1. Exposure

Plasticizers are found in various consumer products, including cosmetics. Numerous phthalates were found in European cosmetics including DnBP and DEHP (DiGangi and Norin, 2002). Thus, the population is exposed to plasticizers through: (i) inhalation of air (e.g. off-gassing from PVC products such as flooring and car dashboard); (ii) ingestion of drinking water and food contaminated with DEHP; (iii) dermal exposure and; (iv) transfer from PVC tubing used in hospitals to administer saline buffer and blood. During the manufacture of plasticizers, workers can be exposed to daily doses 25 times greater than the general population. Extensive uses of plasticizers in medical devices and blood storage bags can expose patients to these compounds. Finally, plasticizers are known to migrate or leach from the plastic product, which can expose humans to these compounds directly through contact with the product or from the compound released in the environment. Few studies have focused on the exposure of DEHP of workers, haemodialysis patients or newborn infants. Latini et al. (2003) confirmed a significant and widespread presence of the DEHP and its metabolite MEHP in the blood of newborn infants. The oxidation of MEHP generates many other metabolites that are studied as potential biomarkers for biological monitoring in human urine and serum (Nuti et al., 2005).

The oral LD_{50} values for DEHP have been reported as 25 g/kg in rats and 30 g/kg in mice. A detrimental dose for humans has been estimated at 69 mg/kg-day. The daily exposure is considered to be much lower, at about 2.3 to 2.8 µg/kg in Europe and 4 µg/kg in the United States. However, these values exclude the indoor air exposure from off-gassing of building materials for which some people could be at greater risks. According to Koch et al. (2003), 12% of the population exceed the daily ingestion rate of DEHP recommended by the EC (37 μg/kg weight per day) and 31% of individuals are exceeding the daily consumption reference dose of 20 μg/kg weight per day recommended by the US Food and Drug Administration.

There are various limiting values set to protect employees against workplace exposures. Limits for air contaminants for all plasticizers are available, but only ten are of great concern. Among them are DMP and DEHP for which recommended exposure limits and permissible exposure limits are set at 5 mg/m^3. The Immediate Danger to Life and Health concentration was set at 2000 mg/m^3 for DMP and at 5000 mg/m^3 for DEHP (NIOSH, 2003). The EU Scientific Committee for Food has allocated tolerable daily intakes of 50 μg DEHP or BBP per kg body weight per day and 100 μg DBO per kg body weight per day. In the United States, the use of materials in the production of food and its packaging is controlled by the Food and Drug Administration. DEHP, DMP and DBP cannot be directly added to food. However, they are allowed in some applications as indirect food additives, meaning that they can be used in materials in contact with food products. DEHP can be used in additives, resinous and polymeric coatings, defoaming agents used in the manufacture of paper and paperboard, acrylic and modified acrylic plastics, semi-rigid and rigid, cellophane and surface lubricants used in the manufacture of metallic articles. DMP is permitted in adhesives, acrylic and modified acrylic plastics and cross-linked polyester resins. DBP is allowed in components of paper and paperboard in contact with aqueous and fatty food, slimicides, cellophane, cross-linked polyester resins and rubber articles intended for repeated use. FDA (2002a,b,c) has issued warnings regarding the use of DEHP in medical devices including intravascular tubing and catheters/cannulae, bags used to store and transport nutrition formulas, and tubing used in internal nutrition. This warning suggests that such devices should be avoided. Research suggests that the migration of a plasticizer from medical devices can be reduced by using trimellitates as replacement (Adams, 2001). Phthalates are also being included in legislation concerning the land application of sewage sludge. Amount of DEHP allowed in treated sludge for land application is proposed at 100 mg/kg d.w. (CEC, 2000).

For DEHP, oral exposures predominate. For certain subsets of the general population, non-dietary ingestion (medical and occupational) or medical exposure is important. Exposure of the general human population to DEHP has been studied more in depth than the other phthalates. It has been estimated to be in the range of 3 to 30 μg/kg of body weight/day (excluding occupational exposure, medical exposures, and non-dietary ingestions in children), the major source being from food residues. These estimates already exceed chronic exposure levels believed to be tolerable for the general population. Preventive limit values, such as reference dose (RfD) of the USEPA and tolerable daily intake (TDI) of the European Union, are 20 μg/kg of body weight/ day and 37 μg/kg of body weight/day, respectively. Higher exposures occur occupationally (up to about 700 μg/kg body weight per day, mainly by inhalation, based on current workplace standards) and in the medical settings through use of plastic tubing, IV bags, etc. (i.e. up to 457 μg/kg body weight/day for haemodialysis patients). Daily exposure to DEHP in medical settings may exceed general population exposures but up to three orders of magnitude (Thickner et al., 2001).

Particular concern has been raised in neonatal care applications because newborns receive among the highest doses of DEHP from blood transfusions, extracorporeal membrane oxygenation and respiratory therapy. Another concern is the presence of phthalates in baby-care products and toys, which represent about 1% of the phthalate plasticizer market. Young children can put plastic toys in their mouth and come in direct contact with plasticizers, but the DEHP exposure via this mode is considerably lower than the critical limit of 69 mg/kg-day (Rahman and Brazel, 2004).

7.3.2 Toxic Effects

The Occupational Safety and Health Administration (OSHA) is a regulatory agency which, among its other duties, issues and enforces regulations that limit exposure to carcinogens in the workplace. OSHA has used two different approaches for limiting exposures: setting permissible exposure limits and requiring specific process technology and procedures. According to the data of the National Toxicology Program, DEHP is the only ester plasticizer found on potential carcinogen's list. It is classified as reasonably anticipated to be a human carcinogen. Numerous plasticizers were shown to have teratogenic properties (Lewis, 1999) while recent data and regulations do not support that most plasticizers are suspected teratogens or mutagens (State of California, 2003). Only DEHP and adipate are considered mutagens.

An estimate of the daily human exposure to a hazardous substance that is likely to have no appreciable risk of adverse health effects over a specified duration of exposure has been developed. It is the Minimal Risk Level. It is evaluated at 0.1 mg/kg-day (acute) and 0.06 mg/kg-per day (chronic) for DEHP and at 0.5 mg/kg-day (acute) for DnBP (ATSDR, 2003).

The toxic effects of plasticizers in human remain unclear, but some studies have been published on impact of phthalates on foetal/neonatal exposure and the relation of maternal nutrition to phthalate teratogenic effects (Latini et al., 2004, 2006; Koch et al., 2006). There is also a possible relation of phthalates exposure and male infertility, especially, their impact on male tract development. Latini et al. (2004, 2006) published interesting reviews on these subjects. Some plasticizers are also believed to mimic oestrogen in vivo and bind to oestrogen receptor blocking the normal endocrine pathways. So, more concerns are raised towards these compounds because they may come in close contact with the human body. In fact, it has been found that phthalate plasticizers leach out of medical plastics such as intravenous bags and dialysis tubing (Tickner et al., 2001; Hildenbrand et al., 2005).

More precisely, the adjuvant effect of phthalate plasticizers has been investigated. An adjuvant effect is an alteration of the immune system giving rise to the development of diseases, including allergy. DEHP in PVC materials was first suspected to promote development of asthma in children, but its adjuvant effect has only been demonstrated in an animal model (Larsen et al., 2002). Latini (2000) documented the DEHP long term toxicity and tissue deposition in animal models and the major factors influencing its action, either dose, time and age. Latini et al. (2004) documented also the fact that infants may represent a population at increased risks of exposure to DEHP. Breast milk, infant formula, baby food, inhalation of indoor air and dermal or oral exposure to indoor dust are major routes of exposure to DEHP for infants. Toys intended for mouthing (bottle nipples, teething rings, pacifiers and rattles) do not contain DEHP in Europe, Canada and United States, although it is still found in toys for older children.

7.3.3 Endocrine Disrupting Properties

The two major plasticizers that are suspected to be endocrine disrupters are DEHP and DBP. These compounds are suspected to stimulate the expression of cellular oestrogen-sensitive endpoints in vitro (Moore, 2000). Most phthalates (BBP, DBP, DiBP, DEP, and DiNP) have endocrine disrupting potencies, six to seven orders of magnitude lower than 17β-estradiol (Harris et al., 1997).

The endocrine disrupting potential of phthalate plasticizers has been reported in mammalian system (Latini et al., 2004; Koch et al., 2006). The adverse effect of PAE as DEHP on male tract development are related to alterations in gene expression of a number of enzymes and transport proteins involved in normal testosterone biosynthesis and transport in the fetal Leydig cell, and to a subsequent reduction in testosterone synthesis. Some PAEs have been shown to disrupt several genes pathways, including, cholesterol transport, steroidogenesis, intracellular lipid and cholesterol homeostasis, insulin signalling, transcriptional regulation and oxidative stress (Latini et al., 2006).

Kim et al. (2003) reported many studies on phthalates exhibiting cytotoxic and estrogenic activity in toxicity-monitoring using recombinant bioluminescent bacteria, MCF-7 cell proliferation, oestrogen receptor-binding in rat uterus, and yeasts transfected with human ER. They also mentioned studies on the development toxicity of DEHP in rats and the teratogenic effects of DEHP in mice and chicks. However, no acute toxicity of plasticizers was observed.

As reported by Hildenbrand et al. (2005), animal experiments have shown that DEHP metabolites impair infertility. The Sertoli cells of the sperm channel involved in spermogenesis, are the most affected cells. Hydrolysis and oxidation of phthalates occur in blood and the oxidative metabolites are toxics.

7.4   Degradation of Toxic Intermediates

Breakdown products of organic pollutants have always been of concern. The pesticide residues are a good example. Organic pesticides tended to replace chemical pesticides because of their biodegradability, but the environmental safety is sometimes compromised by the release of toxic breakdown products (e.g. residues of organophosphate degradation). Other organic pollutants with toxic breakdown products are the non ionic surfactants, alkylphenols. They are degraded into toxic nonylphenol monoethoxylates, nonylphenols and octylphenols. They are present at high levels in environmental samples and classified as endocrine disrupting compounds by governmental authorities (La Guardia et al., 2001; Birkett and Lester, 2003). This important issue also concerns the plasticizers that have different levels of biodegradability in the environment with very few or no information on the metabolic intermediates and their respective toxicity. In fact, DEHP and other PAEs in the environment could be reservoirs of toxic metabolic intermediates (Barnabé et al., 2007; Beauchesne et al., 2007).

Numerous studies have been conducted on biodegradation of plasticizers in the environment (surface waters, soils, sediments, wastewaters, sewage sludges). These studies involved pure microbial culture, addition of microbial enzymes, microbial consortia or activated sludges. Most of these studies on fate of plasticizers have focused on the removal of the parent compounds without considering their partial breakdown products. Table 7.1 presents few studies where metabolites were identified. Phthalic acid, 2-ethylhexanol ($C_8H_{18}O$) and 2-ethylhexanoic acid ($C_8H_{16}O_2$) appear to have been the most frequent toxic metabolites identified during PAE biodegradation studies.

Some of the most extensive studies on the identification of metabolites after partial degradation of DEHP by common soil micro-organisms were performed by Nalli and co-workers. Their works especially demonstrated that the toxic intermediates generated by Rhodococcus rhodochrous include 2-ethylhexanol, 2-ethylhexanal, and 2-ethylhexanoic acid (Nalli et al., 2002; Gartshore et al., 2003; Nalli et al., 2006a). The hydrolysis of ester bonds in DEHP, DEHA and DEHTP leads to a mono-ester and 2-ethylhexanol, which is further oxidized into 2-ethylhexanoic acid. Figure 7.1 illustrates a degradation pathway of DEHP and DEHA in the environment proposed by Nalli and co-workers.

Table 7.1 Studies on metabolites resulting from partial biodegradation of PAEs by pure or mixed cultures or enzymes.

Microbial Culture or Enzyme	Parent Compound	Metabolic Intermediates	Reference
Inocula from diluted and treated municipal solid waste	DEHP	2-ethylhexanol, 2-ethylhexanoic acid	Ejlertsoon and Svenson, 1996
Mycobacterium sp. and other strains[1]	DEHP	2-ethylhexanol, 1,2-benzenedicarboxylic acid	Nakamiya *et al.*, 2005
Pseudomonas fluorescens[1]	DEHP	phthalic acid, benzoic acid, phenol	Zeng *et al.*, 2002
Fungal cutinases, yeast esterases	DEHP	1,3-isonemzofurandione, one unidentified toxic metabolites (produced by yeast esterases)	Kim *et al.*, 2003
Arthrobacter sp., *Sphingolomas paucimobilis*	DMP	mono-ester, phthalic acid	Vega and Bastide (2003)
Mixed culture of *Pseudomonas fluorescens*, *P. aureofaciens*, *Sphingomonas paucimobilis*, *Xanthomonas maltophilia*	DMP	Mono-methyl phthalate, phthalic acid	Wang *et al.*, 2003
Pseudomonas fluorescens B-1	DBP	mono-butyl phthalate, phthalic acid	Xu *et al.*, 2005
Delftia sp.[1]	DBP	phthalic acid, protocatechuate	Patil *et al.*, 2006
Bacillus subtilis[1]	DEHP, DBP	DMP	Quan *et al.*, 2006
Rhodococcus rhodochrous and other common soil microorganisms	DEHP, DEHA, DEHTP[2]	2-ethylhexanol, 2-ethylhexanal, 2-ethylhexanoic acid	Nalli *et al.*, 2002, 2006a; Gartshore *et al.*, 2003

[1] New isolates.

[2] DEHA: di-(2-ethylhexyl) adipate. DEHTP: di-(2-ethylhexyl) terephthalate.

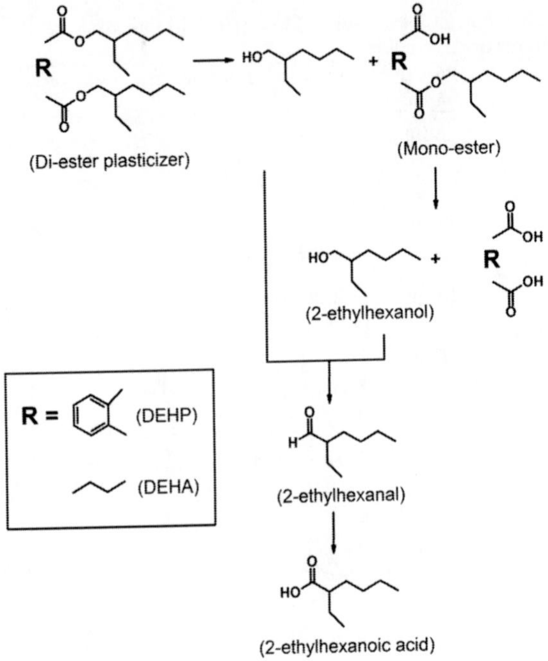

Figure 7.1 Degradation pathway for plasticizers, DEHP and DEHA (Horn et al., 2004; Nalli et al., 2006a).

It has been reported in controlled laboratory studies that these metabolites are more toxic to various species (mammalian models, fishes and arthropods) than the original plasticizers (Nalli et al., 2002; Horn et al., 2004; Nalli et al., 2006a). 2-ethylhexanol is a well known air contaminant. 2-ethylhexanol and 2-ethylhexanal in indoor air are also suspected to be related to sick building syndrome (Nalli et al., 2006a). 2-ethylhexanol is present in breath, urine and serum of patients on haemodyalisis and from patients in intensive care unit and its origin as a metabolite of DEHP degradation, and can have antiproliferative effect. The metabolite 2-ethylhexanal was not reported in most early degradation studies, most likely due to the high volatility of this compound and its tendency to partition into the gas phase. However the production of this compound and its role in the degradation pathway has recently been confirmed (Nalli et al., 2006a). Meanwhile, 2-ethylhexanoic acid is known to be recalcitrant.

The 2-ethylhexanol is not only a degradation product but is also used in the production of plasticizers for PVC resins, and both the alcohol and acid are used as intermediates in the manufacture of inks, paper, rubber, resins, surfactants and lubricants (Staples, 2001). Most environmental releases are presumed to occur during degradation of DEHP in water, soil and sediment (Horn et al., 2004; Barnabé et al., 2007) while they may enter environment during their manufacture, handling or use (Staples, 2001). In air, water, soil and sediment, they are subject to volatilization, adsorption to suspended solids and sediment, biodegradation and photo-degradation, but they have low persistence as noted by Staples (2001).

Field studies also confirmed that DEHP and DEHA and two of its metabolites, 2-ethylhexanol and 2-ethylhexanoic acid were present in surface waters, river sediment, freshly fallen snow, and even in tap water (Horn et al., 2004). Hence, there are reasons to believe that the toxic metabolites are widespread as their parent compounds. DEHA is a less commonly used plasticizer, which is a good additive for polymer flexibility at low temperature (Nalli et al., 2006b). DEHA tends to replace DEHP in some products such as food-grade PVC films (Goulas et al., 2007), floorings and wall coverings (Wypych, 2004) due to its higher biodegradability and less toxicity, but its environmental safety is questionable while considering the toxic intermediates formed during degradation.

In summary, the presence of toxic intermediates of DEHP and other PAEs during degradation with their parent compounds acting as reservoirs justify the importance of considering PAE plasticizers as emerging pollutants. The fate of the plasticizer parent compounds and their toxic metabolites is thus discussed in the next section.

7.5   Fate in the Environment

The understanding of the potential environmental pathways of plasticizers can result in appropriate management of their production and uses, and ensure health and environmental safety. This section will summarize the fate of plasticizers in different environmental compartments: water, soil and sediments, air and organisms. The microbes responsible for degradation of PAEs in the environment will also be discussed.

Briefly, plasticizers enter the environment not only through losses during their manufacture, distribution and through waste disposal, but also by leaching out of the finished product. Hence, the chief routes of environmental releases are: direct transfer (e.g. building materials), urban runoff, industrial air emission/atmospheric deposition, sewage/water distribution pipework deterioration, solid waste disposal (e.g. release of landfill leachate), industrial effluents and sewage treatment plant activities (discharge of effluent, elimination of residues through incineration, land filling and land application). The estimated releases of the three plasticizers found in the toxic release inventory (TRI) maintained by the U.S. Environmental Protection Agency (TRI, 2005) are presented in Table 7.2.

Table 7.2. Estimated releases (in pounds) of three PAEs in 2005 (from TRI).

Plasticizers	Air	Land	Water	Underground Injection	Off-site Waste Transfer
DBP	14,486	47,027	98	89,000	258,678
DEHP	237,093	39,840	429	0	3,820,366
DMP	229,716	1	875	2,294	424,735

Their fate is governed by abiotic and biotic degradation, adsorption to organic matter and volatilization in aquatic and terrestrial systems. The PAEs are degraded in both aerobic and anaerobic environment, but sorption to particles decrease the degradation rate. More precisely, the PAEs undergo the following transformation into the environment:

Abiotic hydrolysis: PAEs are susceptible to hydrolysis, however at slow rates, and it is thought to be negligible in sewage, soils, sediments and surface waters. The products of hydrolysis are an acid and an alcohol. PAE can undergo two hydrolytic steps, initially forming the mono-ester and one free alcohol moiety and a second hydrolytic step creating phthalic acid and a second alcohol. PAEs are hydrolyzed at negligible rates at neutral pH. Acid hydrolysis of PAE is possible, but it is estimated at four orders of magnitude slower than the alkaline hydrolysis rate constants (Ziogou et al., 1989; Staples et al., 1997; Roslev et al., 1998).

Photolysis: Photolysis occurs through UV absorption. The mechanism may be either through direct absorption of UV radiation by the chemicals in air or by absorption of UV radiation by natural substances such as water with the formation of activated species such as singlet oxygen or hydroxy radicals that react with PAE. Photolysis appears to be much more important in the atmospheric fate of PAE than water (Staples et al., 1997).

Biotic degradation: Biodegradation is the main process affecting the environmental fate of PAE. Numerous studies indicate that PAE are degraded by a wide range of bacteria and actinomycetes fungi under both aerobic and anaerobic conditions.

Staples and co-workers in 1997 published an extensive review on the environmental fate of PAE. The authors described important parameters for investigating and understanding the fate and behavior of plasticizers in environment. These parameters are:

Water solubility: This property influences the biodegradation and bioaccumulation potential of PAE, as well as aquatic toxicity. Water solubility is also a determining factor in controlling the environmental distribution of PAE. Losses from wastewater treatment facilities, landfills and sludge-amended soils are partially a function of aqueous solubility.

Octanol/water coefficient (K_{OW}): The equilibrium distribution of an organic chemical between water and octanol is an important physical constant for predicting the tendency of PAEs to partition to water, animal lipids, sediment, and soil organic matter. Relationships exist between Kow and soil sorption, water solubility, bioconcentration, and toxicity.

Vapor pressure (VP): It plays an important role in the fate of PAE released to the atmosphere. The ratio of the vapor pressure to the molar water solubility estimates the Henry's Law constant, which is the measure of the equilibrium distribution coefficient.

Hydrophobicity: Sorption of PAE to soil, sediment, or suspended solids is partially governed by the relative hydrophobicity of the compound. Hydrophobic chemicals adsorb principally to the organic matter associated with the solid.

The most commonly studied plasticizers in environment are PAEs, either DEP, DnBP, DMP, DEHP and DnOP, which are also the most commonly found in the environment (Quan et al., 2006). BBP is also part of some environmental investigations. The values of the parameters governing their environmental fate are presented in Table 7.3.

Table 7.3 Estimated values for the parameters related to environmental fate for the most abundant plasticizers in environment (Staples et al., 1997).

PAE	Log K_{OW}	Aqueous Solubility (mg/ L)	Vapor Pressure (mm Hg, 25°C)	Hydrophobicity
DMP	1.61	4200	2.0E-3	+
DEP	2.38	1100	1.0E-3	+
DnBP	4.45	11.2	2.7E-5	++
BBP	4.59	2.7	5.0E-6	++
DEHP	7.50	0.003	1.0E-7	+++
DnOP	8.06	0.0005	1.0E-7	+++

7.5.1 Air

Plasticizers as PAEs are relatively non volatile as their vapour pressures are very low (see Table 7.3). However, they are still detected in air samples in concentrations ranging from 0.2 to 2453 ng/m^3, as reported in Table 7.4.

Table 7.4. Concentrations of three PAE in various air samples.

PAE	Concentration (ng/m^3)	Sample	Reference
DnBP	0.2-192	Urban air	Weschler, 1984; HSDB, 2001
	1.3-5.0	Rural/Remote	HSDB, 2001
	2453	Apartment	Wensing et al., 2005
DEHP	20-55	Office buildings	Weschler, 1984
	3.6-132	Urban	HSDB, 2001
	0.77-3.6	Rural/Remote	HSDB, 2001
	390	Apartment	Wensing et al., 2005
DMP	0.60-1.74	Office buildings	Weschler, 1984

These concentrations in air are especially attributed to building materials. Major concern has been oriented towards PAE in indoor air because they have been recognized as major indoor pollutants (Bornehag et al., 2005). Potential sources of PAE in indoor air are wall coverings, wall paints, floor coverings and electronic devices (Wensing et al., 2005), which may contain PVC. Wensing et al. (2005) reported many studies on PAE in indoor air and household dust. An apartment indoor air may contain respectively up to 390 ng DEHP m-3 and 2453 ng DBP m-3 (Table 4) while surface dust in children's room may contain about 770 mg DEHP kg-1 and 150 mg DBP kg-1. Fortunately, the analysis of Wensing et al. (2005) of indoor air showed that these values do not result in oral or dermal intake in the range of the Acceptable Daily Intake recommended as stated above.

Wensing et al. (2005) also reported studies on measurement of plastic additives such as flame retardants in vehicle indoor air, but the studies did not assess the PAE levels. Considering that PAEs are a constituent of some vehicle materials and many people pass a major time of their day in their vehicle, it is convenient to investigate the PAE present in vehicle indoor air, especially the so-called "new car odor".

When released to the atmosphere, most plasticizers will be adsorbed on air-borne particulate matter, which is subjected to rain out and gravitational settling. The adsorption of phthalate by dust particles may be responsible for the widespread distribution of the chemicals in an office building (Weschler, 1984). Vapor-phase plasticizers may be subject to degradation by photochemically produced hydroxyl radicals (HO°) which can be formed by photolysis of water vapor (Lyman, 1990a).

Atmospheric fate processes including photooxidation, washout, and vapor-aerosol partitioning has been reviewed by Staples et al. (1997). The authors mentioned that photodegradation via free radical attack is expected to be the dominant degradation pathway in the atmosphere with predicted half-lives of ca. 1 day for most of the investigated PAEs. DEHP may degrade in the atmosphere at relatively significant rates when compared with other plasticizers. DMP may be one of the most persistent PAE in the atmosphere with a half-life as long as about 50 days. These predicted half-lives also suggest that photo-oxidation by hydroxyl radicals could be the major mechanism to transform plasticizers released into the atmosphere (Wypch, 2004). The specific pathway for release of plasticizers into the atmosphere is described in the next sub-sections via volatilization from water and from soil and sediments.

Nalli et al. (2006a) observed a link between the partial biodegradation of plasticizers by microorganisms and presence of some volatile organic compounds in poor indoor air quality. Their observation was especially pointing towards 2-ethylhexanol resulting from the biodegradation of DEHP commonly found in building materials and household dust. The authors concluded that its presence was not as a result of abiotic degradation (hydrolysis of ester bonds promoted by pH of the concrete surface in contact with DEHP; photochemical degradation). Bjork *et al.* (2003) discussed the possibility that the pH of concrete surface can hydrolyze ester bonds of the compounds in contact with it. However, it is not a major pathway of degradation as alkaline hydrolysis of DEHP would have a half-life of 100 years (Nalli *et al.*, 2006a).

7.5.2 *Water Compartments*

Plasticizers have been detected in different water compartments notably, surface waters (streams, rivers, estuaries and marine), rain and snow, groundwater, drinking and tap water, wastewaters and process aqueous streams of sewage treatment plants. Sources of plasticizers in water compartments are enumerated in the beginning of this section. Research demonstrated the widespread occurrence of PAEs in the aquatic environment and higher amount in surface water immediately downstream from sewage treatment plant (Olivier et al., 2005). Table 7.5 presents concentrations of PAE frequently detected at high levels in various water compartments.

As previously mentioned, the fate of plasticizers in aquatic environment is governed by abiotic degradation, biodegradation, volatilization and adsorption reactions. This fate is influenced by the plasticizer solubility that affects the extent of leaching out of plastic products, the movement and fate of dissolved plasticizers in aqueous systems. This further determines the potential of removal of vapour-phase plasticizers from the atmosphere through precipitation (Wypych, 2004). The fate is also influenced by the alkyl chain length: when length increases, the adsorption and bioconcentration potentials increase while the dissolution, evaporation and biodegradation potentials decrease (Birkett and Lester, 2003; Olivier et al, 2005). Plasticizer solubility also influences its toxicity towards aquatic organisms. DEHP generally exhibits very low toxicity due to low water solubility while the lower molecular weight PAE, either DMP, DEP, DBP, DIBP and BBP, can be toxic to aquatic organisms at high concentration.

Table 7.5 Concentrations of three PAE in various environmental water samples.

PAE	Concentration (µg/L)	Location	Reference
DnBP	0.2-42 µg l^{-1}	Surface water, mostly rivers	Fromme *et al.*, 2002; Brossa *et al.*, 2003; Wypych, 2004
	0.003-0.5 µg l^{-1}	Precipitation	Wypych, 2004
	0.45-2.38 µg l^{-1}	Groundwater	Wypych, 2004
DEHP	0.06-180 µg l^{-1}	Surface water, mostly rivers	Fromme *et al.*, 2002; Brossa *et al.*, 2003; Wypych, 2004; Horn et al., 2004
	0.002-0.43 µg l^{-1}	Precipitation	Wypych, 2004
	1.4-8 µg l^{-1}	Groundwater	HSDB, 2001
	130 µg l^{-1}	Melted snow	Horn *et al.*, 2004
DMP	0.002-0.7 µg l^{-1}	Rivers	Wypych, 2004

The solubility of plasticizers in water at 25°C ranges from about 0.1 mg/l to 4 000 mg/l (see Table 7.3). Once dissolved in water, plasticizers can volatilize into the atmosphere or into soil gases. However, this process is relatively slow as they mostly have a low potential for volatilization from water as expressed by Henry's Law constants lower than 10^{-4} atm-m^3/mole. Estimates of half-life of volatilization of plasticizers from water are considered insignificant for DEHP and DMP. It is estimated to be 114 days for DnBP in a lake scenario and 14 days in a river scenario (Wypych, 2004). Other authors estimated half-lives of plasticizers in surface water between one to 56 days and between one day and one year in groundwater (Wypych, 2004). DBP and DMP are particularly considered as biodegradable, in aerobic surface water. More precisely, the estimated half-lives are 1 to 14 days for DnBP and 1 to 7 days for DMP (Howard et al., 1991). In comparison, DEHP biodegradation half-life is estimated between 5 and 23 days.

DEHP partitions into the solid phase (suspended solids or whole sediment) and biota which decreases their presence in the water column (Birkett and Lester, 2003). In fact, adsorption can be a significant mechanism in controlling environmental fate of most of the plasticizers. It is a physicochemical process by which dissolved plasticizers may be removed from water and concentrated at the solid-liquid interfaces. Moreover, plasticizer organic carbon-water partition coefficients (K_{OC}) are usually greater that 1000 l/kg, indicating a moderate affinity for sediments and soil. This coefficient measures the affinity of a compound to partition to organic matter which in turn will control the mobility of the solutes in water. The extent of adsorption of plasticizers is inversely proportional to their solubility in water. More precisely, K_{OC} was estimated to be 1386 l/kg (Russel and McDuffie, 1986) for DBP, 87,429 l/kg (Russel and McDuffie, 1986) or 482,000 l/kg (Williams et al., 1995) for DEHP and between 190 and 1590 l/kg for DMP (Banerjee et al., 1985). All PAEs can partition to suspended and whole sediments with the degree of sorption depending on the alkyl chain length. Contamination of groundwater by PAEs is less important considering the hydrophobic and low solubility nature of these compounds that decreases the percolation due to sorption on organic matter.

Plasticizers in water can undergo abiotic degradation that can transform or degrade the compound into others having different physicochemical properties. Hydrolysis and photodegradation are two processes that may occur in the environment. It has been predicted that DEHP may hydrolyse to form a monoester and then $CO2$, forming also molecules of 2-ethylhexanol. However, it appears that the process may be too slow to have a major impact on the fate of most dissolved plasticizers. In fact, the estimated hydrolysis half-lives are longer than 100 days (HSDB, 2001). Moreover, if solar radiation may react with plasticizers directly or indirectly, it is generally thought that this mechanism does not occur significantly in the environment except in the atmosphere. In groundwater environment, photolysis and volatilization do not occur. Degradation processes are also less important considering the lower concentration of microorganisms and less favorable conditions.

Biodegradation is considered to be the major transformation process in aqueous environment, particularly for PAEs. However, adsorption may prevail for phthalates with higher K_{OW}, lower vapour pressure and longer alkyl chains. Biodegradation is influenced by: dissolved oxygen, oxidation-reduction potential, temperature, pH, bioavailability, presence of particulate matter, concentration of the pollutants, type and concentration of microorganism. Bioavailability is an important factor and its degree depends on: chemical structure and properties related to sorption and persistence; exposure route via biomagnification (uptake via the food chain) or bioconcentration (uptake of the pollutant via the surrounding phase); and type of aquatic life form, either benthic, demersal or pelagic.

Horn et al. (2004) detected the toxic metabolites of DEHP and DEHA biodegradation, either 2-ethylhexanol and 2-ethylhexanoic acid, in river and snow samples, and even in tap water. 2-ethylhexanoic acid is the most stable form of the breakdown products (Nalli et al., 2002; Nalli et al., 2006b), so it is expected to be ubiquitous in the environment where micro-organisms and appreciable quantities of plasticizers are present.

7.5.3 Soil and Sediment

As adsorption is a major fate mechanism for many plasticizers, sediments can act as a sink for dissolved plasticizers, depending on their carbon content. Various plasticizers have been detected in different sediment and soil samples. The largest reported concentration is 1100 mg/kg for DnBP in river sediments, the range of measured concentrations being 100 mg/kg to 1100 mg/kg. Other reported concentrations are 0.18 to 70 mg/kg of DEHP in coastal sediments (river) and 0.2 to 150 mg/kg of DMP in canal sediments (HSDB, 2001). However, the detected concentrations may reflect point sources rather than the background levels (Kohli, 1989).

Volatilization from dry soil is not a major mechanism that influences the fate of plasticizers in the environment. In fact, estimated half-lives of plasticizers spilled on dry soil are inversely proportional to vapour pressure. As measured vapour pressures are $3.08 \times 10\text{-}4$ mmHg at 25°C for DMP (HSDB, 2001), 2.0×10^{-5} mmHg at 25°C for DnBP (HSDB, 2001; ATSDR 1999) and 7.2×10^{-8} mm Hg at 25°C for DEHP (HSDB, 2001; ATSDR, 1991), the estimated soil-evaporation half-lives for plasticizers spilled onto dry soil showed that most plasticizers would be relatively persistent if biodegradation is not considered. The half-lives are estimated at 19 days for DnBP, 31 days for DMP and greater than one year for DEHP (Thomas, 1990).

Biodegradation was cited as the most rapid process available to degrade plasticizers in biologically active soil. PAE may be used by aerobic and anaerobic soil microbes as a source of carbon and energy, and that their half-lives in soils may range from less than one week to several months (Staples et al., 1997). Half-lives are likely to be longer under anaerobic conditions, and in cold, nutrient-poor environments. The extent of primary biodegradation was studied, yielding to half-lives between 1.9 days for DMP and 15.4 days for DnBP in natural environments (Sugatt et al., 1984), suggesting that there is a potential for phthalates to biodegrade in natural environment. Another study (Howard et al., 1991) estimated biodegradation half-lives for different plasticizers, indicating that they were of 1 to 7 days for DMP, 2 to 23 days for DnBP and 5 to 23 days for DEHP.

However, it was found that compounds having a Koc greater than 100 000 would be expected to be adsorbed significantly, thus decreasing the probability of biodegradation as the compound was less available to organisms. Hence, it was found that although DEHP can degrade readily as its estimated half-life is less than 15 days, the plasticizer was more persistent in a sandy, clay loam soil where after 70 days, only 10% was degraded (Cartwright et al., 2000).

Numerous authors have reported works on the presence of phthalates in sediments. Suzuki et al. (2001) have found phthalate in sediments. Kao et al. (2005) concluded that plasticizers tend to accumulate in freshwater sediments because their biodegradation occur slowly. Horn et al. (2004) have also detected the DEHP and high amount of toxic metabolites in sediments of an important river in an urban area. Olivier et al. (2005) reported a higher amount of phthalates in sediment of receiving water downstream of sewage treatment plant. Fromme et al. (2002) found DEHP in 135 samples of sediment, measured concentrations being about 0.21-8.44 mg DEHP/kg and 0.06-2.08 mg DBP/kg. Finally, Ziogou et al. (1989) reported works on aerobic and anaerobic degradation of phthalates in sediments and soils. Sediments contain normally PAE in the range of over 3 to 4 orders of magnitude in comparison with content in surface water. Zurmühl et al. (1991) also studied the transport of phthalate-esters in undisturbed and unsaturated soil columns.

Sediments constitute an important compartment for hydrophobic and low solubility organic pollutants such as, DEHP that favor sorption. Biodegradation is also affected as a consequence of less bioavailability. Fate is not only determined by sorption and degradation, but also by transport of the suspended particulate sediments that redistribute the bound pollutants to great distances with the possibility of being eventually released in the water column.

7.5.4 Biosolids-Amended Soils

Land application of biosolids is another way for organic pollutants to be reintroduced in the environment. PAEs are hydrophobic compounds known to accumulate in sewage sludge. Once sludge is applied, non-specific van der Waals interactions are expected to dominate the sorption of DEHP in soils (de Jonge et al., 2002). DEHP can be mineralized in sludge and sludge–soil mixtures (Roslev et al., 1998), but part of the DEHP is resistant to biomineralization due to sorption interactions, and may therefore be subject to subsequent leaching.

Organic pollutants having endocrine disrupting properties can be potentially hazardous to the health of farm animals grazing the treated pasture (Meijer et al., 1999; Boerlan et al., 2002; Rhind et al., 2002). Humans can also be affected through the consumption of products derived from grazing animals and contaminated vegetation (Rhind et al., 2005). Although most of DEHP ingested by rats may be degraded in the rat gut, other studies have shown that significant quantities are accumulated in the tissues of sheep. However, studies on land application of sludge and its effect on farm animals suggested that this practice is unlikely to cause large increases of DEHP in animal tissues (Rhind et al., 2005). Besides, one report stated that sheep in soil-amended sewage sludge exhibited more feminine behavior, while other study reported no changes at all in sheep or traces of these chemicals in their organs (Erhand and Rhind, 2004; Rhind et al., 2005).

Roslev et al. (1998) studied the degradation of DEHP by indigenous and microbial additives (inoculum of adapted aerobic microbes) in sludge-amended soil. Indigenous microorganisms were found to be more efficient in DEHP degradation. The kinetics of this mineralization was reported in Madsen et al. (1999). The estimated mineralization was 32% of the initial DEHP content (1.6 mg/kg d.w.) in sludge-amended soil after one year of incubation under anaerobic conditions at 20°C.

A major factor that influences the enzymatic degradation of hydrophobic substances as phthalates is the bioavailability of the substrate in soil and sediment, which is strongly affected by sorption, partitioning and diffusion (Roslev et al., 1998). Presence of other organic substrates also explains the low rate of DEHP biodegradation in soils. Roslev et al. (1998) suggested that immobilization of DEHP and its metabolites in sludge matrix will limit the degradation of DEHP in sludge-amended soil. In fact, de Jonge et al. (2002) have shown that DEHP was strongly bound to the sludge phase.

7.5.5 Bioconcentration in Organisms

Another pathway for plasticizers in the environment is related to aquatic organisms. Staples et al. (1997) reported bioconcentration studies of PAEs involving fishes, bacteria, algae, crustacean, insecta, molluscans, among others. The partitioning behavior of plasticizers between water and organisms is evaluated by K_{ow} as octanol is believed to imitate the fatty structures in plants and the aqueous phase of living tissue. More precisely, K_{ow} greater than 10,000 indicates that the chemical is strongly hydrophobic and should partition significantly to organic phases such as fish tissue (Lyman, 1990b). Log K_{ow} for different PAE plasticizers are presented in Table 7.3, yielding values up to 7.60 for DEHP. Hence, partitioning to aquatic organisms may be an important pathway for many of the plasticizers.

Apart from the K_{ow} value, the bioconcentration factor can indicate the degree to which a chemical accumulates in an aquatic organism at equilibrium when compared with its concentration in the water. Bioconcentration factor was measured for different plasticizers yielding values of 2.125 for the fathead minnow for DNP, 114 to 137 for the bluegill sunfish and 155 to 900 for the fathead minnow for the DEHP, and 57 for the bluegill fish and 5 for the sheephead minnow for the DMP (Staples et al., 1997; HSDB, 2001). As values are below 1000 ml/g, it suggested that bioconcentration by fish is not a significant environmental pathway.

Trophic transfers are also thought to be insignificant because of intensive biotransformation (Staples et al., 1997). Bioconcentration of DEHP by fescue, lettuce, carrots and Chile peppers was minimal from a sewage applied soil (Aranda et al., 1989). This observation was related to the adsorption of phthalate by the sewage and soil which could have limited the amount of DEHP available to be absorbed by plant roots.

7.5.6 Microbial Degraders

Microorganisms found in environment can degrade partially and/or totally the PAE. Metabolic breakdown by microorganisms is considered as a major way of environmental degradation of plasticizers. Although some individual microbes are capable of completely mineralizing PAE, more efficient metabolism appears to result from mixed microbial populations, typically found in the environment. A review of the microbial degradation of PAE was published by Staples et al. (1997) on this subject. The authors reported that biodegradation involves acclimated inocula (e.g. wastewater, activated sludge from sewage treatment plant, and landfill leachate) added to a media spiked with the PAE as sole carbon source. They also reported works on experiments in shake flask or microcosm containing the test chemical and freshwater, marine water, soil or sediments. Since 1997, many microorganisms have been isolated from various environmental compartments for their ability to degrade

PAE, mostly under aerobic conditions. Some of these works are presented in Table 7.6.

Table 7.6 Works published in the last decade[1] on new or common microorganisms able to degrade phthalate esters under aerobic conditions.

Microorganisms	Target Compound(s)	Culture Media or Environmental Sample, and Specific Culture Conditions	Reference
New isolates			
Delftia sp. TBKNP-05	DBP	Mineral salts with the plasticizer as sole carbon source	Patil *et al.*, 2006a
Pseudomonas fluorescences FS1	DMP, DEP, DBP, DIBP, DOP, DEHP	Mineral salts with the plasticizers as sole carbon source	Zeng *et al.*, 2004
Pseudomonas fluorescences B-1	DBP	Mineral salts with the plasticizer as sole carbon source	Xu *et al.*, 2005
Bacillus subtilis subsp.	DEHP	Soil, corn steep liquor medium	Quan *et al.*, 2005
Fusarium oxysporum f. sp. *pisi*	DEHP, BBP	Purified enzyme solution; addition of compound in enzyme solution	Kim *et al.*, 2002, 2003
Gordonia sp.	DEHP	Machine oil-contaminated soil	Chatterjee *et al.*, 2003; Nishioka *et al.*, 2006
Mycobacterium sp. and other strains	DEHP	Garden soil	Nakamiya *et al.*, 2005
Bacillus sp.	DMP	Synthetic, spiked with the target plasticizer; cell-free extracts of esterase	Niazi *et al.*, 2001
Flavobacterium sp.	Phthalic acid	Soil, carbon-limited growth medium	Tanaka *et al.*, 2006
Bacillus sp. NCIM 5220	DMP, DBP	Synthetic, spiked with the target plasticizers; immobilized cells	Patil *et al.*, 2006b
Unidentified strain SD3	DEHP	Spiked wastewater sludge	Roslev *et al.*, 2007

[1] Consult Staples *et al.* (1997) for past works. Chang and Zystra (1998), Eaton (2001) and other works related to those presented in this Table. Xu *et al.* (2005) reported some studies on DMP degradation by pure cultures (*Pasteurella mutocida*, *Sphingomonas paucimoblis*) and mixed culture (*Klebsialla oxytoca*, *Methylobacterium mesophilium*). Niazi *et al.* (2001) also reported other studies involving *Bacillus*, *Pseudomonas*, *Micrococcus*, *Moraxella* and *Comamonas* spp..

[2] Di-octyl terephthalate.

Table 7.6 continued

Microorganisms	Target Compound(s)	Culture Media or Environmental Sample, and Specific Culture Conditions	Reference
Common strains			
Bacillus subtilis, Rhodococcus sp., *Aspergillus niger, A. punicens, Actinomucor elegans, Phnaerochaete chrysosporium*	DEHA, DOP, DOTP[2]	Mineral salt medium spiked with target plasticizers	Nalli *et al.*, 2006b
Rhodococcus rhodocrous, Rhodotura rubra	DEHA, DOP, DOTP	Mineral salt medium spiked with target plasticizers	Nalli *et al.*, 2002; Gartshore *et al.*, 2003
Rhodococcus ruber	DBP	Rubbish landfill soil	Li *et al.*, 2006
Pure or mixed cultures (*Pseudomonas fluorescens, P. aureofaciens, Sphingomonas paucimobilis, Xanthomonas maltophilia*)	DBP, phthalic acid, DMP	Wastewater sludge, minimal salts medium	Wang *et al.* (2003a, 2003b, 2004); Wang, 2004

PAE with shorter alkyl chains are known to be very easily biodegraded in comparison with longer alkyl chains. The initial step of the biodegradation is the hydrolysis of the ester linkage between each alkyl chain and the aromatic rings to form a monoester and subsequently phthalic acid. Some microorganism can only hydrolyze one of the two ester bonds (Xu et al., 2005). The phthalic acid can then be mineralized into CO_2 and H_2O. There are two principal catabolic pathways: (1) some strains partially degrade the compounds by hydrolyzing selectively one of the two ester linkages to generate a mono-alkyl phthalate and an alcohol; (2) other strains can totally degrade and mineralize the mono-alkyl or di-alkyl phthalates (Quan et al., 2006). Partial degradation by microorganisms can lead to the production of toxic metabolites. Enzymes studied or involved in PAE degradation are bacterial esterases (Niazi et al., 2001; Patil et al., 2006a), bacterial dioxygenases (Patil et al., 2006a), fungal cutinases of *Fusarium oxysporum* subsp. piti and yeast esterase of *Candida cylindracea* (Kim et al., 2002, 2003), hydrolases like lipases (Kurane et al., 1984), pork liver esterase (Gavala et al., 2004) and deshydrogenases/decarboxylases (reported by Roslev et al., 1998). These enzymes catalyze hydrolytic reactions as well as esterification and transesterification reactions that lead to detoxification and/or partial or complete degradation. Immobilized cells of a *Bacillus sp.* strain in alginate and polyurethane have been shown to increase the degradation of DBP and phthalic acid (Patil and Karegoudar, 2005; Patil et al., 2006). Such information on potential of microorganisms and enzymes to degrade PAE could be exploited for bioremediation in environmental compartments. Roslev et al. (1998) investigated this possibility in soil while other authors instead targeted on wastewater treatment (Patil and Karegoudar, 2005; Patil et al., 2006).

7.5.7 Landfill Sites

Various types of materials are disposed in solid waste landfill sites and plastic products represent one category. As a number of PVC products have been used in construction industries, this type of waste material disposed of in landfills may also pose a threat to the aquatic environment. Phthalates are so far the most used plasticizers in plastic products and fraction of PVC that ends in municipal solid waste is about 1% (Mersiowsky et al., 2001). Therefore, they may be released to the environment along with the landfill leachate. Asakura et al. (2004) showed that DEHP is the most abundant phthalate in raw leachate (9.6-49 µg 1-1) and processed leachate (4.9-62 µg 1-1). Based on their data, the authors mentioned that these amounts of DEHP were not necessarily related to the quantities of plastic in landfill sites and that the treatment of leachate (by aeration, coagulation/sedimentation and biological treatment) could not result in efficient removal of DEHP.

Horn et al. (2004) studied the presence of the toxic metabolites of DEHP in a landfill leachate, but did not report any concentrations. On the contrary, Fromme et al. (2002) reported 0.12-10.2 µg DEHP 1-1 in runoff water samples from a domestic waste dump. Likewise, Marttinen et al. (2002) reported PAE concentrations in 11 untreated or treated landfill leachates. The authors reported about 1-60 µg DEHP 1-1 in untreated leachates and 0-4 µg DEHP 1-1 in treated leachates. Bauer and Herrmann (1997) reported the publications of several working groups on occurrence of phthalates in landfill leachates. The authors estimated a DEHP concentration of about 2.6% in household wastes, with about 90% of contribution from DEHP. They suggested that there may be a constant output of phthalate from municipal landfills to the surrounding groundwater if no protection system against leakage is installed. Bauer and Herrmann (1998) reported that phthalate esters were mainly present in the dissolved organic carbon part of the leachate. The results indicated that in municipal landfill leachates, the dissolved organic carbon is much more important as a transport vehicle for hydrophobic phthalate esters than the suspended particles present. Further, Bauer and Herrmann (1998) confirmed that dissolved organic macromolecules, mainly humic-like substances, enhance the solubility of PAEs. In the biochemical environments of municipal landfills, short chain PAEs can be degraded by base-catalyzed hydrolysis or by microorganisms aided by enzymes which split the side chains.

Landfill environment is characterized by anaerobic and reductive conditions and the absence of UV radiation. Waste constituents are generally subject to leaching by infiltration water. Biodegradation may also occur as well as adsorption and retention of phthalate esters in the solid waste matrix (Mersiowasky et al., 2001) is possible. Mersiowasky et al. (2001) detected phthalic monoesters and phthalic acid in the leachate obtained with landfill simulation assays. Due to microbial transformation, the concentrations in the leachate are, however, not correlated with the losses. Phthalates and their degradation products may occur transiently at low concentrations. The occurrence of phthalic compounds in landfill leachate cannot conclusively be attributed to plasticizers and/or PVC products. There are a number of other possible sources. The shorter-chained phthalates are not or only partially applied as plasticizers; and may thus be introduced by other materials. Examples include perfumes and cosmetics (DMP, DEP), cellulose products (DEP), as well as inks, polymer dispersions and coatings (DBP, BBP). Jonsson et al. (2003) analyzed leachates from 17 different landfills in Europe and found that phthalic esters were present in the majority of leachate. In some cases, the range of target compounds exceeds the range of PVC additives so as to include transformation products and related chemicals as well. It is therefore necessary to settle for a limited selection which is considered representative. These observed occurrences of degradation products, of all diesters studied, supported that they are degraded under the landfill conditions covered by this study.

The crucial question is the quantity of plasticizers introduced to landfills and how much is available for complete biodegradation. Taking into consideration all facts, the contribution of flexible PVC products in landfills to greenhouse gas emissions may be estimated to be negligible. While the segregation of putrescibles from landfilled waste will cause an increased relative importance of all residual waste components, a significance of flexible PVC products in this respect appears unlikely.

7.5.8 Summary of the Environmental Fate of Plasticizers

Plasticizers are released into the environment by industrial sources, but there are no mandatory requirements to track their release. They have been detected in air samples collected in rural, remote and urban environments, and in samples of indoor air. They have also been detected in trace quantities in precipitation, surface water, groundwater and drinking water samples. Plasticizers have also been detected in sediment and soil samples.

The environmental fate of these plasticizers is controlled by their chemical properties. There are two main reasons as to why plasticizers released into the environment do not accumulate in a water system. DEHP is adsorbed on soil and does not leach into water when released to land. It is also biodegradable and atmospheric bound vapors and mists will travel long distances and will be ultimately deposited on land with rain water.

Furthermore, the plasticizers are largely hydrophobic and have limited volatility. Volatilization from water and dry soil may be slow processes. Moreover, hydrolysis and photodegradation may not be major mechanisms in the degradation of dissolved plasticizers. Biodegradation in water and soil may be the most important processes that transform plasticizers. It appears that adsorption by soil and sediment may be a major sink for plasticizers. It also appears that bioconcentration of most of the plasticizers by aquatic organisms may not be significant. Hence, biodegradation and adsorption appear to be the major mechanisms that control the fate of plasticizers released into the environment.

7.6   Fate in Wastewater Treatment Plants

Sewage treatment plants (STPs) effluents and residues are also major sources of plasticizers in environment. The plasticizers enter principally the STPs through the sewage. Sources of plasticizers in sewage are: urban runoff, sewage pipework deterioration, water from snow melt, industrial air emission/atmospheric deposition and wastewaters from domestic, commercial, institutional and industrial activities (Rule et al., 2006). Plasticizer compounds that eventually reach STPs tend to be concentrated in sewage sludges. This, in turn, may add constraints to the ultimate disposal of these sludges and/or possibilities for their beneficial uses (CEC, 2000; Birkett and Lester, 2003). Processes for wastewater and sludge treatment can remove organic pollutants to a certain extent, either through biological and chemical degradation, and result in their transformation into volatile compounds (Birkett and Lester, 2003). These transformations may also result in the formation of toxic metabolites that may be introduced in the environment through effluent discharge and residues disposal. The plasticizer parent compounds can act as pollutant reservoirs that slowly biodegrade to produce the more toxic compounds including 2-ethylhexanol and 2-ethylhexanoic acid.

The occurrence of some important PAEs in process streams and residues of STPs is shown in Table 7.7. A lot of data are available for effluent and sludge. However, there are data gaps for other residues as grits from pre-treatment systems and ashes from incineration system.

Understanding the fate and behavior of phthalates is important as the knowledge can be used to improve the design and operation of sewage treatment systems and to improve effluent and sludge quality. This requires the building of new STPs with innovative treatment systems. Besides, given that plasticizers and their degradation products may pose a threat to ecosystems after they are released into receiving waters, a study of the fate of these compounds is warranted. In particular, it is essential to evaluate their prevalence in various treatment plant compartments and to identify processes that are most effective in reducing the environmental impacts associated with these compounds (Barnabé et al., 2007).

Among the works published on fate of organic pollutants in STPs, some working groups carried out a detailed study (with mass balance) on the occurrence of PAE, especially DEHP, in many process streams and residues of STPs. These studies are summarized in Table 7.8.

Table 7.7 Concentration ranges of some PAEs in sewage, effluent and sludge of STPs.

PAE	Concentration[1]	Sample	Reference
DBP	3-9 µg l[-1]	Sewage	Marttinen et al., 2002
	n.d. - 10.4 µg l[-1]	Effluent	Marttinen et al., 2003a; Fromme et al., 2002
	n.d. - 3.1 mg kg[-1]	Sludge	Marttinen et al., 2003a,b; Cai et al., 2007
DEHP	28 - 122 µg l[-1]	Sewage	Marttinen et al., 2002
	1.74 - 182 µg l[-1]	Effluent	Marttinen et al., 2003a; Fromme et al., 2002
	4.4 - 346 mg kg[-1] d.w.	Sludge	CEC, 2000; Cheng et al., 2000; Fromme et al., 2002; Marttinen et al., 2003a,b; Beauchesne et al., 2007; Cai et al., 2007
DMP	n.d. - 1 µg l[-1]	Sewage	Marttinen et al., 2002
	n.d. – 0.237 µg l[-1]	Effluent	Roslev et al., 2007
	n.d. - 2.0 mg kg[-1] d.w.	Sludge	Cai et al., 2007; Roslev et al., 2007

[1]   n.d. = not detected.

Studies on fate as presented in Table 7.8 give sufficient information on the behavior and treatment of PAE in STPs. However, they are limited to particular STPs or do not consider the formation of toxic metabolites. Other studies or reviews on occurrence and behavior of PAE in specific wastewater or sludge treatment systems are available. The next sub-sections present and discuss the more recent data available on PAE in sewage and each step of its treatment.

7.6.1 Sewage

Available data on PAE in sewage give less information on their sources as sewage is rarely described in details by the researchers. Knowledge of the sewage system (e.g. age of buildings near wastewater catchments or presence of new buildings or many domestic households; age and deterioration level of urban wastewater catchments; levels of domestic, institutional, commercial and industrial activities in the deserved area...) leads to a better understanding of the sources of PAE in sewage. Two working groups studied the occurrence of PAE in sewages of different composition (Marttinen et al., 2002; Rule et al., 2006). Marttinen et al. (2002) observed that DEHP concentrations were at the same level in sewage containing household wastewater and stormwater runoff when compared to sewage containing industrial discharges and landfill leachates. Rule et al. (2006) published detailed study on presence of DEHP in urban wastewater catchments. Among the studied sources, the authors observed that DEHP concentrations were higher in new domestic discharges than the old ones. As PVC is still used in construction material (e.g. pipes, floor and wall covering). DEHP can leach from this "new" PVC and can be an important source in sewage.

Table 7.8 Summary of published work on fate of PAEs in different STPs.

STP Characteristics	Target PAE and Sample Investigated	Salient Results	Reference
Danish STP serving a population of 80,000; industrial load of 20-25%; equipped with pre-treatment system (screening and grit removal), primary settler, activated sludge with biological nitrogen and phosphorus removal systems, sludge digester	DOP, DnNP, DnBP, BBP in sewage, effluent, sludge (primary and secondary).	Between 60-70% of phthalates were removed by microbial degradation and 20-35% were sorbed to primary and secondary sludge.	Fauser et al., 2003
Finnish STP serving a population of 250000 with a flow rate of 75000 m³ per day; equipped with pre-treatment system (screening and grit removal), settlers, activated sludges including a dephosphoration/nitrification step, sludge thickener and anaerobic digester	DMP, DEP, DnBP, BBP, DEHP, DnOP in sewage, effluent, returned supernatant and filtrate, sludge (primary, secondary, thickened, treated)	DEHP removal from liquid phase was on average 94%; main removal process was sorption to sludge. On average 29% of DEHP was removed by activated sludge. Return flows contained high DEHP.	Marttinen et al, 2003a
English STP serving a population of 11000 and equipped with sedimentation tank, secondary treatment (trickle filters, humus tanks and reedbeds), sludge thickener and anaerobic digester	Many PAEs in sewage and raw sludge as well as in primary tank, trickle filter, humus tank and reedbed systems	Trickle filter system removes 94-99% of DBP and <1-44% of DEHP.	Olivier et al., 2005
Canadian STP serving a population of 1800000 with a flow rate of 2.2x10⁶ m³ per day ; equipped with pre-treatment system (screening and grit removal), primary treatment only addition of FeCl₃+alum, following by addition of acrylamide polymer and sedimentation), sludge homogenizer (no treatment before dewatering)	DEHP, DEHA plus related toxic metabolites (2-ethylhexanol, 2-ethylhexanal, 2-ethylhexanoic acid) in influent, pre-treated wastewater, effluent, grit residues, sludges (homogenized, dewatered, dried)	Most DEHA (98%) and some DEHP (20%) were removed. Metabolites were detected in all samples, except the dried sludge. All analyzed residues contain DEHP and DEHA, including the grit residues and the scums at surface of sedimentation tanks.	Barnabé et al., 2007
Danish STP serving a population of 100000 similar to the STP studied by Fauser et al., 2003.	DMP, BBP and DEHP in influent, effluent, aeration tank and digester	About 93%, 91%, 90% and 81% or, respectively, DMP, DBP, BBP and DEHP were removed by the treatment.	Roslev et al., 2007

7.6.2 Pre-Treatment Processes

Residues from screening (mostly large objects or putrescible matter) are usually incinerated or landfilled, when residues from grits (mostly inorganic materials) are dumped on site and sometimes sent to landfills. Toxic organic pollutants may bind to putrescible matter or inorganic particles. Apart from Marttinen et al. (2003a) and Barnabé et al. (2007), no study has been carried out to verify effect of pre-treatment on organic pollutant removal: only assumptions are made on weak removal of such compounds at this stage. Marttinen et al. (2003a) mentioned that approximately 65% of the sewage DEHP were present in the WW following typical mechanical pre-treatment and sedimentation. According to Barnabé et al. (2007), the pre-treatment system decreased concentration of DEHP, DEHA and DEHTP and their related partial breakdown products, either 2-ethylhexanol, 2-ethylhexanal and 2-ethylhexanoic acid.

Pre-treatment systems may involve separation of oils and fats, which could be a way to improve DEHP removal as described by Olivier et al. (2005). Barnabé et al. (2007) detected DEHP, DEHA and DEHTP in grit residues produced by a grit removal process in a STP. These plasticizers were likely adsorbed onto fats, oils and greases. The authors also observed relatively low concentrations of 2-ethylhexanal and 2-ethylhexanoic acid, and no 2-ethylhexanol. The residual plasticizers in the grit are expected to contribute to the plasticizer loads in landfill leachates and, ultimately, will contribute to the production of toxic degradation products. Indeed, landfill leachates are already known to contain large amounts of plasticizers (Marttinen et al., 2002), which are likely to originate from wastewater treatment plant residues and plastic wastes.

7.6.3 Primary Treatment Processes

During primary treatment, chemical degradation and volatilization of toxic organic compounds may occur in the upper part of the basins, but losses are small. Biotic degradation can also occur, but sorption is likely to be a more important way of removal. In Birkett and Lester (2003), it is mentioned that removal degree of organic pollutants is largely dependent on suspended solids (SS) removal, which is controlled by settling characteristics of particles (density, size and ability to flocculate), sludge retention time (SRT) and surface loading. In fact, when toxic organic compounds have log K_{OW} value greater than 4, the major removal process would be adsorption to the settled sludge (or its SS) (Birkett and Lester, 2003). Organic carbon content of solids as well as polarity and composition of the organic matter, are major factors influencing sorption processes. Moreover, sorption to organic matter can be assessed by log K_{OC} value, so that toxic organic compounds with high value tend to sorb and be accumulated in the primary sludge. These sludges are usually mixed and thickened with sludges from secondary treatment for further decontamination, stabilization, dewatering and final disposal (by incineration, landfilling or land application) and then contribute to the total load of organic pollutants in the STP residues.

Very few works present exact data on fate of few toxic organic compounds during primary treatment of WW. Most works presume the fate of toxic organic compounds based on their behaviour related to their physicochemical properties. In the works described in Table 8, Olivier *et al.* (2005) and Barnabé *et al.* (2007) presented data on the presence of PAE in primary sedimentation tanks. Olivier *et al.* (2005) observed a variable removal of DEHP by sedimentation and sometimes higher concentrations of DEHP than the raw sewage entering the system. This variability could be due to the sludge retention time, the sampling regime and the desorption of DEHP from solids. For DBP, 98% of the observed removal was attributed to the secondary treatment. Barnabé *et al.* (2007) studied DEHP, DEHA and DEHTP in a STP equipped with a physico-chemical treatment system. The authors observed a decrease of 20% of the DEHP found initially in the influent of the STP and low concentrations of 2-ethylhexanol and 2-ethylhexanoic acid, but more of 2-ethylhexanal. Chemical reactions such as hydrolysis and reductive dehalogenation can be responsible for the breakdown of the parent compounds into toxic metabolites. The chemical reactions may favour subsequent biodegradation by microorganisms. Temperature, pH and moisture are major factors responsible for removal of organic pollutants by chemical and biological degradation during primary treatment. Finally, the scums at surface of primary treatment basin contain fat, oils, grease or other hydrophobic compounds and sometimes biofilms offer surface for sorption of some organic compounds. In fact, Barnabé *et al.* (2007) analyzed the scums at the surface of a primary physico-chemical treatment basin and found DEHP, DEHA and DEHTP as well as toxic metabolites, either 2-ethylhexanol and 2-ethylhexanal, but no 2-ethylhexanoic acid.

7.6.4 Secondary Treatment Processes

Secondary wastewater treatment processes include activated sludge process, sequential biological reactor (SBR), biofiltration, and aerated ponds. Most of the reported work on fate of PAE has focused on activated sludge systems (Marttinen et al., 2003a; Fauser et al., 2003; Roslev et al., 2007) and less common systems such as trickling filter (Olivier et al., 2005). Activated sludge systems are known to biodegrade a significant fraction of DEHP entering STPs. Marttinen et al. (2003a) mentioned that approximately 6% of the sewage DEHP were present in the secondary effluent and calculated that 29% were removed in activated sludges while 65% was attributed to the primary treatment. Olivier et al. (2005) studied the treatment of DEHP and DEP in sewage by trickle filter and observed removal percentages of 94-99% for DEP and <1-44% for DEHP. The authors attributed most of the removal to the trickle filter treatment process.

During biological or secondary treatment, the pathways for organic pollutants removal are (Birkett and Lester, 2003): (i) sorption onto solids, microbial flocs, extracellular polymeric substances and/or biofilms; (ii) biological or chemical degradation and; (iii) volatilization (during aeration). Degradation and sorption are the major removal pathways. For all types of biological treatment, the sorption of organic pollutants onto suspended matter (solids, dead or live cells) lead to accumulation in secondary sludge. In fact, biological sludges from secondary treatment are mostly constituted by dead or alive cells, which provide a large surface area for organic pollutant adsorption. It is the major removal pathway during secondary treatment for organic pollutants having high K_{OW} and K_{OC} values. Biodegradation of organic pollutants to metabolites and transformation into volatile compounds occur aerobically. Chemical reactions are occurring, but biodegradation predominate by degrading chemically modified organic compounds. So, production of toxic metabolites during secondary treatment and potential release into ecosystem through effluent discharge can be of concern, but this issue cannot be assessed because of the lack of data. When aeration and agitation are involved, volatilization of low molecular mass and vapour pressure organic compounds by air stripping can happen. Therefore, sorption and degradation are still major modes of removal. As for primary treatment, temperature, pH and moisture are factors to consider for organics removal by chemical or biological degradation. High hydraulic retention time and sludge retention time in biological treatment systems such as activated sludges or SBR are known to enhance sorption and degradation processes and are sometimes necessary to degrade some organic pollutants (Birkett and Lester, 2003). Finally, recirculation of acclimated biomass in activated sludge systems is also known to increase the bodegradation rate of organic pollutants such as plasticizers (Fang and Zheng, 2004). It increases the chance of biodegradation to occur within the works because of longer retention times (Olivier et al., 2005). Fang and Zheng (2004) showed that DEP and DBP are adsorbed onto activated sludge and its extracellular polymeric substances, suggesting that removal can be also due to recirculation of biomass in this type of system.

7.6.5 Tertiary Treatment Processes

Tertiary treatment is principally applied to reduce pathogens during summer or remove phosphate to achieve water quality regulations. It may be disinfection (disinfectant addition, UV lights), dephosphatation, denitrification and filtration processes. Removal and fate of organic pollutants after tertiary treatment of wastewater is less discussed in literature (Birkett and Lester, 2003). Olivier et al. (2005) discuss about the possibility of improving DEHP removal with carbon filters, but there are data gaps on this subject in literature.

7.6.6 Biosolids Processing

Adsorption on suspended solids is an important way of removing many organic compounds during conventional wastewater treatment (Birkett and Lester, 2003). DEHP, which is characterized by a high log K_{OW} of 7.5, will tend to accumulate in sludges (Fountoulakis et al., 2006) and proved to be the most abundant plasticizer in these STP residues (see Table 7.7). Abundance of DEHP varies with the type of sludge and treatment (Beauchesne et al., 2007).

Depending on the type of sludge treatment and dewatering processes, biological and chemical degradation and transformation into volatile compounds are expected to be major removal pathways for organic pollutants (Birkett and Lester, 2003). Most of organic pollutants present are considered as adsorbed on organic or inorganic matter and microbial cells (dead or alive), but remain available for biodegradation. Organic pollutants biodegradation occurs during anaerobic digestion processes (by reductive halogenation, nitroreduction and sulfoxide reduction) and during aerobic digestion (mostly through enzymatic reactions that partially or totally degrade the compound). Anaerobic biodegradation may occur during sludge homogenization, thickening, anaerobic digestion and dewatering. Aerobic biodegradation can occur while the sludge is being aerated during digestion or agitated during homogenization, thickening and dewatering. Volatilization occurs mostly by air stripping when sludge is aerated (e.g., aerobic digestion, dewatering) or agitated (during sludge thickening), but it is less significant for loss of organic pollutants than biodegradation. The temperature, pH, water content and retention time are factors to consider for organics removal by chemical or biological degradation during sludge treatment.

Among sludge treatment processes, effect of sludge digestion on PAE has been intensively investigated. Many studies have been reported by Birkett and Lester (2003) on removal of PAE during sludge digestion. Table 7.9 reports few recent works on PAE biodegradation during sludge pre-treatment, treatment and composting.

In sludge digestion processes, higher removal percentages of PAE are normally observed at longer periods (Birkett and Lester, 2003; Benabdallah El-Hadj et al., 2006). DEHP is reported to be persistent under anaerobic conditions and is removed at high percentage under aerobic conditions, which could be enhanced at high temperature (Fauser et al., 2003; Benabdallah El-Hadj et al., 2006).

If a sludge pre-treatment (blending, sonification, chemical or/and thermal hydrolysis) is applied prior to treatment, it will favour organic pollutants degradation by making organic matter more available for degradation. It can break organic matter aggregate or cells and liberate compounds within or bounded. In Birkett and Lester (2003), a short review on effects of some WWS pre-treatment (sonification, lime stabilization, thermal hydrolysis) was attempted and two other articles are reported in Table 9. Ultrasonic treatment of DBP, BBP and DEHP was found to be efficient (Yim et al., 2002; Psillakis et al., 2004). Chang et al. (2007) recently demonstrated that an ultrasonic pre-treatment applied prior to aerobic digestion can efficiently improve the degradation rate, in order, of DBP, BBP, DEP and DEHP. Gavala et al. (2004) studied the effect of combination of thermal and enzymatic pre-treatment (with pork liver esterases) on DEHP, DEP and DBP. DEHP, DEP and DBP were slowly degraded during mesophilic anaerobic digestion of thermally pre-treated (70°C) primary sludge. Enzymatic pre-treatment with commercial esterase (from pork liver crude), combined or without a thermal pre-treatment increased the degradation rate of DEHP, DEP and DBP to two orders of magnitude faster than under normal mesophilic anaerobic conditions. Considering DEHP as example, it was considered to be the most recalcitrant compound among the commonly used PAEs and anaerobic mesophilic digestion of primary sludge showed that DEHP had a half life of 198-70 days (Gavala et al., 2003a,b). However, with enzymatic pre-treatment using pork liver esterases, the half-life can be reduced to 1 day (Gavala et al., 2004).

Table 7.9 Recent works (2003-2007)[1] on biodegradation of plasticizers during sludge pre-treatment, digestion and composting.

Process	Plasticizer	Salient Results	Reference
Pre-treatment			
Ultrasonic treatment	DEP, BBP, DBP, DEHP	Order of biodegradation rates: DBP > BBP > DEP > DEHP. Ultrasonic treatment accelerated chemical reactions, increased the activity of microorganisms during bio-reaction and thus enhanced biodegradation.	Chang et al., 2007
Thermal	DEP, DBP, DEHP	Biodegradation rates after the thermal pretreatment was proportional to the PAE solubility in water: the higher the solubility, the higher the percentage of the reduction (DEP > DBP > DEHP).	Gavala et al., 2003a
Digestion			
Anaerobic	DEHP	Modeling using ADM1 model demonstrated that DEHP removal was limited by the transfer of DEHP within the solid fraction.	Fountoulakis et al., 2006
Mesophilic and thermophilic anaerobic	DEHP	Higher removal was achieved at thermophilic conditions. High retention time enhances the biodegradation of DEHP.	Benabdallah El-Hadj et al., 2006, 2007
Mesophilic anaerobic	DEHP, DBP	After 28 days of incubation, 95.3 and 94.7 % of DBP and DEHP, respectively remained in petrochemical sludge whereas, 94.1 and 97.3% of DBP and DEHP, respectively were reported in sewage sludge. Degradation rates were affected by alterations in pH value, temperature, substrate concentration, electron donors and acceptors, surfactants and heavy metals	Chang et al., 2005
Aerobic	DEHP	The DEHP removals were 33–41% and 50–62% in 7 and 28 days, respectively. Aeration reduced DEHP present in sewage sludges to levels acceptable for agricultural use.	Marttinen et al., 2003b
Anaerobic	DEHP	The DEHP removal efficiency from the water phase was on average 94% of sewage DEHP, principal processes being sorption to primary and secondary sludges.	Marttinen et al., 2003a

Table 7.9 continued

Process	Plasticizer	Salient Results	Reference
Anaerobic	DEHP, DBP	A combination of recalcitrant (DEHP) and biodegradable (DBP) PAEs into the anaerobic sewage sludge digesters showed that high levels of DEHP in wastewater sludge affected methanogenesis and removal of biodegradable PAEs.	Alatriste-Mondragon *et al.*, 2003
Anaerobic	DEHP, DBP, DEP	Batch and continuous experiments showed that DEP and DBP present in sludge are rapidly degraded under mesophilic anaerobic conditions. Accumulation of high levels of DEHP (more than 60 mg/l) in the anaerobic digester has a negative effect on DBP and DEHP removal rates as well as on the biogas production.	Gavala *et al.*, 2003b
Thermophilic anaerobic		High temperature (68oC) during 5 days enhanced DEHP removal from 9.6% to 34-53%.	Hartmann and Ahring, 2003
Composting			
Lagooning and activated sludges	DEHP, DBP, DMP	The appearance and accumulation of PAEs with a short alkyl side-chain was observed. Thus, alkyl side-chain degradation precedes aromatic ring-cleavage during composting stabilization of sewage sludge.	Amir *et al.*, 2005
Raw and digested sludges	DEHP	Composting removed 58% of the DEHP content of the raw sludge and 34% of that of the anaerobically digested sludge during 85 days stabilization in compost bins.	Marttinen *et al.*, 2003b

[1] Please consult Staples *et al.* (1997) and Birkett and Lester (2003) for past works.

Literature is scarce on PAE degradation or removal after sludge conditioning with coagulants and polyelectrolytes for further dewatering. However, Marttinen et al. (2003a) have reported data on fate of BEHP after sludge treatment and conditioning. Approximately half of the BEHP introduced to sludge treatment was returned to the influent in forms of supernatant and filtrate. According to Marttinen *et al.* (2003a, 2004), the reject water can contain large amount of hydrophobic organic pollutants sorbed to dissolved solids or fine particles. In fact, the authors observed DEHP concentration in the range of 360-2400 µg/l in the reject water. Besides, 32% of BEHP in the influent were founded in treated sludge (after thickening, anaerobic digestion and dewatering). It was also assumed that another 32% were biologically removed during anaerobic digestion and 2% after dewatering.

There are also data gaps on the effect of sludge drying and incineration processes on PAE. Barnabé et al. (2007) detected DEHP and DEHA in dried sludge samples, produced by drum dryers that heat dewatered sludge at temperature over

450oC. For sludge incineration processes, heating temperature can reach approximately 900oC, which mineralizes organic components. Thus, it is presumed that plasticizers would be absent from ash residues (Barnabé et al., 2007).

As mentioned previously, biotic degradation occurs during sludge treatment. Therefore, it is expected that metabolites would be produced throughout treatment and dewatering as the plasticizers undergo biodegradation. According to Beauchesne et al. (2007), the metabolites, 2-ethylhexanol, 2-ethylhexanal and 2-ethylhexanoic acid were found in almost all types of sludge, except the dried sludge. Thus, it is apparent that mixed consortia of microorganisms in sewage systems accomplishes neither the complete transformation of the plasticizers nor the complete breakdown of the metabolites. In dried sludge, it is likely that the metabolites would have been either thermally oxidized or volatilized during the drying process. Presence of plasticizers in treated, dewatered or dried sludge can be of concern for beneficial uses such as land application. However, large quantities of DEHA in sludges would be expected to degrade fairly rapidly in the environment. This degradation will ultimately result in the production of the toxic metabolites. Meanwhile, land applied sludges can act as reservoirs for the production of toxic metabolites by soil microorganisms.

7.7   Fate in Drinking Water and Reuse Water

Drinking water can contain plasticizers at concentrations ranging from 0.03 to 470 µg/l. Wypych (2004) reported DMP at concentrations of 0.1 to 470 µg/l in drinking and tap waters. DEHP concentrations of 4.26 to 86 µg/l in drinking and tap waters have been reported (Brossa et al., 2003; Horn et al., 2004; Wypych, 2004). Even the toxic metabolites of DEHP have been found in tap water.

Plasticizer contamination in aquifers used as sources of drinking water may originate from diffuse sources such as percolation/infiltration and waste disposal practices, or point sources. Percolation of these compounds is attenuated due to sorption on soil organic matter, but still they may find way into groundwater aquifers further contaminating the groundwater sources.

Organic pollutants as PAE are removed during drinking water treatment. Adsorption processes as granular activated carbon are known to remove DEHP (USEPA, 2001). The activated carbon is generally derived from wood and coal. The base carbon material is dehydrated followed by carbonization through slow heating in the absence of air. It is then activated by oxidation at high temperatures (200 to 1000°C), resulting in a highly porous, high surface area per unit mass material. Likewise, coagulation/filtration processes have also been involved in removal of DEHP from potable water. The coagulation/filtration process involves the addition of chemicals like iron salts, aluminum salts, with and without anionic, cationic, or anionic- cationic polymers that coagulate and destabilize particles suspended in the water. Meanwhile, DEHP removal is not possible by the conventional lime softening process. The usual chlorination in fact leads to more oxidized forms of DEHP which are potentially more toxic than the parent compound.

Materials in contact with drinking water supplies may be an additional source of contamination, especially if they are made of plastics. However, there is a limited use of phthalates in small areas which are actually in contact with water. Meanwhile, DEHP was included by USEPA in a group of specified organic chemicals, presence of which is tested by local authorities. Maximum contaminant level is established at 6 µg/l. As previously stated, data showed that some amount of plasticizers are detected in aqueous media in the environment, but these are usually in parts per billion, which is well below the maximum contaminant level established by the USEPA. However, presence of such contaminants as DEHP in water can be of significant concern for water reuse. Moreover, as the toxicity component will be getting accumulated leading to long term effects further restricting reuse.

7.8 Conclusion

The discharge of plasticizers into the environment must be priorities in order to minimize human and environmental health impacts. Industries must make efforts to limit the use of toxic plasticizers and especially those that can potentially act as reservoir for toxic metabolites such as 2-ethylhexanol. Thus, new and eco-friendly plasticizers, leading to no toxic intermediates after biodegradation, could be developed. Other considerations mentioned by Rahman and Brazel (2004) must be taken for plasticizer development in order to reduce the potential to leach out from the product or migrate to other polymeric substances and surroundings. Research must also be conducted on the fate and transport of other less frequently used plasticizers. These studies could assess the possibility of these plasticizers to partially degrade into toxic compounds once released in the environment.

Meanwhile, as proposed by some authors (Marttinen et al., 2004; Olivier et al., 2005), improving the design and operation of STP in order to increase PAE removal could be a solution to minimize intrusion of plasticizers in environment. For this, it could be important to gain knowledge on other biological treatment systems other than activated sludges or trickling filters as well as the diurnal and seasonal occurrence and removal of phthalates in all types of treatment systems. Detailed investigations on fate of plasticizers in STPs are required. In fact, following the work of Marttinen et al. (2003a), the same working group initiated an interesting study on the removal of DEHP contained in the reject water of a STP (consisting of supernatant and filtrate of sludge thickening, treatment and dewatering processes), which is normally returned in the influent of the station and can cause an internal pollution load (Marttinen et al., 2004). The authors used a sequential batch reactor for treatment and were able to remove DEHP from this reject water. This study suggested that the treatment of the reject water from sludge processes can reduce the amount of hydrophobic organic pollutants such as DEHP in the process streams of a STP. The invasion of plasticizers in environment through STP effluent and residues as a source can also be addressed by developing approaches to minimize the release of plasticizers into sewer systems or improving or optimizing processes for the stabilization of sludges to achieve the removal of plasticizer-related contaminants. All these considerations must also be taken account during the construction of new STPs. If new systems for wastewater, sludge and reject water treatment remove efficiently PAE and their relative toxic metabolites, there may be a possibility that other priority and emerging organic pollutants are degraded during the process.

Additionally, in order to favour the beneficial uses of sludge, it is recommended that plasticizer and toxic metabolite removal must be an important consideration during the development or the optimization of sludge treatment processes. It seems reasonable to fix a limit for phthalate concentrations in sludge for land application as proposed by CEC (2000). This agency selected DEHP for limiting their content in sludge along with other priority organic pollutants found in high amount in sludge, either poly aromatic hydrocarbons, polychlorinated biphenyls, the nonylphenol ethoxylates, linear alkylbenzene sulphonates and dibenzodioxins/furans. A limit of 10-100 mg DEHP kg-1 d.w was proposed by the CEC and it could be important to assess the potential environmental threat of the toxic metabolites at this range.

Finally, research is mandatory to fill gaps in the environmental fate and movement of plasticizers, especially for plasticizers with large molecular weights (Wypych, 2004). There is also a lack of information on the identification, toxicity, occurrence, fate and transport of plasticizer metabolites in the environment and STP. As major sources of plasticizers in environment, it is particularly important to evaluate the prevalence of the parent compounds and toxic metabolite in different process streams of STPs. This may allow to: (i) identify the treatment processes that are most effective in reducing environmental impacts associated with these compounds; (ii) select wastewaters and sludges that are safe for beneficial uses such as land application or bioconversion into value-added products; (iv) elucidate the need for eco-friendly alternatives to conventional plasticizers; and (v) assess the need to refine water quality monitoring protocols by including plasticizers and their toxic intermediates.

References

Adams, R.C. (2001). A comparison of plasticizers for use in flexible vinyl medical products. Medical Device Diagnostic Ind., April.
Alatriste-Mondragon, F., Iranpour, R., Ahring, B.K. (2003). Toxicity of di-(2-ethylhexyl) phthalate on the anaerobic digestion of wastewater sludge. Water res., 37(6): 1260-1269.
Amir, S., Hafidi, M., Merlina, G., Hamdi, H., Jouraiphy, A., El Gharous, M., Revel, J.C. (2005). Fate of phthalic acid esters during composting of both lagooning and activated sludges. Process Biochem., 40 : 2183-2190.

Aranda, J.M., G.A. O'Connor, Eiceman, G.A. (1989). Effects of sewage sludge on diethylhexyl phthalate uptake by plant. J. Environ. Qual. 18:45-50.

Asakura, H., Matsuto, T., Tanaka, N. (2004). Behavior of endocrine-disrupting chemicals in leachate from MSW landfill sites in Japan. Waste Manag., 24(6): 613-22.

ATSDR (1991). Toxicological Profile for di-(2-ethylhexyl) phthalate. Agency for Toxic Substances and Disease, Atlanta, Georgia, United States.

ATSDR (1999). Toxicological Profile for di-n-butyl phthalate. Agency for Toxic Substances and Disease, Atlanta, Georgia, United States.

ATSDR (2003). Minimal risks levels, MRLs, for hazardous substances – January. Agency for Toxic Substances and Disease Registry.

Banerjee, P., Piwoni, M.D., Ebeid, K. (1985) Sorption of organic contaminants to a low carbon subsurface core. Chemosphere, 14: 1057-1067

Barnabé, S., Beauchesne, I., Cooper, D.G., Nicell, J.A. (2007). Plasticizers and their degradation products in the process streams of a large urban physicochemical sewage treatment plant. Water Research, in press.

Bauer, M.J., Herrmann, R. (1997). Estimation of the environmental contamination by phthalic acid esters leaching from household wastes. Sci. Tot. Environ., 208: 49-57.

Bauer, M.J., Herrmann, R. (1998). Dissolved organic carbon as the main carrier of phthalic acid esters in municipal landfill leachates. Waste Manage. Res., 16 (5): 446-454.

Beauchesne, I., Barnabé, S., Cooper, D.G., Nicell, J.A. (2007). Plasticizers and Related Toxic Degradation Products in Wastewater Sludges. In proceedings of IWA specialist conference – Moving Forward, Wastewater Biosolids Sustainability : Technical, Managerial and Public Synergy, June 24-27, Moncton, New Brunswick, p.231-238.

Benabdallah El-Hadj, T., Dosta, J., Mata-Alvarez, J. (2006). Biodegradation of PAH and DEHP micro-pollutants in mesophilic and thermophilic anaerobic sewage sludge digestion. Water Sci. Technol., 53 (8): 99-107.

Birkett, J.W., Lester, J.N. (2003). Endocrine disrupters in wastewater and sludge treatment processes. CRC Press LLC & IWA publishing, London, UK, 295 pages.

Bjork, F., Eriksson, C.-A., Karlsson, S., Khabbaz, F. (2003). Degradation of components in flooring systems in humid and alkaline environments. Construction and Building Materials, 17: 213–221.

Boerjan, M.L., Freijnagel, S., Rhind, S.M., Meijer, G.A.L. (2002). The potential reproductive effects of exposure of domestic ruminants to endocrine disrupting compounds. Animal Science, 74: 3-12.

Bornehag, C.-G., Lundgren, B., Weschler, C.J., Sigsgaard, T., Hagerhed-Engman, L., Sundell, J. (2005). Phthalates in indoor dust and their association with building characteristics. Environ Health Perspect., 113(10): 1399–1404.

Brossa L, Marcé RM, Borrull F, Pocurull E. (2003). Determination of endocrine-disrupting compounds in water samples by online solid-phase extraction-programmed-temperature vaporization–gas chromatography-mass spectrometry. J. Chromatography A, 998: 41-50.

Cai, Q.-Y., Mo, C.-H., Wu, Q.-T., Qiao-Yun Zeng, Q.-Y., Katsoyiannis, A. (2007). Occurrence of organic contaminants in sewage sludges from eleven wastewater treatment plants, China. Chemosphere, 68 (9): 1751-1762.

Cartwright, C.D., Owen, S.A., Thompson, I.A., Burns, R.G. (2000). Biodegradation of diethyl phthalate in soil by a novel pathway. FEMS Microbio,l Lett., 186: 27–34.

CEC (2000). Working document on sludge – 3rd draft. Commission of the European Communities Directorate-General Environment, ENV.E.3/LM, Brussels, 27 April 2000.

Chang, H.K., Zylstra, G.J. (1998). Novel organization of the genes for phthalate degradation from Brukholderia cepacia DBO1. J. Bacteriol., 180: 6529-6537.

Chang, B.V., Liao, G.S., Yuan, S.Y. (2005). Anaerobic Degradation of Di-n-butyl Phthalate and Di-(2-ethylhexyl) Phthalate in Sludge. Bull. Environ. Contam. Toxicol., 75 (4) : 775-782.

Chang, B.V., Wang, T.H., Yuan, S.Y. (2007). Biodegradation of four phthalate esters in sludge. Chemosphere, 69(5): 769-775.

Chatterjee, S., Dutta, T.K. (2003) Metabolism of butyl benzyl phthalate by Gordonia sp. strain MTCC 4818. Biochem. Biophys. Res. Commun., 309: 36–43.

Cheng, H.F., Chen, S.Y., Lin, J.G. (2000). Biodegradation of di-(2-ethylhexyl) phthalate in sewage sludge. Water Sci. Technol., 41 (12): 1–6.

de Jonge, H., de Jonge, L.W., Blicher, B.W. Moldrup P. (2002). Transport of di(2 ethylhexyl)phthalate (DEHP) applied with sewage sludge to undisturbed and repacked soil Columns. J. Environ. Qual., 31: 1963-1971.

Di Gangi, J., Norin, H. (2002). Pretty Nasty – Phthalates in European Cosmetics Products. Published by Healthcare Without Harm, USA and in association with Women's Environmental Network and Swedish Society for Nature Conservation.

Eaton, R.W. (2001). Plasmid-Encoded Phthalate Catabolic Pathway in Arthrobacter keyseri 12 B. J. Bacteriol., 183 (12): 3689-3703.

Ejlertsson, J., Johansson, E., Karlsson, A., Meyerson, U., Svensson, B.H. (1996a). Anaerobic degradation of xenobiotics by organisms from municipal solid waste under landfilling conditions. Antonie van Leeuwenhoek, 69: 67–74.

Ejlertsson, J., Meyerson., U., Svensson, B.H. (1996b). Degradation of phthalic acid esters in municipal solid waste under methanogenic conditions. Biodegradation, 7: 345–352.

Erhard, H.W., Rhind, S.M. (2004). Prenatal and postnatal exposure to environmental pollutants in sewage sludge alters emotional reactivity and exploratory behavior in sheep. Sci. Total Environ., 332: 101-108.

Fang, H.H.P, Zheng, H. (2004). Adsorption of phthalates by activated sludge and its biopolymers. Environ. Technol., 25: 757-761.

Fauser, P., Vikelsoe, J., Sorensen, P.B., Carlsen, L. (2003). Phthlates, nonylphenols and LAS in an alternately operated wastewater treatment plant – Fate modelling based on measured concentrations in wastewater and sludge. Water Res., 37: 1288-1295.

FDA - U.S. Food and Drug Administration (2002a). Code of federal regulations. Title 21. Food and drugs. Chapter 1. Food and drug administration. Department of health and human services.

FDA - U.S. Food and Drug Administration (2002b). Center for Devices and Radiological Health. FDA Public Health Notification: PVC Devices Containing Plasticizer DEHP – July.

FDA - U.S. Food and Drug Administration (2002c). Center for Devices and Radiological Health. FDA Public Health Notification: PVC Devices Containing Plasticizer DEHP - September.

Fountoulakis, M.S., Stamatelatou, K., Batstone, D.J., Lyberatos, G. (2006). Simulation of DEHP biodegradation and sorption during the anaerobic digestion of secondary sludge. Water Sci. Technol., 54(4): 119-128.

Fromme, H., Küchler, T., Otto, T., Pilz, K., Müller, J., Wenzel, A. (2002). Occurrence of phthalates and bisphenol A and F in the environment. Water Res., 36: 1419-1438.

Gartshore, J., Cooper, D.G., Nicell, J.A. (2003). Biodegradation of plasticizers by Rhodotorula Rubra. Env. Toxicol. Chem., 22 (6): 1244-1251.

Gavala, H.N., Yenal, U., Skiadas, I.V., Westermann, P., Ahring, B.K. (2003a). Mesophilic and thermophilic anaerobic digestion of primary and secondary sludge. Effect of pre-treatment at elevate temperature. Water Res., 37 : 4561–4572.

Gavala, H.N., Alatriste-Mondragron, F., Iranpour, R., Ahring, B.K. (2003b). Biodegradation of phthalate esters during anaerobic digestion of sludge. Chemosphere, 52: 673-682.

Gavala, H.N., Yenal, U., Ahring, B.K (2004). Thermal and enzymatic pretreatment of sludge containing phthalate esters prior to mesophilic anaerobic digestion. Biotechnol. Bioeng., 85 (5) : 561-567.

Gomez, H.A., Aguilar, C.M.P. (2003). Social and economic interest in the control of phthalic acid esters. Trends Anal. Chem., 22 (11): 848-857.

Goulas, A.E., Zygoura, P., Karatapanis, A., Georgantelis, D., Kontominas, M.G. (2007). Migration of di(2-ethylhexyl) adipate and acetyltributyl citrate plasticizers from food-grade PVC film into sweetened sesame paste (halawa tehineh): Kinetic and penetration study. Food Chem. Toxicol., 45 (4): 585-591.

Harris, C.A., Henttu, P., Parker, M.G., Sumpter, J.P. (1997). The Estrogenic Activity of Phthalate Esters in Vitro. Environmental Health Perspectives, 105(8): 802-811.

Hartmann, H., Ahring, B.K. (2003). Phthalic acid esters found in municipal organic waste: enhanced anaerobic degradation under hyper-thermophilic conditions. Water Sci. Technol., 48 : 175–183.

Hildenbrand,L., Lehmann, H.D., Wodarz, R., Ziemer, G., Wendel, H.P. (2005). PVC-plasticizer DEHP in medical products: do thin coatings really reduce DEHP leaching into blood? Perfusion, 20 (6): 351-357.

Horn, O., Nalli, S., Cooper, D.G., Nicell, J.A. (2004). Plasticizer metabolites in the environment. Water Res., 38: 3693-3698.

Howard, P.H., Boethling, R.S., Jarvis, W.F., Meylan, W.M., Michalenko, E.M. (1991). Handbook of environmental degradation rates. Lewis Publishers, Chelsea, Michigan.

HSDB - Hazardous Substances Data Bank (2001). National Library of Medicine [Online].. http://www.toxnet.nlm.nih.gov/cgi_bin/sis/htmlgen?HSDB. Consulted on September 10th 2007.

Jonsson, S., Ejlertsson, J., Ledin, A., Mersiowsky, I., Svensson, B.H. (2003). Mono- and diesters from o-phthalic acid in leachates from different European landfills. Water Res., 37: 609–617.

Kao, P.H., Lee, F.Y., Hseu, Z.Y. (2005). Sorption and Biodegradation of Phthalic Acid Esters in Freshwater Sediments. J. Environ. Sci. Health, Part A, 40 (1): 103-115.

Kim, Y.H., Lee, J., Ahn, J.Y., Gu, M.B., Moon, S.H. (2002). Enhanced degradation of an endocrine-disrupting chemical, butyl benzyl phthalate, by Fusarium oxysporum f. sp. pisi cutinase. Appl. Environ. Microbiol., 68: 4684–4688.

Kim, Y.H., Lee, J., Moon, S.H. (2003). Degradation of an endocrine disrupting chemical, DEHP [di-(2-ethylhexyl)-phthalate], by Fusarium oxysporum f. sp. pisi cutinase. Appl. Environ. Microbiol., 63 (1): 75–80.

Koch, H.M., Drexler, H., Angerer, J. (2003). An estimation of the daily intake of di(2 ethylhexyl)phthalate (DEHP) and other phthalates in the general population. Int. J. Hyg. Environ. Health., 206: 77–83.

Koch, H.M., Preuss, R., Angerer, J. (2006). Di(2-ethylhexyl)phthalate (DEHP): human metabolism and internal exposure - an update and latest results. Int. J. Androl., 29 (1), 155–165.

Kohli, J., Ryan, J.F., Afghan, B.K. (1989). Phthalate esters in the aquatic environment. In: Analysis of trace organics in the aquatic environment, Afghan, B.K., Chau, A.S.Y. (eds), Chapter 7, p.243-281, CRC Press. Boca Raton, Florida.

Koo, H.J., Lee, B.M. (2004). Estimated exposure to phthalates in cosmetics and risk assessment. J. Toxicol. Environ. Health A, 67 (23-24): 1901-1914.

Kurane, R., Suzuki, T., Fukuoka, S, (1984). Purification and some properties of a phthalate ester hydrolysing enzyme from Nocardia erythropolis. Appl. Microbiol. Biotechnol., 20: 378–383

La Guardia, M.J., Hale, R.C., Harvey, E., Mainor, T.M. (2001). Alkylphenol ethoxylate degradation products in land-applied sewage sludge (biosolids). Environ. Sci. Technol., 35: 4798-4804.

Larsen, S.T., Lund, R.M., Nielsen, G.D., Thygesen, P., Poulsen, O.M. (2002). Adjuvant effect of di-n-butyl-, di-n-octyl-, di-iso-nonyl- and di-iso-decyl phthalate in a subcutaneous injection model using BALB/c mice. Pharmacol. Toxicol., 91(5): 264-72.

Latini, G. (2000). The potential hazard of exposure to DEHP in babies: a review. Biol. Neonates, 78 (4): 269-276.

Latini, G., De Felice, C., Presta, G. (2003). In utero exposure to Di-(2-Ethylhexyl)-phthalate and duration of human pregnancy. Environ. Health Perspect., 111: 1783–1785.

Latini, G. (2005). Monitoring phthalate exposure in humans. Clinica Chimica Acta, 361: 20–29.

Latini, G., De Felice, C., Verrotti, A. (2004). Plasticizers, infant nutrition and reproductive health. Reproductive Toxicol., 19: 27-33.

Latini, G., Vecchio, A.D., Massaro, M., Verrotti, A., De Felice, C. (2006). Phthalate exposure and male infertility. Toxicology, 226: 90-98.

Lewis, R.J. (1999). Sax's dansgerous properties of industrial materials - 10th edition. John Wiley, New York.

Li, J., Chen, J., Zhao, Q., Li, X., Shu, W. (2006). Bioremediation of environmental endocrine disruptor di-n-butyl phthalate ester by Rhodococcus rubber. Chemosphere, 65 (9): 1627-1633.

Lopez-Espinosa, M.J., Granata, A., Araque, P., Molina-Molina, J.M., Puertollano, M.C., Rivas, A., Fernández, M., Cerrillo, I., Olea-Serrano, M.F., Lopez, C., Olea, N. (2007). Oestrogenicity of paper and cardboard extracts used as food containers. Food Additives and Contaminants, 24(1): 95-102.

Lyman, W.J. (1990a). Atmospheric residence time In: Lyman, W.J., Reehl, W.F., Rosenblatt, D.H. (eds), Handbook of chemical property estimation methods, American Chemical Society, Washington D.C., Chapter 10.

Lyman, W.J. (1990b). Octanol/water partition coefficient. In: Lyman, W.J., Reehl, W.F., Rosenblatt, D.H. (eds), Handbook of chemical property estimation methods, American Chemical Society, Washington D.C., Chapter 1.

Madsen, P.L., Thyme, J.B., Henriksen, K., Møldrup, P., Roslev, P. (1999). Kinetics of di(ethylhexyl)phthalate (DEHP) mineralization in sludge-amended soil. Environ. Sci. Technol., 33 : 2601-2606.

Marttinen, S.K., Kettunen, R.H., Rintala, J.A. (2002). Occurrence and removal of organic pollutants in sewages and landfill leachates. Sci. Total Environ., 301: 1-12.

Marttinen, S.K., Kettunen, R.H., Sormunen, K.M., Rintala, J.A. (2003a). Removal of bis (2-ethyhexyl) phthalate at a sewage treatment plant. Water Res., 37: 1385-1393.

Marttinen, S.K., Hanninen, K., Rintala, J.A. (2003b). Removal of DEHP in composting and aeration of sewage sludge. Chemosphere, 54: 265-272.

Marttinen, S.K., Maria Ruissalo, M., Rintala, J.A. (2004). Removal of bis (2-ethylhexyl) phthalate from reject water in a nitrogen-removing sequencing batch reactor. J. Environ. Manag., 73 (2): 103-109.

Meijer, G.A.L., Bree, J.A., Wagenaar, J.A., Spoelstra, S. F. (1999). Sewerage overflows put production and fertility of dairy cows at risk. J. Environ. Qual., 28: 1381-1383.

Mersiowsky, I., Weller, M., Ejlertsson, J. (2001). Fate of plasticized PVC products under landfill conditions: a laboratory-scale landfill simulation reactor study. Water Res., 35 (13): 3063–3070.

Moore, N.P. (2000). The estrogenic potential of the phthalate esters. Reprod. Toxicol., 14: 183-192.

Nakamiya, K., Hashimoto, S., Ito, H., Edmonds, J.S., Yasuhara, A., Morita, M. (2005). Microbial Treatment of Bis (2-Ethylhexyl) Phthalate in Polyvinyl Chloride with Isolated Bacteria. J. Biosci. Bioeng., 99 (2): 115-119.

Nalli, S., Cooper, D.G., Nicell, J.A. (2002). Biodegradation of plasticizers by Rhodococcus rhodochrous. Biodegradation, 13: 343-352.

Nalli, S., Horn, O.J., Grochowalski, A.R., Cooper, D.G., Nicell, J.A. (2006a). Origin of 2-ethylhexanol as a VOC. Environ. Pollut., 140 (1): 181-185.

Nalli, S., Cooper, D.G., Nicell, J.A. (2006b). Metabolites from the biodegradation of di-ester plasticizers. Sci. Total Environ., 366: 286-294

Niazi, J.H., Prasad, D.T., Karegoudar, T.B. (2001). Initial degradation of dimethyl phthalate by esterases from Bacillus species. FEMS Microbiology Letters, 196 : 201–205.

NIOSH (2003). Pocket guide to chemical hazards. NIOSH publication no.97-140 – January.

Nishioka, T., Iwata, M., Imaoka, T., Mutoh, M., Egashira, Y., Nishiyama, T., Shin, T., Fuji, T. (2006). A mono-2-ethylhexyl phthalate hydrolase from a Gordonia sp. that is able to dissimilate di-2-ethylhexyl phthalate. Appl. Environ. Microbiol., 72 (4): 2394-2399.

Nuti, F., Hildenbrand, S., Chelli, M., Wodarz, R., Papini, A.M. (2005). Synthesis of DEHP metabolites as biomarkers for GC-MS evaluation of phthalates as endocrine disrupters. Bioorganic Medicinal Chem., 13: 3461-3465.

Olivier, R., Eric, M., William, J. (2005). The occurrence and removal of phthalates in a trickle filter STW. Water Res., 39: 4436-4444.

Patil, N.K., Karegoudar, T.B. (2005). Parametric Studies on Batch Degradation of a Plasticizer Di-n-Butylphthalate by Immobilized Bacillus sp.. World J. Microbiol. Biotechnol., 21: 1493-1498.

Patil, N.K., Kundapur, R., Shouche, Y.S., Karegoudar, T.B. (2006a). Degradation of a Plasticizer, di-n-Butylphthalate by Delftia sp. TBKNP-05. Curr. Microbiol., 52: 369-374.

Patil, N.K., Veeranagouda, Y., Vijaykumar, M.H., Nayak, S.A., Karegoudar, T.B. (2006b). Enhanced and potential degradation of o-phthalate by Bacillus sp. immobilized cells in alginate and polyurethane. Int. Biodeteriorat. Biodegrad., 57: 82-87.

Psillakis, E., Mantzavinos, D., Kalogerakis, N. (2004). Monitoring the sonochemical degradation of phthalate esters in water using solid-phase microextraction. Chemosphere, 54 : 849–857.

Quan, C.S., Q. Liu, W.J. Tian, J. Kikuchi and S.D. Fan (2005). Biodegradation of an endocrine disrupting chemical, di-ethylhexyl phthalate, by Bacillus subtitlis no. 66. Appl. Microbiol. Biotechnol., 66 : 702-710.

Quan, C.S., Zheng, W., Liu, Q., Ohta, Y., Fan, S.D. (2006). Isolation and characterization of a novel Burkholderia cepacia with strong antifungal activity against Rhizoctonia solani. Appl. Microbiol. Biotechnol., 72 (6) : 1276-1284.

Rahman, M., Brazel, C.S. (2004). The plasticizer market: an assessment of traditional plasticizers and research trends to meet new challenges. Prog. Polym. Sci., 29 : 1223-1248.

Rhind, S.M., Smith, A., Kyle, C.E., Telfer, G., Martin, G., Duff, E., Mayes, R.W. (2002). Phthalate and alkyl phenol concentrations in soil following applications of inorganic fertilizer or sewage sludge to pasture and potential rates of ingestion by grazing ruminants. J. Environ. Monitor., 4: 142–148.

Rhind, S.M., C.E. Kyle, G. Telfer, E.I. Duff, Smith, A. (2005). Alkyl phenols and diethylhexyl phthalate in tissues of sheep grazing pastures fertilized with sewage sludge or inorganic fertilizer. Environ. Health Perspectives, 113:447-453.

Roslev, P., Madsen, P.L., Thyme, J.B., Henriksen, K. (1998). Degradation of phthalate and di-(2-ethylhexyl)phthalate by indigenous and inoculated microorganisms in sludge-amended soil. Appl. Environ. Microbiol., 64: 4711–4719.

Roslev, P., Vorkamp, K., Aarup, J., Frederiksen, K., Nielsen, P.H. (2007). Degradation of phthalate esters in an activated sludge wastewater treatment plant. Water Res., 41: 969-976.

Rule, K.L., Comber, S.D.W., Ross, D., Thornton, A., Makropoulos, C.K., Rautiu, R. (2006). Sources of priority substances entering an urban wastewater catchment-trace organic chemicals. Chemosphere, 63 (4): 581-591.

Russell, D.J., McDuffie, B. (1986). Chemodynamic properties of phthalate esters: Partitioning and soil migration. Chemosphere, 15 (8): 1003-1021.

Sate of California (2003). Chemicals known to the state to cause cancer or reproductive toxicity – March 14. Environmental Protection Agency, Office of Environmental Health Hazard Assessment, Safe Drinking Water and Toxic Enforcement Act of 1986.

Staples, C.A., Peterson, D.R., Parkerton, T.F., Adams, W J. (1997). The environmental fate of phthalate esters: a literature review. Chemosphere, 35: 667–749.

Staples, C.A. (2001). A review of the environmental fate and aquatic effects of a series of C4 and C8 oxo-process chemicals. Chemosphere, 45: 339-346.

Sugatt, R.H., O'Grady, D.P., Banerjeem, S., Howard, P.H., Gledhill, W.E. (1984). Shake Flask Biodegradation of 14 Commercial Phthalate Esters. Appl. Environ. Microbiol., 47(4): 601–606.

Suzuki, T., Yaguchi, K., Suzuki, S., Suga, T. (2001). Monitoring of Phthalic Acid Monoesters in River Water by Solid-Phase Extraction and GC-MS Determination. Environ. Sci. Technol., 35 (18) : 3757 -3763.

Tanaka, T., Yamada, K., Iijima, T., Iriguchi, T., Kido, Y. (2006). Complete Degradation of the Endocrine-Disrupting Chemical Phthalic Acid by Flavobacterium sp. J. Health Sci., 52 (6): 800-804.

Thomas, R.G. (1990). Volatilization from soil. In: Lyman, W.J., Reehl, W.F., Rosenbault, D.H. (eds), Hanbook of chemical property estimation methods, American Chemical Society, Washington, D.C. Chapter 16.

Tickner, J.A., Schettler, T., Guidotti, T., McCally, M., Rossi, M. (2001). Health risks posed by use of Di-2-ethylhexyl phthalate (DEHP) in PVC medical devices: A critical review. Am. J. Ind. Med., 39 (1): 100-111

TRI-Toxic Release Inventory (2005). National Library of Medicine [Online]. Available at http://www.toxnet.nlm.nih.gov/cgi_bin/sis/htmlgen?TRI. Consulted on September 10th 2007.

USEPA (2001). Removal of endocrine disrupting compounds using drinking water treatment processes. EPA/625/R-00/015, Washington, DC.

Vega, D., Bastide, J. (2003). Dimethylphthalate hydrolysis by specific microbial esterase. Chemosphere, 51: 663–668.

Wang, J. (2004). Effet of DBP on activated sludge. Process Biochem., 39: 1831-1836.

Wang, Y., Fan, Y., Gu, J.-D. (2003). Microbial degradation of the endocrine-disrupting chemicals phthalic acid and dimethyl phthalate ester under aerobic conditions. Bulletin of Environmental Contamination and Toxicology, 71: 810–818.

Wang, Y., Fan, Y., Gu, J.-D. (2004). Dimethyl phthalate ester degradation by two planktonic and immobilized bacterial consortia. Int. Biodeterioration Biodegrad., 53: 93–101.

Weschler, C.J. (1984). Indoor-Outdoor Relationships for Nonpolar Organic Constituents of Aerosol Particles. Environ. Sci. Technol., 18: 648-652.

Wensing, M., Uhde, E., Salthammer, T. (2005). Plastics additives in the indoor environment; flame retardants and plasticizers. Sci. Total Environ., 339 (1-3) : 19-40.

Williams, M.D., Adams, W.J., Parkerton, T.F., Biddinger, G.R., Robillard, K.A. (1995). Sediment sorption coefficient measurement for four phthalate esters: experimental results and model theory. Environ. Toxicol. Chem., 14 (9):1477-1486.

Wypych, G. (2004). Handbook of Plasticizers. ChemTec Publishing, Toronto, 687 pages.

Xu, X.-R., Li, H.-B., Gu, J.-D. (2005). Biodegradation of an endocrine-disrupting chemical di-n-butyl phthalate ester by Pseudomonas fluorescens B-1. Int. Biodeterioration Biodegrad., 55 (1): 9-15.

Yim, B., Nagata, Y., Maeda, Y. (2002). Sonolytic Degradation of Phthalic Acid Esters in Aqueous Solutions. Acceleration of Hydrolysis by Sonochemical Action. J. Phys. Chem. A, 106 (1): 104 -107.

Zeng, F., Cui, K.Y., Fu, J.M., Sheng, G.Y., Yang, H.F. (2002). Biodegradability of di(2-ethylhexyl) phthalate by Pseudomonas fluorescens FS1. Water, Air, Soil Pollut., 140: 297–305.

Zeng, F., Cui, K., Li, X., Fu, J.M., Sheng, G.Y (2004). Biodegradation kinetics of phthalate esters by Pseudomonas fluoresences FS1. Process Biochem., 39: 1125-1129.

Ziogou, K., Kirk, P.W.W., Lewter, J.N. (1989). Behaviour of phthalic acid esters during batch anaerobic digestion of sludge. Water Res., 23: 743–748.

Zurmühl, T., Durner, W., Herrmann, R. (1991). Transport of phthalate-esters in undisturbed and unsaturated soil columns. J. Contam. Hydrol., 8: 111–133.

CHAPTER 8

Surfactants

S. Yan, B. Subramanian, S. Barnabe, R.D. Tyagi and R.Y. Surampalli

8.1. Introduction

Surfactants (surface active agents or wetting agents) are organic chemicals that reduce surface tension in water and other liquids. Surfactants represent a major, multi-purpose groups of organic compounds. It had a volume of nearly $3x10^{10}$ kg per year all over the world (Berna et al., 1998). The most familiar use for surfactants is in soaps, dishwashing liquids, laundry detergents and shampoos. Other important uses are in industrial applications such as lubricants, emulsion polymerisation, textile processing, mining flocculates, petroleum recovery, and a variety of other products and processes. Surfactants are also used as dispersants after oil spills (Scott and Jones, 2000; Petrovic et al., 2002). These compounds can enter the environment after their application, after use they are usually discharged into municipal sewer systems and afterwards treated in wastewater treatment plants, where they are completely or partially removed by a combination of sorption and biodegradation.

There are hundreds of compounds that can be used as surfactants. These are usually classified by their ionic behaviour in solutions: anionic, cationic, non-ionic or amphoteric (zwiterionic). Some examples of major commercial and industrial surfactants are summarized in Table 8.1. Each surfactant class has its own specific properties (Petrovic and Barcelo, 2004) (CCME, 1992). The two major groups of surfactants are the anionics and non-ionics with a global production of around 2.5 and 0.5 million tons per year, respectively (Lara-Martin et al., 2006). Their main components are linear alkylbenzene sulfonates (LAS) for the anionics and alkylphenol polyethoxylates (APEOs) for the non-ionics,.

LAS is used in the formulation of detergents and other cleaning products. The environmental behavior of LAS, as one of the most widely-used xenobiotic organic compounds, has aroused considerable interest and study (Alvarez-Munoz et al., 2007). APEO applications also include pesticide formulations and industrial products, with 80% as nonylphenol polyethoxylates (NPEO) and 20% as octylphenol ethoxylates (OPEO) with around 20%. However, the use of APEOs is banned or restricted in Europe because their degradation products are toxic and estrogenic to aquatic organisms (Jobling et al., 1996). NPEO are widely used in a number of commercial and household formulations, including detergents, cosmetic products, water-based paints, inks and textiles (Birkett and Lester, 2003). Thus, this chapter will focus on discussing the occurrence and behavior (fate and transport) of the main categories of LAS and APEO and their metabolites in natural and engineered systems.

Table 8.1 Some examples of major commercial and industrial surfactants.

Type	Commercial and domestic examples	Major industrial examples
Non-ionic	Dodecyl dimethylamine oxide; coco diethanol-amide alcohol ethoxylates; linear primary alcohol polyethoxylate	alkylphenol ethoxylates; alcohol ethoxylates; EO/PO polyol block polymers; polyethylene glycol esters; fatty acid alkanolamides
Cationic	Stearalkonium chloride; benzalkonium chloride	quaternary ammonium compounds; amine compounds
Anionic	Sodium linear alkylbenzene sulphonate (LABS); sodium lauryl sulphate; sodium lauryl ether sulphates	Petroleum sulphonates; linosulphonates; naphthalene sulphonates, branched alkylbenzene sulphonates; linear alkylbenzene sulphonates; alcohol sulphates
Amphoteric	Cocoamphocarboxyglycinate; cocamidopropylbetaine	Betaines; imidazolines

8.2 LAS and APEO Surfactants

8.2.1 Introduction

LAS formulations consist of a mixture of homologs with linear alkyl chains of different lengths, usually in the range of 10 to 13 C atoms, increasing in hydrophobicity with increasing chain length (higher homologs) (Figure 8.1a). Their aerobic biodegradation in the aquatic environments takes place by the formation of sulphophenyl carboxylic acids (SPCs) (Fig. 8.1b). The metabolic biodegradation route of LAS is shown in Figure 8.2. Under anaerobic conditions, the degradation of LAS remains under study with researchers reporting that either LAS is not degraded anaerobically (Sarracin et al., 1999) or that only primary biodegradation occurs (León et al., 2001; Sanz et al., 2003).

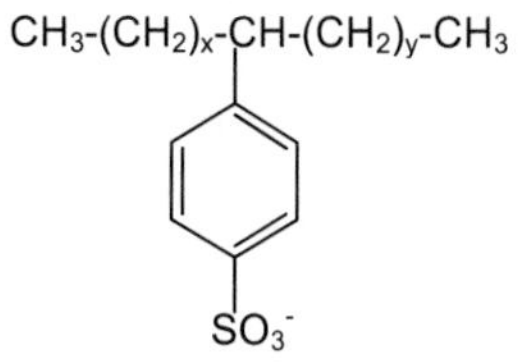

$$CH_3\text{-}(CH_2)_x\text{-}CH\text{-}(CH_2)_y\text{-}CH_3$$

SO$_3^-$

(a) Linear alkylbenzene sulphonate (LAS): x,y: 0-10; x+y: 7-10

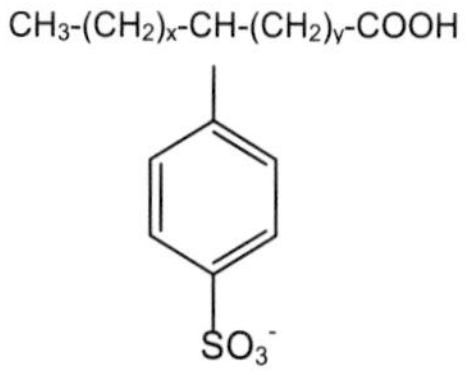

$$CH_3\text{-}(CH_2)_x\text{-}CH\text{-}(CH_2)_y\text{-}COOH$$

SO$_3^-$

(b) Sulfophenyl carboxylates (SPC): x,y: 0-10; x+y: 0-10

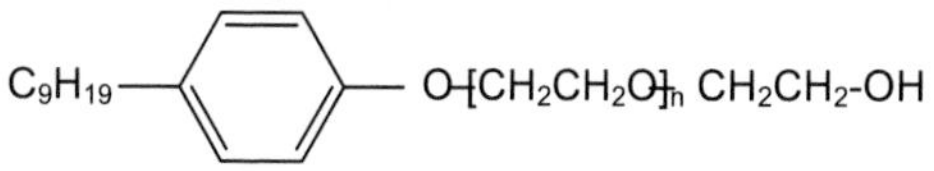

$$C_9H_{19}\text{——}\bigcirc\text{——} O\text{-}[CH_2CH_2O]_n\, CH_2CH_2\text{-}OH$$

(c) NPnEO, nonylphenol polyethoxylates (*n* indicates the number of ethoxy units)

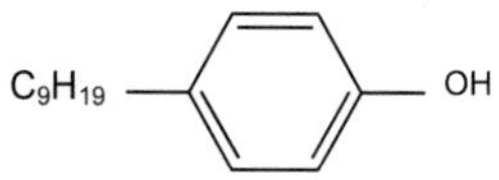

$$C_9H_{19}\text{——}\bigcirc\text{——} OH$$

(d) NP, nonylphenol.

Figure 8.1 Chemical structures of LAS, SPC, NPnEO and NP.

APEO are a class of surfactants which are manufactured by reacting alkylphenols (APs) with ethylene oxide. An APEO molecule consists of two parts: the AP and the ethoxylate (EO) moiety. This structure makes APEOs soluble in water and helps disperse dirt and grease from soiled surfaces into water (Ying et al., 2002). In contrast to most of the other types of surfactants, APEOs are biotransformed starting at the hydrophilic part of their molecules (Swisher, 1987). Such a biotransformation pathway results in formation of a number of stable metabolic products including nonylphenol (NP), short-chain nonylphenol polyethoxylates (NP1EO, NP2EO) and nonylphenoxy carboxylic acids (NPnECs) (Reinhard et al., 1982; Giger et al., 1984;

Giger et al., 1987; Abel, 1987; Ball et al., 1989; Holt et al., 1992) (Figure 8.3). Some of these metabolites, notably NP, are more lipophilic and consequently much more toxic than the parent compounds themselves (Granmo et al., 1989; Naylor et al., 1992a). Therefore, increasing attention has been focused on the ecotoxicological effects of APEO in the aquatic environment.

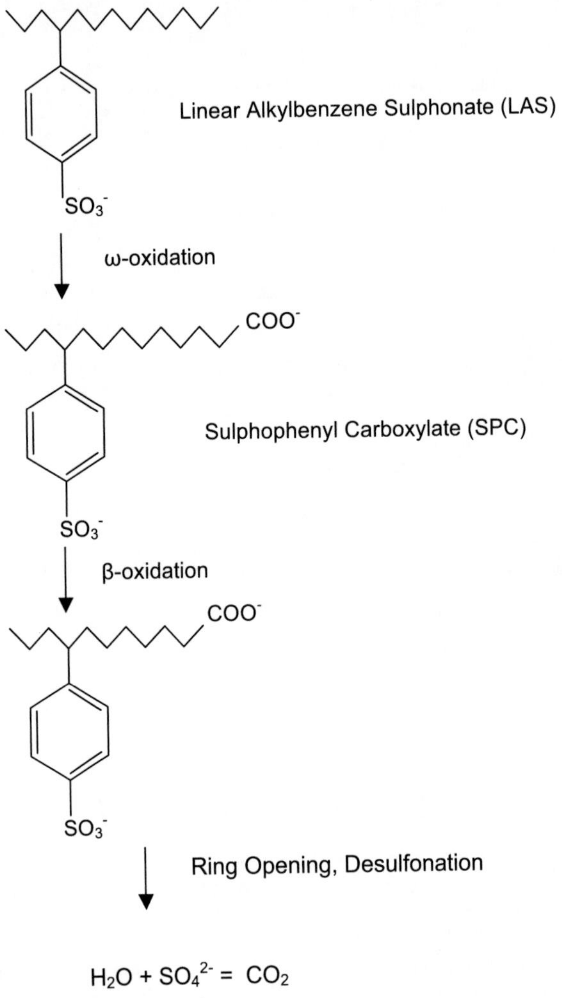

$$H_2O + SO_4^{2-} = CO_2$$

Figure 8.2 Metabolic biodegradation route of LAS.

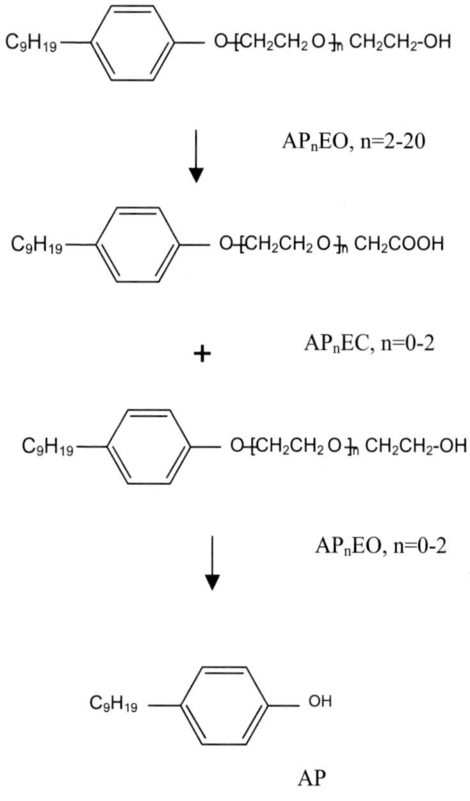

$$C_9H_{19} - \langle \text{benzene ring} \rangle - O\!-\!(CH_2CH_2O)_n\,CH_2CH_2\text{-}OH$$

↓ AP$_n$EO, n=2-20

$$C_9H_{19} - \langle \text{benzene ring} \rangle - O\!-\!(CH_2CH_2O)_n\,CH_2COOH$$

+ AP$_n$EC, n=0-2

$$C_9H_{19} - \langle \text{benzene ring} \rangle - O\!-\!(CH_2CH_2O)_n\,CH_2CH_2\text{-}OH$$

↓ AP$_n$EO, n=0-2

$$C_9H_{19} - \langle \text{benzene ring} \rangle - OH$$

AP

N: nonyl;   O: octyl

Figure 8.3 Mechanism of APEO biodegradation (AP: Nonylphenol).

Commercial NPEO is a mixture with varying numbers of ethoxylate groups and variations in the degree of alkyl chain branching. The ethoxylate chain is easily degraded and hence NP is found in the environment partly as a degradation product of NPEO and partly as the parent compound. The chemical structures of NPEO and NP are shown in Figure 8.1c & d.

8.2.2 Properties

Pure LAS is a solid at ambient temperatures. The melting point for LAS has been experimentally determined. The melting and boiling points increase with increasing alkyl chain length. The boiling point for LAS has not been determined experimentally due to decomposition. Vapor pressure and log Kow also increases with increasing alkyl chain length. Since surfactants such as LAS preferentially partition to the octanol-water interface, it is impossible to accurately measure a log Kow. The most reliable calculated value for C11.6 LAS (log Kow = 3.32) takes into account the various phenyl position isomers of LAS. LAS is water soluble, with a critical micelle concentration (CMC) value of 0.1 g/L and forms a clear solution in water at concentrations up to 250 g/L.

The physicochemical properties of NP and NPEO are summarized in Table 8.2. The solubility of an APEO surfactant depends on the number of polar groups forming the hydrophilic part of the molecule. Lower APEO oligomers (EO< 5) are usually described as "water-insoluble" or lipophilic, whereas the higher oligomers are described as "water-soluble" or hydrophilic (Ahel and Giger, 1993a). The physicochemical data can help predict the partitioning behaviour of these substances among different phases (air, water and sediment/soil) in the environment.

Table 8.2 Summary of the properties for LAS and APE metabolites (NP, NPEO).

Name	Molecular Weight (g mol^{-1})	Water Solubility (mg/l at 20°C)*	Log Kow **	Koc(1/kg)	Half-Life (days)	Reference
LAS	297-339 (C10-C13)	35-90	0.6-2.7		7-22 ***	(McAvoy et al., 1994)
NP	220	5.43	4.48	245,470	30-58	(Ahel and Giger, 1993)
Nonylphenol monoethoxylate (NPE1O)	264	3.02	4.17	288,403		
Nonylphenol diethoxylate (NPE2O)	308	3.38	4.21	151,356		
Nonylphenol triethoxylate (NPE3O)	352	5.88	4.20	74,131		

* Ahel and Giger (1993a) ; ** Ahel and Giger (1993b) ; *** Holt et al. (1989); **** McAvoy et al., 1994)

8.3 Occurrence and Behavior in Natural Systems

After use, surfactants are ultimately discharged into aquatic ecosystems through both treated and untreated wastewater discharges, and are deposited into agricultural soils as part of sludges from wastewater treatment plants (WWTPs). Comprehensive field studies have shown that degradation of surfactant in WWTPs can be incomplete and some residues of the surfactant and its aerobic breakdown intermediates can enter the receiving waters (Schoberl, 1995; Tabor and Barber, 1996). If domestic wastewater is discharged directly into natural streams, the levels can be considerably higher although aqueous phase concentrations may be reduced by sorption onto river sediments (Trehy et al., 1990), or by biodegradation through indigenous bacterial communities (Perales et al., 1999). However, the water solubility of these compounds can also enable their convectional transport over relative long distances. High mobility of these compounds also make them potentially hazardous with respect to contamination of groundwater resources, and is, therefore, of concern with respect to drinking water quality (Reemtsma, 1996; Eichhorn et al., 2002a). Significant levels of the intact surfactant and its aerobic breakdown intermediates have also been reported in terrestrial and aquatic environments (Gonzalez-Mazo et al., 1998; Lara-Martin et al., 2007).

8.3.1 Surface Water and Groundwater

8.3.1.1 Surface Water

After treatment in WWTPs, the effluent containing residual surfactants is discharged into aquatic ecosystems (Feijtel et al., 2000). A continuous input of these compounds is responsible for their presence and the presence of their metabolites in receiving waters (González-Mazo et al., 1997, Marcomini et al., 2000, Yadav et al., 2001 and León et al., 2002). Many studies have evaluated the concentrations of these metabolites in surface waters (rivers, lakes and coastal waters as well as aquatic biota), and a wide range of concentrations has been reported from around the world, the location and the concentration of LAS and their intermediate in the surface water are summarised in Table 8.3, those of APEO and their intermediate in the surface water are also summarised in Table 8.4.

Table 8.3 Concentration of LAS and their intermediate in the surface water.

Surfactant	Concentration (µg/L)	Location	Reference
LAS	53	River, USA	(Trehy et al., 1996)
LAS	0.24-9.71	River, USA	(Sanderson et al., 2006)
LAS	0.94-11.1	sediment pore water, USA	(Sanderson et al., 2006)
LAS	14 - 155	river Rio Macacu	(Eichhorn et al., 2002b)
SPC	1.2 - 14	river Rio Macacu	(Eichhorn et al., 2002b)
LAS	1.2-102	Laguna de Bay, Philippines	(Eichhorn et al., 2001)
LAS.	0.5 to 3.1	rivers in Taiwan	Ding et al., 1999)
LAS	100	river Guadalete (Cadiz, SW of Spain),	(Eichhorn et al., 2002)
LAS	1.8	Rhine river, Germany	(Eichhorn et al., 2002)
LAS	5.0	Llobregat river, Spain	(Eichhorn et al., 2002)
SPC	5.0	Llobregat river, Spain	(Eichhorn et al., 2002)
SPC	1.8	the Rhine river, Germany	(Eichhorn et al., 2002)
LAS	up to 120	a tidal channel in Cadiz, Spain	(González-Mazo et al., 1997)
LAS	2.6 to 420	a lagoon in Venice, Italy	(Marcomini et al., 2000)

Table 8.4 Concentration of APEO and their intermediate in the surface water.

Surfactant	Concentration (μg/L)	Location	Reference
NP	<ND-0.92	Canada	(Bennie et al., 1997)
NP	<ND-1.19	USA	(Snyder et al., 1999)
NP	<0.1–1.4	in Italy	(Vitali et al., 2004)
NPs, NP1EOs, and NP2EOs	up to 76	Rivers, the Aire and Lea, U.K.	(Bester et al., 2001)
NP	0.0007 to 0.0044	German Bight of the North Sea	(Bester et al., 2001)
NP	0.033	Elbe estuary	(Bester et al., 2001)
NP	1.7 to 7.3	Changjiang River, China	(Shao et al., 2005)
NPEOs	2.5 to 97.6	Changjiang River, China	(Shao et al., 2005)
NP	0.1 to 0.55	Haihe River, China	(Jin et al., 2004)
NP	n.d. to 41.3	in Lake Shihwa in Korea	(Li et al., 2004b)
NP	from <0.02 to 0.3	eight rivers in Japan	(Tsuda et al., 2002)
NPEO	up to 0.5	eight rivers in Japan	(Tsuda et al., 2002)
NP1EC and NP2EC	from 0.11 to 2.8	Tokyo, Japan	(Isobe and Takada, 2004)
NP and NP1EO	up to 3.4	Tokyo, Japan	(Isobe and Takada, 2004)
NP	n.d. to 5.1	18 major rivers of Taiwan	(Cheng et al., 2006)
NP1EO	n.d. to 0.5	18 major rivers of Taiwan	(Cheng et al., 2006)
NP1EC + NP2EC + NP3EC	n.d. to 63.6	18 major rivers of Taiwan	(Cheng et al., 2006)
CAP1EC+CAP2EC	n.d. to 94.6	18 major rivers of Taiwan	(Cheng et al., 2006)

8.3.1.2 Groundwater

Since groundwater represents one of the most important reservoirs for drinking water supplies in many countries, contamination by numerous organic pollutants is a major concern. To date, only a few publications have reported on APnEOs and their metabolites in groundwater (Schaffner et al., 1987; Barber et al., 1988; Abel, 1991; Field et al., 1992). Barber et al. (1988) studied transport of NP in an aquifer that was contaminated by rapid infiltration of secondary sewage effluent. Their field data suggested that NP was retarded during subsurface transport which was in agreement with its predicted retardation factor of 3.4. Field et al. (1992) were able to identify at the same field site carboxylated residues of APnEO surfactants. Ahel (1991) reported occurrence of NP, NP1EO, NP2EO and NP2EC in groundwater from three different field sites in the alluvial aquifer of the Sava River, Croatia. Groundwater contamination by APs and APEOs has also been reported in Switzerland, Israel, and the United States (Zoller, 1993; Tamage, 1994).

8.3.2 Sediments

LAS Sorption of LAS has been studied for a number of different types of solids, including river and marine sediments, soils, humic substances and kaolinite (House and Farr, 1989; Knaebel et al., 1996; Traina et al., 1996). Some studies with natural sediments have indicated positive correlation between organic carbon and the adsorption of LAS (Urano et al., 1984; Matthijs and De Henau, 1985; McAvoy et al., 1994), whereas others have found a poor correlation (Hand and Williams, 1987; Ou et al., 1996). Marchesi et al. (1991) presented the results of adsorption of LAS on two river sediments in terms of a hydrophobic bonding mechanism. Hand and Williams (1987) examined structure–activity relationships for the sorption of various homologues and mixtures of C10–C14 LAS by riverine sediments. It was found that sorption increased with increasing chain length, indicating that non-polar sorption mechanisms were responsible for LAS retention. Traina et al. (1996) studied the association of C10, C12 and C14 LAS with dissolved humic substances. The data indicated the significance of non-polar forces in LAS–organic matter interactions. However, some studies (McAvoy et al., 1994; Ou et al., 1996) suggested that the mechanism of LAS sorption on soils and sediments is not based on hydrophobic interactions. Brownawell et al. (1991) observed non-linear sorption of C10, C12 and C14 LAS to soils and sediments and concluded that the sorption was due to non-polar, electrostatic, and specific chemical interactions. Ou et al. (1996) observed linear isotherms at low concentrations of LAS whereas cooperative adsorption was observed at higher concentrations of LAS. In the study of the sorption of LAS to surfaces and sediment particles, evidence for both hydrophobic and specific or electrostatic interactions was reported by Westall et al. (1999). Table 8.5 shows the concentration of LAS in sediments.

Table 8.5 Concentration of LAS and their intermediate in the sediments.

Surfactant	Concentration ($\mu g/L$)	Location	Reference
LAS	0.94 to 11.1	Lowell, USA Bryan, USA	(Sanderson et al., 2006)
LAS	0.2 to 69 $\mu g.g^{-1}$	in sediments from the Bay of Tokyo	Kikuchi et al. (1986)
LAS	0.5 and 24 $\mu g.g^{-1}$	sediments from the Tamagawa Estuary in the Bay of Tokyo	Takada and Ogura (1992)
LAS	2 to 20 $\mu g.g^{-1}$	Venice Lagoon	Marcomini et al. (1988)
LAS	0.04–0.11 $\mu g/g$	sediment samples from the	Bester et al. (2001)

		German Bight of the North Sea	
LAS	86 mg/kg	sediments at sampling points near the discharge of WWTP effluents	Gonzalez et al. (2004).

Research by Lara-Martin et al. (2007) showed for the first time the degradation of LAS under anaerobic conditions, together with the presence of metabolites and the identification of microorganisms involved in this process. Laboratory experiments performed with anoxic marine sediments spiked with 10-50 ppm of LAS demonstrated that its degradation reached 79% in 165 days via the generation of sulfophenyl carboxylic acids (SPCs). Almost all of the added LAS (>99%) was found to be attached to the sediment while the less hydrophobic SPCs were predominant in solution, as their concentration increased progressively up to 3 mg/L during the full course of the experiment. Average half-life for LAS has been estimated to be 90 days, although higher values should be expected when the LAS concentration exceeds 20 mg/L, due to inhibition of the microbial community (Lara-Martin et al., 2007).

APEO Vitali et al. (2004) found that levels of NP in sediments of the Rieti district, Italy, showed a relatively low contamination and a good autodepuration capacity of the studied aquatic environments. The concentrations of NPs in sediment were in the range of 44-567 $\mu g kg^{-1}$, accumulation factors in sediment samples ranged from 10^2 to 5×10^3.

Levels up to 1.7 mg/kg of NP, 400–760 µg/L for NPEOs and NPECs were detected in sediments at sampling points near the discharge of WWTP effluents, clearly showing that in coastal areas receiving WWTP effluents, surfactants and their degradation products are widespread contaminants (Gonzalez et al., 2004). Khim et al. (1999) found 0.100–3.80 µg/g of Alkylphenol (AP) in an estuary in Korea. de Voogt et al. (1997) found nonyl- and octylphenol-ethoxylates in coastal sediments obtained in Great Britain in slightly lower concentrations ranging from 0.002 to 0.03 µg/g. This may be due to different pollution patterns in Britain as well as to different geologies and hydrodynamics of the respective coastlines. Concentrations of alkylphenolic substances in surficial sediments in Ontario, Canada, ranged from ND for many of the compounds to 1.75 $\mu g\ g^{-1}$ for 4-NP (Mayer et al., 2007).

Nonylphenols in samples taken from the Lake Shihwa in Korea and its surrounding creeks flowing through municipal and industrial areas, and into the lake were detected at the concentration ranges of 0.0–116.6 and 0.3–31.7 mg/kg in suspended particle and sediment samples, respectively (Li et al., 2004a).

8.3.3 Air and Atmosphere

Data for LAS and APEO in air and atmosphere is limited. Vejrup and Wolkoff (2002) measured the content of LAS in the particle fraction of floor dust sampled from seven different public buildings varied between 34 µg LAS g^{-1} dust and 1500 µg LAS g^{-1}, while the content of the fibre fractions was generally higher with up to 3500 µg

$LASg^{-1}$ dust. The use of a cleaning agent with LAS resulted in increases of approximately 30% of the amount of LAS in the floor dust after floor wash relative to just before floor wash. However, the most important source of LAS in the indoor floor dust appeared to be residues of detergent in clothing. Thus, a newly washed shirt contained 2960 µg $LASg^{-1}$ clothing. The analysis of the office dust samples indicated that LAS (and probably other detergents) might be of importance for the indoor environment (Vejrup and Wolkoff, 2002).

The occurrence of NPs in coastal and urban atmospheres was reported for the first time by Dachs et al. (1999). Water-to-air volatilization of NPs from estuarine waters is a source of NPs to the estuarine atmosphere. The high concentrations ranging of 2.2-70 ng m^{-3} found in the coastal atmosphere of the New York-New Jersey Bight suggests that NP occurrence in the atmosphere may be an important human and ecosystem health issue in urban, industrial, and coastal-impacted areas receiving treated sewage effluents (Dachs et al., 1999).

Van Ry et al. (2000) studied the seasonal trends of nonylphenols and tert-octylphenol (tOP) in the atmosphere of the Lower Hudson River Estuary. Gas phase concentrations of NPs at a coastal site (Sandy Hook) ranged from below the detection limit to 56.3 ng/m^{3}, while concentrations at a suburban site (New Brunswick) ranged from 0.13 to 81 ng/m^{3}. NPs and t-OP exhibited seasonal trends with higher gas phase concentrations during summer than during fall and early winter (Van Ry et al., 2000).

The atmospheric occurrence of NPs and tOP was assessed at three sites in the lower Hudson River Estuary (LHRE). Gas-phase NP concentrations at a coastal site (Sandy Hook) ranged from n.d. to 56.3 ng m^{-3}, while concentrations at a suburban site (New Brunswick) ranged from 0.13 to 81 ng m^{-3}. Gas-phase concentrations of tOP ranged from n.d. to 1.0 ng m^{-3} at Sandy Hook and from 0.01 to 2.5 ng m^{-3} at New Brunswick. NPs and tOP exhibited seasonal dependence with higher gas-phase concentrations during summer than during fall and early winter. Temperature explained 40-62% of the variability in the log (gas phase) NP and tOP concentrations. Assessment of the influence of local wind direction on atmospheric NP concentrations provided evidence for the predominance of local sources rather than long-range transport. Based on simultaneous water and over-water gas-phase samples and subsequent estimation of air-water exchange fluxes, volatilization and advection to the Atlantic Ocean accounted for 40 and 26% of the removal of NPs from the water column of the LHRE, respectively. The estimated half-life of NPs in the water column of the LHRE was 9 days (Van Ry et al., 2000).

8.4 Fate and Transformation in Engineered Systems

The major routes of entry of LAS and APEOs into the environment are via sewage treatment process where these compounds are subjected to physical and biological treatment. Other routes are direct discharge of sewage to rivers, lakes, and seas (Painter, 1992; Malz, 1997). The extent of the sorption of APEOs and LAS to particles significantly influences their fate, bioavailability and toxicity (Knaebel et al., 1996; Traina et al., 1996; Wolf and Feijtel, 1998).

8.4.1 Water Treatment

A conventional drinking water treatment system consists of flocculation and clarification, rapid or slow sand filtration, granular activated carbon (GAC) filtration and post chlorination (Tanghe and Verstraete, 2001). Chlorination and ozonation are also important treatment processes of water supply systems. The removal efficiencies for NP, NP1EO and NP2EO in drinking water by the utilisation of ozone (O_3), sodium hypochlorite (NaClO) and chlorine dioxide (ClO_2) in the laboratory tests were studied and are summarized in Table 8.6 (Lenz et al., 2004). Various other studies on treatment of surfactants in water treatment are reported (Eichhorn et al., 2002; Petrovic et al. 2003; Shao et al., 2005; Tanghe and Verstraete, 2001). A summary of the fate of surfactants in different water treatment processes is presented in Table 8.7. OP, NPs, NP1EOs, and NP2EOs have all been detected in drinking water, up to 34 ng/L (Kuch and Ballschmiter, 2001). Eichhorn et al. (2002) compared the fate of SPCs during drinking water preparation from the different treating stages the Spanish waterworks and found that in, neither prechlorination nor flocculation followed by rapid sand filtration had an impact.

Table 8.6 The fate of NP and NPnEO in water treatment process

Treatment ng/l	ozone		NaClO		ClO_2		Reference
	Initial Conc.	End Conc.	Initial Conc.	End Conc.	Initial Conc.	End Conc.	
NP	<10	<10	76	<10	78	<20	(Lenz et al., 2004)
NP1EO	570	410	390	330	440	<24	(Lenz et al., 2004)
NP2EO	1,580	1,110	1370	1125	1430	120	(Lenz et al., 2004)
NP1EC	<20	<10	41	<10	20	23	(Lenz et al., 2004)
NP1EC	<20	<20	<20	<20	28	30	(Lenz et al., 2004)

8.4.2 Municipal and Industrial Wastewater Treatment

LAS After application, LAS is usually discharged into municipal sewer systems. The concentration of LAS in the influents of domestic wastewater treatment plants typically ranges from 1 to 7 mg/l (Clara et al., 2007). Sanderson et al. (2006) reported influent wastewater concentrations from 2.75 to 3.96 mg l^{-1} with the highest in Lowell and lowest in Bryan, USA. Clara et al. (2007) calculated an averaged specific loading rate of 367 mg $l{-}1$ p.e.$^{-1}$ (population equivalent) (135–650 mg l^{-1} p.e.$^{-1}$)

LAS are generally considered to be biodegradable.An extensive study conducted on the fate of LAS during wastewater treatment indicated that they are efficiently removed by physical, chemical and biological processes (Eichhorn et al., 2002). Apart from precipitation and adsorption onto suspended solids, which can produce 30 to 70% removal (Berna et al., 1989), microbial degradation generally accounts for the major elimination route (~ 80%) resulting in 95 to 99.5% removal of LAS in activated sludge systems (Painter and Zabel, 1989). More than 99% removal of LAS in aerobic WWTPs has been reported by many researchers (Leon et al., 2004; Temmink and Klapwijk, 2004).

Table 8.7 The fate of surfactant in different water treatment process.

Surfactant	Treatment	Efficiency	Reference
SPC	ozonation, granular activated carbon filtration, and final chlorination	The final SPC level was about 2 μgL^{-1} in drinking water.	(Eichhorn et al., 2002)
SPC	rapid sand filtration	Diminished the SPC concentration by >85%.	(Eichhorn et al., 2002)
SPC	slow sand filtration prior to the closing chlorination	SPC down to a level of <0.05 μgL^{-1}	(Eichhorn et al., 2002)
NPEO, NP, NPEC	Prechlorination settling and flocculation ozonation, GAC filtration, and final disinfection with chlorine	Entering waterworks ranged from 8.3 to 22 $\mu g/L$. Prechlorination reduced the concentration of NPECs and NPEOs by about 25- 35% and of NP by almost 90%. settling and flocculation followed by rapid sand filtration (7%), ozonation (87%), GAC filtration (73%), and final disinfection with chlorine (43%), resulting in overall elimination ranging from 96 to 99%. The residues detected were generally below 100 ng/L, with one exception for NP2EC with a concentration of 215 ng/L was detected.	(Petrovic et al., 2003)
NPEOs	conventional water treatment process	The 4-NP removal efficiency varied in a range of 62% to 95% with final NP of 0.1 to 2.7 $\mu g/L$ in drinking water.	Shao et al. (2005)
NP	granular activated carbon (GAC)	With contact times of 4 d and 24 hr and GAC dosages of 1 and 0.1 g L^{-1} no saturation of the GAC could be obtained with NP total contaminant loadings up to 10 000 $\mu g\ L^{-1}$. Higher NP concentrations could not be applied due to its low water solubility (~5 mg L^{-1}). The sorption capacity of GAC for NP was at least 100 mg g^{-1} GAC. A full-scale GAC filter unit will be sufficient to remove environmentally relevant NP concentrations of 10 $\mu g\ L^{-1}$.	(Tanghe and Verstraete, 2001)

Although degradation pathways of LAS are poorly understood, it is generally accepted that the main aerobic breakdown intermediates of LAS are SPC which may appear in the effluent of treatment plants. However, in anaerobic environments the degradation is less likely to occur (Madsen et al., 1997). The removal mechanism of LAS in sewage treatment plants depends on the system used (Berna et al., 1989).

A study was conducted to evaluate the removal of LAS in a membrane bioreactor (MBR) and a conventional activated sludge (CAS) system. Removal efficiencies of over 97% were achieved in both reactor systems. Another study (Gonzalez et al., 2007) showed similar elimination of LAS in the both systems. The results indicated that LAS removal was due to biodegradation, rather than sorption. The effect of operational variables, such as hydraulic retention time, LAS composition and hydrophobicity of the membrane used in the MBR, was negligible in the range tested (De Wever, et al. 2004). While Terzic et al. (2005) compared the degradation of LAS in a CAS system and MBR based on hollow fiber membranes, and obtained higher elimination rates using the MBR. Trehy et al. (1995) and McAvoy et al. (1993) reported that removals obtained by a rotating biological contactor were similar to those observed in the activated sludge WWTPs. Effluent concentrations of LAS in WWTP effluent range between 7.9–71 µg l^{-1} . Comparable influent and effluent concentrations are also reported in the literature (Clara et al., 2007; Matthijs et al., 1999; Feijtel et al., 1995; Waters and Feijtel, 1995).

It has been reported that LAS at higher concentrations is not biodegradable (Zhang et al., 1999). As a result, chemical processes may be used to degrade aqueous LAS. Among these processes, advanced oxidation technologies (AOTs), such as ultraviolet with the wavelength of 254 nm (UV-254) and the combination of ultraviolet with the wavelength of 254 nm and hydrogen peroxide (UV-254/H_2O_2) are attractive alternatives for the treatment of wastewater containing bioresistant compounds (Tabrizi and Mehrvar, 2006). A summary of studies on the chemical treatment of LAS is shown in Table 8.8.

Table 8.8 Chemical degradation of LAS by different advanced oxidation processes

Initial Concentration (mg/L)	Oxidation Scheme	Efficiency	References
1600	Wet air oxidation	Increased the biodegradability under higher temperature	(Patterson et al., 2002)
125	TiO2/UV365	Optimum concentration of TiO2 = 3 g/L gave r_o = 1.7 mg LAS/L min	(Venhuis and Mehrvar, 2004)
0.25–2	Fenton's reaction	90 mg FeSO4/L, 60 mg H2O2/L, 95% degraded	(Lin et al., 1999)
1000	Fenton's reaction	30 mg FeSO4/L, 60 mg H2O2/L, 38% degraded	(Cuzzola et al., 2002)
1000	Wet air oxidation	LAS was readily oxidized under mild condition	(Mantzavinos et al., 2000)
100	UV-254 nm	by UV-254 alone was capable of degrading LAS in 120 min by only 40%.	(Tabrizi and Mehrvar, 2006)
	H_2O_2	The optimum concentration of H_2O_2 was found to be 720 mg/L.	
	Combination of UV-254	The degradation of LAS by	

	nm and H_2O_2	combination of UV-254 and H_2O_2 was up to 95% in 120 min.	
138	UV-254 nm	in 300 min, 70% removal	(Venhuis and Mehrvar, 2004)
2.8×10^{-5} mol/L	TiO2/UV-C	Conversion rate can be up to 93% in 120 min	(Saien et al., 2003)

APEO The biodegradation of surfactants in the WWTP has been studied in numerous papers and in general most of the surfactants are well eliminated by conventional wastewater treatment. Under optimized conditions more than 90–95% can be eliminated, although the percentage of elimination can vary depending on the operating characteristics of the WWTP (i.e. plant size, sludge retention time, hydraulic retention time, temperature) (Gonzalez et al., 2007). It was observed that low removal rates were achieved in highly loaded plants. Clara et al. (2005a) as well as Ahel et al. (1994) concluded that NPEO removal is better in WWTPs that are loaded lightly, due to variations of redox conditions relevant for the degradation of those compounds.

During wastewater treatment, the hydrophilic ethoxylate chains are shortened and transformed to monoethoxylate (NP1EO) and diethoxylate (NP2EO). NP1EO and NP2EO are further biodegraded to completely deethoxylated NP, which is more lipophilic and toxic one and more resistant to biodegradation compared with the long chain ethoxylates. NP, NP1EO and NP2EO have been reported to cause a number of estrogenic responses on aquatic organisms and thus they have been classified as endocrine disruptors (EDCs) by several organizations. Nonylphenol, which has been listed as a priority pollutant in the Water Framework Directive (European Union), is a mixture of different branched and linear chain isomers (ortho-, meta- or para-), with the most common ring isomers being the para isomers (4-NPs), represented by 4-n-nonylphenol.

Studies investigating the biodegradation of NP have focused principally on wastewater treatment based on activated-sludge processes (Hernandez-Raquet et al., 2007). Langford et al. (2005) found that with increasing sludge age, there is an increase in mixed liquor solids concentration in activated sludge which results in greater bacterial numbers and the potential for greater species diversity which therefore increases compound degradation. However, increased degradation of long chain compounds resulted in an accumulation of shorter chain compounds and nonylphenol, which are more resistant to degradation (Langford et al., 2005). Tanghe et al. (1998) reported that in lab-scale activated-sludge reactors, NP removal rates varied as a function of temperature. Complete NP degradation was observed at 28 °C. In contrast, high levels of NP, mainly concentrated in the sludge fraction, were found at 10 °C. This sludge sorption phenomenon was accentuated when the influent was supplemented with organic material. Staples et al. (2001), studying the ultimate biodegradation of NPE and NP, demonstrated that, after 35 days at 22 °C, even if NP was degraded, it was not completely removed, demonstrating lengthy persistence in the environment .

In order to reduce the concentration of surfactants and metabolites in the wastewater effluents new technologies are being applied. Membrane bioreactor (MBR) treatment is an emerging technology based on the use of membranes in combination with the traditional biological treatment Gonzalez et al. (2007). In the CAS system, 87% of parent NPEOs were eliminated, but their decomposition yielded persistent acidic and neutral metabolites which were poorly removed. The elimination of short ethoxy chain NPEOs (NP1EO and NP2EO) averaged 50%, whereas nonylphenoxy carboxylates (NPECs) showed an increase in concentrations with respect to the ones measured in influent samples. Nonylphenol (NP) was the only nonylphenolic compound efficiently removed (96%) in the CAS treatment. On the other hand, MBR showed good performance in removing nonylphenolic compounds with an overall elimination of 94% for the total pool of NPEO derived compounds (in comparison of 54%-overall elimination in the CAS). The elimination of individual compounds in the MBR was as follows: 97% for parent, long ethoxy chain NPEOs, 90% for short ethoxy chain NPEOs, 73% for NPECs, and 96% for NP. Consequently, the residual concentrations were in the low µg/l level or below it (Gonzalez et al., 2007).

Li et al. (2000) compared conventional treatment with the membrane assisted biological wastewater treatment, and showed that membrane treatment improved the elimination of nonylphenol ethoxylates (NPEOs) but did not entirely stop their discharge in the permeate. Wintgens et al. (2002, 2004) reported 70–99% removal of nonylphenol (NP) from a waste dump leachate plant in different systems using nanofiltration or reverse osmosis membranes. Terzic et al. (2005) compared the degradation of NPEOs and their degradation products in a CAS system and MBR based on hollow fiber membranes, and obtained higher elimination rates using the MBR (Gonzalez et al., 2007).

The concentration and distribution of nonylphenol polyethoxylates (NPEOs represents the mixture, and NPnEO represents the monomer) and its metabolites in influent and effluent of four municipal WWTPs in the north of China were measured. The concentration and distribution of these chemicals in the sludge of two WWTPs were also determined, and the transfer and fate of NPEOs in the sewage treatment process were discussed by analyzing the distribution of the products in the effluent and the sludge. Results showed that NPEOs and their metabolites existed in all samples of the influent, effluent, and sludge. NPEOs were degraded in the sewage treatment process with the removal efficiency in the range of 23.38%-77.11%, or an average of 52.86%. However, the large analogs of NPnEO were only degraded to small ones, whose degradation rate was rather slow, and consequently the degradation was not complete. Hence, the concentrations of some small metabolites, such as nonylphenol (NP), nonylphenol monoethoxylate (NP1EO), and nonylphenol diethoxylate (NP2EO) were elevated in the effluent. These small metabolites are more toxic than the large NPnEO analogs, and some of them were reported to exhibit environmental endocrine disrupting activity. From this point of view, the process of sewage treatment does not reduce but elevate the risk of NPEOs, which becomes the main source of these small NPnEO in the environment. The sludge exhibited good adsorption ability for NPEOs, especially for the small analogs, which led to the high level of NPEOs in the sludge. Hence, reasonable disposal of the surplus sludge to avoid re-pollution is very important (Hou and Sun, 2007).

Due to the interactions between the different fractions, nonylphenol, nonylphenol ethoxylates and nonylphenol carbocylates cannot be treated separately, but an integrated evaluation of the behaviour of those compounds during wastewater treatment has to be performed. As NP1EO was the dominant fraction in the influent of the sampled WWTPs, at least partial NPnEO degradation seemed to occur already in the sewer system. In the influent of the nine sampled WWTPs, in the average 27% of the cumulative load was present in the form of NP, approximately 61% as NP1EO and about 12% as NP2EO. In the effluent in the average 46% of the cumulative load were present in form of NP, approximately 36% as NP1EO and about 18% as NP2EO (Clara et al., 2005a).

The efficiency of six WWTPs in the Catalonian region, Spain, to remove several classes of ionic and non-ionic surfactants was investigated (Gonzalez et al., 2004). Occurrence and distribution of nonylphenol ethoxylates (NPEOs) and their degradation products were studied in coastal areas receiving the WWTP effluents. Concentrations of NPEOs in raw water entering WWTPs ranged from 60 to 190 µg/L. In effluents, concentrations ranged from 2.8 to 6.6 µg/L, which corresponded to an average primary elimination of 93–96%. Nonylphenol (NP) was found in concentrations from 0.2 to 18 µg/L in WWTP influents and up to 5 µg/L in the treated water, showing a clear declining trend with respect to concentrations.

The effluent of 17 WWTPs across Norway, Sweden, Finland, The Netherlands, Belgium, Germany, France and Switzerland was studied for the presence of nonylphenol (NP). NP could be detected in the effluent of all 14 STW where this measurement was attempted, with a median of 0.31 µg/l and values ranging from 0.05 to 1.31 µg/l (Johnson et al., 2005).

The investigations at three Northeast Kansas wastewater treatment plants (WWTPs) found NPnEOs and NP in influent wastewater at levels from nondetectable to more than 200 µg/L. Conventional unit processes at these WWTPs were not completely effective in removal of these organic wastewater contaminants. Low levels (up to 23 µg/L) of NPnEOs and NP were detected in the WWTP effluents that are discharged into the Kansas River (US) (Keller et al., 2003).

8.4.3 Solids Treatment and Disposal

Sewage sludge is one of the potential routes for introducing surfactants and their metabolites into environment. Among a number of waste substances discharged through the sewers, surfactants are one of the classes of organic compounds with the greatest tendency to accumulate in sewage sludge (Petrovic and Barcelo, 2004). Typically, 25% of surfactants entering a WWTP via influent are removed during primary treatment via primary sludge. Surfactants adsorbed at the biological sewage sludge are carried out in original or partly degraded form when excess sludge is removed and subsequently treated. Because of restricted metabolic pathways, the majority of common surfactants are not degradable under anaerobic conditions and sludge after the anaerobic digestion process is rich in surfactants (Petrovic and Barcelo, 2004).

LAS The presence of LASs in sewage sludge greatly depends upon the type of treatment (aerobic or anaerobic), content of the raw sewage, water hardness, age of the sludge, and so on. As an average 15–35% of LASs is physically removed by precipitation during the primary settling step and are, therefore, not exposed to biological treatment. This precipitate (primary sludge) follows the anaerobic digestion process with the bulk of the sludge produced. LAS concentration in anaerobically digested sludge is generally higher than in aerobically stabilized sludge. A summary of the fate of LAS and their degradation products in sludge is presented in Table 8.8. There are significant differences in the quantity of LAS in different kinds of sludges. Generally, sludges exposed to aerobic conditions contain far less LAS than primary sludge or anaerobically digested sludge (Giger et al., 1989). The great variations from different studies also can be attributed to the different measuring techniques used to quantify LAS (Painter and Zabel, 1989). It was reported that they are easily degraded under aerobic conditions, but in anaerobic environments the degradation process does not occur (Madsen et al., 1997). On the other hand, as they absorb on to sewage solids during primary settlement of sewage, they bypass the aeration tank and hence are not degraded in regular treatment processes (De Wolfe and Feijtel, 1997). This is considered the most important removal mechanism of LAS loading to the terrestrial environment through sludge amendment (Abad et al., 2005).

Although LAS show lower toxicity to aquatic organisms when compared to nonylphenol, they generally occur at higher concentrations in sewage sludges. It has been suggested that LAS biodegradation is severely inhibited when sorptive phases are available and this could explain the high concentrations found in sewage sludges (Abad et al., 2005).

Anaerobic treatment of sludge is widely used but it must be investigated with respect to its ability to remove certain organic contaminants which may be toxic for many organisms at certain concentrations. In the literature their behavior is well documented under aerobic treatment and they are known to biodegrade rapidly under these conditions (Schöber, 1989; Connell, 1997; IPCE, 1996; Berna et al., 1991, 1995; Romano and Ranzani, 1992; Giger et al., 1989; Prats et al., 1997; Holt et al., 1995; Greiner and Six, 1997). However there are uncertainties about their biodegradation under anaerobic conditions. Furthermore at WWTPs, at least 20% of the mass load on LAS entering the plant will be present on the suspended solids.

The subsequent fate of these organic chemicals following sewage sludge disposal is a topic of current concern and there is a requirement for such basic information as concentrations of specific compounds in sludge in order to assess the environmental impact of specific sludge disposal practices. Gomez-Rico et al. (2007)

reported, for LAS concentration in the seventeen sewage sludges samples had 130 to 32000 mg kg^{-1}, 82% presented <10,000 mg kg^{-1}, 12% presented 10,000 to 20,000 mg kg^{-1} and 6% presented >20,000 mg kg^{-1}. It can be emphasized that 12% (only two samples) presented <2600 mg kg^{-1} LAS (the limit) (Union, 2000). Jensen (1999) collected information about different types of sludges from several countries between 1986 and 1997 and showed that typical LAS levels of aerobic digested sludge were in the range of 100 to 500 mg kg^{-1} dry weight (dw), whereas anaerobically digested sludge might contain on average 5000 to 15,000 mg kg^{-1} dw. In a large Danish survey (Torslov et al., 1997), aerobically and anaerobically treated sludges from the same plant were found to contain 11 and 13,600 mg LAS kg^{-1} dw, respectively. Under aerobic conditions, total mineralization of LAS proceeds through degradation of the alkyl group via o-oxidation, b-oxidation, desulfonation, and finally degradation of the phenyl ring (Haagensen et al., 2002).

Other studies have also shown that in anaerobically treated sludge, relatively high concentrations of LAS can be found (5–10 g/kg dry solids) whereas activated sludge and aerobically treated sludge, contain low LAS concentrations (0.1–0.8 g LAS/kg dry matter) (Berna et al., 1989, McAvoy et al., 1993 and Waters and Feitjel, 1995). Carballa et al. (2007) studied the effect of an oxidative pre-treatment with ozone on the removal of LAS in the digested sludge. The total LAS content of the digested sludge (55–125 mg kg^{-1}) was much lower than the values reported in other literature (Table 8.9). The removal efficiencies (ranging from 50% to 90%) of the different LAS homologues were similar in both digesters during all the experiments. The removal efficiencies reported in literature for full-scale anaerobic digesters vary from 18% (Prats et al., 1997) to 35% (Osburn, 1986). However, it is not clear to which process (binding, humification, co-metabolism and anaerobic desulphonation) these can be ascribed (Carballa et al., 2007).

Table 8.9 Typical concentration of LAS in various sludges.

Type of Sludge	LAS Concentration (g/kg dry weight)	Reference
Primary sludge	5.34-6.31	(Painter and Zabel, 1989)
Activated sludge	0.09-0.86	(Painter and Zabel, 1989)
Anaerobically digested sludge	5-15	(Jensen, 1999)
	5.2-30.2	(Painter and Zabel, 1989)
	2-10	(Giger et al., 1989)
	2.1-2.9	(Painter and Zabel, 1989)
	0.1-0.5	(Jensen, 1999)
Digested sewage sludges	12	(McEvoy and Giger, 1985)
Six municipal sewage sludges	0.02-0.43	(Eganhouse et al., 1988)
Sewage sludges with predominantly industrial and domestic catchments	0.125-0.180	(Sweetman et al., 1992)
	0.153-0.176	

After conventional and pre-ozonation treatment of sewage sludge	0.8–2.4 mg/kg^{-1} (C10), 13.6–20.8 mg/kg^{-1} (C11), 20.2–43.7 mg/kg^{-1} (C12) and 19.1–39.7 mg/kg^{-1} (C13).	(Carballa et al., 2007)
Air-dried digested sludge	0.15-0.16	(Painter and Zabel, 1989)
Agricultural soils amended with anaerobic digested sludge	0.0002-0.02	(Jensen,1999)
in non-digested sludge	8.4–14.0 (average 12.6)	(Prat et al., 1993)
After anaerobic digestion	range 12.1–18.8 (average 15.8)	(Prat et al., 1993)
After aerobic fermentation process	dropped to 6.0	(Prat et al., 1993)
	0.13 to 32	(Gomez-Rico et al., 2007)
Aerobically treated sludges	0.011	(Torslov et al., 1997
Anaerobically treated sludges	13.6	(Torslov et al., 1997

To date published experimental evidence of LAS being degraded under strict anaerobic conditions is rare. Recent publications have indicated that some removal/primary degradation of LAS in anaerobic treatment using UASB reactors is possible (Haggensen et al., 2002; Mogensen and Ahring, 2002; Sanz et al., 1999). It was also reported that LAS was degraded using NO$_3^-$ as the electron acceptor in the acidogenic step of a 2 stage UASB reactor setup (Almendariz et al., 2001). Thus, the few recent reports of LAS degradation under anaerobic conditions seem to conflict with other publications which indicate no degradation. However the experiments and the scenarios investigated by the different authors did not have many similarities.

Anaerobic degradation of sludge amended with LAS was tested in a one-stage continuous stirred tank reactor (CSTR) and a two-stage reactor system consisting of a CSTR as first step and upflow anaerobic sludge bed (UASB) reactor as the second step. Anaerobic removal of LAS was only observed at the second step but not at the first step. Removal of LAS in the UASB reactors was approximately 80% where half was due to absorption and the other half was apparently due to biological removal as shown from the LAS mass balance. In batch experiments, it was found that LAS at concentrations higher than 50 mg/l is inhibitory for most microbial groups of the anaerobic process. Therefore, low initial LAS concentration is a prerequisite for successful LAS degradation. The results suggested that anaerobic degradation of LAS is possible in UASB reactors when the concentration of LAS is low enough to avoid inhibition of microorganisms active in the anaerobic process (Angelidaki et al., 2004).

Sanz et al. (2006) studied the effect of temperature on the biodegradation of LAS during the composting of anaerobically digested sludge from a wastewater treatment plant. It was shown that the optimum temperature for the biodegradation of LAS was around 40 °C (Sanz et al., 2006).

APEOs From an ecotoxicological point of view, the most critical sludge-bound-surfactant-derived compounds are weakly estrogenic NPs and short ethoxy chain NPEOs. The overall rate of biodegradation of the parent NPEOs was found to be limited because of the formation of biorefractory metabolites (ultimate biodegradation <40%). Approximately 63% of all nonylphenolic compounds introduced into WWTPs are discharged into the environment in the form of lipophilic NPEO1 and NPEO2 and NP, carboxylated derivatives (NPECs), or untransformed NPEOs. Forty percent of the total output of nonylphenolic compounds is via digested sewage sludge. Consequently, nonylphenolic compounds are frequently detected in substantial amounts in digested sludges .

NPEOs were detected in concentrations ranging from values below mg/kg range to over 500 mg/kg (maximum concentration of 2 g/kg was reported for sludge from an WWTP receiving industrial wastewaters), NPEO1 and NPEO2 being the predominant species. Generally, higher contamination is found in WWTPs using anaerobic digestion (Minamiyama et al., 2006).

NPs were found in concentrations ranging from the lower mg/kg range to the lower g/kg range (Table 8.10). Adsorption of NPs onto primary and secondary sewage sludge, and its formation from NPEOs during anaerobic stabilization of sludge, results in extremely high concentrations of NPs in anaerobically digested sludge. The high levels of NPs (more than 1 g/kg) are consistently found in anaerobically stabilized sludge. Giger et al. (1989) reported a 15-fold increase of NP concentration during anaerobic digestion resulting in extremely high concentrations of NPs in anaerobically digested sewage sludge (450–2530 mg/kg, mean 1010 mg/kg). The concentrations in activated sewage sludge, in mixed primary and secondary sludge, and in aerobically stabilized sludge were substantially lower, suggesting that the formation of NPs is favored under mesophilic anaerobic conditions.

The investigations at three Northeast Kansas wastewater treatment plants (WWTPs) found that a large portion of NPnEOs and NP appeared to adsorb to the biosolids, a phenomenon that likely prevented their degradation in the bioreactors. As much as 898 mg/kg NP was measured in biosolids from one WWTP. Onsite composting appeared to reduce NP, nonylphenol mono-ethoxylate (NP1EO), and nonylphenol di-ethoxylates (NP2EOs) in the biosolids (Keller et al., 2003).

In most activated sludge treatment plants, the suspended solids are removed via primary settling and are directed to an anaerobic digester. Minamiyama et al., (2006) carried out a study to determine the fate of APEO in sewage sludge treatment processes, especially in an anaerobic digestion process. The following key results were obtained from the anaerobic digestion experiments: (1) Approximately 40% of NP1EO was converted to NP whenNP1EO was injected to an anaerobic digestion testing apparatus operated at a retention time of approximately 28 d and a temperature of 35 °C with thickened sludge sampled from an actual WWTP; (2) Conversion of NP1EC to NP and negligible conversion of NP2EC within 20 days when these compounds were injected to an anaerobic digestion testing apparatus with thickened sludge.

Hawrelak et al. (1999) studied the primary degradation products of alkyphenol ethoxylate surfactants in recycled paper sludge, the data shown in Table 8.9. NP was persistent and high NP concentrations (20 mg/g dw) may inhibit anaerobic gas production (Battersby and Wilson, 1989; Ejlertsson et al., 1999; Hernandez-Raquet et al., 2007). On the contrary, in a more recent study, Patureau et al. (2006) demonstrated that NP can be partially eliminated under methanogenic conditions in continuous reactors. However, in those experiments, the resulting digested sludge contained higher levels of NP per gram of dw than the sludge prior to anaerobic treatment. Performance of aerobic thermophilic sludge processes has also been studied in batch and continuous cultures (Banat et al., 2000; Patureau et al., 2006), showing that in both conditions complete removal of NP was not achieved.

Table 8.10 Typical concentration of APEO and their degradation products in various sludge.

Type of sludge	NP and OP concentration (µg/g dry weight)				Reference
	4-NP	NP1EO (i.e.NPE1)	NP2EO (i.e.NPE2)	NPEO	
	4.61	1.21	0.39		Hawrelak et al. (1999)
	3.98	0.17	0.14		Hawrelak et al. (1999)
	2.35	0.07	0.08		Hawrelak et al. (1999)
in the sludges of municipal sewage treatment plants in Canada	100-500	100-500	5-150		(La Guardia et al., 2001)
In a Danish study of 20 sludge samples	15			average NPEO concentrations of 15	(TørCshløevm et al., 1997).
anaerobic digestion,	0.450-2500				Gomez-Rico et al. (2007)
six municipal sewage sludges	0.02-0.43				Eganhouse et al. (1988)
sewage sludges with predominantly industrial and domestic catchments.	0.125-0.180 0.153-0.176				(Sweetman et al., 1992)
Swedish sewage sludge	3.9				(Samsøe-Petersen, 2003)

| Air-dried digested sludge | 0.15-0.16 | Painter and Zabel (1989) |
| Agricultural soils amended with anaerobic digested sludge | 0.0002-0.02 | Jensen (1999) |

Hernandez-Raquet et al. (2007) studied that removal of NP and its estrogenic activity in different sludge treatment processes. It was found that under anaerobic conditions, no degradation of NP and its estrogenic activity was observed. Indeed, an accumulation of the compound occurred. In contrast, high removal of NP was achieved in aerobic conditions as well as in aerobic post-treatment of anaerobically pre-digested sludge, with a concomitant reduction of the sludge's estrogenic potency. Hernandez-Raquet et al. (2007) also found that in sludge treatment, NP and total NPEO elimination only occurred when aerobic treatment of sludge was included.

NP concentration in sludge from different WWTPs showed concentrations varying with the treatment process used. Secondary sludge from WWTP using the activated sludge process displayed NP concentrations of about 128 mg/kg dw. Stabilised sludge by aerobic or anaerobic processes showed NP concentrations ranging, respectively, 80–650 and 450–2530 mg/kg dw (Ahel and Giger, 1985; Bennie, 1999; Brunner et al., 1988; Giger et al., 1984; Lee et al., 2004). NP concentrations in sludge vary regionally, depending on the use of NPEO surfactants. NP concentrations range from 0.3 to 67 mg/kg dw in Denmark and from 14.3 to 3150 mg/kg dw in Spain. The mean NP concentrations reported in California and New York were, respectively, 754 and 1500 mg/kg dw (Abad et al., 2005; Hernandez-Raquet et al., 2007; La Guardia et al., 2001; Pryor et al., 2002; Törslöv et al., 1997).

Total OP, NPs, NP1EOs, and NP2EOs ranged from 6.1 to 981 mg/kg, NP concentrations ranged from 5.4 to 887 mg/kg, with a mean of 491 mg/kg. Eleven U.S. WWTPs examined contained detectable levels of OP, NPs, NP1EOs, and NP2EOs. Nine exceeded the current Danish land application limit (30 mg/kg; sum of NPs, NP1EOs, and NP2EOs) by 6 to 33 times. NPs were the major component, and their concentrations therein ranged from 5.4 to 887 mg/kg dw. OP, reportedly 10 to 20 times more estrogenic than NP, was detected in these same nine biosolids at levels up to 12.6 mg/kg. Three biosolids were also subjected to the U.S. Environmental Protection Agency Toxicity Characteristic Leaching Procedure Method 1311. NPs and NP1EOs were both detected in the leachate; the former at concentrations from 9.4 to 309 g/L (La Guardia et al., 2001).

Gomez-Rico et al. (2007) analyzed nonylphenolic compounds, including NP and NP1EO + NP2EO in 17 sewage sludges and observed that NP + NP1EO + NP2EO exceeded the maximums exceeded the EU's proposed limit in the third draft of the Working Document on Sludge for NP + NP1EO + NP2EO in most samples. It was also found that the ratio of NP1EO + NP2EO to NP was related to sludge treatment at the WWTP. At the same time, it was seen that the higher values of organic pollutants belonged to digested sludges (Gomez-Rico et al., 2007). For NP +

NP1EO + NP2EO, the samples had total contents of 190 to 3500 mg kg^{-1} dw, whereas the concentration limit proposed by the EU (2000) is 50 mg kg^{-1}. Paulsrud et al. (2000) found NP + NP1EO + NP2EO in high concentrations in sludge samples from all the WWTPs in Norway (22 to 650 mg kg^{-1} dw), and similar experiences have been reported from Switzerland (44 to 7214 mg kg^{-1} in 1989 to 1991 and 23 to 171 mg kg^{-1} in 1993.

La Guardia et al. (2001) studied 11 different biosolids from the United States, and they found 6.1 to 974 mg kg^{-1} dw NP + NP1EO + NP2EO, and the composted sludges had the lowest values. The values obtained by Abad et al. (2005) for sludges from Spain ranged from 14.3 to 3150 mg kg^{-1} NP + NP1EO + NP2EO, and most samples presented concentrations higher than the EU proposed limit. The Danish results (Jensen & Jepsen 2005) were lower than those observed in other countries, 1.5 to 133 mg kg^{-1} dw NP + NP1EO + NP2EO in 1997 and 1 to 25 mg kg^{-1} in 2002. This difference may be related to the lower alkylphenol ethoxylate consumption rates in Denmark as a result of the voluntary restrictions spurred on by alkylphenol ethoxylate byproduct environmental impact concerns. The Danish EPA set a limit of 10 mg kg^{-1} for NP + NP1EO + NP2EO in 2000. As seen in the literature, most countries present high values, and it is difficult to comply with the limit.

Composting has been extensively investigated as a bioremediation technique for a variety of pollutants (Semple et al., 2001) but there are few reports on the fate of organic pollutants during composting of sewage sludge. Composting accelerates decomposition of organic material and is a useful technique to decrease the concentration of organic contaminants including NPE in waste products. As results of the biodegradation process, the distribution of NPE oligomers were observed to change and after 100 days of composting only the degradation products nonylphenol monoethoxylate (NPE1), nonylphenol diethoxylate (NPE2) and 4-nonylphenol (NP) remained. These degradation products are more toxic to the environment than the parent NPE surfactant itself (Bennie, 1999; Hawrelak et al., 1999)

Moeller and Reeh (2003) investigated the degradation of NPEOs in sewage sludge as well as in municipal solid waste under aerobic composting conditions. It was found that at fixed temperature of 35, 50 and 55°C: at the beginning of the composting process 4-NP was present in concentration around 2 mg/kg dw and NPEO1 as well as NPEO2 were practically absent. 4-NP was then degraded to concentrations below 1 mg/kg dw within 28 to 90 days of composting. While the degradation of 4-NP was reduced at 65 °C, possibly due to the degrading organisms could not function at high temperatures.

Gibson et al. (2007) investigated the impact of pilot-scale composting and drying of sludge on the concentrations of NP. Concentrations of 4-NPs during composting decreased from 114 mg/kg ash in the starting material to less than 15 mg/kg ash in the mature compost. Losses during the first 9 weeks were slow (18%) but increased thereafter so that total losses were 88%. While the concentrations of 4-NPs during drying decreased from an average of 345 mg/kg ash to 212 mg/kg ash in the final material, a total loss of 39%.

Enhanced treatments of sewage sludge produce a more manageable product for agricultural use by stabilizing the material, removing water, and reducing the possibility of pathogen transfer. We investigated the impact of pilot-scale composting and drying of sludge on physicochemical characteristics and on the concentrations of some organic contaminants. During the 143 day composting procedure, Concentrations of 4-nonylphenols (4-NPs) fell by 88%. Losses of 4-NPs (39%) were less than in composting and stopped when moisture content became constant. Both treatments are simple, practical procedures that reduce the volume of waste and are applicable in situ on farms. Composting would be the method of choice for reducing organic contaminants but requires much longer times than drying (Gibson et al., 2007).

8.4.4 Soil and Groundwater Remediation Systems

LAS The sorption of LAS to soil is a combination of several mechanisms and sorption to both the organic and inorganic fraction of the soil has been demonstrated. There is no clear correlation between the type and content of soil components and the extent of LAS sorption (Jacobsen et al., 2004). It was found that the sorption isotherms in soils for LAS were linear for concentrations below 90 mg L^{-1} (Ou et al., 1996)

Holt et al., (1989) described a program of monitoring work to determine the fate of LAS in sludge amended soil. The concentrations of LAS in soils are given for a large number of locations (24 farms and 51 fields) in the Thames Water Authority (TWA), U.K. The sites selected provide a range of soil types, frequency and concentration of sludge applications and agricultural uses (pasture/arable). In addition, the disappearance of LAS from soil with time was shown at three selected sites following different sludge application practices. The concentrations of LAS found in the sludge amended soil were generally less than 1 μg LAS/g soil. When this data was compared with the estimated total cumulative load based on known sludge applications, the majority of the sites showed losses of LAS >98%. In nine fields, the concentration of LAS in soil were in the range <0.2-20 μg g-1 representing losses of LAS between 70 and 99% of the estimated total cumulative load. The time course studies confirmed the rapid removal of LAS from sludge amended soils at three locations (5 fields) and for three different methods of application. The homologue distributions determined for LAS in soil suggested that microbial breakdown rather than leaching was the prime mechanism for its removal. Overall, the data indicated that an adequate safety margin existed between the concentrations of LAS in sludge amended soils and those likely to affect the growth of crop plants (Holt et al., 1989).

Application of sewage sludges is the most significant source of LASs in the terrestrial environments. LASs degrade rapidly in aerobic terrestrial environments as a result of microbial activity (Litz et al., 1987; Schoberl et al., 1988; Knaebel et al., 1990). For LAS, half-lives of 1 to 22 days in soil have been reported by Litz et al. (1987), de Henau et al. (1986b), Holt et al. (1989), Waters et al. (1989), and HERA (2004). It has been documented (Venhuis and Mehrvar, 2004) that LAS which makes up 0–16,000 mg/kg of the total dry weight of sewage sludge may influence microbial activity of the soils. Short-term effects of LAS on agricultural soils that were amended by sewage sludge varied with LAS concentration and incubation time. It was also suggested that LAS has a potential to inhibit biological activity. LAS in concentration greater than 40–60 mg kg^{-1} dw may have toxic effects on reproduction and growth of soil invertebrates. Earthworms and enchytracids were found to be four times more sensitive to LAS than springtails and mites (Mungray and Kumar, 2007).

Doi et al. (2002) determined the sorptive and biodegradable characteristics of LAS in a soil below a Florida, USA, septic system drainfield. Three distinct soil samples were collected from the septic system drainfield study site. These soils were used in laboratory sorption and biodegradation studies. The sorption test was designed to determine the partitioning of LAS between groundwater and soil in each sample. Results indicated that the sorption distribution coefficient (Kd) decreased from 4.02 to 0.43 L/kg and that the rate of ultimate biodegradation (first-order rate constant, k1) decreased from 2.17 to 0.08/d with increasing distance (0.7–1.2 m vertically below ground surface [BGS] and 0 to 6.1 m horizontally) from the drainfield. The three soils showed 49.8 to 83.4% LAS mineralization (percentage of theoretical CO_2) over 45- or 59-d test periods. These results demonstrate that subsurface soils in the systems studied possessed the potential to sorb and biodegrade LAS (Doi et al., 2002).

Eichhorn et al. (2005) identified for the first time SPC with 5–13 carbon atoms in the aliphatic side chain in agricultural soils treated with sewage sludge. Quantification of LAS and SPC in soil from 10 field sites, which differed in the history of sludge application, gave total concentrations of 120–2840 µg kg^{-1} for LAS and of 4–220 µg kg^{-1} for SPC. It was concluded that the anionic surfactant LAS was rapidly biodegraded in agricultural soils with high conversion rates immediately after the sludge amendment. Residues of LAS persisted over longer periods of time in soil, presumably due to very low bioavailability (Eichhorn et al., 2005).

For example, Berna et al. (Berna et al., 1989) detected ~1 mg kg^{-1} LAS in soil 90 days after sludge application, while Holt et al. (Holt et al., 1989) reported 0.7 mg kg^{-1} after 60 days and <0.2 mg kg^{-1} after 21 days, respectively, in experiments carried out at different field sites. In contrast, de Ferrer et al. (De Ferrer et al., 1997) reported significantly higher residual LAS concentrations of 16.7 mg kg^{-1} after a period of 62 days. Giger et al. (Giger et al., 1987) also reported a relatively high LAS concentration of 5.0 mg kg^{-1} on a permanent pasture site 104 days after sludge application. However, the initial concentrations directly after sludge amendment varied by more than order of magnitude among the studies cited above (Eichhorn et al., 2005).

LAS leaving the plants adsorbed to sludges continued to biodegrade during the soil amendment operation and no accumulation effect was detected during sewage treatment or on the sludge amended soils (Berna et al., 1989). The kinetics of LAS breakdown in sludge-amended soils were demonstrated to approximate first-order kinetics. In the studies of Holt and Bernstein (Holt and Bernstein, 1992) the LAS level dropped from 145 to 20 mg kg^{-1} within 10 days and then further to 16 mg kg^{-1} after another 10 days. A similar trend was reported by de Ferrer et al. (De Ferrer et al., 1997) with a starting concentration of 155 mg kg^{-1} and a decrease to 56 mg kg^{-1} after 6 days; an LAS content of 28 mg kg^{-1} soil was measured after a total elapsed time of 15 days. In the work performed by Giger et al. (Giger et al., 1987) LAS biodegradation was reported to amount to 62% after only 10 days (reduction from 45 to 17 mg kg^{-1}), and after 23 days a further 8.4% of the initially present LAS had disappeared. Spreading of sewage sludge on arable land resulted in an LAS concentration of 2.6 mg kg^{-1} (Waters et al., 1989), which had declined to 1.5 mg kg−1 after 8 days and further to 0.6 mg kg^{-1} after another 10 days. Regarding the frequency of sludge application and the LAS levels measured in the soil samples (ignoring the field at St. Joan), our data confirm an extensive removal of LAS through biodegradation (confirmed by the presence of metabolites; see below). No evidence was obtained that the anionic surfactant was accumulated as a result of repetitive sludge application (up to five applications in consecutive years).

Transport and biodegradation of LAS in sewage-contaminated groundwater were investigated over a range of dissolved oxygen concentrations (Krueger et al., 1998). Both laboratory column and an 80-day continuous injection tracer test field experiments were conducted. The rates of LAS biodegradation increased with increasing dissolved oxygen concentrations and indicated the preferential biodegradation of the longer alkyl chain LAS homologues (i.e., C12 and C13) and external isomers (i.e., 2- and 3-phenyl). However, for similar dissolved oxygen concentrations, mass removal rates for LAS generally were 2-3 times greater in laboratory column experiments than in the field tracer test. Under low oxygen conditions (<1 mg/L) only a fraction of the LAS mixture biodegraded in both laboratory and field experiments. Biodegradation rate constants for the continuous injection field test (0.002-0.08 day^{-1}) were comparable with those estimated for a 3-h injection (pulsed) tracer test conducted under similar biogeochemical conditions, indicating that increasing the exposure time of aquifer sediments to LAS did not increase biodegradation rates (Krueger et al., 1998).

The soil and groundwater conditions at a single home septic system near Jacksonville (Florida), USA represent a worst case for demonstrating the treatment of septic tank effluent (Nielsen et al., 2002). Despite these adverse circumstances, the septic treatment system (aerobic infiltration surface/unsaturated soil and saturated soil) effectively removes the septic tank effluent (STE) components. Most of the treatment occured at the infiltration surface since about > 96% of LAS, and >99% of alcohol ethoxylate (AE) and alcohol ether sulfate (AES) were removed when a 0.4-m unsaturated zone existed. Under the worst conditions, when very little (0.01 m) or no unsaturated treatment zone was present, LAS and AES surfactant residues were detected in the groundwater; AE was not detected. LAS was detected only up to 11.7 m horizontally and 3.7 m vertically from the drainfield during the wet season. Since it was possible for the surfactants to have migrated as much as 260 m if they had moved uninhibited with the wastewater plume, it is clear that removal was occurring. The most likely removal mechanisms for these surfactants are biodegradation and sorption (Nielsen et al., 2002).

The occurrence of LAS in a series of soil samples originating from the municipality of Roskilde (Denmark) was studied (Carlsen et al., 2002). The study included soil samples from eight different locations with different histories: a preserved natural area that has not been cultured for 50-100 years, a soil that has been ecologically cultured for 40 years, a soil sustainably manured and ecologically cultured for 5 years (formerly conventionally cultured) and a soil that has been conventionally cultured using artificial fertilizer. In addition, a soil was studied that had been sludge amended by applying medium amounts of sludge as well as a soil that has been amended with high amounts of sludge for a 25-year period. In the latter case, the sludge amendment was abandoned 6 years before the first sampling, followed by the application of artificial fertilizers. Finally, a meadow in the run-off zone from a sludge storage area was included in the investigations. In addition to the soil samples, selected samples of the applied sludge and other fertilizers were analyzed for their possible content of LAS. Apart from the location where the soil had been heavily sludge-amended and the location situated in the run-off zone of the sludge storage, concentrations of LAS in all soil samples were found to be below approximately 1 mg/kg, which is well below the proposed preliminary soil quality criteria for LAS of 5 mg/kg. On the other hand, the study unambiguously disclosed that in the case of heavy sludge amendment, the proposed soil quality criteria might well be exceeded (Carlsen et al., 2002).

APEO The sorption mechanism for NP is much simpler. Under most environmental conditions, the molecule has no electrical charge and therefore primarily interacts with the soil by hydrophobic sorption to the organic fraction. The compound is only deprotonated when pH exceeds the pKa value for phenols of approximately 10. However, sorption by hydrogen bonding also has been demonstrated (John et al., 2000; Jacobsen et al., 2004).

Degradation and mobility of NP were investigated in a lysimeter study using a sandy loam soil and 45-cm soil columns (Jacobsen et al., 2004). Anaerobically digested sewage sludge was incorporated in the top-15-cm soil layer to an initial content of 0.56 mg NP kg-1 dry wt., respectively. It was found that the concentrations in the top-15-cm soil layer declined to 25 and 45% of the initial contents for LAS and NP, respectively, within the first 10 d of the study. At the end of the study, after 110 d, NP content was below the detection limit. Assuming first-order degradation kinetics, half-lives of 37 d were estimated for NP. In addition, no NP was measured in concentrations above the detection limits of 150 and 50 µg kg-1 dry wt., respectively, in soil layers below the 15 cm of sludge incorporation, indicating negligible downward transport of the surfactants in the lysimeters (Jacobsen et al., 2004).

Ahel et al. (1996) studied the behaviour of various persistent metabolites derived from nonylphenol polyethoxylate (NPnEO) surfactants during infiltration of river water to groundwater at two field sites situated in the northern part of Switzerland (Glatt River and Sitter River). Nonylphenol (NP), nonylphenol monoethoxylate (NP1EO), nonylphenol diethoxylate (NP2EO), nonylphenoxy acetic acid (NP1EC) and nonylphenoxy(ethoxy) acetic acid (NP2EC) were observed in the two investigated rivers at relatively high concentrations with average values of the individual types of nonylphenolic compounds ranging from 1.8 to 25 g/l. The average concentrations of NP, NP1EO and NP2EO in groundwater were significantly lower (range < 0.1-1/~g/1) suggesting an efficient elimination of these compounds during infiltration. In contrast, the elimination of nonylphenoxy carboxylic acids was less efficient. Most of the observed elimination occurred in the first 2.5 m of the aquifer, while further decrease in concentration was rather slow. In one sampling period, residual concentrations of nonylphenolic compounds up to 7.2 g/l were detected in a pumping station used for drinking water supply which is situated 130 m from the Glatt River bed. Concentrations of NP, NP1EO and NP2EO in both river water and groundwater showed a pronounced seasonal variability with higher values observed during winter. The data suggest that low temperatures, which prevail in winter, significantly reduce the elimination efficiency of NP and to a lesser extent of NP1EO, while the behaviour of NP2EO was not affected. Such behaviour indicates biological transformation as the responsible elimination process. A comparison of average elimination efficiencies of nonylphenolic compounds was the following sequence: NP2EO > NP1EO > NP > NP1EC = NP2EC (Ahel et al., 1996).

Swartz et al. (2006) monitored NP, NP1EC and NP2EC, in a residential septic system and in downgradient groundwater in Cape Cod, MA, USA. The apparent persistence of NP1EC and NP2EC in groundwater over the distance investigated was consistent with a study of APEO behavior in river water infiltrating to groundwater over a distance as far as 14 m, although unlike the present study, Ahel et al. (Ahel et al., 1996) measured substantially lower concentrations of the lower homologues (NP1-2EO) and NP over the first several meters of transport in the aquifer. Differing oxygenation conditions, sorption capacities, and microbial activity may account for these apparent differences in behavior of NP and the lower homologues between these two study sites.
(Swartz et al., 2006)

A Canadian study indicated that 60% of the original NPs and 30% of OP remained in the soil 60 days after application but decreased to nondetectable levels 90 days after application. A Danish study also suggested that soil concentrations of NPs, NP1EOs, and NP2EOs remained constant during a 28-day testing period.

8.5 Regulations

Public concern about the many unregulated chemicals, including the surfactants and their degradation products discussed in this paper, is increasing, as new insights concerning environmental and health issues associated with the production and use of chemicals continue to emerge, highlighted, for instance, by the increasing focus on endocrine disrupters over the past few years (Lauridsen and Røpke, 2005).

Gomez-Rico et al. (2007) reported, for LAS, the samples had 130 to 32000 mg kg^{-1}, and the EU limit (Union, 2000) is 2600 mg kg^{-1}. Therefore, some samples exceeded the maximum. Furthermore, there were values ≥10 of this. Of the samples, 82% presented <10,000 mg kg^{-1}, 12% presented 10,000 to 20,000 mg kg^{-1} and 6% presented >20,000 mg kg^{-1}. It can be emphasized that 12% (only two samples) presented <2600 mg kg^{-1} LAS (the limit) (Union, 2000). Several articles can be found about LAS contents in biosolids from sewage-treatment plants showing that many countries have high values, surpassing the limits proposed by the EU in 2000 in many cases.

In a subsequent work, Jensen and Jepsen (2005) found that LAS contents had substantially decreased in Denmark from 1995, when the range was 13 to 13,725 mg kg^{-1} dw, until 2002, when the range was 50 to 1507 mg kg^{-1}. This decrease, which makes LAS levels in Denmark one order of magnitude lower than those found elsewhere in the EU, has been facilitated by effective control from authorities, voluntary phasing-out agreements with industry, improved source-identification tools, better handling and post treatment methods (aerobic stabilization better than anaerobic), and higher waste taxes (Jensen & Jepsen 2005). The cut-off value in this country is 1300 mg kg^{-1} (Gomez-Rico et al., 2007).

In accordance with the requirements of the European Water Framework Directive (2000/60/EEC), Austria defined environmental quality criteria regarding the good chemical status of surface waters. For LAS the limit value is set with 270 μg L^{-1} (BGBl. II 96/2006). The effluents of the investigated WWTPs keep to this limit value even undiluted, being in the average ten times below the required river concentrations.

The European Commission proposal for a Directive on environmental quality standards (EQS) in the field of water policy (COM(2006)397 final) defines a limit concentration for NP of 0.3 μg L^{-1}. Only the effluent of WWTP 2 reached an effluent concentration below this limit value. In order to keep to the defined quality criteria for all the other sampled WWTPs a three- to six-fold dilution of the effluent in the receiving waters is necessary. Besides the EQS, which refer to annual average values, the Commission proposal also defines a maximum allowable concentration for NP which amounts to 2 μg L^{-1}. All measured effluent concentrations complied with this maximum value.

8.6 Conclusion

Surfactants and their degradation products persist in the aquatic environment due to their widespread usage and discharge of sewage effluents into surface waters. They are also measured in sludge, soil sediment and groundwater. Conventional unit processes at wastewater treatment plants (WWTPs) are not completely effective in removal of these organic contaminants in wastewater. A large portion of these organic contaminants appeared to adsorb to the biosolids, a phenomenon that likely prevented their degradation in the bioreactors. Therefore, application of surfactants (LAS,

NP2EO, NP1EO, NP, NP1EC and etc.) contaminated sludges is a possible source of ground and surface water contamination as well as through direct exposure from soil ingestion, more work is needed to understand the fate of these compounds in agricultural settings. Little work has been reported investigating the fate of surfactants (LAS, NP2EO, NP1EO, NP, NP1EC and etc.) in high-temperature composting environments or during heat treatment of contaminated sludge. Further work therefore needs to be done to assess the fate of these organic contaminants during processing of sludges, to prevent harm to the environment, in particular to soil when sludge is used for land application. The environmental risks posed by surfactants and their degradation products can be assessed based on the comparison of the environmental concentration, however, the relevant regulations and more toxicity data are needed for terrestrial risk assessment of surfactants and their degradation products.

References

Ahel, M., and Giger, W. (1993) Aqueous solubility of alkylphenols and alkylphenol polyethoxylates. Chemosphere 26: 1461-1470.

Ahel, M., Schaffner, C., and Giger, W. (1996) Behaviour of alkylphenol polyethoxylate surfactants in the aquatic environment--III. Occurrence and elimination of their persistent metabolites during infiltration of river water to groundwater. Water Research 30: 37-46.

Alvarez-Munoz, D., Lara-Martin, P.A., Blasco, J., Gomez-Parra, A., and Gonzalez-Mazo, E. (2007) Presence, biotransformation and effects of sulfophenylcarboxylic acids in the benthic fish Solea senegalensis. Environment International 33: 565-570.

Angelidaki, I., Torang, L., Waul, C.M., and Schmidt, J.E. (2004) Anaerobic bioprocessing of sewage sludge, focusing on degradation of linear alkylbenzene sulfonates (LAS). Water Science and Technology 49: 115-122.

Berna, J.L., Moreno, A., and Bengoechea, C. (1998) J. Surfact. Detergents 1: 263.

Berna, J.L., Ferrer, J., Moreno, A., Prats, D., and Ruiz Bevia, F. (1989) The fate of LAS in the environment. TENSIDE SURFACTANTS DETERG. 26: 101-107.

Bester, K., Theobald, N., and Schroder, H.F. (2001) Nonylphenols, nonylphenolethoxylates, linear alkylbenzenesulfonates (LAS) and bis (4-chlorophenyl)-sulfone in the German Bight of the North Sea. Chemosphere 45: 817-826.

Birkett, J.W., and Lester, J.N. (2003) Endocrine Disrupters in Wastewater and Sludge Treatment Processes. CRC Press, Boca Raton, FL

Carballa, M., Manterola, G., Larrea, L., Ternes, T., Omil, F., and Lema, J.M. (2007) Influence of ozone pre-treatment on sludge anaerobic digestion: Removal of pharmaceutical and personal care products. Chemosphere 67: 1444-1452.

Carlsen, L., Metzon, M.B., and Kjelsmark, J. (2002) Linear alkylbenzene sulfonates (LAS) in the terrestrial environment. Science of the Total Environment 290: 225-230.

Cheng, C.-Y., Wu, C.-Y., Wang, C.-H., and Ding, W.-H. (2006) Determination and distribution characteristics of degradation products of nonylphenol polyethoxylates in the rivers of Taiwan. Chemosphere 65: 2275-2281.

Cuzzola, A., Bernini, M., and Salvadori, P. (2002) A preliminary study on iron species as heterogeneous catalysts for the degradation of linear alkylbenzene sulphonic acids by H2O2. Appl. Catal. B: Environ. 36: 231–237.

Dachs, J., Van Ry, D.A., and Eisenreich, S.J. (1999) Occurrence of estrogenic nonylphenols in the urban and coastal atmosphere of the lower Hudson River estuary. Environmental Science and Technology 33: 2676-2679.

De Ferrer, J., Moreno, A., Vaquero, M.T., and Comellas, L. (1997) Monitoring of LAS in direct discharge situations: Linear alkylbenzene sulfonate in untreated sewage and on sludge modified soils. Tenside, Surfactants, Detergents 34: 278-282.

Doi, J., Marks, K.H., DeCarvalho, A.J., McAvoy, D.C., Nielsen, A.M., Kravetz, L., and Cano, M.L. (2002) Investigation of an onsite wastewater treatment system in sandy soil: Sorption and biodegradation of linear alkylbenzene sulfonate. Environmental Toxicology and Chemistry 21: 2617-2622.

Eichhorn, P., Lopez, O., and Barcelo, D. (2005) Application of liquid chromatography-electrospray-tandem mass spectrometry for the identification and characterisation of linear alkylbenzene sulfonates and sulfophenyl carboxylates in sludge-amended soils. Journal of Chromatography A 1067: 171-179.

Eichhorn, P., Flavier, M.E., Paje, M.L., and Knepper, T.P. (2001) Occurrence and fate of linear and branched alkylbenzenesulfonates and their metabolites in surface waters in the Philippines. Science of the Total Environment 269: 75-85.

Eichhorn, P., Knepper, T.P., Ventura, F., and Diaz, A. (2002) The behavior of polar aromatic sulfonates during drinking water production: a case study on sulfophenyl carboxylates in two European waterworks. Water Research 36: 2179-2186.

European Union Decision No 2455/2001/EC of the European Parliament and of the Council of 20 November 2001 establishing the list of priority substances in the field of water policy and amending directive 2000/60/EC, Off. J. L331, 15/12/2001.

Gibson, R.W., Wang, M.-J., Padgett, E., Lopez-Real, J.M., and Beck, A.J. (2007) Impact of drying and composting procedures on the concentrations of 4-nonylphenols, di-(2-ethylhexyl)phthalate and polychlorinated biphenyls in anaerobically digested sewage sludge. Chemosphere 68: 1352-1358.

Giger, W., Brunner, P.H., and Ahel, M. (1987) Detergent derived organic chemicals in sewage and sewage sludge (Organische waschmittelinhaltsstoffe und deren abbauprodukte in abwasser und klarschlamm). GAS WASSER ABWASSER 67: 111.

Gomez-Rico, M.F., Font, R., Aracil, I., and Fullana, A. (2007) Analysis of organic pollutants in sewage sludges from the Valencian community (Spain). Archives of Environmental Contamination and Toxicology 52: 306-316.

Gonzalez-Mazo, E., Forja, J.M., and Gomez-Parra, A. (1998) Fate and distribution of linear alkylbenzene sulfonates in the littoral environment. Environmental Science and Technology 32: 1636-1641.

Gonzalez, S., Petrovic, M., and Barcelo, D. (2004) Simultaneous extraction and fate of linear alkylbenzene sulfonates, coconut diethanol amides, nonylphenol ethoxylates and their degradation products in wastewater treatment plants, receiving coastal waters and sediments in the Catalonian area (NE Spain). Journal of Chromatography A 1052: 111-120.

Gonzalez, S., Petrovic, M., and Barcelo, D. (2007) Removal of a broad range of surfactants from municipal wastewater - Comparison between membrane bioreactor and conventional activated sludge treatment. Chemosphere 67: 335-343.

Hernandez-Raquet, G., Soef, A., Delgenes, N., and Balaguer, P. (2007) Removal of the endocrine disrupter nonylphenol and its estrogenic activity in sludge treatment processes. Water Research 41: 2643-2651.

Holt, M.S., and Bernstein, S.L. (1992) Linear alkylbenzenes in sewage sludges and sludge amended soils. Water Research 26: 613-624.

Holt, M.S., Matthijs, E., and Waters, J. (1989) The concentrations and fate of linear alkylbenzene sulphonate in sludge amended soils. Water Research 23: 749-759.

Hou, S., and Sun, H. (2007) Pollution of NPEOs in four municipal sewage treatment plants in the north of China. Frontiers of Environmental Science and Engineering in China 1: 196-201.

Isobe, T., and Takada, H. (2004) Determination of degradation products of alkylphenol polyethoxylates in municipal wastewaters and rivers in Tokyo, Japan. Environ. Toxicol. Chem. 23: 599–605.

Jacobsen, A.M., Mortensen, G.K., and Hansen Bruun, H.C. (2004) Degradation and Mobility of Linear Alkylbenzene Sulfonate and Nonylphenol in Sludge-Amended Soil. Journal of Environmental Quality 33: 232-240.

Jin, X.L., Jiang, G.B., Huang, G.L., Liu, J.F., and Zhou, O.F. (2004) Determination of 4-tert-octylphenol, 4-nonylphenol and bisphenol A in surface waters from the Haihe River in Tianjin by GC–MS with selected ion monitoring. Chemosphere 56: 1113–1119.

John, D.M., House, W.A., and White, G.F. (2000) Environmental fate of nonylphenol ethoxylates: Differential adsorption of homologs to components of river sediment. Environmental Toxicology and Chemistry 19: 293-300.

Johnson, A.C., Aerni, H.R., Gerritsen, A., Gibert, M., Giger, W., Hylland, K. et al. (2005) Comparing steroid estrogen, and nonylphenol content across a range of European sewage plants with different treatment and management practices. Water Research 39: 47-58.

Keller, H.; K. Xia; Bhandari, A. "Occurrence and distribution of nonylphenol polyethoxylates and 4-nonylphenol in northeast Kansas wastewater treatment facilities." Practice Periodical of Hazardous, Toxic and Radioactive Waste Management. 7:203-213, 2003.

Krueger, C.J., Radakovich, K.M., Sawyer, T.E., Barber, L.B., Smith, R.L., and Field, J.A. (1998) Biodegradation of the surfactant linear alkylbenzenesulfonate in sewage- contaminated groundwater: A comparison of column experiments and field tracer tests. Environmental Science and Technology 32: 3954-3961.

Kuch, H.M., and Ballschmiter, K. (2001) Determination of endocrine-disrupting phenolic compounds and estrogens in surface and drinking water by HRGC-(NCI)-MS in the picogram per liter range. Environmental Science and Technology 35: 3201-3206.

La Guardia, M.J., Hale, R.C., Harvey, E., and Matteson Mainor, T. (2001) Alkylphenol ethoxylate degradation products in land-applied sewage sludge (biosolids). Environmental Science and Technology 35: 4798-4804.

Lara-Martin, P.A., Petrovic, M., Gomez-Parra, A., Barcelo, D., and Gonzalez-Mazo, E. (2006) Presence of surfactants and their degradation intermediates in sediment cores and grabs from the Cadiz Bay area. Environmental Pollution 144: 483-491.

Lara-Martin, P.A., Gomez-Parra, A., Kochling, T.K., Sanz, J.L., Amils, R., and Gonzalez-Mazo, E. (2007) Anaerobic degradation of linear alkylbenzene sulfonates in coastal marine sediments. Environmental science & technology 41: 3573-3579.

Lauridsen, P.V., and Røpke, I. (2005) Experience with Chemicals Regulation - Lessons from the Danish LAS Case. The Journal of Transdisciplinary Environmental Studies 4: 1-15.

Lenz, K., Beck, V., and Fuerhacker, M. (2004) Behaviour of bisphenol A (BPA), 4-nonylphenol (4-NP) and 4-nonylphenol ethoxylates (4-NP1EO, 4-NP2EO) in oxidative water treatment processes. Water Science and Technology 50: 141-147.

Li, D., Kim, M., Oh, J.-R., and Park, J. (2004a) Distribution characteristics of nonylphenols in the artificial Lake Shihwa, and surrounding creeks in Korea. Chemosphere 56: 783-790.

Li, D., Kim, M., Oh, J.R., and Park, J. (2004b) Distribution characteristics of nonylphenols in the artificial Lake Shihwa, and surrounding creeks in Korea. Chemosphere 56: 783-790.

Lin, S.H., Lin, C.M., and Leu, H.G. (1999) Operating characteristics and kinetic studies of surfactant wastewater treatment by Fenton oxidation. Water Res. 33: 1735–1741.

Mantzavinos, D., Burrows, D.M.P., Willey, R., Biundo, G.L., Zhang, S.F., Livingston, A.G., and Metcalfe, I.S. (2000) Wet air oxidation of aqueous solution of linear alkylbenzene sulfonate. Ind. Eng. Chem. Res. 39: 3659–3665.

Mayer, T., Bennie, D., Rosa, F., Rekas, G., Palabrica, V., and Schachtschneider, J. (2007) Occurrence of alkylphenolic substances in a Great Lakes coastal marsh, Cootes Paradise, ON, Canada. Environmental Pollution 147: 683-690.

McAvoy, D.C., White, C.E., Moore, B.L., and Rapaport, R.A. (1994) Chemical fate and transport in a domestic septic system: Sorption and transport of anionic and cationic surfactants. Environmental Toxicology and Chemistry 13: 213-221.

Minamiyama, M., Ochi, S., and Suzuki, Y. (2006) Fate of nonylphenol polyethoxylates and nonylphenoxy acetic acids in an anaerobic digestion process for sewage sludge treatment. Water Science and Technology 53: 221-226.

Mungray, A.K., and Kumar, P. (2007) Degradation of anionic surfactants during drying of UASBR sludges on sand drying beds. Journal of Environmental Management doi:10.1016/j.jenvman.2007.05.006.

Nielsen, A.M., DeCarvalho, A.J., McAvoy, D.C., Kravetz, L., Cano, M.L., and Anderson, D.L. (2002) Investigation of an onsite wastewater treatment system in sandy soil: Site characterization and fate of anionic and nonionic surfactants. Environmental Toxicology and Chemistry 21: 2606-2616.

Patterson, D.A., Metcalfe, I.S., Xiong, F., and Livingston, A.G. (2002) Biodegradability of linear alkylbenzene sulfonates subjected to wet air oxidation. Journal of Chemical Technology and Biotechnology 77: 1039-1049.

Petrovic, M., and Barcelo, D. (2004) Analysis and fate of surfactants in sludge and sludge-amended soils. TrAC Trends in Analytical Chemistry 23: 762-771.

Petrovic, M., Diaz, A., Ventura, F., and Barcelo, D. (2003) Occurrence and removal of estrogenic short-chain ethoxy nonylphenolic compounds and their halogenated derivatives during drinking water production. Environmental Science and Technology 37: 4442-4448.

Petrovic, M., Fernandez-Alba, A.R., Borrull, F., Marce, R.M., Mazo, E.G., and Barcelo, D. (2002) Occurrence and distribution of nonionic surfactants, their degradation products, and linear alkylbenzene sulfonates in coastal waters and sediments in Spain. Environmental Toxicology and Chemistry 21: 37-46.

Prat, D., Ruiz, F., Vazquez, B., Zarzo, D., Berna, J., and Moreno, A. (1993) Environ Toxicol Chem. 12: 1599.

Saien, J., Ardjmand, R.R., and Iloukhani, H. (2003) Photocatalytic decomposition of sodium dodecyl benzene sulfonate under aqueous media in the presence of TiO2. Phys. Chem. Liq. 41: 519–531.

Samsøe-Petersen, L. (2003) Organic contaminants in Sewage Sludge. Swedish Environmental Protection Agency. Report no. 5217. ISBN 91-620-5217-9.

Sanderson, H., Dyer, S.D., Price, B.B., Nielsen, A.M., van Compernolle, R., Selby, M. et al. (2006) Occurrence and weight-of-evidence risk assessment of alkyl sulfates, alkyl ethoxysulfates, and linear alkylbenzene sulfonates (LAS) in river water and sediments. Science of The Total Environment 368: 695-712.

Sanz, E., Prats, D., Rodriguez, M., and Camacho, A. (2006) Effect of temperature and organic nutrients on the biodegradation of linear alkylbenzene sulfonate (LAS) during the composting of anaerobically digested sludge from a wastewater treatment plant. Waste Management 26: 1237-1245.

Scott, M.J., and Jones, M.N. (2000) The biodegradation of surfactants in the environment. Biochimica et Biophysica Acta (BBA) - Biomembranes 1508: 235-251.

Swartz, C.H., Reddy, S., Benotti, M.J., Yin, H., Barber, L.B., Brownawell, B.J., and Rudel, R.A. (2006) Steroid estrogens, nonylphenol ethoxylate metabolites, and other wastewater contaminants in groundwater affected by a residential septic system on cape cod, MA. Environmental Science and Technology 40: 4894-4902.

Tabrizi, G.B., and Mehrvar, M. (2006) Pilot-plant study for the photochemical treatment of aqueous linear alkylbenzene sulfonate. Separation and Purification Technology 49: 115–121.

Tamage, S.S. (1994) Environmental and Human Safety of Major Surfactants, Alcohol Ethoxylates and Alkylphenol Ethoxylates; Lewis Publishers: Chelsea, MI.

Tanghe, T., and Verstraete, W. (2001) Adsorption of nonylphenol onto granular activated carbon. Water, Air, and Soil Pollution 131: 61-72.

TørCshløevm, J., L.Samsøe-Petersen, Rasmussen, J.O., and Kristensen, P. (1997) Use of waste products in agriculture. Environ. Project 366. Ministry of Environ. and Energy, Copenhagen.

Tsuda, T., Suga, K., Kaneda, E., and Ohsuga, M. (2002) 4-Nonylphenol, 4-nonylphenol mono- and diethoxylates, and other 4-alkylphenols in water and shellfish from rivers flowing into lake Biwa. Bulletin of Environmental Contamination and Toxicology 68: 126-131.

Union, E. (2000) Working document on sludge, 3rd draft, 19 p. Available at: http://www.ewaonline.de/downloads/sludge_en.pdf.

Van Ry, D.A., Dachs, J., Gigliotti, C.L., Brunciak, P.A., Nelson, E.D., and Eisenreich, S.I. (2000) Atmospheric seasonal trends and environmental fate of alkylphenols in the lower Hudson River Estuary. Environmental Science and Technology 34: 2410-2417.

Vejrup, K.V., and Wolkoff, P. (2002) Linear alkylbenzene sulfonates in indoor floor dust. The Science of The Total Environment 300: 51-58.

Venhuis, S.H., and Mehrvar, M. (2004) Health effects, environmental impacts, and photochemical degradation of selected surfactants in water. International Journal of Photoenergy 6: 115-125.

Vitali, M., Ensabella, F., Stella, D., and Guidotti, M. (2004) Nonylphenols in freshwaters of the hydrologic system of an Italian district: association with human activities and evaluation of human exposure. Chemosphere 57: 1637-1647.

Waters, J., Holt, M.S., and Matthijs, E. (1989) Fate of LAS in sludge amended soils. Tenside, Surfactants, Detergents 26: 129-135.

Ying, G.G., Williams, B., and Kookana, R. (2002) Environmental fate of alkylphenols and alkylphenol ethoxylates - A review. Environment International 28: 215-226.

Zhang, C., Valsaraj, K.T., Constant, W.D., and Roy, D. (1999) Aerobic biodegradation kinetics of four anionic and nonionic surfactants at sub- and supra-critical micelle concentrations (CMCs). Water Research 33: 115-124.

Zoller, U. (1993) Groundwater contamination by detergents and polycyclic aromatic hydrocarbons - a global problem of organic contaminants: Is the solution locally specific? Water Science and Technology 27: 187-194.

CHAPTER 9

Brominated Fire Retardants

Shankha Banerji, Rao Surampalli, Tian Zhang, and R.D. Tyagi

9.1 Introduction

Fire retardants are substances that can delay or prevent combustion. Water is the most common fire retardant used. Here we are interested in chemicals used as fire retardants, that eventually enter the environment and cause possible harm.

There are more than 175 different types of compounds used as fire retardants. These compounds can be placed in four categories - halogenated organics, phosphorus-based, nitrogen-based, and other inorganic compounds. The halogenated compounds (chlorinated and brominated) are the most common fire retardants as they are less expensive and are quite effective. In this group, brominated compounds are more common than the chlorinated types. The phosphorus and nitrogen based fire retardants are monoammonium phosphate, diammonium phosphate, ammonium polyphosphate and ammonium sulfate. These compounds individually or in combination are mixed with other chemicals such as corrosion inhibitors, alcohol, gum thickeners and surfactants in a fire retardant formulation. In the other inorganic compounds category, compounds like aluminum hydroxide, antimony oxides and chlorides may be included.

In this chapter, brominated fire retardants (BFRs) will be covered since they are the most used fire retardants and have serious environmental effects. There are more than 75 different brominated fire retardants (Birnbaum and Staskal, 2004). BFRs contribute to about 38% of the global bromine demand. The electronic industry is the major consumer of BFRs. The four main applications of BFRs in computers are: printed circuit boards, connectors, plastic covers and cables. BFRs are also used in many other products such as TV plastic covers, carpets, paints, upholstery and domestic kitchen appliances. Its effectiveness for fire prevention and hindrance to the spread of fire makes it a very useful fire retardant.

Reactive BFRs are chemically bound to the polymeric material used for the manufacture of the electronic or household items mentioned earlier. In contrast, the additive flame retardants are blended with polymeric materials along with additives

315

like plasticizers to form the product. They are not chemically bound to the polymer materials, which make them more liable to enter the environment by leaching. In addition, there are synergistic flame retardants like antimony trioxide or zinc borate which are used with the additive flame retardants to improve their action (D'Silva et al., 2004).

While benefits achieved through enhanced fire safety are critical, they should be achieved in a manner that minimizes risk to human health and the environment. From an environmental point of view, BFRs are a group of potential endocrine disrupting chemicals (EDCs) causing increasing concern (Vos et al., 2003; Kucher and Purvis, 2004). The importance of BFRs has been highlighted recently by the findings that their concentrations in the environment and in humans are rapidly increasing (Alaee and Wenning, 2002). In addition, some BFRs have been shown to be toxic and bioaccumulative compounds (Darnerud, 2003). Like polychlorinated biphenyls (PCBs), they have been found in remote areas like the Arctic and in deep oceans (Sellstrom et al., 1993). Some BFRs may exhibit toxicity to wildlife and humans through various mechanisms, including neurobehavioral development, thyroid hormone levels, fetal toxicity / teratogenicity, and liver and kidney morphology (Darnerud 2003). Women accumulate polybrominated diphenyl ethers (PBDEs) in their bodies throughout their lifetime. During pregnancy, the mother's blood transports nutrients to her embryo and removes waste products through transfer across the placenta. A number of bioaccumulative compounds like (PBDEs) have been shown to move freely across the placenta into the infant's bloodstream, as though the placenta were transparent to the chemicals (Lione, 1988). It is, therefore, possible for fetuses to be exposed to PBDEs during vulnerable periods of brain development.

In this chapter, the main categories of BFRs and their properties will be introduced, followed by their occurrences and behavior (fate and transport) in natural and engineered systems. Details on toxicity of BFRs will not be included in this chapter. There are several reviews on the subject that can provide needed information (de Wit, 2001; Darnerud, 2003; Birnbaum and Staskal, 2004; Kucher and Purvis, 2004). Issues related to BFRs' characteristics as potential EDCs, regulations, U.S. toxic policy, and the current status of research in this area will be discussed.

9.2 General Information on BFRs

9.2.1 Main Categories of BFRs

Brominated fire retardants can be classified into five categories – bisphenol, diphenyl ethers, cyclohydrocarbons, phenols and phthalic acid derivatives. Among these BFRs, the industrial production of the first three is the highest and consequently has major impacts on the environment. Thus, only these three classes of compounds will be included in this chapter – brominated bisphenols (BBP), polybrominated diphenyl ethers (PBDE) and polybrominated cyclohydrocarbons. Figure 9.1 shows the chemical structures of some representative BFRs.

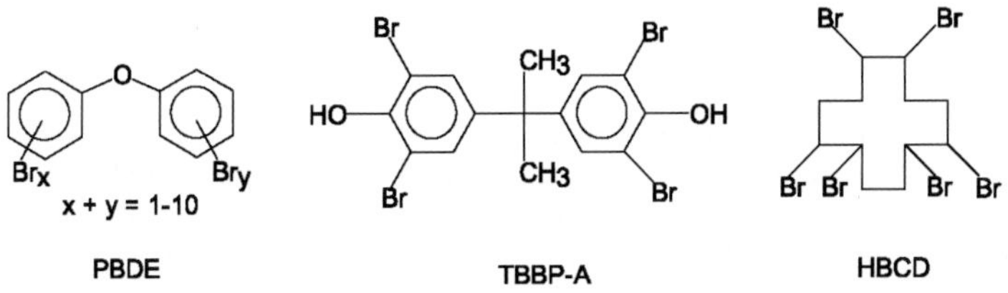

PBDE   **TBBP-A**   **HBCD**

Figure 9.1 Chemical structures of polybrominated diphenyl ethers (PBDE), tetrabromobisphenol A (TBBP-A), and hexabromocyclododecane (HBCD).

Bromobisphenol: In this group the most important compound is tetrabromobisphenol A (TBBPA), which is a reactive fire retardant with the highest global consumption for fire retardation in the world in 2003 (Alaee et al., 2003). It has also been used as a fire retardant additive in some polymers. The majority of the TBBPA is used in the production of epoxy and polycarbonate resins used in printed circuit boards to make them fire resistant. It is chemically bound to the resins of the printed circuit boards. In addition, a small amount of TBBPA is used as an additive for fire retardation in acrylonitrile-butadiene-styrene resin and high-impact styrene. A derivative of TBBPA
(bis 2-hydroxy ethyl ether) is also used as a flame retardant for paper and textiles adhesives and coatings.

Polybrominated diphenyl ethers (PBDE): PBDEs are used as additive flame retardants for a variety of plastics, foams, surface coatings and synthetic fibers. In commercial formulation, there is a mixture of PBDEs with various degrees of bromination. Like PCBs, PBDEs have many possible congeners formed during manufacture. However, most commercial PBDE products have only a few congeners at significant levels. The three most common PBDEs commercial products are: penta-BDE, octa-BDE and deca-BDE (i.e., $x + y$ = 5, 8, and 10, respectively, in Fig. 1). Penta-BDE is used in the manufacture of flexible polyurethane foams and polyurethane elastomers. These polyurethanes are used to make upholstery and furnishings. Octa-BDE is used as a flame retardant in the manufacture of acrylonitrile-butadiene-styrene plastics (ABS), which are used as covers or casings for electrical or electronic equipments. It is also used in the manufacture of nylon, polycarbonate, phenol formaldehyde resins, low density polyethylene, adhesives and coatings. Deca-BDE is the largest volume PBDE used. It is used in electronic equipments.

Polybrominated cyclohydrocarbons: The most important compounds in this group are the hexabromocyclododecanes (HBCDs). They are additive flame retardants used in extruded and high impact polystyrene foams. They are used as thermal insulation in buildings, in upholstery textiles and to some extent in electrical equipment housings. Since the production restrictions in US and Europe and the discovery of its persistence and toxicity, HBCDs are being considered as replacements to PBDEs (D'Silva et al., 2004).

9.2.2 Sources

As shown in Table 9.1, release of BFRs to the environment may occur during their manufacture, incorporation into resins and polymer products, while the incorporated products are in use, and finally as the products are disposed or recycled (Danish Environmental Protection Agency, 1999; USEPA, 2005).

Table 9.1 Sources of BFRs in the environment

BFRs	Sources	References
Tetrabromobisphenol (TBBPA)	TBBPA manufacturing plants	Zweidinger et al., 1979.
	- dust on TV sets	Watanabe & Sakai, 2003.
	- electrical and electronic recycling plants	Morf et al., 2005. Sjodin et al., 2001
Polybrominated diphenyl Ethers (PBDEs)	- home and work furnishings plastic coverings, etc.	Hale et al., 2002.
	- dust on TV sets	Watanabe & Sakai, 2003
	- electrical & electronic recycling plants	D'Silva et al., 2004 ; Morf et al., 2005.
	- wastewater spray irrigation	Goel et al., 2006.
Hexabrominated Cyclododecanes (HBCDs)	- home and office dust	Greenpeace, 2004.
	- HBCD manufacturing plants	Covaci, et al., 2006.
	-electrical & electronic recycling plants	Morf et al., 2005.

A study in Sweden reported high amounts of BFRs in river sediments near a factory manufacturing these compounds (Sellstrom and Jansson, 1995). Stream sediments near a textile mill using several BFRs also had higher levels of these compounds downstream compared to upstream (Sellstrom et al., 1998). Other possible sources of these compounds in the environment are from its release from consumer products that have BFRs and from sites where the consumer products are disposed. Fine dust in off gases from a waste electrical and electronic equipment recycling plant in Switzerland had BFRs present which could be a source of these compounds if not managed properly (Morf et al., 2005). It was reported that by the year 2004 over 315 million computers would become obsolete and have BFRs present in the casings and circuit boards that could be a source of these compounds in the environment. Incineration of these BFR containing products may increase the amount of brominated and mixed bromochlorodioxins and furans that reach the environment (D'Silva et al., 2004).

Tetrabromobisphenol A (TBBPA): TBBPA is a reactive fire retardant that is incorporated in polymers comprising epoxy and polycarbonate resins. In the final plastic product there may be some unpolymerized TBBPA which may leak out and enter the environment. In addition, the additive use of this BFR can release it from the polymers. In the environment both additive- and reactive-TBBPA and its metabolites have been found. At a plant manufacturing TBBPA, the air samples had 1.8 μg TBBPA/m^3 (Zweidinger et al., 1979). Work place air in an electronic recycling plant in Sweden had higher amounts of TBBPA compared to other work place environments (Sjodin et al., 2001). Dust collected from the back covers of TV sets containing BFRs had about 21 mg TBBPA /g (Watanabe and Sakai, 2003). Disposal of discarded products containing BFRs can also release these chemicals in the environment. A study by Morf et al. (2005) indicated that in a Swiss recycling plant, the waste electrical and electronic equipment had about 1420 mg/kg of TBBPA, some of it could be potentially released into the environment.

Polybrominated diphenyl ethers (PBDEs): As PBDEs are also used as additive BFR, their release from plastics, textiles and other equipments where they have been used, is possible. They have been found in home and work environments presumably being released from furnishings, plastic coverings, etc. It has been suggested that polyurethane foam flame retardants is one of the most important sources of emission of penta-BDE in the environment in North America (Hale et al., 2002). Watanabe and Sakai (2003) reported that dusts collected from back cover of TV sets whose cabinets contained BFRs had about 0.32mg PBDEs /g. Electronic recycling plants can also be a source of these compounds (D'Silva et al., 2004). The Swiss study reported the amounts of penta-BDE , octa-BDE and deca-BDE in electrical and electronic recycling plant to be 34 mg/kg, 530 mg/kg and 510 mg/kg, respectively. Potentially some of these compounds could be released into the environment (Morf et al., 2005). Goel et al. (2006) investigated the potential for spray irrigation of municipal wastewater as a source of atmospheric PBDEs. They found the level of PBDEs in air at a site in Maryland and Delaware near spray irrigation fields to be 5-10 times higher than background levels. The average concentration BDE-47 and BDE-99 at the sites representing background levels in air were 10-17pg/m^3 and 5.3 - 7.7 pg/m^3, respectively, while at the site near the spray irrigation operation it was 175 pg/m^3 and 26 pg/m^3, respectively.

Hexabrominated cyclododecanes (HBCDs): As it is used as an additive BFR, it can be released from the materials where it has been added, e.g. polystyrene insulation materials. Thus it can contaminate the localized area where it is present, e.g. homes, offices, etc. Dust samples from homes and offices in Belgium had up to 58μ HBCD/g

(Greenpeace report, 2004). Factories that produce HBCDs are a ready source of this compound also. Concentrations up to 28.5 µg HBCD/m³ in the air in a plant producing extruded polystyrene foam with HBCD flame retardant was reported by Covaci et al. (2006). The study by Morf et al. (2005) reported 17 mg HBCD/g in an electrical and electronic recycling plant, which may be an environmental source for these compounds.

9.2.3 Properties

Important properties of TBBPA, PBDEs, and HBCDs are described below (USEPA, 2005).

Tetrabromobisphenol A (TBBPA): TBBPA is a crystalline colorless powder at room temperature with a melting point of 180°C and a boiling point of 316°C. It is not a volatile compound with a vapor pressure of less than 1mm Hg at 20 °C. It is a highly lipophilic compound with a log K_{ow} = 4.5, and it has low water solubility (0.72 mg/L) (Birnbaum and Staskal, 2004). However, it is very soluble in acetone and methanol (D'Silva et al., 2004).

Polybrominated diphenyl ethers (PBDEs): As mentioned earlier PBDEs have many congeners depending on the number and location of the bromine atom on the ring. The properties of the congeners depend on their structure and vary considerably from congener to congener. Most PBDEs have high K_{ow} values (log K_{ow} varying from 5.7 to 7.9), indicating lipophilic properties. The commercial PBDEs have boiling points between 310 to 425°C. Most PBDEs have a very low vapor pressure. The hexa-BDE (BDE-153) has a vapor pressure of 7.6 x 10^{-6} mmHg. The more brominated BDEs have even lower vapor pressures. However, tri-BDE (BDE-28) has a vapor pressure of 2 x10^{-3} mmHg (Orazio, 2005). These chemicals are thermally labile which allows them to act as flame retardants. In general they are relatively stable to chemical and biochemical degradation which accounts for their persistence in the environment (D'Silva et al., 2004). Commercial penta-BDE generally consists of a mixture of about 24-38% tetra congener, 50-60% penta congener and 4-8 % hexa congener. It is a viscous liquid at ambient temperatures. Commercial octa-BDE is a white powder. It has several congeners present: 10-12 % hexa-, 44 % hepta-, 31-35 % octa- , 10-11% nona- and <1% deca –congener. Commercial deca-BDE is a white powder and is a relatively pure compound with only a few congeners. It has about 97 % deca-, < 3% nona- and a very small amount of octa-congener (Birnbaum and Staskal, 2004).

Hexabrominated cyclododecanes (HBCDs): Like other BFRs it is also very lipophilic compound with a log K_{ow} of 5.6. Its water solubility is very low (0.0034 mg/L). It is not volatile as it has a vapor pressure of 4.7 x 10^{-7} mm Hg. In commercial HBCD, there are three stereoisomers- γ-HBCD (75-89 %), α-HBCD (10-13%) and β-HBCD) (1-12 %). The HBCD molecules can be thermally rearranged at temperatures > 160° C, with differing amounts of these stereoisomers. Such rearrangements may occur during processing of materials containing HBCDs, i.e. extrusion of polystyrene (Covaci et al., 2006).

9.3 Occurrence and Behavior in Natural Systems

In general, it has been reported that PBDEs are extremely stable in the environment. However, PBDEs accumulate in environmental biota and are widely found in the aquatic food chain. PBDEs also occur in deep-sea food chains as they have been reported in the harbor seal and the sperm whale (de Boer et al., 1998). Studies indicate that fish (e.g., juvenile carp, rainbow trout, sunfish, etc.) exposed to deca-BDE can metabolize it into the less fully brominated compounds associated with the penta- and octa- BDE. However, it is unclear whether this breakdown happens due to UV light exposure, microbial activity, or other processes. Moreover, it is not very clear how chemical alterations or biodegradation affect their persistence in the environment.

Different congeners are found at different levels in environmental media and wildlife. Generally the highest measured concentrations are for the tetra (> 50%), penta (20-30%), hexa (15-20%) and hepta and octa brominated (< 20%) congeners (USEPA, 2005). The congener pattern in the environment is often different from that in the commercial product which indicates that they are not totally stable. Reported half lives ($T_{1/2}$) of PBDE congeners in air, water and soil are > 2days, 2 months and 6 months, respectively (Birnbaum and Staskal, 2004). However, there have been reports about their photolytic breakdown and microbial decomposition.

Information on the generic chemical manufacturing process of fire retardants, the life cycle for a fire-retardant chemical can be found in many references (e.g., USEPA, 2005). Figure 9.2 shows possible fates of BFRs in the environment. They are generally less mobile in the environment because of their physical properties, such as low volatility and low water solubility. However, BFRs with lower brominated products have some degree of mobility as they can be volatilized to some extent. The environmental behavior of BFRs is similar to polychlorinated biphenyls (PCBs) (Watanabe and Sakai, 2003).

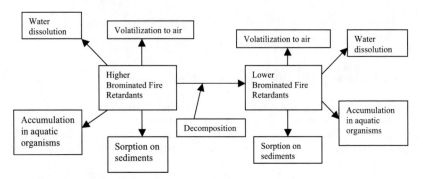

Figure 9.2 Fate of brominated fire retardants in environment.

9.3.1 Surface and Groundwater

The flame-retardant chemical may present an aquatic exposure if the water solubility of the compound is greater than 1×10^{-6} g/L or the compound is dispersible in water (USEPA, 2005).   BFRs are lipophilic compounds having low water solubilities, so it is not expected that these compounds will be found in large concentrations in surface and ground waters. Watanabe and Sakai (2003) reported practically no BFRs in water samples in Japan in 1977 and in 1987-88, with only one sample having TBBPA at < 30-50 ng/L. In 1999 the level of PBDE in surface waters in Lake Ontario was reported to be 4 to 13 pg/L. The congeners BDE -47 and BDE - 99 were the predominant BDEs present (Luckey, et.al, 2001). Lower brominated congeners like BDE-47 (tetra-BDE) and BDE-99 (penta-BDE) are more water soluble than higher brominated congeners like BDE-209 (octa-BDE) and expected to be more likely to be present in water. In coastal waters of the Netherlands, the concentration of BDE-47, BDE-99 and BDE-153 (hexa-BDE) were found to be 1, 0.5 and 0.1 pg/L, respectively (Booij et al., 2000).   A report on Lake Michigan water column study found the total PBDEs to increase from 0.031 in 1997 to 0.158 ng/L in1999. Kuch et al. (2001) found BDE-47 in German surface waters at concentrations ranging from < 0.2 to 0.71ng/L.

Buckus et al. (2005) reported the levels of some BFRs in wet-only precipitation collected in the Great Lakes Basin. The average concentrations of a congener BDE-209 (deca-BDE) and other PBDEs were found to be 44 ng/L and 0.21 to 4.2 ng/L, respectively.   Streets et al. (2006) reported that the Lake Michigan dissolved phase PBDE congener concentrations of 0.2 to 10 pg/L were similar to dissolved phase PCBs. PBDE congener concentrations were from nondetected to 13 pg/L.

No reported data on groundwater concentrations of BFRs are available at the present time.

9.3.2 Municipal and Industrial Wastewater

Wastewater: Most data on BFRs in wastewater have been for sludge samples with very limited data on wastewater itself. Kuch et al. (2001) reported BDE-47 and BDE-99 concentrations in wastewater plant effluent of 1.1 ng/L. Most of the PBDEs were associated with the suspended solids. In the Netherlands deBoer et al. (2003) also noted that BDE-47 and BDE-209 were in the particulate fraction of the wastewater plant effluent with concentration ranging from 11 to 35 and 310 to 920 ng/g dry weight (dw), respectively. Song et al. (2006) reported the fate and partitioning of PBDEs in a wastewater plant in Ontario, Canada. The total (dissolved plus in the solid fraction) average sum of 8 PBDE congeners (BDE-28, -47, -99, -100, -138, -153, -154, -183) in the influent wastewater was 265 ($\pm$ 210) ng/L, but 85 % of the PBDEs were the congeners BDE-47 and -99. The average effluent of the 8 PBDE congeners was 36 ($\pm$ 10) ng/L with BDE-47 and -99 the predominant congeners.

Morris et al. (2004) reported some limited HBCD and TBBPA concentration data in wastewaters in the Netherlands and England. The average (5 samples) influent and effluent wastewater HBCDs concentrations (on a dry weight basis) in The Netherlands were 954 ng/g and 4.9 ng/g, respectively. The average influent and effluent values (5 samples) for HBCDs in wastewaters in England were 6.3 ng/g and < 3.9 ng/g, respectively. The corresponding average TBBPA values in Netherlands wastewater influent and effluent were < 6.9 ng/g and 42 ng/g, respectively. The higher TBBPA concentration in the effluent was possibly due to a lower size of particles in the effluent than in the influent. The average TBBPA data for influent and effluent of wastewater in England were 7.5 ng/g and < 3.9 ng/g, respectively.

Sludge: Most data on BFRs from wastewater plants are for the sludge. There is some concern about spread of BFRs in the environment and biota through land application of sludge containing BFRs (de Wit, 2002). Hale et al. (2001) analyzed 11 sludge samples from Virginia, Maryland, New York and California and found high levels of BFRs in these sludge samples. The levels of penta-BDE (congeners BDE-47, -99, -100) were 1,100 ng/ g dw regardless of sludge pretreatment and location. These values were much higher than European sludge, which was expected as penta-BDE use in US is also higher. The deca-BDE (BDE-209) concentrations varied widely in these sludge (84 to 4,890 ng/g dw). Sludge from another wastewater plant in south-central Virginia had the same penta-BDE congeners as the sediments in a study reported by Hale et al. (2002). The concentration of total PBDE was 1,370 ng/g dw. The concentrations of PBDEs in sludge in 1988 from a Swedish wastewater plant were reported to be around 20-30 ng/g dw (de Wit, 2002). Sludge data from 1997-98 showed that PBDE values were higher in several Swedish wastewater plants, with BDE-209 varying from 230 to 320 ng/g dw and BDE -99 varying from 56-100 ng/g dw.

In a study comparing BFRs in sewage sludge from the Nertherlands, England and Ireland, Morris et al. (2004) reported that HBCD levels in sludge in Ireland and

England was higher than those in the Netherlands. The range of HBCDs in UK sludge was 500 to 2700 ng/g dw, while that in the Netherlands the range was 600 to 1300 ng/g dw. Swedish wastewater sludge had HBCD concentration varying from 19-54 ng/g dw (Selstrom et al., 1999). The maximum concentration of TBBPA in sludge from Cork, Ireland was found in dewatered sludge and secondary sludge at 192 ng/g dw and 9100 ng/g dw, respectively. The maximum concentration of TBBPA in the sludge from the Netherlands was 600 ng/g dw. Selstrom et al. (1999) reported lower concentrations of TBBPA in Swedish sewage sludge which were in the range of 3.6 to 45.0 ng/g dw. Song et al. (2006) looked into the partitioning of various congeners of PBDEs in both primary and waste activated sludge from Ontario, Canada. The average sum of 8 congeners (BDE - 28 through BDE -183) in primary sludge was 1626 (± 576) ng/g dw, but 2698 (± 1141) ng/g dw in waste activated sludge. The predominant congeners in both sludge were BDE-47, -99 and -100, which paralleled the composition of PBDE in the wastewater.

9.3.3 Soils and Sediments

It has been mentioned that soil may be a substantial sink for some BFRs (Mueller et al., 2006), but only a few reports are available to quantify the extent of contamination in the soils environment. There is a lot of information about BFRs in river sediments as compared to soils.

Harrad and Hunter (2006) reported the surface soil PBDE congener concentrations in West Midlands in UK. The congeners BDE-47 and BDE-99 were the main components of the PBDEs found in the soils. The sum of five congeners (BDE-28, -47, -99, -100, -153 and -154) varied from 0.241 to 3.89 ng/g dw in these soils.

In a study of surface soils (0-5 cm) from rural woodlands and grasslands in UK and Norway, Hussanin et al. (2004) reported total PBDEs concentration from 0.065 to 12 ng/g dw. Predominant congeners were: PBDE-47, -99, -100, -153 and -154, which are the major constituents of commercial penta-BDE. The reported congener mix in soil closely matched that in the commercial product. This led to the conclusion that there was little weathering or degradation of the BFR product during transportation from the source through air-soil route. A congener BDE-183 which is a marker for the octa-BDE was also found at concentrations with a median value of 50ng/g dw. Generally, it was a minor component of soil PBDEs, but it was high in soils of northern England. A report on the soil outside a recently closed polyurethane foam manufacturing facility in North Carolina found penta-BDE to be 76 ng/g dw (Hale et al., 2002). Studies in Japan reported the levels of TBBPA in soils to range between 0.5 and 140 ng/g dw (Watanabe et al., 1983).

Levels of total PBDE in river sediments in the vicinity of Osaka, Japan in the 1980s were found ranging from 33 to 410 ng/g dw (Watanabe et al., 1987). The deca-BDE was the predominant congener, with its concentration one order higher than the tetra – to hexa- PBDEs. Japan Environmental Agency reported the PBDE levels in river, estuarine and marine sediments (Watanabe and Sakai, 2003) throughout Japan in 1987 and 1988. The levels of deca-BDE varied from 10 to1370 ng/g dw in 1987 and 4 to 6,000 ng/g dw in 1988. In a more recent report, Sakai et al. (2002) found PBDE levels in sediments collected at the mouth and coastal areas of Osaka Bay, Japan, ranged from none to 910 ng/g dw. The predominant congener was deca-BDE in these samples.

The river and marine sediment PBDE data in Europe and North America was reviewed by de Wit (2002). It varied from a few ng/g dw to μg/g dw. Sellstrom et al (1999) reported the highest level of BDE-47 and BDE-99 in the sediments from River Humber in UK. The total for the two congeners was 13.1 ng/g dw. The highest level of BDE-209 at 1700 ng/g dw was found in sediments from the River Mersey in UK. Another UK study reported on the PBDE concentrations in sediments in the River Skerne and Tees estuary. The BDE-47 and BDE 99 concentrations in the sediments were up to 368 ng/g dw and 898 ng/g dw, respectively (Allchin et al., 1999).

Dodder et al. (2002) presented PBDE data of surficial sediment samples from Hadley Lake in Indiana. The lake is close to a research and development facility of a PBDE producer. The major congeners detected were BDE-209 followed by BDE-99, -153, -154, -47 and -100. The concentrations of BDE-209 ranged from 19 to 36 ng/g dw. The concentrations of other PBDEs were < 5 ng/g dw.

In another US study Hale et al. (2001) reported PBDEs in surface sediments from 137 lakes in Virginia. PBDEs were detected at >0.5 ng/g dw in 22% of the samples. The congeners in order of predominance were: BDE-47, BDE-99 and BDE-100. The highest PBDE concentration in the sediment was 52.3 ng/g dw. Sediments from a stream nearby a closed polyurethane foam manufacturing plant in North Carolina had penta-BDE concentration ranges of < 1 to 132 ng/g dw (Hale et al., 2002). A recent study measured the PBDEs in the surface sediments of Great Lakes (Li et al., 2006). They found the predominant congener to be BDE-209 at a concentration of 50 ng/g dw and the total PBDEs to be 53 ng/g dw. In Lake Erie surface sediments, the predominant PBDE congeners were BDE-209, BDE-47 and BDE- 99 (Chu et al., 2005). The sum of 8 congeners Σ_8 PBDE (BDE-28, -47, -99,- 100, -138, -153, -154, -183) were quite low (0.25 -0.5 ng/g dw) in the eastern and central Lake Erie basins, but in the western basin it was between 0.5 – 5 ng/g dw, with a highest value of 11.7ng/g dw in a location on the west-central side.

Morris et al. (2004) reported that HBCDs were found in all river and estuarine sediment samples in UK, Belgium and the Netherlands. The highest total HBCD value of 1,700 ng/g dw was measured in the sediments from the River Skerne in England. This site was close to a factory manufacturing BFRs. River Skerne is a tributary of River Tees. River Skerne sediments also had PBDEs as noted earlier. HBCDs were also evident in River Tees at concentrations ranging from 295 to 511ng/g dw. The total HBCD concentrations in the Netherlands river sediments were generally lower than that found in rivers in England. The HBCD concentrations in sediments in River Rhine in Holland varied from 2.3 to 34 ng/g dw. The sediments from Scheldt basin in Belgium had HBCD values ranging from 38 to 950 ng/g dw which could be due to the impact of textile industry utilizing HBCD in their products.

Morris et al. (2004) also reported the levels of TBBPA in the sediments of the three countries – UK, Belgium and the Netherlands. The sediments from River Skerne in UK had the highest TBBPA value of 9,750 ng/g dw. The sediments from River Tees had mean TBBPA concentrations of 25 ng/g dw. Other rivers in UK had lower sediment TBPPA values of < 2.4 ng/g dw. The level of TBBPA in Dutch and Belgium river sediments were in the range of 0.5 to 24 ng/g dw. The TBBPA in Lake Erie sediments were found in only two sites, one at the mouth of the Detroit River at a concentration of 0.51 ng/g dw and at detection level in the mid-eastern basin (Chu et al., 2005).

9.3.4 Air and Atmosphere

Air is one of the important transport media for many BFRs to locations far from the source. BFRs have been found both in indoor and outdoor air samples. The volatilization of BFRs depends on vapor pressure of the compound. The BFRs reviewed in this chapter all have a very low vapor pressure so it is not expected that their levels in the air will be high. As most BFRs are quite hydrophobic, they are associated with particulates (dust) in air.

Window organic films (an 11-100-nanometer-thich coating) were collected from indoor and outdoor window surfaces, along an urban-rural transect extending northward from Toronto, Canada by Butt et al. (2004). They found the total air concentrations of PBDEs of indoor window films (42.1 [g/m^3]) were 1.5 to 20 times greater than outdoor films (4.8 pg/m^3), indicating that urban indoor air is a source of PBDEs to urban outdoor air and the outdoor regional environment. Congener profiles were dominated by BDE-209 (51.1%), followed by BDE-99: 13.6% and -47: 9.4%). They also found that urban total PBDE concentrations were ~10 times greater than rural concentrations, indicating an urban-rural gradient and greater PBDE sources in urban areas.

In southern Taiwan in 1991, total PBDE concentrations (tri- to hexa-BDEs) in air samples ranged from 100 to 190 pg/m^3 (Watanabe et al., 1992). Air borne dust samples in Osaka, Japan in 1993-94 mainly contained deca-BDE (BDE-209) ranging from 83-3,100 pg/m^3, while the other congeners were less than 46 pg/m^3. In a more recent paper Ohta et al. (2002) reported the total PBDE concentration in air samples in Osaka, Japan to vary from 104 to 347 ng/m^3. The congener BDE-209 was 96 % of the total PBDEs.

Dodder et al. (2000) reported levels of PBDEs in air in urban, rural and remote shorelines of the Great Lakes. All samples contained congeners BDE- 47, -99, -100, - 153 and -154, but the predominant congeners were BDE -47 and - 99. The highest concentrations were found near the City of Chicago. The total PBDEs concentration ranged from 6.9 to 77 pg/m^3. They also found that higher brominated

BDEs were in the particulate phase in the air samples. About 80 % of BDE-47 was in vapor phase but only 20% of BDE-153 was in vapor phase. Sjodin et al. (1999) reported that > 99 % of the BDE-209 was in the particulate phase in the air samples at a Swedish electronic dismantling plant. In a later study Sjodin et al. (2001) reported air particle-associated and semi- volatile PBDEs, TBBPA and other BFRs in a plant where electronic products were being recycled. In the dismantling room the BDE-209 and BDE -183 were the two predominant PBDEs in the particulate and semi volatile fractions at a mean concentration of 36 and 19 ng/m^3, respectively, while other congeners were at much lower levels. Most of these compounds were associated with the particles in the air. The mean concentration of total TBBPA in air was 30 ng/m^3. In circuit board assembly room BDE -47 was the highest PBDE in the air at a mean concentration of 0.35 ng/m^3; the concentration of TBBPA in the room was only 0.2 ng/m^3. These air concentrations dropped off considerably in the offices with computers (BDE-47 < 0.1 ng/m^3; TBBPA 0.036 ng/m^3) and in other rooms.

Levels of PBDEs in the air were reported by Alaee et al. (2001) in Arctic and Great Lakes region in 1994. In the Arctic region the total PBDEs ranged from 10 to 700 pg/m^3, while in the Great Lakes area it was 31 to 2000 pg/m^3. The predominant congeners in the air were BDE-99 followed by BDE-47. Congener BDE-209 and higher brominated congeners were not found in these samples. The profile of PBDE congeners in the US air was different than in Japan where BDE-209 was the predominant species present.

The PBDE concentration in air samples in a rural site in southern Ontario, Canada in spring of 2000 ranged from 88 to1,250 pg/m^3 (Gouin et al., 2002). The highest levels of PBDEs were thought to be released from a snowpack during the spring melt. The PBDE levels in the air dropped to 1 and 230 pg/m^3 after the bud burst. The major congeners in the air were BDE-17, -28 and -47.

Air samples from 10 locations in West Midland in UK were analyzed for PBDEs by Harrad and Hunter (2006). The total PBDEs varied from 2.84 to 23.3 pg/m^3. The major congener found in the air samples was BDE-47. The PBDE congener pattern in the air was different than in soils with the ratio of congener 47:99, higher in air than in soil. The lower brominated congeners like BDE-47 volatilize from products more. The higher brominated like BDE-99 once in the air have a higher atmospheric deposition rate and retention by soils after deposition compared to congeners like BDE-47.

Pozo et al. (2006) reported global PBDE levels in the air. Consistently, only three out of 17 targeted congeners (BDE-47, 99,- 100) were found in most samples. The sum of these three congeners ranged from below the detection limit (3.7 pg/m^3) to 24 pg/m^3, which was at an agricultural site in Georgia. The reported levels of these compounds in California and Kuwait City were 19 and 17 pg/m^3. To some extent, many PBDEs were associated with fine particles in the air.

Air samples in urban and rural areas in Sweden showed the presence of HBCDs ranging from 2 to 610 pg/m^3 (Covaci et al., 2006). Air samples from plants producing HBCDs or where extruded polystyrene foam flame retardants with HBCDs were being manufactured, had values of HBCDs as high as 28,500 ng/m^3. Dust samples in offices and homes in Europe and the US had measurable quantities of HBCDs. In Belgium air dust samples from homes and offices had HBCD levels up to 58,000 ng/m^3 (Greenpeace report, 2004).

Several studies have been conducted to model the fate and transport of BFRs and atmospheric vegetation uptake of PBDEs. Air-particulate distribution revealed that penta- and higher BDE congeners were mainly associated with particulates even in warmer temperatures, whereas for the tri- and tetra-BDE congeners, a significant temperature dependence was observed. Particulate-bound deposition velocity to vegetation was calculated to be 3.8 m/h. Net gaseous transfer velocities ranged from 2.4 to 62.2 m/h (St-Mand et al., 2007).

9.3.5 Aquatic and Terrestrial Organisms

BFRs have been found in aquatic as well as in terrestrial organisms. PBDEs are of major environmental and health concern as it resembles structurally to polychlorinated biphenyls (PCBs). Other BFRs (HBCD and TBBPA) can biomagnify in many species (Morris et al., 2004). In general, a log K_{ow} of 3.5 to 5 corresponds to BCFs (bioconcentration factors) of approximately 1,000 (moderate) to 5,000 (high bioaccumulation potential) (USEPA, 2005). As the log K_{ow} increases above 8, the bioaccumulation potential decrease. Many BFRs have a log $K_{ow} < 8$, and therefore, have,a tendency to concentrate in lipid-rich tissues of the organisms, so many of the reported data are normalized for lipid weight (lw). Only a few selected references are being presented here.

Aquatic Organisms: In Holland, deBoer et al. (2000) found in fresh water mussels that were hung in nets in water, BDE-209 levels below detection levels. The concentration ranges for congeners BDE-47, -99 and -153 were 0.7-0.17, 0.4-11 and <0.1- 1.5 ng/g dw, respectively. BDE-47 levels in eels from Dutch rivers and lakes were reported to be < 20 to 1700 ng/g lw and BDE-47 was the major component (70%) of the total PBDE (de Boer et al., 2000).

High levels of PBDEs (tri-BDEs and hexa- BDEs) were found in fish samples in River Viskan in Sweden in 1979 -1980 where many textile mills are located (Andersson and Blomkvist, 1981). The PBDEs range in fish was 950-27,000 ng/g lw. The major congener found was BDE-47. Earlier no PBDEs were found at this location in 1977. Later in 1987, higher levels of BDE-47,-99 and -100 were found here in fish (de Wit, 2002). Different varieties of fish were collected from estuaries and rivers in UK to determine the amount of PBDEs in them (Allchin et al., 1999). In the Tees Bay, the fish (plaice, flounder and dab) BDE-47 and BDE-99 concentration ranges were 520-9500 and 83-370 ng/g lw, respectively. Fishes (flounder) from River

Humber had BDE-47 and BDE-99 concentrations of 1600 and 160 ng/g lw, respectively.

Asplund et al. (1999) found tri- and hexa-BDEs in steelhead trout from Lake Michigan in 1995. The total PBDE (sum of BDE-47, -99 and -100) was 2700 ng/g lw. In studies from Virginia, US, fish muscle samples from 50 sites were analyzed for PBDEs (Hale et al., 2000). The predominant congener was BDE-47 with concentrations greater than 1000 ng/g lw at 9 of the 50 sites. The highest level of PBDE measured (about 57,000 ng/g lw) was in carp downstream from a textile and furniture factory.

Ikonomou et al. (2002) presented data on the levels of PBDEs in aquatic organisms in a pristine harbor and near a paper mill location in British Columbia, Canada. The total PBDE concentrations in the liver of English sole ranged from 12 to 340 ng/g lw and the corresponding value in the hepatopancreas of Dungeness crab was 4.2 to 480 ng/g lw.

Streets et al. (2006) reported that the mean log BAFs (= bioaccumulation factors = the observed ratio of fish tissue concentration to dissolved water concentration expressed in the equivalent units) of PBDE congeners for Lake Michigan trout were 7.3 for BDE-47, 7.3 for BDE-66, 6.7 for BDE-99, and 7.5 for BDE-100, respectively. They also found that the BAFs for PBDEs were similar to or slightly lower than PCBs. However, Gustafsson et al. (1999) reported that PBDE, BAFs were 5-8 times higher than the respective values for PCBs (Katsoyiannis, 2007).

Morris et al. (2004) reported the HBCDs and TBBPA concentrations in marine organism in North Sea. Muscle of whiting fish had total HBCD and TBBPA concentrations of < 73 and 136 ng/g lw, respectively. Harbor porpoise blubber had mean concentrations of total HBCD and TBBPA of 312 and 83 ng/g lw, respectively. In the biota the α-HBCD was the predominant species present, which could be due to higher hydrophilic property of this isomer and possibly because of metabolism of the γ- and β- isomer by the organism (Law et al., 2005).

Terrestrial Organisms: PBDEs were found in birds in Baltic and North Sea at concentrations ranging from 80 to 350 ng/g lw (Jansson et al.,1987). In the coast of England the total PBDEs in cormorants were from 300-6400 ng/g lw in liver with BDE-47 being the main congener (Allchin et al., 2000). The level of BDE-47 in glaucous gulls in Norway varied from 290 to 643 ng/g lw (de Wit 2002). McKinney et al. (2006) reported the levels of PBDEs and other organohalogen compounds in bald eagles in British Columbia, Canada and Santa Catalina islands in the US. The sum of 8 congeners of PBDEs ranged from 1.78 -8.49 ng/g. The dominant congeners were BDE-47, -99 and 100. HBCDs were also reported in birds by many investigators (Morris et al., 2004, Lindberg et al. 2004, Sellstrom et al. 2003). The HBCDs in cormorant liver from Holland, in peregrine falcon eggs from Sweden and in guillemont eggs from Baltic Sea varied two orders of magnitude with a maximum value of 7100ng/ g lw.

Voorspoels et al. (2007) reported that the sum of 7 PBDE congeners (BDE 28 to 183) in rodents (wood mice and bank vole) ranged from 2.4 to 23 ng/g lw in liver , from 2.2 to 30 ng/g lw in muscle and 0.23 to 4.9 ng/g lw in brain. BDE-209 was not detected in any tissues above the detection limit. The levels of PBDEs in passerines (great tits) were higher, ranging from 50 to 500 ng/g lw in body fat and 87 to 540 ng/g lw in eggs. The biomagnification ranged from 2 to 34 (lipid-normalized basis) in the predatory bird food chain.

BFRs have been found in human milk and blood throughout the world (Lunder and Sharp, 2003; Watanabe and Sakai, 2003). In Japan PBDEs, TBBPA and other organocholrine compounds were found in the blood of adults (Nagayama et al., 2000). All blood samples tested had PBDEs with a range of 1.2 to 18 ng/g lw. The blood PBDEs level in workers in the electronics-dismantling industry was one order of magnitude higher than in the general population (Sjodin et al., 1999). The PBDE congener profile in the blood of workers occupationally exposed to these fire retardants was quite different than in general population; they had higher brominated congeners such as BDE-183 and BDE-209 in their blood.   The TBBPA concentrations in blood of these workers were about 1 ng/g lw. BDE-47 concentrations in serum from women in California increased with time (Petreas et al., 2003). In the 1960s BDE-47 was about 2 ng/g lw compared to 50 ng/g lw in the 1990s. These results are quite similar to the experiences in Canada, and it shows that much higher body burdens for PBDEs are occurring in North America than in Europe or Japan.

The breast milk from 12 nursing women one month after delivery in Japan was tested for PBDEs concentration by Ohta et al (2002). The concentration of total PBDEs were in the range of 0.67 to 2.8 ng/g lw. The breast milk samples in North America were reported to be much higher than in Europe or Japan (Lunder and Sharp, 2003). Comparing the PBDEs concentration in human milk in Sweden, Japan, Canada and the US, it was found that contamination levels in the breast tissue of California women and in the breast milk of women throughout America are up to 75 times higher than those found in European counrties (Kucher and Purvis, 2004). There was a large difference in the median levels. The Swedish and Japanese median levels for PBDEs were 3.2 and 1.4 ng/g lw, respectively while the Canadian and US median levels were 25 and 41 ng/g lw, respectively (Betts 2002, Ryan et al., 2002).

Another important factor is the congener composition in the human blood or milk samples. Human tissue samples from California had a different congener mix compared to the commercial product. In the commercial penta- BDE the ratio of BDE-99 to BDE-47 is about 2:1 but in human samples the BDE-47 is 2.5 times higher than BDE-99 (Mazdai et al., 2003, Petreas et al., 2003). In addition in commercial formulations BDE-100 is found in small quantities, but in human samples the amount of BDE-100 and BDE-99 are about equal. The congener profile in wildlife was somewhat different but it still was very much like that in human samples.

In summary, BFRs are found in the environment at different locations in all types of media, water, air, soil, biota including human beings. The level of some BFRs is increasing in the environment and in human tissues which are of concern since some of these compounds have toxic properties.

9.4 Fate and Transformation in Engineered Systems

As indicated earlier, most of the BFRs have relatively low volatility and water solubility, so they are associated with the sediments or in solid phases. They have been found in surface waters at low concentrations, but their presence in groundwater has not been reported.

9.4.1 Water Treatment

It is clear from the data presented here that rivers and lake waters have low but measurable amounts of BFRs, particularly TBBPA and PBDEs. The concentrations in lake and river waters are in the ng /L range. These chemicals are not listed in the contaminant candidate list of USEPA and as such are not being monitored (Scharfenaker 2005). In addition, as yet no systematic study has been reported about the removal of BFRs from drinking water treatment processes.

These chemicals are often associated with the sediments and particles, so with the removal of the turbidity in a water treatment plant, they will be removed from the water phase and end up in the sludge part, which could have negative environmental consequences.

9.4.2 Municipal and Industrial Wastewater Treatment

There are a few reports on the concentration of BFRs in the influent and effluent of wastewater plants. La Guardia et al. (2003) reported that a wastewater treatment plant below the Virginia border released large quantities of deca-BDE into the adjacent stream. The level of deca-BDE in the effluent discharged from the wastewater plant was about 12,000 ng/L, much higher than the most common components of penta-BDE at 12 ng/L (for BDE-47) and 8 ng/l (for BDE-99). The stream contained 50 parts per billion of deca-BDE. In addition, high levels of deca-BDE and smaller quantities of the other PBDEs were found in soil and sediments as far as 6.7 miles away from the treatment plant. Therefore, wastewater treatment plants can be another source of release of these chemicals (Kucher and Purvis, 2003). In general, however, it is expected that BFRs will be removed from wastewater by partitioning on the solids present. Thus the primary and secondary sludge will have a large fraction of the incoming BFRs in the wastewater.

Song et al. (2006) reported the data on the fate and partitioning of PBDEs in a wastewater treatment plant. This plant was a conventional secondary plant with activated sludge process as the secondary treatment process. The summarized data from their study is shown in Table 9.2. It can be seen that total PBDE (8 congeners) removal in the primary portion of the plant was about 66% and total removal through the plant was 86 %. In addition the predominant congeners present in the wastewater were BDE-47, -99 and -100.

Table 9.2 Mean PBDEs concentrations in wastewater treatment plant flows in Ontario, Canada (Song et al., 2006).

PBDE Congeners	Wastewater in ng/L			Sludge in ng/g dw	
	influent	primary effluent	final effluent	primary sludge	waste activated sludge
BDE-28	1.3	0.6	<0.7	8.0	14
BDE-47	102	36	14	586	963
BDE-99	121	41	16	757	1247
BDE-100	19	6.7	2.8	122	167
BDE-138	1.0	<1.2	<1.2	9.1	17
BDE-153	11	3.5	1.6	84	109
BDE-154	7.6	2.5	1.1	49	71
BDE-183	1.7	<1.9	<1.9	12	22
Σ_8PBDE	265	90	36	1626	2698

The data on sludge PBDEs were also included in Table 9.2. As expected the PBDE removed from the wastewater was incorporated in the sludge. It can be seen that the total PBDE (for 8 congeners) in the primary sludge was less than that in the waste activated sludge, which could be due to a higher solids content in the secondary process which caused a higher partitioning of these compounds on the solids. The congener mix in the sludge was quite similar to that in the wastewater, indicating no specific debromination occurring during the treatment process. The PBDE sludge data presented were similar to that reported by others in the US but higher than that reported in Europe (Hale et al., 2001). The authors pointed out that land application of this sludge could be a source of PBDEs to the environment. They also calculated the mass loading of PBDEs to the Little River, that receives the final effluent from the plant, was about 0.7 kg/yr.

PBDEs in methanol/water have been found to be photodecomposed by natural sunlight to lower brominated compounds (Eriksson et al., 2004). They found that the photodecomposition rate seemed to be dependent on the degree of bromination. The reaction rate was also dependent on the solvent. In methanol/water mixture the reaction rate was 1.7 times lower than in pure methanol. The photodecomposition half-life ($T_{1/2}$) of deca-BDE in methanol/water was about 0.5 h, while it was 12 d for the tetra-BDE. The products of photodecomposition were lower brominated compounds. Rayne et al. (2003) examined the photolytic degradation of BDE-15 (4,4´-dibromodiphenyl ether) in organic (methyl cyanide and methanol) and aqueous (water plus methyl cyanide mixture, 1:1) solvents. The reductive debromination was the only reaction observed under the test conditions with no evidence of cleavage of C-O bonds. Similar results were also reported by Sellstrom et al. (1998) who found that deca-BDE in toluene and on silca gel was successively debrominated by UV light to lower brominated products such as tetra-BDEs. The $T_{1/2}$ for the degradation in toluene was 15 min. Sand treated with deca-BDE and exposed to UV radiation or sunlight had a similar effect as in toluene solution or on silica gel. The deca-BDE was debrominated but at a slower rate. The $T_{1/2}$ for UV and sunlight degradation was 12 h and 37 h, respectively.

Anaerobic microbial reductive debromination of BDE-15 (4,4'-dibromodiphenyl ether) was reported by Rayne et al. (2003) in fixed –film plug flow reactor. The final products of the debromination reaction were BDE-3 (4-bromodiphenyl ether) and diphenyl ether (DE). They found no hydroxylation or methoxylation metabolites during this degradation process. The degradation of DE to other products was possible but not clearly identified. Earlier, Schaefer et al. (2001) had found that deca-BDE did not biodegrade at 21 to 25°C in 32 weeks in an anaerobic sediment/water microcosm. The first reported anaerobic degradation of deca- BDE (BDE-209) was by Gerecke et al. (2005). They used sewage sludge from a mesophilic digester as a seed and tried five primer compounds (4-bromobenzoic acid; 2,6-dibromodiphenyl; tetrabromobisphenyl A; hexabromocyclododecane; decabromobiphenyl) to hasten the biodegradation rate of the compound. Parallel experiments without the primer compounds were also conducted. The experiments were continued up to 238 days. The results showed clearly the reductive debromination of the parent compound to nona-BDEs and octa-BDEs. In the presence of primers the BDE-209 concentration decreased by 30% in 238 days with the formation of lower brominated products. In systems without the primers, the rate was 50 % lower. The nona-BDEs formed could be further debrominated under the experimental conditions. There was also some evidence of BDE-209 biodegradation to lower brominated products in a full scale anaerobic digester.

In another study reporting the anaerobic degradation of deca-BDE and octa-BDE by the bacteria *Sulfurospirillum multivorans* and *Dehalococcoides* species, it was found that *S. multivorans* could produce the lower brominated products like octa- and hepta-BDEs with deca-BDE as a substrate, but it could not debrominate the octa-BDE (He et al., 2006). However, the *Dehalococcoides* cultures were able to debrominate anaerobically the octa-BDEs to hepta- through di- BDEs products, but they were incapable of degrading the deca-BDE. Some of the products identified were more toxic PBDE species such as hexa-BDE-154, penta- BDE-99, tetra-BDE-49 and tetra-BDE-47, which may have some negative environmental consequences.

Very few reports have been published on the degradation of TBBPA and HBCDs. TBBPA can be photodegraded in aqueous media. The reported $T_{1/2}$ for the photodegradation process varied from 6.6 to 80.7 days depending on the season. However, the photodegradation half life of the compound was only 0.12 days when adsorbed on silica gel (WHO, 1995). Eriksson and Jakobsson (1998) also reported the photodegradation of TBBPA both in the presence and absence of hydroxyl radicals. The main product of degradation was 2,4,6-tribromophenol, with several other minor products. A laboratory study reported that TBBPA could be reductively debrominated by microorganisms via a two-step process (Ronen and Abeliovich, 2000). In the first step under anaerobic conditions, TBBPA is converted to bisphenol-A, which is then mineralized under aerobic conditions in the next step. The seed culture for the first step came from sediments in a stream that received chemical wastes. Fackler (1989a, 1989b and 1989c) reported the partial aerobic and anaerobic biodegradation of TBBPA in environmental media such as soil, sediments and water. The $T_{1/2}$ under the test conditions was about 2 months.

The removal of total HBCDs in a wastewater plant in the Netherlands was reported to be > 99 % by Morris et al. 2004. Most of the removed HBCDs were concentrated in the sludge solids with a range from 0.5 to 2.7 mg/kg dw.

9.4.3 Other Systems

There has been very little information on the transformation of BFRs from solids treatment and disposal systems. Only one report documented the presence of HBCD and TBBPA in landfill leachate in Holland (Morris et al., 2004). In this particulate phase of the leachate, a mean HBCD concentration was found to be 6 µg/g dw. In another Dutch landfill leachate, the maximum total HBCD was 36 µg/g dw, with γ HBCD being the predominant isomer present. The TBBPA concentration in the leachate varied from < 0.3 to 320 ng/g dw, with a mean concentration of 54 ng/g dw.

To the authors' knowledge, no information on soil and groundwater remediation system has been reported for any BFRs at the present time. Currently, researchers from the United States Geological Service are conducting studies on ecological effects of fire retardant chemicals and fire suppressant foams on growth and community characteristics of terrestrial vegetation and insects that rely on the vegetation as a food source, which would generate information on phytoremediation of fire retardants (Northern Prairie Wildlife Research Center, 2006).

9.5 Issues Related to Regulation, Policy, and Future Actions

As early as 1986, industrial users in Germany agreed to phase out PBDEs. The European Union has developed a policy banning the use of all PBDEs (penta, octa, and deca) in consumer electronic beginning in mid-2006 and banning the marketing and use of penta and octa products in all sectors beginning in mid-2004 (Directive 2003/11/EC, 2003). Legislation has been passed to restrict its use in Hawaii and California in 2006 (January 1 and June 1, respectively) (USEPA, 2005). Maine requires electronic manufacturers or importers to phase out all BFRs by 2006. Michigan has introduced bills to ban BFRs by 2006.

The only US manufacturer of penta- and octa-BDE, Great Lakes Chemical Corp., has voluntarily phased out the production of these chemical in response to the reports that these chemicals are accumulating in humans and wildlife and are quite toxic. They have been called the "new PCBs" by some. EPA has taken action to prevent the manufacture or import of these two BFRs after January 1, 2005, without further evaluation of potential risks (USEPA, 2007). However, there are many manufactured items still in use or in storage containing these BFR compounds before the production curtailment in Europe and US to cause environmental concerns for a long time in the future, as is the case with PCBs. Currently, finding alternatives to PBDEs and stimulating innovation for next-generation, safer chemical fire retardants and safer non-chemical technologies are critical priorities for the industry and all parties involved (USEPA, 2005).

The case of PBDEs illustrates the shortcomings of federal and state chemical regulatory policies. In 1976, congress passed the primary law regulating toxic chemicals, the Toxic Substances Control Act (TSCA), which grandfathered all existing chemicals on the market into use without health effects testing or analysis. Since the law's inception, USEPA has never used its authority to ban a chemical and has only offered regulations on five different chemicals, including PCBs. TSCA clearly failed to effectively regulate toxic chemicals.

Issues related to BFRs have taught us another lesson. Chemicals should never be regulated based on the "innocent until proven guilty" policy. Instead, a new approach based on a "Principle of Precautionary Action" should be followed. The principle says the following: when an activity raises threats of harm to human health or the environment, precautionary measures should be taken even if some cause and effect relationships are not fully established scientifically. In this context the proponent of an activity, not the public, should bear the burden of proof (O'Brien, 2000).

9.6 Conclusion

The most common BFRs are tetrabrominated bisphenol A (TBBPA), polybrominated diphenyl ethers (PBDEs) and hexabromocyclodo decanes (HBCDs). Brominated fire retardants are finding their way into the environment in ever increasing concentrations and in remote areas. The release of these compounds to the environment occurs during their manufacture, while being incorporated into resins or polymers to make them fire resistant, during the use of the products containing these compounds and when the products are disposed off or recycled.

All BFRs are lipophilic with low water solubilities, which generally makes their presence in the environment scarce. In the environment (air or water) they are often associated with particulate matter because of their lipophilic properties. Because of their persistence and low biodegradation profile, several PBDE congeners have accumulated in biota and are widely found in the aquatic food chain. Unlike

PCBs and DDT, the levels of some BFRs in the environment and in humans have increased during the last decades.

Some of the BFRs are toxic to man and other animals and can bioaccumulate in fatty tissues of animals. The major health effects of PBDEs include: (a) endocrine disruption (some PBDEs are closely resemble thyroid hormones, bond to thyroid hormone transfer proteins, and disrupt thyroid hormone balance); (b) neurobehavioral effects (PBDEs can cause learning and motor deficits); (c) reproductive toxicants (e.g., exposure to PBDEs can delay the onset of puberty in males and decrease the weight of male rats' reproductive organs and sperm count, etc.), and (d) probable carcinogens (with mixed evidence of certain congeners). However, most of studies are based only on in vitro or short-term experiments. Therefore, significant gaps in knowledge exist for the situation of chronic and low-level exposure of humans and wildlife.

Although considerable studies have been conducted on fate and transport of BFRs, sufficient information still is not available on occurrence, distribution, extent of uptake and transformation of BFRs in the environment and by humans and wildlife. Further research is needed. In the absence of adequate data, measures must be taken to prevent human exposures to these chemicals when there is clear evidence of potential adverse health effects

References

Alaee, M.; Arias, P.; Sjodin, A.; Bergman, A. (2003) An overview of commercially used brominated flame retardants, their applications, their use patterns in different countries/regions and possible modes of release. Environ International, 29, 683-689.

Alaee, M.; Cannon.; Muir, D.; Blanchard, P.; Bruce, K.; Fellin, P. (2001) Spatial distribution and seasonal variation of PBDEs in Arctic and Great lakes air. Organohalogen Compounds, 52, 26-29.

Alaee, M.; Wenning, R.J. (2002) Significance of brominated flame retardants in the environment: current understanding, issues and challenges. Chemosphere, 46, 579-582.

Allchin, C.R.; Morris, S.; Bennett, M.; Law, R.J.; Russell, I. (2000) Polybrominated diphenyl ethers in sediments and biota downstream of potential sources in the UK. Organohalogen Compounds, 47,190-193.

Allchin, C.R.; Law, R.J.; Morris, S. (1999) Polybrominated diphenyl ethers in sediments and biota downstream of potential sources in the UK. Environ. Poll., 105, 197-207.

Andersson, O.; Bloomkvist, G. (1981) Polybrominated aromatic pollutants found in fish in Sweden. Chemosphere, 10, 1051-1060.

Asplund, L.; Hornung, M.; Petersoan, R.E.; Turesson, K.; Bergman, A. (1999) Levels of polybrominated diphenyl ethers (PBDEs) in fish from Great lakes and Baltic Sea. Organohalogen Compounds, 40, 351-354.

Betts, K.S. (2002) Rapidly rising PBDE levels in North America. Environ. Sci. Technol., 36, 50A-52A.

Birnbaum, L.S.; Staskal, D.F. (20045) Brominated flame retardants: cause for concern? Environ. Health Perspective, 112 (1), 9-17.

Booij, K.; Zeagers, B.N.; Boon, J.P. (2000) Levels of some polybrominated diphenyl ether (PBDE) flame retardants along Dutch coast as derived from their accumulation in SPMDs and blue mussels (Mytilus edulis). Organohalogen Compounds, 47, 253-255.

Buckus, S.; Archer, M.; Harrison, B.; Williams, D.; Muir, D.C.G.; Alaee, M. (2005) Spatial and temporal distribution of brominated flame retardants (BFRs) in wet-only precipitation collected in the Great Lakes Basin. Paper presented at the US Society of Environmental Toxicology and Chemistry (SETAC) conference in Baltimore, MD.

Chu, S.; Shahmiri, S.; Haffner, G.; Ciborowski, J.; Hamaed, A.; Drouillard, K.; Letcher, R. (2005) Polybrominated diphenyl ethers including decabromodiphenyl ether (BDE-209) and tetrabromobisphenol-A (TBBPA) and degradation products in sediments from Lake Erie. Paper presented at the US SETAC conference in Baltimore, MD.

Covaci, A.; Gerecke, A.C.; Law, R.J.; Voorspoels, S.; Kohler, M.; Heeb, N.V.; Leslie, H.; Allchin, C.R.; de Boer, J. (2006) Hexabromocyclododecanes (HBCDs) in the environment and humans: a review. Environ. Sci. Technol., 40, 3679-3688.

Danish Environmental Protection Agency (1999) Brominated Flame Retardants-Substance flow analysis and assessment of alternatives. Environment Project No. 494.

Darnerud, P.O. (2003) Toxic effects of brominated flame retardants in man and wildlife. Environ International, 29, 841-853.

de Boer, J.; van der Horst, A.; Wester, P.G. (2000) PBDEs and PBBs in suspended particulate matter, sediments, sewage plant treatment plants and in effluents and biota from the Netherlands. Organphalogen Compounds, 47, 85-88.

de Boer, J.; Wester, P.G.; Klamer, H.J.; Lewis, W.E.; Boon, J.P. (1998) Do flame retardants threaten ocean life? Nature, 394, 28-29.

de Wit, C. (2002) An overview of brominated flame retardants in the environment. Chemosphere, 46, 583-624.

Directive 2003/11/EC of the European Parliament and the council of 6 February 2003 amending for the 24th time Council Directive 76/769/EEC relating to restrictions on the marketing and use of certain dangerous substances and preparations (pentabromodiphenyl ether, octabromodiphenyl ether). Off. J. Eur. Union L 2003, 42 (Feb 15), 45-46.

Dodder, N.G.; Strandberg, B.; Hites, R.A. (2002) Concentrations and spatial variations of polybrominated diphenyl ethers and several organochlorine compounds in fishes from northeastern United States. Environ. Sci. Technol., 36, 146-151.

Dodder, N.G.; Strandberg, B.; Hites, R.A. (2000) Concentration and spatial variations of polybrominated diphenyl ethers in fish and air in northeastern United States. Organohalogen Compounds, 47, 69-72.

D'Silva, K.; Fernandes, A.; Rose, M. (2004) Brominated organic micropollutants-igniting the flame retardant issue. Critical Reviews in Environ. Sci. Technol., 34, 141-207.

Eriksson, J.; Jakobsson, E. (1998) Decomposition of tetrabromobisphenol-A in the presence of UV-light and hydroxyl radicals. Organohalogen Compounds, 35, 419-422.

Eriksson, J.; Green, N.; Marsh, G.; Bergman, A. (2004) Photochemical decomposition of 15 polybrominated diphenyl ether congeners in methanol/water. Environ. Sci. Technol., 38, 3119-3125.

Fackler, P.H. (1989a) Bioconcentration and Elimination of ^{14}C-Residues by Eastern Oysters (Crassostrea Virginica) Exposed to Tetrabromobisphenol A. Report No.89-1-2918, Springborn Life Sciences Inc., Wareham, MA.

Fackler, P.H (1998b) Bioconcentration and Elimination of C-Residues by Fathead Minnows (Pimephales Promelas) Exposed to Tetrabromobisphenol A. Report No.89-3-2952, Springborn Life Sciences Inc., Wareham, MA.

Fackler, P.H. (1989c) Determination of the Biodegradability of Tetrabromobisphenol A in Soil under Aerobic Conditions. Report no.88-11-2848. Springborn Life Science Inc., Wareham MA.

Gerecke, A.C.; Hartmann, P.C.; Heeb, N.V.; Kohler, H.P.E.; Giger, W.; Schmid, P.; Zennegg, M.; Kohler, M. (2005) Anaerobic degradation of decabromodiphenyl ether. Environ. Sci. Technol., 39, 1078- 1083.

Goel, A.; McConnell, L.L.; Torrents, A.; Scudlark, J.R.; Simonich, S. (2006) Spray irrigation of treated municipal wastewater as a potential source of atmospheric PBDEs. Environ. Sci. Technol., 40, 2142-2148.

Gouin, T.; Thomas, G.O.; Cousins, I.; Barber, J.; Mckay, D.; Jones, K.C. (2002) Air-surface exchange of polybrominated diphenyl ethers and polychlorinated biphenyls. Environ. Sci. Technol., 36, 1426-1434.

Greenpeace Report (2004) Hazardous Chemicals in Belgian House Dust. http://www.greenpeace.org/raw/content/belgium/nl/press/reports/rapport-hazardous--chemicals-in.pdf .

Gustafsson, K.; Bjork, M.; Burreau, S.; Gilek, M. (1999) Bioaccumulation kinetics of brominated flame retardants (polybrominated diphenyl ethers) in blue mussels (Mytilus edulis). Environ. Toxicol. Chem., 18(6), 1218-1224.

Hale, R.C.; La Guardia, M.J.; Harvey, E.P.; Mainor, T.M.; Duff, W.H.; Gaylor, M.O.; Jacobs, E.M.; Mears, G.L. (2000) Comparison of brominated diphenyl ether fire retardant and organochlorine burdens in fish in Virginia rivers (USA). Organohalogen Compounds, 47, 65-68.

Hale, R.C.; La Guardia, M.J.; Harvey, E.P.; Gaylor, M.O.; Mainor, T.M.; Duff, W.H. (2001) Flame retardants: persistent pollutants in land applied sludges. Nature, 412, 140-141.

Hale, R.C.; La Guardia, M.J.; Harvey, E.; Mainor; T.M. (2002) The potential role of fire retardant-treated polyurethane foam as a source of brominated diphenyl ethers to the US environment. Chemosphere, 46, 729-735.

Harrad, S.; Hunter, S. (2006) Concentrations of polybrominated diphenyl ether in air and soil on a rural-urban transect across a major UK conurbation. Envion. Sci. Technol., 40, 4548-4553

Hassanin, A.; Brevik, K.; Meijer, S.N.; Steinnes, A.J.; Jones, K.C. (2004) PBDEs in European background soils: levels and factors controlling their distribution. Environ. Sci. Technol., 38, 738-745..

He, J.; Robrock, K.R.; Alverez-Cohen, L. (2006) Microbial reductive debromination of polybrominated diphenyl ethers (PBDEs). Environ. Sci. Technol., 40, 4429-4434

Ikonomou, M.; Reyne, S.; Fisher, M.; Fernandez, M.; Cretney, W.; Occurrence and congener profiles of polybrominated diphenyl ethers (PBDEs) in environmental samples from coastal British Columbia, Canada. Chemosphere, 46, 649-663.

Jansson, B.; Aspplund, L.; Olsson, M. (1987) Brominated fire retardants- ubiquitous environmental pollutants? Chemosphere, 16, 2343-2349.

Katsoyiannis, A. (2007) Comment on "partitioning and bioaccumulation of PBDEs and PCBs in Lake Michigan." Environ. Sci. Technol, 41, 3391.

Kuch, B.; Hagenmaier, H.; Korner, W. (2001) Determination of brominated flame retardants in sewage sludges and sediments in south-west Germany. Paper presented at 11th annual European SETAC meeting in Madrid, Spain,

Kucher, Y.; Purvis, M. (2004) Boby of Evidence: New Science in the Debate over Toxic Flame Retardants and our Health. PIRGIM Education Fund, 2004.

La Guardia, M.; Hale, R.C.; Harvey, E. (2003) Are wastewater treatment plants sources for polybrominated biphenyl ethers? Paper presented at the Annual Meeting of the Society of Environmental Toxicology and Chemistry (SETAC), 2003.

Law, R.J.; Kohler, M.; Heeb, N.V.; Gerecke, A.C.; Schmidt, P.; Voorspoels, S.; Covaci, A.; Becher, G.; Janak, K.; Thomsen, C. (2005) Hexabromocyclododecane challenges scientists and regulators. Environ. Sci. Technol., 39, 281A -287A.

Li, A.; Rockne, K.J.; Sturchio, N.; Song, W.; Ford, J.C.; Buckley, D.R.; Mills, W.J. (2006) Polybrominated diphenyl ethers in the sediments of the Greta Lakes 4-Influencing factors, trends and implications. Environ. Sci. Technol., 40, 7528-7534.

Lindberg, P.; Sellstrom, U.; Haggberg, L.; de Wit, C. (2004) Higher brominated diphenyl ethers and hexabromocyclododecane found in eggs of peregrine flacons (Falco peregrinus) breeding in Sweden. Environ. Sci. Technol., 38, 93-96.

Lione, A. (1988) Plolychlorinated biphenyls and reproduction. Reproductive Toxicology, 2, 83-89.

Lucky, F.; Fowler, B.; Litten, S. (2001) Establishing baseline levels of polybrominated diphenyl ethers in Lake Ontario surface waters. Presented at the 2nd workshop on Brominated Flame Retardants. Stockholm Sweden, 2001. The Swedish Chemical Society, 337-340.

Lunder, S.; Sharp, R. (2003) Mothers' Milk: Record levels of toxic fire retardants found in American mothers' breast milk. Environmental Working Group, Sept. 2003.

Mazdai, A.; Dodder, N.G.; Abernathy, M.; Hites, R.; Bigsby, R. (2003) Polybrominated diphenyl ethers in maternal and fetal blood samples. Environ. Health Perspect., 111, 1249-1252.

McKinney, M.A.; Cesh, L.S.; Elliott, J.E.; Williams, T.D.; Garcelon; D.K.; Lectcher, R.J. (2006) Brominated flame retardants and halogenated phenolic compounds in

North American west coast bald eaglet (Haliaeetus leucocephalus) plasma. Environ. Sci, Technol., 40, 6275 -6281.

Morf, L.S.; Tremp, J.; Gloor, R.; Huber, Y.; Stengele, M.; Zennegg, M. (2005) Brominated flame retardants in waste electrical and electronic equipment: substance flows in a recycling plant. Environ. Sci. Technol., 39, 8691-8699.

Morris, S.; Allchin, C.R.; Zegers, B.N.; Hafta, J.J.H.; Boon, J.P.; Belpaire, C.; Leonards, P.E.G.; van Leewen, S.P.J.; de Boer, J. (2004) Distribution and fate of HBCD and TBBPA brominated flame retardants in North Sea estuaries and aquatic food webs. Environ. Sci. Technol., 38, 5497-5504.

Mueller, K.E.; Mueller-Spitz, S.R.; Henry, H.F.; Vonderheide, A.P.; Soman, R.S.; Kinkle, B.K.; Shann, J.R. (2006) Fate of pentabrominated diphenyl ethers in soil: abiotic sorption, plant uptake and the impact of interspecific plant interactions. Environ. Sci.Technol., 40, 6662-6667

Northern Prairie Wildlife Research Center. (2006) Ecological effects of fire retardants chemicals and fire suppressant foams. http://npwrc.usgs.gov/resources/habitat/fireweb/index.htm.

O'Brien, M. (2000) Making Better Environmental Decisions: An alternative to Risk Assessment. The MIT Press, Cambridge, Massachusetts, 2000.

Ohta, S.; Ishizaki, D.; Nishimura, H.; Nakano, T.; Aozasa, O.; Shimidzu, Y. (2002) Comparison of polybrominated diphenyl ethers in fish, vegetables, meat and levels in human milk of nursing women in Japan. Chemosphere, 46, 689-696.

Orazio, C. (2005) Brominated flame retardants" USGS Case study III. www.rnrf.org/2005cong/orazio.pdf

Petreas, M.; She, J.; Brown, R.; Winkler, J.; Winham, G.; Rogers, E. (2003) High body burdens of 2,2´,4,4´- tetrabromodiphenyl ether (BDE-47) in California women. Environ. Health Perspect., 111, 1175-1180.

Pozo, K. (2006) Towards a global network for persistent organic pollutants in air: results from a GAPS study. Environ. Sci. Technol., 40, 4867-4873.

Rayne, S.; Ikonomou, M.G.; Whale, M.D. (2003) Anaerobic microbial and photochemical degradation of 4, 4´- dibromodiphenyl ether. Wat. Res., 37, 551-560.

Ryan, J.; Patry, B.; Mills, O.; Beaudoin, N. (2002) Recent trends in levels of brominated diphenyl ethers (BDEs) in human milks from Canada. Organohalogen Compounds, 58, 173-176.

Sakai, S.; Hayakawa, K.; Okamoto, K.; Takatuki, H. (2002) Time trends and horizontal distribution of polybrominated diphenyl ethers (PBDEs) in sediment cores from Osaka Bay, Japan. Organohalogens Compounds, 58, 189-192.

Schaefer, E.C.; Flaggs, R. (2001) Potential for Biotransformation of Radiolabeled Decabromodiphenyl Oxide (DBDPO) in Anaerobic Sediment. Report by Wildlfe International, Ltd. Easton, MD.

Scharfenaker, M. (2005) USEPA finalizes second contaminant candidate list. J. Am. Water Works Assoc., 97(3), 12- 21.

Sellestrom, U. (1999) Determination of Some Polybrominated Flame Retardants in Biota, Sediments and Sewage Sludge. Ph.D. dissertation, Stockholm University, Stockholm, Sweden.

Sellstrom, U.; Jansson, B. (1995) Analysis of Tetrabromobisphenol A in a product and environmental samples. Chemosphere, 31, 3085-3092.

Sellstrom, U.; Kierkegaard, A.; de Wit, C.; Jansson, B. (1998) Polybrominated diphenyl ethers and hexabromocyclododecane in sediments and fish in Swedish river. Environ. Toxicol. Chem., 17, 1065-1072.

Sellstrom, U.; Bignert, A.; Kierkegaard, A.; Haggberg, L.; de Wit, C.A.; Olsson, M.; Jansson, B. (2003) Temporal trend studies on tetra- and penta-brominated diphenyl ethers and hexabromocyclododecane in guillemont egg from the Baltic Sea. Environ. Sci. Technol., 37, 5496-5501.

Sjodin, A.; Thuresson, K.; Hagmar, L.; Klasson-Wehler, E.; Bergman, A. (1999) Occupational exposure to polybrominated diphenyl ethers at dismantling of electronics- ambient air and human serum analysis. Organohalogen Compounds, 43, 447 -451.

Sjodin, A.; Carlsson, H.; Thuresson, K.; Sjolin, S.; Bergman, A.; Ostman, C. (2001) Flame Retardants in indoor air at an electronic recycling plant and at other Work Environments. Environ. Sci. Technol., 35, 448-454.

Song. M.; Chu, S.; Letcher, R.J.; Seth, R. (2006) Fate, partitioning, and mass loading of polybrominated diphenyl ethers (PBDEs) during treatment processing of municipal sewage. Environ. Sci. Technol., 40, 6241-6246.

Streets, S.S.; Henderson, S.A.; Stoner, A.D.; Carlson, D.L.; Simcik, M.F.; Swackhamer, D.L. (2006) Partitioning and bioaccumulation of PBDEs and PCBs in Lake Michigan. Environ. Sci. Technol., 40, 7263-7269.

USEPA. (2005) Furniture Flame Retardancy Partnership: Environmental Profiles of Chemical Flame-Retardant Alternatives for Low-Density Polyurethane Foam. USEPA, 742-R-05-002A, September 2005.

USEPA. (2007) Pollution Prevention and Toxics- Polybrominated diphenylethers (PBDEs). http://www.epa.gov/oppt/pbde.

Voorspoels, S.; Covaci, A.; Jaspers, V.L.B.; Neels, H.; Schepens, P. (2007) Biomagnification of PBDEs in three small terrestrial food chains. Environ. Sci. Technol., 41, pp. 411-416.

Vos, J.G.; Becher, G.; van den Berg, M.; de Boer, J.; Leonards, P.E.G. (2003) Brominated flame retardants and endocrine disruption. Pure Appl. Chem., 75, 2039-2046.

Watanabe, I.; Kashimoto, T.; Tatsukawa, R. (1983) The flame retardant tetrabromobisphenol A and its metabolite found in river and marine sediments in Japan. Chemosphere, 12, 1533-1539.

Watanabe, I.; Kashimoto, T.; Tatsukawa, R. (1987) Polybrominated biphenyl ethers in marine fish, shellfish and river and marine sediments in Japan. Chemosphere, 16, 2389-2396.

Watanabe, I.; Kawano, M.; Wang, Y.; Chen, Y.; Tatsukawa, R. (1992) Polybrominated dibenzo-p dioxin (PBDDs) and –dibenzofurans (PBDFs) in atmospheric air in Taiwan and Japan. Organohalogen Compounds, 8, 309-312.

Watanabe, I.; Sakai, S. (2003) Environmental release and behavior of brominated flame retardants. Environ. International, 29, 665-682.

WHO. (1995) Tetrabromobisphenol A and derivatives.   Environmental Health Criteria 172, World Health Organization, Geneva.

Zweidinger, R.A.; Cooper, S.D.; Erickson, M.D.; Pellizzair, E.D. (1979) Sampling and analysis for semivolatile brominated organics in ambient air. In: Monitoring Toxic Substances. Schuetz, D. (Ed.), ACS Symposium Series, 94, 217-231. American Chemical Society, Washington, DC.

CHAPTER 10

Pesticides

Tian C. Zhang, Keith C. K. Lai, and Rao Y. Surampalli

10.1   Introduction

The word "pesticide" is a generic term, which covers any substance or mixture of substances used for preventing, destroying, repelling or mitigating pests, or intended for use as plant regulators, defoliants or dessicants (USEPA, 2007a). Pests are considered to be any living organism unwanted or causing damage to crops, humans or other animals. An example of pest includes insects, mice, unwanted plants (weeds), fungi, bacteria and viruses. Pesticides are not necessarily natural or synthetic chemicals, but may include microorganisms or their components such as endotoxins from *Bacillus thuringiensis*, or macroorganims such as predatory wasps specifically bred to control caterpilliars and aphids (Tyagi et al., 2002; Hamilton and Crossly, 2004). Currently, the most common types of the pesticides being used are herbicides, insecticides, fungicides and bactericides. Other pesticides being applied include nematicides for controlling parasitic microscopic worms living in soil, avicide for birds, molluscicide for snails and slugs, piscicide for fishes, algicides, rodenticides, and miticides (AGCare, 2007).

Human society has a long history of applying pesticides for pest control. In ancient China, chalk and wood ash were used to control insects in enclosed spaces; plant extracts were used for treatment stored grain, and arsenic sulfide was used to control human lice (Perry et al., 1998). Ancient Romans killed insects by burning sulfur and controlled weeds with salt. In the 1600s, a mixture of honey and arsenic was used to control ants (Delaplane, 2000). Early in the 20th century, arsenical pesticides including lead arsenate were widely used in the United States for controlling insect pests in fruit orchards, vegetable fields, golf courses and turf farms (Shepard, 1951). After World War II, organochlorine pesticides such as dichloro diphenyl trichloroethane (DDT), aldrin and dieldrin emerged, and the use of the arsenical pesticides began to be phased out. This is because the organochlorine pesticides were more effective at the lower application rates than the arsenical pesticides and thereby are less expensive to use. Another advantage of the organochlorine pesticides is stemmed from their broad-spectrum activity against insect pests.

In this chapter, the sources, the main categories of pesticides and their properties will be introduced, followed by their occurrences, fate and transport in natural and engineered systems. Issues relating to pesticides' characteristics as potential EDCs, regulations, U.S. toxic policy, and current status of research in this area will be discussed; conclusions will be given. However, details on toxicity of pesticides are not included in this chapter. There are several reviews and databases on the subject that can provide needed information, such as Merck Index (10th ed., 1983), the EPA integrated Risk Information Service (IRIS) database (USEPA, 2008a), the EPA Pesticide Product Information System Databases (USEPA, 2008b), the EPA Region III Risk-Based Concentration (RBC) Table (USEPA, 2008c), and the EXTOXNET database (Miller, 2008).

10.1.1 Sources

Pesticides are commonly used in agriculture, forestry, transportation (for roadside weed control), in urban and suburban areas, in lakes and streams (for control of aquatic flora and fauna) and in various industries. The amount of pesticide used in the world is tremendous. In 2001, over 5 billion pounds of pesticides were used, which corresponded to a world pesticide expenditure of about 32 billion US dollars. In the U.S., a steady increase in pesticide consumption was observed between mid-1960 and the end of 1970s because of increasing uses of herbicides. Currently, the total pesticide amount used has remained relatively constant at about 1.2 billion pounds per year. In 2001, the amount of pesticide used in the U.S. and related expenditure accounted for about 24% and 35% of the world market, respectively. Generally, agricultural sectors account for the majority of the total pesticide consumption, while non-agricultural sectors such as industry/commercial/government and home/garden sectors account for less than 25% of the total uses. Nowadays, most of the pesticides used are herbicides, which account for about 65% of the total pesticides used in the U.S. agricultural sectors. Insecticides, miticides, fungicides, nematicide and fumigant only occupy approximately 6% to 15% (Kiely et al., 2004). In comparison to herbicides, insecticides generally are applied more selectively and at lower rates. In response to environment concerns and development of more effective alternative pesticides, there have been some major changes in insecticide uses over the years. For instance, DDT was widely applied to control insect pests on crop and forest lands, around homes and gardens, and for industrial and commercial purposes between 1940s and 1960s; use of DDT was banned in the U.S. in 1972 (USEPA, 2007b). However, when the use of more persistent pesticides dropped, the use of other less persistent pesticides increased.

Generally, environmental pesticide pollution can be classified into two categories, which are point-source and nonpoint-source pollution. By definition, point-source pollution comes from specific and identifiable places or sources (Müller et al., 2002). These sources include the wash water generated and spills released during cleanup of pesticide-sprayed equipment, improper pesticide container rinsing and disposals, leaks and spills at pesticide storage sites, back-siphoning and spills generated during mixing and loading of pesticides, and the pesticide residues washed out from impervious areas (e.g., farmyards, streets, roofs, etc.) during storm events (Pesticide Action Network UK, 2000). Point-source pollution generally involves small areas with high pesticide concentrations, which can contaminate large areas of water over time and most likely causes acute incidents. The amount and type of pesticides released from the point-source are regulated by laws.

Nonpoint-source pollution is more widespread and more difficult to identify and quantify than the point-source pollution. It comes from many diffuse sources including pesticides from agricultural and residential lands. It starts with precipitation falling on the ground. When the resulting runoff moves over and through the soil, it picks up and carries away pesticides, and finally transports them to streams, rivers, wetlands, lakes and groundwater. Atmosphere, which people commonly overlook, is also one of the nonpoint-source pollution. In the U.S., nearly every pesticide has been detected in air, rain, snow or fog across the nation. Such airborne pesticides can be transported far from their origin, and atmospheric deposition of the airborne pesticides is not evenly distributed on the ground. Nowadays, the pesticide pollution issues mainly come from legitimate application of pesticides by farmers rather than the illegal uses and accidental spills.

Pesticide pollution of indoor environments is also a serious problem. It is reported that approximately 80% of people's exposure of pesticides occurs indoor. The source of pesticides in indoor environments mainly comes from insecticides, termiticides, rodenticides, fungicides, and disinfectants used to kill household pests; they are sold as sprays, liquids, sticks, powders, crystals, balls, and foggers. Other possible sources include pesticides used on lawns and gardens that drift or are tracked inside houses, and stored in pesticide containers (USEPA, 2007c).

10.1.2 Regulations and Future Tendency

The USEPA regulates all pesticides, including both active ingredients and pesticide formulation, under the Federal Insecticide, Fungicide, and Rodenticide Act (FIFRA) to ensure that pesticide use does not cause unreasonable adverse effects on humans and the environment. Under FIFRA, the registrant of a pesticide must submit specific data to the EPA to support the conclusion that the product meets this standard before being marketed and sold. As part of the registration process, the EPA differentiates between general-use and restricted-use pesticides (GUPs and RUPs), which is based on toxicity classes shown in Table 10.1. The EPA requires manufacturers to put information on the label about when and how to use the pesticide.

Table 10.1 EPA pesticide toxicity classes.

Toxicity Class	Toxicity Rating	Signal Word on Label
I	Highly toxic	DANGER-POISON
II	Moderately toxic	WARNING
III	Slightly toxic	CAUTION
IV	Practically non-toxic	CAUTION

While pesticides have played an important role in improving public health through disease vector reduction and increased food production, pesticide use and disposal have resulted in undesirable releases of these toxic chemicals to the environment (Carson, 1966; Shrivastava, 1987). Pesticides can harm public health, livestock and the environment, cause acute (e.g., stinging eyes, rashes, blisters, blindness, nausea, dizziness, diarrhea, and death) and chronic (e.g., cancers, birth defects, reproductive harm, neurological and developmental toxicity, immunotoxicity and disruption of the endocrines system) adverse health effects. In general, children, farm workers and pesticide applicators are more vulnerable to pesticide exposure. Therefore, pesticides (e.g., Fonofos, Dimethoate, Terbufos, Sulfone, Terbacil, and many others) have been considered as contaminants of emerging concerns.

The full environmental, public health and social costs of pesticide use might be about \$24 billion per year (Pimentel, 2005). Currently, there is a need to devise ways to reduce pesticide use in crop production while still maintaining crop yields. There are proven alternatives to conventional pesticide use (which sees pests as intruders that must be removed with pesticides). These approaches consider pest problems within a broad context that considers many factors, including the presence of antural enemies, the trend of the pest population, the time and weather patterns that can influence whether the pest will flourish or simply die away. Currently, there is an increased tendency to combine these alternatives with organic methods and other non-pesticide techniques to control pests and grow crops (CCG, 2001; PCC, 2002; Miller, 2004; CPR, 2008; USEPA, 2008d).

10.2 Physical and Chemical Properties

The following characteristics are used to described the physical and chemical properties of pesticides: molecular weight, color, form and odor; water solubility; partition coefficient (K_{ow}); soil sorption coefficient (K_{oc}); vapor pressure; EPA toxicity classification; ACGIH threshold limit values-time-weighted average (TLV-TWA); NIOSH recommended exposure limits; OSHA permissible exposure limits; EPA oral reference doses and inhalation reference concentrations; and carcinogenicity. In this section, properties and related information on several major insecticides and herbicides are described. Information on characteristics of a pesticide can be found in the aforementioned databases.

10.2.1 Insecticides

Insects are known to have existed on the earth for more than 250 million years, while humanoids have only existed for about 3 million years (Ware and Whitacre, 2004). Human beings have already learned to live and compete with the insect world. Insecticides are chemical or biological agents used to kill harmful or destructive insects, or prevent the insects from engaging in behaviors which are unfavorable to human beings. Among various insecticides, organochlorines, organophosphorous insecticides, and carbamate esters are the main groups of the insecticides commonly applied (Ballantyne et al., 1999). Hence their general structures, specific mode of action, toxicology, and the major factors affecting their chemical behaviors will be briefly discussed as follows.

The mode of action of most insecticides involves their influence on nervous systems, cuticle production systems or water balance of insects, or their inhibition to energy production systems or endocrine systems of insects. The impact on the nervous system is the basic mode of action of most insecticides, which can alter the signal transfer along a nerve fiber and across the synapse from one nerve fiber to another or from a nerve fiber to a muscle fiber. The insecticides with the mode of action of influencing the cuticle production systems behave as chitin synthesis inhibitors (CSIs) to prevent insects from producing chitin, which is a major component of insect exoskeleton, for new cuticle syntheses, thereby hindering insects from successfully molting to the next stage. Some insecticides can affect the insect water balance by removing the protective waxy covering from the insect body, which helps prevent water loss from the insect cuticular surfaces. Therefore, the insects poisoned by these insecticides consequently suffer a rapid water loss from cuticles and eventually die from desiccation. The insecticides with the mode of action of the energy production system inhibition can bind to protein called cytochrome in electron transport systems of mitochondria, thereby blocking the production of adenosine triphosphate (ATP) inside an insect body. Some insecticides can act as insect grow regulators (IGRs) by affecting the endocrine or hormone systems of insects. The insecticides with this mode of action can mimic the action of juvenile hormones produced in an insect brain to keep the poisoned insects in an immature stage and thereby are not able to molt to an adult stage successfully and cannot reproduce normally (Valles and Koehler, 2003).

10.2.1.1 Organochlorine Insecticides

Organochlorines are insecticides composed of carbon, chlorine, hydrogen and sometimes oxygen atoms and with a number of carbon-chlorine bonds. Insecticides belonging to this group are characterized by the presence of cyclic carbon chains (including benzene ring), lack of any particular active intramolecular sites, apolarity, lipophilicity, and chemical unreactivity (Matsumura, 1985). Organochlorine insecticides are also known as chlorinated hydrocarbons, chlorinated organics, chlorinated insecticides and chlorinated synthetics. There are four main types of the organochlorine insecticides, which are diphenyl aliphatics (dichlorodiphenyl ethane), hexachlorocyclohexane, cyclodienes, and polychloroterpenes (Ware, 1983).

Diphenyl Aliphatics

Diphenyl aliphatics, which consist of an aliphatic carbon chain with two phenyl rings attached, are the oldest group of the organochlorine insecticides. Major examples of the diphenyl aliphatics include DDT and five DDT analogues which are dichlorodiphenyl dichloroethane (DDD), dicofol, ethylan, chlorobenzilate and methoxychlor (Reigart and Roberts, 1999). Their chemical structures are illustrated in Figure 10.1. DDT, probably the most well-known insecticide, has been the most widely used pesticidal agent for many years. Its success after syntheses is stemmed from its high insecticidal activity, low acute mammalian toxicity, simple manufacturing and handling, low price, and long duration activity. It is very effective on controlling disease-carrying pests, and it is estimated that almost 1 billion people in the world have been saved from malaria by the use of DDT (Büchel, 1983). The mode of action of DDT has never been clearly established, but in some complex manner it prevents normal transmission of nerve impulse in the insect nervous system by destroying the delicate balance of sodium and potassium ions within the axons of the neuron. The affected neurons fire impulses spontaneously, causing the insect muscles to twitch followed by convulsions and death eventually. Unlike other chemicals generally show a high activity at a high temperature, DDT has a negative temperature correlation. It becomes more toxic to insects when the surrounding temperature decreases (Ware and Whitacre, 2004).

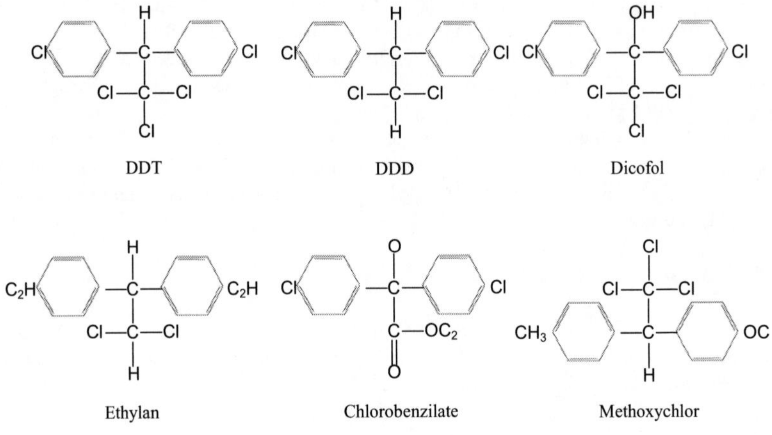

Figure 10.1 Chemical structure of DDT and its analogues.

Despite the fact that DDT is effective in controlling disease-carrying pests and has a low acute mammalian toxicity, it was categorized as an environmental hazard, and the use of DDT was completely banned on January 1, 1973. DDT is persistent in soil and aquatic environments, and in animal and plant tissues. It is not susceptible to being broken down by microorganisms, enzymes, heat or ultraviolet light. Since DDT is quite soluble in fatty tissues and resistant to metabolism, it is readily stored in the fatty tissues of any animals ingesting DDT alone or DDT-contaminated food. It was reported that about 50 to 95% of DDT ingested is absorbed in animal bodies. It can accumulate in animals which prey on other animals or eat plant tissues bearing even trace amounts of DDT (Ware, 1983). Animals poisoned by DDT first become nervous and hyperexcitable with excessive blinking, cold skin, ruffled fur, lack of appetite, and muscular weakness. It is then followed by the onset of fine tremors due to muscular fibrillation. The advanced stage of poisoning culminates in paralysis, clonic convulsion and death. Accumulation of DDT in birds, such as osprey, falcon, golden eagle, seagull and pelican, may result in the reduction of eggshell thickness. This is because DDT can inhibit carbonic anhydrase activity, which is generally acknowledged to play an important role in forming eggshells (Matsumura 1985). A human poisoned by a low level of DDT becomes weak and giddy, but can recover within 48 hrs. A human poisoned by a high DDT level suffers liver damage which can be lethal (Matsumura, 1985).

Hexachlorocyclohexane

Hexachlorocyclohexane (HCH), which is synthesized from chlorinated benzene with six chlorine atoms and was erroneously called benzenehexachloride, was first discovered in 1825 (O'Brien, 1967). However, its insecticidal properties were not recognized at that moment. In 1942, the French and British entomologists found that the insecticidal properties of HCH are stemmed from its γ-isomer, which is named lindane.

As illustrated in Figure 10.2, three chlorine substituents of lindane take axial and the other three take equatorial conformations on the chair form of cyclohexane (Perry et al., 1998). Theoretically, there are many isomers of HCH in which seven of them (α, β, γ, δ, ε, η and θ) are known. Generally, a normal mixture of HCH contains only about 12% of lindane and the other isomers are left as inert materials or insecticidally inactive ingredients (Ware 1983). Similar to DDT, lindane is also a neurotoxicant, but it is about 5 to 10 times more effective against insects and is also a more acute nerve poison in comparison to DDT (Perry et al., 1998). Insects poisoned by lindane show the symptoms of tremors, ataxia (loss of coordinated movements), convulsions, falling on their back, prostration, paralysis and death. Because of the nervous and muscular activity induced by the lindane poisoning, the respiratory rate of the poisoned insects is largely increased (Ware and Whitacre, 2004). Lindane was banned in 1976.

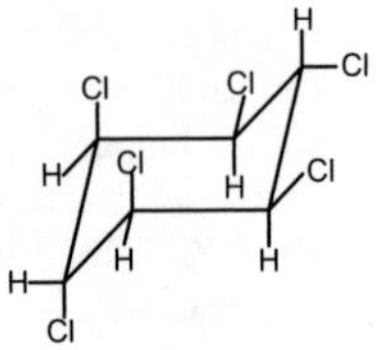

Figure 10.2 Spatial configuration of γ-isomer of HCH (Lindane).

The isomers of HCH are stable to light, high temperature, hot water and acid, but they are dechlorinated by alkali. It is approximately 100 times more volatile than DDT. Crude HCH has a characteristic musty odor imparting an off-flavor to some edible crops; while the pure Lindane is odorless (Matsumura, 1985). In analogy to DDT, Lindane also shows a negative correlation between temperature and toxicity. Mammals poisoned by Lindane suffer a rise in blood pressure and a fall in their heart beat because of the stimulation of the central nervous system. There is an increase in respiration followed by restlessness, coarse tremors of the whole body, salivation, grinding of the teeth and convulsions. Eventually, the poisoned mammals suffer depression of respiration and death from cardiac arrest. Human beings poisoned by Lindane show the symptom of dizziness, headache, weakness, diarrhea and epileptiform attacks (Perry et al., 1998). Ingestion of HCH-treated wheat was reported resulting in human dermal toxicity diagnosed as porphyria cutanea tarda. The skin blisters become very sensitive to sunlight and heal poorly, which results in scarring and contracture formation (Reigart and Roberts, 1999).

Cyclodienes

Cyclodienes are the collective group of synthetic cyclic hydrocarbons produced by the condensation reaction called Diels-Alder reaction. This reaction involves the combination of a compound containing the diene group (i.e., -CH=CH-CH=CH-) with another double-bonded compound, as shown in Eq. (10.1) (O'Brien, 1967). Despite their generic name, only a few cyclodienes possess two double bonds.

$$\text{(Eq. 10.1)}$$

The main examples of cyclodienes are chlordane, heptachlor, aldrin, dieldrin and endosulfan. As illustrated in Figure 10.3, they all have a fully chlorinated ring system showing a chlorinated "endomethylene bridge" bridging the ends of the ring. Some of the cyclodienes possess another ring, but it is usually not chlorinated. Cyclodienes are a group of highly active insecticides, and they are stable in soil and the ultraviolet light of sunlight. Hence cyclodienes were widely applied as soil insecticides for controlling termites and soil-borne insects, and as termiticides for protecting wood and wooden structures (Ware and Whitacre, 2004). However, their use was cancelled by the USEPA in 1975 because of their persistence in environment leading to bioconcentration in wildlife food chain, and the resistance developed in several soil insects (Perry et al., 1998).

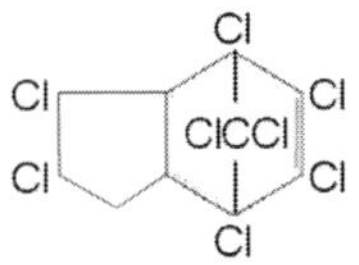

Chlordane 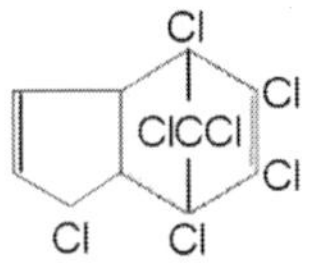   Heptachlor 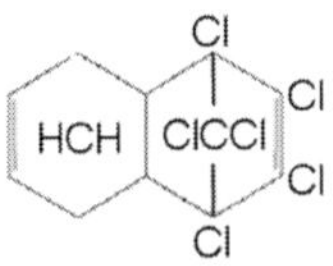   Aldrin

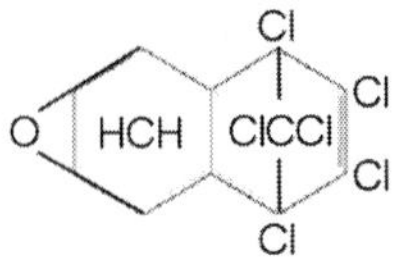

Dieldrin 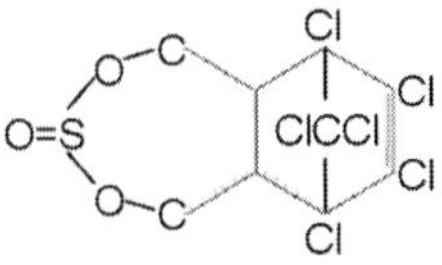   Endosulfan

Figure 10.3 Chemical structure of chlordane, heptachlor, aldrin, dieldrin, and endosulfan.

In analogy to most organochlorine insecticides, cyclodiene compounds also act as neurotoxicants. Although their mode of action in killing insects is not fully understood, it is known that the primary action involves an inhibition of GABA (γ-aminobutyric acid) receptors, which can increase chloride permeability of neurons. Cyclodienes can prevent chloride ions from entering the neurons, and thereby antagonize the "calming" effects of the GABA. Unlike DDT and HCH, cyclodienes show a positive temperature correlation in which their toxicity to insects increases with an increase in the ambient temperature (Ware and Whitacre, 2004). Insects poisoned by cyclodienes resemble symptoms as those poisoned by DDT including strong tremors, short trains of impulses and an increase in respiration rates. For instance, housefiles and blowflies poisoned by cyclodienes have a period of normal activity followed by a more quiescent period after which wing tremors become frequent and increasing in severity, and the poisoned insects show spasms of wing beat without flying. After a while, this culminates in uncoordinated flight, loss of ataxic gait and falling on the back. Mammals poisoned by cyclodienes first show the prominent symptom of convulsion, which may be accompanied by confusion, excitability, incoordination and a coma. Unlike DDT, cyclodiene compounds can be absorbed through the skin. Moreover, they can produce changes in liver cells. Some are carcinogenic to human beings (Perry et al., 1998).

Polychloroterpenes

There are only two types of polychloroterpenes, which are toxaphene (introduced in 1947) and strobane (discovered in 1951). Strobane was relatively insignificant, while toxaphene was used on cotton first in a combination with DDT, since toxaphene itself has a low toxicity to insects. In 1965, toxaphene was formulated with an organophosphorus insecticide called methyl parathion because several types of cotton insects became resistant to DDT.

Toxaphene is a mixture of 177 polychlorinated derivatives, which are 10-carbon compounds with Cl_6, Cl_7, Cl_8, Cl_9 or Cl_{10} constituents. Figure 10.4 shows the general structure of toxaphene. Similar to cyclodienes, toxaphene is also persistent in soils, but not as persistent as cyclodienes. After introducing onto the surface of plant tissues, toxaphene generally disappears in 3 to 4 weeks due to volatilization rather than metabolism or photolysis by ultraviolet light in sunlight. Toxaphene is readily metabolized by mammals and birds, and is not stored in body fat in an extent similar to DDT, HCH and cyclodienes. In spite of the low toxicity to insects, birds and mammals, fish are highly susceptible to toxaphene poisoning. The mode of action of toxaphene and strobane involves an impact on neurons causing an imbalance in sodium and potassium ions (Ware, 1983; Ware and Whitacre, 2004).

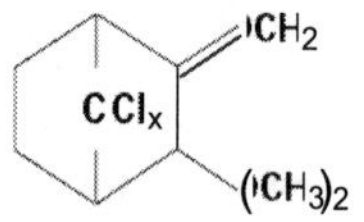

Toxaphene

Figure 10.4 Chemical structure of toxaphene.

10.2.1.2. Organophosphorus Insecticides

Organophosphorus compounds are among the largest groups of insecticides currently being used. The discovery of organophosphorus insecticides largely substituted and finally weeded out the persistent organochlorine insecticides. This is because organophosphorus insecticides are generally more toxic than the organochlorine insecticides, but they are chemically nonpersistent and quite biodegradable by most organisms in natural environment. All organophosphorus insecticides are derived from phosphoric acid. They are neutral ester or amide derivatives of phosphorous acids carrying a phosphoryl (P-O) or thiophosphoryl (P-S) group. Thus, organophosphorus insecticides are also named as phosphorus esters. Generally, they are the most toxic insecticides to insects and vertebrates (Ware, 1983).

The insecticidal qualities of organophosphorus compounds were first observed in Germany during World War II in the study of materials closely related to the nerve gases such as sarin, soman, and tabun. Because of the similarity of chemical structures of the organophosphorus compounds to the "nerve gas", their mode of action is analogous. This is also the reason why organophosphorus insecticides are also called nerve gas relatives. Organophosphorus insecticides exert their toxicity by inhibiting certain enzymes of the nervous system called cholinesterase. Basically, the transfer of a signal along a nerve fiber happens by changing the electrical potential across the nerve cell membrane through the movement of ions in and out of the nerve cells. At the terminal end of a nerve fiber, the signal is transferred across the synapse to the next nerve cell by the release of neurotransmitters such as acetylcholine. Subsequently, the neurotransmitter must be eliminated by an enzyme called cholinesterase in order to restore the sensitivity of the synapse. Inhibition of the function of cholinesterase by organophosphorus insecticides leads to an accumulation of acetylcholine at the neuron/neuron and neuron/muscle junctions or synapses, thereby resulting in rapid twitching of the voluntary muscles and finally paralysis (O'Brien, 1967).

Depending on the structure, the organophosphorus insecticides can be categorized as aliphatic, phenyl and heterocyclic derivatives (Ware and Whitacre, 2004). Aliphatic organophosphorus insecticides are phosphoric acid derivatives, which possess short linear carbon chains. Tetraehtyl pyrophosphate (TEPP), malathion, trichlorfon, monocrotophos, dichlorvos and mevinphos are examples of this group of the organophosphorus insecticides (Figure 10.5). TEPP is the first organophosphorus insecticide introduced in agriculture. It is very unstable in water and it hydrolyzes quickly (12-24 hrs) after being sprayed in crops. Since TEPP is very toxic, it was never available for home use. Unlike TEPP, malathion is safe for home uses; it can even be used on human beings for controlling head, body and crab lice. Malathion is the most heavily used aliphatic organophosphorus insecticide and was quickly adopted by agriculture for use on most vegetables, fruits and forage crops. Trichlorfon is an aliphatic organophosphorus insecticide containing three chlorine atoms. It has been used for crop pest control and fly control around barns and farm buildings. Monocrotophos is a nitrogen-containing organophosphorus insecticide. It is highly toxic to mammals so it had a limited use in agriculture and never will be available to home gardeners. Dichorvos has a very high vapor pressure and gives strong fumigant quality for insect control at home and in other closed areas. Mebinphos has a very short insecticidal life; it generally leaves no residue on the vegetables. Thus, it has been widely used in commercial vegetable production even though it is a highly toxic organophosphorus insecticide (Ware, 1983).

Phenyl organophosphorus insecticides are generally more stable than the aliphatic organophosphorus insecticides so that their residue lasts longer than those of the aliphatic organophosphorus insecticides. Phenyl organophosphorus insecticides contain a benzene ring with one of the ring hydrogen atom displaced by an attachment to the phosphorus moiety and the others are frequently substituted by Cl, NO_2, CH_3, CN or S. Examples of the phenyl organophosphorus insecticides include parathion (ethyl and methyl), stirofos, profenophos, sulprofos and isofenphos (Figure 10.6). Parathion is the second organophosphorus insecticide introduced into agriculture after TEPP. Ethyl parathion was the first phenyl organophosphorus insecticide commercially used. However, it has not been available for home use because of its hazard. Unlike ethyl parathion, methyl parathion has a lower toxicity to human and domestic animals, shorter insecticidal life, broader spectrum of the insecticidal activity than ethyl parathion, and thereby, is more useful than ethyl parathion. Stirofos is a home-safe insecticide for control of the home and livestock pests. Profenophos and sulprofos are only used on field crops. They have a wide range of insect control. Isofenphos is mainly used as a soil insecticide in field crops and vegetables (Ware and Whitacre, 2004).

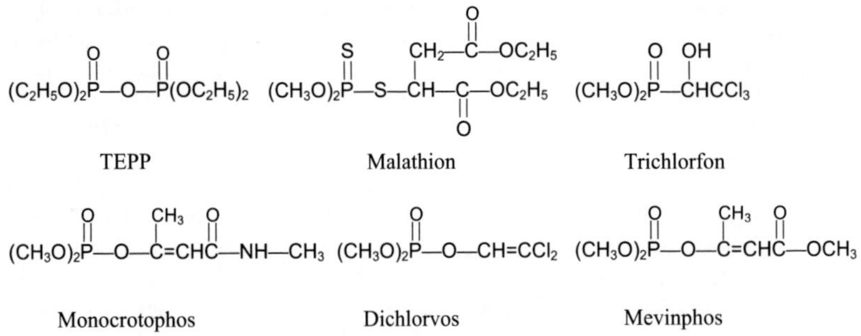

Figure 10.5 Chemical structure of the aliphatic organophosphorus insecticides including TEPP, malathion, trichlorfon, monocrotophos, dichlorvos and mevinphos.

Similar to the phenyl organophosphorus insecticides, heterocyclic organophosphorus insecticides also contain a ring structure. However, one or more carbon atoms in the ring of the heterocyclic derivatives are displaced by unlike or different atoms such as oxygen, nitrogen or sulfur. Heterocyclic organophosphorus insecticides are highly complex molecules and their residues generally last longer than those of the aliphatic and phenyl derivatives. Diazinon, azinphosmethyl, chlorpyrifos, methidathion, phosmet and dialifor are the main examples of the heterocyclic organophosphorus insecticides. Their chemical structure is shown in Figure 10.7. Diazinon is the first heterocyclic organophosphorus insecticides being introduced and it is a relatively safe organophosphorus insecticide. It has been widely used to control insects in home, garden, lawn and ornamentals, and for fly control in stables and pet quarters. Azinphosmethyl is the second oldest heterocyclic organophosphorus insecticide and it is used as both insecticides and acaricides in cotton production. Chlorpyrifos is the most popular insecticide being applied for control of cockroaches and other household insects. Methidathion is mainly used on forage, field crops, true fruits, and nut crops for control of a wide variety of insect and mite pests. Phosmet is the insecticide mainly used for controlling weevil pests and the use of dialifor is generally limited to apples, grapes, pecans and citrus (Ware and Whitacre, 2004).

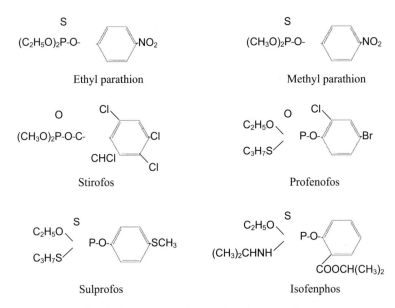

Figure 10.6 Chemical strcuture of the phenyl organophosphorus insecticides including ethyl and methyl parathion, stirofos, profenophos, sulprofos and isofenphos.

As alluded to above, the mode of action of the organophosphorus insecticides is mainly stemmed from their inhibition to the enzyme called cholinesterase. However, this inhibition also exerts toxic impacts on vertebrates including human beings. Vertebrates acutely poisoned by the organophosphorus insecticides show the symptoms of sweating, salivation, urination, defecation, lachrymation, nausea, vomiting, abdominal cramps, constriction of the pupil, slowing of the heart, and a drop in blood pressure. Accumulation of acetylcholine caused by the cholinesterase inhibition at the junction of motor nerves and skeletal muscle, and in ganglia of the autonomic nerve system consequently leads to involuntary twitching of the muscles, muscle spasms, and weakness of the respiratory muscle of the vertebrate. Moreover, accumulation of acetylcholine in the central nervous system causes restlessness, anxiety, insomnia, neurosis, tremor, ataxia, convulsions, depression of the respiratory center, a coma, and fatal asphyxia.

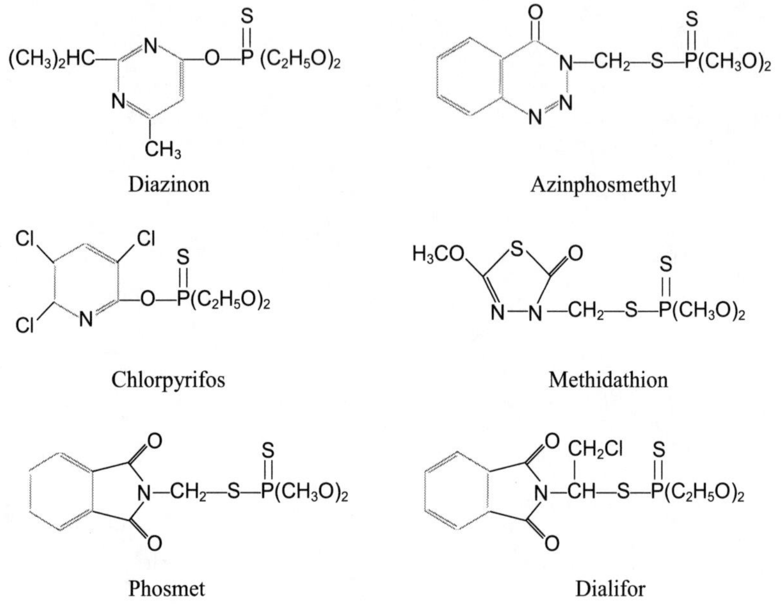

Figure 10.7 Chemical structure of heterocyclic organophosphorus insectides including diazinon, azinphosmethyl, chlorpyrifos, methidathion, phosmet and dialifor.

A man poisoned by the organophosphorus insecticides suffers headache, giddiness, nervousness, blurred vision, weakness, nausea, cramps, diarrhea, slowing of the heart, and a drop in blood pressure. Other signs of the poisoning include sweating, salivation, lachrymation, vomiting, constriction of the pupils, uncontrollable muscular twitches, and flaccid paralysis. In severe cases, the poisoned man suffers convulsions, a coma, loss of reflexes, and even fatal respiratory failure (Perry et al., 1998).

10.2.1.3. Carbamate Insecticides

Carbamate insecticides are the latest group of cholinesterase-inhibited insecticides being established. In general, carbamate insecticides are the synthetic derivatives of physostigmine (Figure 10.8), which is a principal alkaloid of the plant called *Physostigma venenosum*. Despite the fact that physostigmine and its synthetic analogues such as prostigmine are effective in inhibiting insect cholinesterase, they are not effective insecticides. This is because their high polarity and low lipid solubility properties prevent them from penetrating the insect cuticle and the ion-impermeable sheath of the insect nerve. Hence the currently available carbamate insecticides have been modified by eliminating the polar moiety of physostigmine to increase their lipid solubility. The general carbamate structure is illustrated in Figure 10.8 in which R_1 and R_2 are either hydrogen, methyl, ethyl, propyl or other short-chain alkyls, and R_3 is phenol, naphthalene or other cyclic hydrocarbon rings (Matsumura, 1985). Common examples of carbamate insecticides include carbaryl, methomyl, oxamyl, aldicarb, carbofuran, bufencarb, methiocarb, aminocarb and propoxur (Figure 10.8).

The mode of action of carbamate insecticides in controlling insects is similar to organophosphorus insecticides in which they also inhibit the insect cholinesterase and thereby lead to the accumulation of acetylcholine at the neuron/neuron and neuron/muscle junctions or synapses. However, the selectivity of carbamate insecticides against the cholinesterase of different species is more pronounced than organophosphorus insecticides. Furthermore, the cholinesterase inhibition by carbamate insecticides is reversible, while the inhibition by organophosphorus insecticides is irreversible. Thus, all the inhibited cholinesterase in a synapse can be fully recovered after several hours of the carbamate insecticides exposure if no further carbamate insecticides enter the synapse (Perry et al., 1998).

Carbaryl (or Sevin), which was introduced in 1956, is the best known carbamate insecticide. It is a naphthyl carbamate and an N-methyl compound, and is widely used as a lawn and garden insecticide (O'Brien, 1967). Its prevalence is stemmed from its low mammalian oral and dermal toxicity, and wide spectrum of insect control (100-150 species) (Matsumura, 1985). Methomyl, oxamyl, aldicarb and carbofuran, are systemic insecticides since they have relatively high water solubility and are also not easily metabolized by plants. They are generally taken into the roots of plants and translocated to the aboveground parts, thereby killing any sucking insect feeding on the plant juices. These systemic insecticides are usually used as soil insecticides and/or nematicides. Occasionally, aldicarb can be detected in shallow groundwater. Bufencarb is mainly applied in agriculture as a soil insecticide. Methiocarb and aminocarb are effective insecticides for the control of foliage- and fruit-eating insects, especially slug and snail in flower gardens and ornamentals. Moreover, methiocarb is used as a seed dressing and a bird repellent for cherries and blueberries. Propoxur is a popular insecticide used by pest control operators for controlling cockroaches and other household insects in restaurants, kitchens and homes. It is highly effective against the cockroaches, which are resistant to the organochlorine and organophosphorus insecticides (Matsumura, 1985; Perry et al., 1998; Ware, 1983).

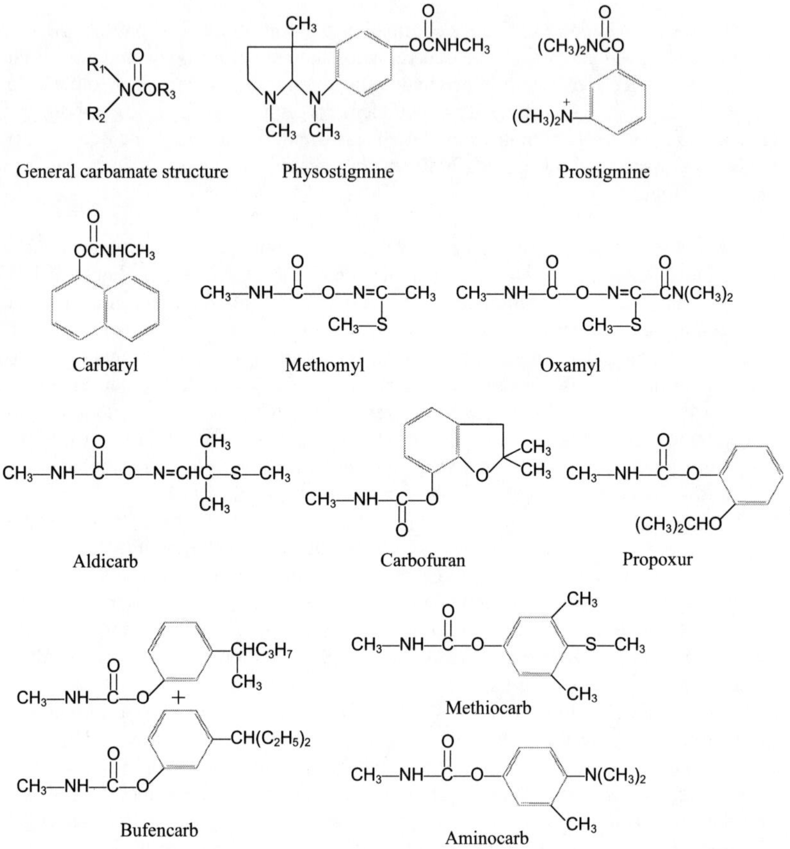

Figure 10.8 General carbamate structure, and chemical structure of physostigmine, prostigmine, carbaryl, methomyl, oxamyl, aldicarb, carbofuran, propoxur, bufencarb, methiocarb, and aminocarb.

Mammals poisoned by carbamate insecticides generally show analogous symptoms as those poisoned by the organophosphorus insecticides, which include constriction of the pupil, blurred vision, weakness, nausea, vomiting, sweating, salivation, tearing, urination, pulmonary edema, and convulsion. Fatal poisoning by carbamate insecticides may also happen due to respiratory failure, bronchial constriction, lowered blood pressure, neuromuscular block of respiratory muscles, and failure of the respiratory center of the brain. Human exposure of a sublethal dose of carbamate insecticides suffers headache, salivation, nausea, vomiting, muscular weakness and diarrhea. Customarily, a human poisoned by carbamate insecticides can be fully recovered in only a few hours after exposure, which is much faster than organophosphorus insecticide poisoning. In addition, unlike the organophosphorus insecticide poisoning, the carbamate insecticide poisoning does not lead to a delayed neurotoxic effect culminating in flaccid paralysis (Perry et al., 1998).

10.2.2 Herbicides

Herbicides are a more effective and economical means of weed control in comparison to hoeing and hand pulling. In the past, weeds are thought of as pests only in farmers' crops. Nowadays, the definition of the weeds has been broadened to cover all unwanted plants including those that obscure the view along roads, produce allergic reactions, and constitute fire hazards on vacant lots. Moreover, low value trees in forests, aquatic plants growing in irrigation systems, in fish hatcheries and recreational lakes, annuals which invade an airstrip making landing hazardous, perennials that heave the pavement, as well as perennial grasses which invade parks, golf greens and private gardens are all classified as weeds (Crafts, 1961). Hence herbicides have been extensively used away from the farm such as industrial sites, roadsides, ditch banks, irrigation canals, fence lines, recreational areas, railroad embankments, and power lines (Ware, 1983).

Currently available herbicides can be either selective or nonselective. Selective herbicides are used to kill weeds without harming surrounding crops, while nonselective herbicides kill all vegetation. In addition, herbicides can be applied to the leaves of plants (foliar application) or to soils (soil application) in which the herbicides ultimately reach underground plant organs such as roots, rhizomes, seeds, and emerging seedlings. Herbicides can be applied to areas before crops are planted, which is called preplanting application. They can also be applied prior to or after the emergence of crops or the weeds from the soils, which are named as preemergence and postemergence application, respectively.

A contact type of herbicides, which is named as contact herbicides, initially kill only the parts of the plants to which the herbicides are applied, but the whole plants may ultimately wither because of the failure of the important plant organs by the contact action. Contact-type herbicides are specifically effective against annuals; a complete coverage by contact herbicides is important in weed controls. Unlike contact herbicides, translocated herbicides are effective on impairing distant plant organs which do not receive a direct contact with herbicides. This is because they can be transported throughout the plants once being absorbed either by roots or aboveground parts of the plants. The translocated herbicides are effective against all weeds especially perennials. Moreover, uniform application of herbicides rather than complete coverage is only needed for applying translocated herbicides for weed controls (Ashton and Crafts, 1981; Ware, 1983).

According to the chemical structure of herbicides, there are currently two major types of herbicides: inorganic and organic herbicides. Inorganic herbicides include different inorganic acids and salt such as sodium chlorate and sulfuric acid. Organic herbicides includes petroleum oil, organic arsenicals, phenoxyaliphatic acids, substituted amides, nitroanilines, substituted ureas, carbamates, thiocarbamates, heterocyclic nitrogens, aliphatic acids, substituted benzoic acids, phenol derivatives, substituted nitriles, bipyridyliums, and myco-herbicides. The chemical structure, mode of action and toxicology of some inorganic and organic herbicides will be briefly discussed in the following section.

10.2.2.1 Inorganic Herbicides

Inorganic herbicides are the first chemicals used for weed control. Copper sulfate ($CuSO_4$) has been used to selectively kill weeds in grain fields. Sodium arsenite solutions were the commercially standard herbicides between 1906 and 1960 used for soil sterilization (Ware, 1983). Ammonium sulfamate ($NH_4SO_3NH_2$) has been applied for brush control. Other salts of inorganic herbicides include ammonium thiocyanate (NH_4SCN), ammonium nitrate (NH_4NO_3), ammonium sulfate (NH_4SO_4) and iron sulfate ($FeSO_4$). Their modes of action generally involve dessication and plasmolysis in which the protoplasm of a plant cell shrinks away from its cell wall due to the removal of water from its large central vacuole.

Another group of inorganic herbicides is borate herbicides, which include sodium tetraborate ($Na_2B_4O_7 \cdot 5H_2O$), sodium metaborate ($Na_2B_2O_4 \cdot 4H_2O$), and amorphous sodium borate ($Na_2B_8O_{13} \cdot 4H_2O$). The mode of action of borate herbicides for weed control is not fully understood at the moment, but is related to the accumulation of boron in reproductive structures of plants. Borate herbicides are nonselective and persistent herbicides. They are first absorbed by plant roots and then translocated to aboveground parts. Sodium chlorate ($NaClO_3$) is a nonselective herbicide, which has been widely used as a soil sterilant for more than 40 years. Sulfuric acid was mainly used as a foliar herbicide. In analogy to ammonium salts, the mode of action of both $NaClO_3$ and sulfuric acid also involves dessication and plasmolysis. Although some inorganic herbicides are still being used for weed and brush control, they are gradually being replaced by organic herbicides because of their persistence in soils.

10.2.2.2 Organic Herbicides

Phenoxyaliphatic Acids

Phenoxyaliphatic acids are used as herbicides in the forms of parent acids, or more commonly as salts and esters. The first phenoxy herbicide being synthesized is 2,4-D [(2,4-dichlorophenoxy)acetic acid], which was introduced in 1944. 2,4,5-T [(2,4,5-trichlorophenoxy)acetic acid] and MCPA [(4-chloro-*o*-tolyl)oxy]acetic acid] are another two important herbicides of this group. Other phenoxy herbicides include 2,4-DB [4-(2,4-dichlorophenoxy)butyric acid], MCPB [4-((4-chloro-*o*-tolyl)oxy)butyric acid], silvex [2-(2,4,5-trichlorophenoxy)propionic acid], dichlorprop [2-(2,4-dichlorophenoxy)propionic acid], mecoprop [2-((4-chloro-*o*-

tolyl)oxy)proionic acid], acifluorfen [sodium 5-(2-chloro-4-(trifluoro-methyl)-phenoxy)-2-nitrobenzoate], and diclofop methyl [2-(4-(2,4-dichlorophenoxy)-phenoxy)-methyl-propanoate]. Chemical structures of some phenoxy herbicides are illustrated in Figure 10.9. Phenoxy herbicides are highly selective for broad-leaf weeds and can be translocated throughout the plants. At optimum dosage, foliar application of phenoxy herbicides kills the root systems of perennial weeds. An overdose of phenoxy herbicides may cause excessive contact injury to foilage, which may lead to little translocation. However, high dosage of phenoxy herbicides can be applied as preemergence herbicides through soil application and as postemergence herbicides to flooded rice by application through water (Ashton and Crafts, 1981; Ware, 1983).

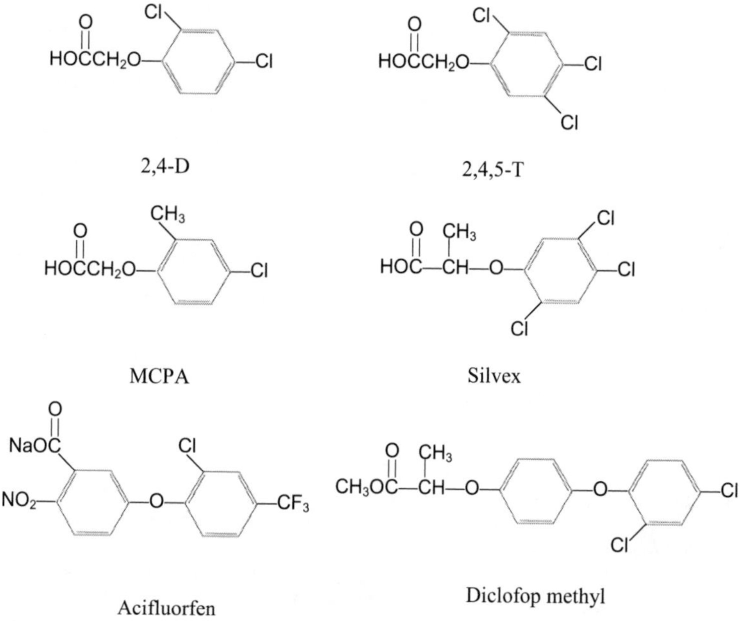

Figure 10.9 Chemical structure of 2,4-D, 2,4,5-T, MCPA, silvex, acifluorfen, diclofop methyl.

The mode of action of phenoxy herbicides is similar to that of growth hormones (i.e., by mimicking auxins, which are growth regulators produced by plants). The herbicidal actions influence cellular division, activate phosphate metabolism and modify nucleic acid metabolism of plants. Immediately after foliar application of phenoxy herbicides, there is an epinastic bending of the plants. In susceptible species of weeds, tumors, secondary roots, and fasciated structures are subsequently developed. In mature cells, phenoxy herbicides lead to dedifferentiation and initiation of cell division, while they inhibit cell division in primary meristem. A low level of phenoxy herbicides abnormally stimulates plant growth by stimulating the synthesis of nucleic acid and protein. They also induce cell enlargement by increasing the activity of autolytic and synthetic enzymes, which are responsible for synthesis of new cell wall materials and cell wall loosening. On the other hand, a high level of phenoxy herbicides inhibits the synthesis of nucleic acid, protein and new cell wall materials, thereby inhibiting plant growth. Consequently, the plants wither because of the abnormalities in plant structure and growth induced by phenoxy herbicides (Ashton and Crafts, 1981).

Phenoxy herbicides are mildly irritating to eyes, skin, and respiratory and gastrointestinal linings. They are not readily absorbed from lung and significantly stored in fat. Moreover, cutaneous absorption seems to be insignificant. However, they are easily absorbed from the gastrointestinal tract. Animals poisoned by phenoxy herbicides suffer vomiting, diarrhea, anorexia, weight loss, mouth and pharynx ulcer, as well as toxic injury to kidneys, liver, and the central nervous system. Humans who ingest a high dosage of phenoxy herbicides suffers severe metabolic acidosis (i.e., low arterial pH and bicarbonate content), injury of striated muscle, hyperthermia, and fatally renal and multiple organ failure. In addition to phenoxy herbicides, dioxin compounds, including polychlorinated dibenzodioxins and polychlorinated dibenzofurans (impurities formed during the manufacturing processes of 2,4-D and 2,4,5-T), are highly stable and toxic. They are exceptionally toxic to multiple mammalian tissues. Humans poisoned by these dioxin compounds shows the symptom of chloracne (a chronic and disfiguring skin condition) and soft-tissue sarcoma (Regait and Roberts 1999). As a result, the use of some phenoxys, particularly 2,4,5-T and silvex has been restricted by the USEPA.

Substituted Amides

Amide herbicides are comprised of a diverse group of chemicals. They have various biological properties and are readily degraded in soils and by plants. Chloroacetamides, which possess a monochlorinated methyl group ($Cl-CH_2-$), are a major subdivision of amide herbicides. Examples of chloroacetamides include allidochlor, propachlor, and metolachlor. Instances of other amide herbicides include diphenamid, propanil, and napropamide. Allidochlor is one of the earliest amide herbicides being synthesized. It is a preemergence herbicide and is selective for grass control by inhibiting the germination or early seedling growth through alkylation of the –SH group of protein. Diphenamid is also a preemergence herbicide. It is applied to soils and exerts little contact impact to plants. Diphenamid only affects seedling so that most established plants are tolerant to it. Napropamide is a soil-applied herbicide, which has been used for control of grasses and broad-leaf weeds in vineyards and orchards, and in strawberries, ornamentals and tobacco. Propanil is a selective postemergence herbicide for control of broad-leaf weeds in rice fields. The chemical structures of some amide herbicides are illustrated in Figure 10.10 (Ashton and Crafts, 1981; Ware, 1983).

The mode of action of amide herbicides generally involves growth inhibition and especially inhibition of root elongation. Soil-applied amide herbicides, such as allidochlor, diphenamid, and napropamide, inhibit root elongation, seed germination and/or early seedling growth of weeds by interfering cell division, cell enlargement, nucleic acid and protein synthesis. Therefore, the affected seedlings emerging from the soils are either severely stunted and/or malformed. Certain soil-applied herbicides

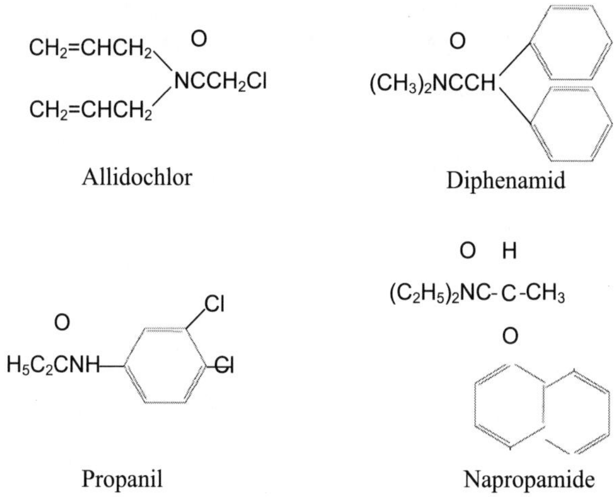

Figure 10.10 Chemical structure of allidochlor, diphenamid, propanil, and napropamide.

were reported to be able to inhibit stem, coleoptile, shoot, and/or leaf growth due to the translocation of the herbicides absorbed by the roots to other parts of plants. The prime effect of foliage-applied amide herbicides, such as propanil, is the inhibition of photosynthesis and localized necrosis of leaves, which is associated with alternation of the cellular membrane. Moreover, they may exert growth inhibition in other parts (root and coleoptile) of plants (Ashton and Crafts, 1981).

Allidochlor is severely irritating to eyes and skin. Rats poisoned by allidochlor are characterized by apparent discomfort, coma, salivation, convulsion, and liver degeneration. Human exposure to allidochlor suffers severe contact dermatitis including persistent swelling and violaceous edmema of the exposed parts, hemorrhagic bullae, crusting, and an exudative intertrigo. Rats and dogs poisoned by propanil show the symptoms of growth depression, reduction in feed consumption and hemoglobin concentrations, and central nervous system depression. Human exposure to propanil was reported suffering chloracne (Hayes and Laws, 1991).

Substituted Ureas

Thousand of substituted ureas have been tested to be able to control weeds and many of them are still in use today. Major examples of substituted ureas include DCU [1,3-bis(2,2,2-trichloro-1-hydroxyethyl)urea], fenuron (3-phenyl-1,1-dimethylurea), monuron [3-(*p*-chlorophenyl)-1,1-dimethylurea], diuron [3-(3,4-dichlorophenyl)-1,1-dimethylurea], and neburon [1-*n*-butyl-3-(3,4-dichlorophenyl)-1-methylurea]. As shown in Figure 10.11, the hydrogen atoms of the urea functional group of the substituted ureas are replaced by various carbon chains and rings. In general, most substituted ureas are relatively nonselective, are most effective against broadleaf, and are used primarily as preemergent weed killers; some are foliar-applied herbicides. In addition, by taking the advantages of water solubilities and adsorptive properties of substituted ureas by soils, selectivity may be obtained in certain crops. For instance, shallow-rooted weeds may be controlled in deep-rooted crops by using substituted ureas with low water solubility and high adsorptive properties in a soil of relatively high adsorption capacity, or vice versa.

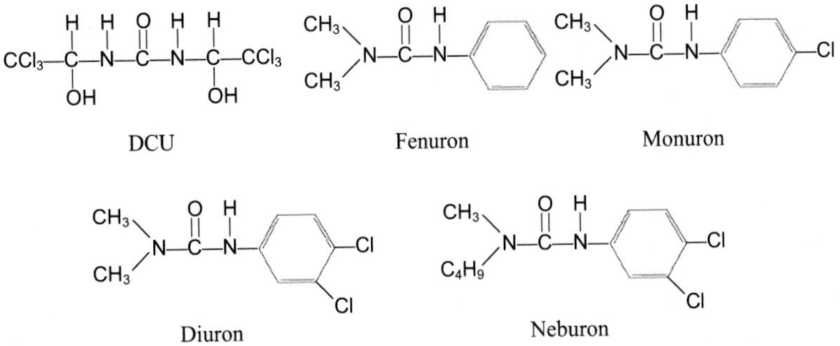

Figure 10.11 Chemical structure of DCU, fenuron, monuron, diuron, and neburon.

DCU was the first substituted ureas being used commercially for weed control. It is toxic to grasses and selective in certain broad-leaf plants. Currently, DCU is not registered for use in crops in the U.S. Fenuron, monuron, diuron, and neburon are all soil sterilants; weeds are killed by first root absorption of these herbicides which are then transported upward in transpiration stream. Fenuron has a solubility of about 2900 ppm; it is readily leached into deep soils by rainfall or irrigated water. Fenuron has been proven to be effective for control of medium- to deep- rooted weeds in regions of moderate rainfall and against shallow-rooted weeds in arid regions. Monuron has a solubility of about 230 ppm. In comparison to fenuron, monuron tends to remain in the topsoil layer and is leached slowly. It is commonly used as an industrial and agricultural soil sterilant in the region of moderate rainfall, and has been widely applied in citrus orchards, vineyards, and cranberry fields. Diuron has a solubility of 42 ppm. It is safer than monuron because of its low solubility and limited leaching. Diuron is recommended for industrial and agricultural soil sterilization in regions of medium to high rainfall. Neburon has even lower solubility (4.8 ppm) than diuron. It usually stays on the top of soil and attacks the roots of seedlings as they emerge. Thus, neburon should be applied in spring prior to the germination of the seeds of the weeds.

The primary mode of action of substituted ureas for weed control is the inhibition of photosynthesis. This involves blockage of electron transport, and thereby, subsequently prevent the formation of ATP and NADPH required for carbon dioxide fixation. Consequently, plants starve from lack of photosynthate. In addition to the photosynthesis inhibition, foliar application of substituted ureas also induces some symptoms in nonphotosynthetic tissues of the plant. At a high concentration of substituted ureas in leaves, light green areas initially appear and the leaves finally become necrotic. The symptoms in leaves appearing at a low foliar concentration include wilting of leaves, the appearance of silver and/or indeterminate grey blotches, and rapid yellowing. It was postulated that the appearance of these symptoms is due to the formation of secondary phytotoxic substance, alternation of the protective carotenoid related reactions, and/or happening of the photooxidative pigment-destructive reactions by substituted ureas (Crafts, 1961; Ashton and Crafts, 1981; Ware, 1983).

Diuron has a low acute toxicity to mammals. It was reported that rats poisoned by diuron became drowsy and ataxic, and were prostrate and breathed slowly. They also suffered diarrhea and diuresis along with reduced intake of food and water, leading to weight loss, had symptoms of hypothermia, glycosuria, proteinuria, and aciduria, as well as suffered lethally respiratory failure depending on the dosage of diuron being taken. However, rats that survived from diuron poisoning could evidently recover at 48 hrs, and disappearance of most signs of toxicity was observed after 72 hrs. In addition, diuron is irritating to human skin, eyes, or nose, but humans who accidentally ingested diuron-containing herbicide showed no signs of intoxication (Hayes and Laws, 1991).

Triazines

Triazines are the most familiar group of heterocyclic nitrogen because of their heavy uses as herbicides. In the U.S, they are the second largest group of herbicides sold in the market (Hayes and Laws, 1991). They are characterized by six-member rings each contains three nitrogen atoms. Triazines are strong photosynthesis inhibitors, and are used for nonselective weed control and selectively in certain crops based on the differences in the ability of tolerant plants to degrade or metabolize triazines. The greatest uses of triazines today are as selective herbicides in maize and as nonselective herbicides on industrial sites. They are generally applied to soils for their postemergence activity. Chlorazine, simazine, atrazine, cyanazine, prometon, propazine, metribuzin, and hexazinone are some of the examples of triazines (Figure 10.12).

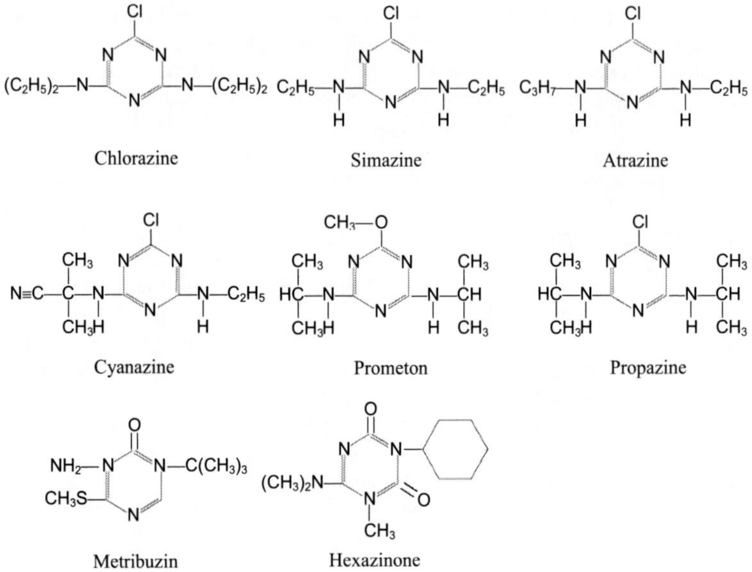

Figure 10.12 Chemical structure of chlorazine, simazine, atrazine, cyanazine, prometon, propazine, metribuzin, and hexazinone.

The first triazine herbicide being synthesized is called chlorazine. It displayed varying selectivity and is effective against weeds in maize, cotton, and snapbeans for both preemergence and postemergence application. Simazine was introduced in 1955. It was proven that simazine is more toxic and less selective than chlorazine. Simazine does not have a strong contact action, but kills weeds by a slow dying back of leaves. Atrazine is also a preemergence and postemergence herbicide commonly used on maize, pineapples, and nursery conifer, as well as for selective control of pond weeds especially submerged plants. Metribuzin is mainly used on soybeans, wheat, sugarcane, and a few vegetables. Hexazinone is usually applied for control of annual, biennial, and perennial weeds, and woody plants on non-cropland or in conifer plantings (Crafts, 1961; Ware, 1981).

The generally phytotoxic symptoms of triazines are foliar chlorosis followed by necrosis at high concentration and an increase in the greening of the leaves at low triazine concentration. Past studies of the toxic impacts of atrazine on weeds indicated that they could largely alter the fine structure of chloroplasts, accelerate vacuolation of the cells of developing leaves, reduce airspace-system of mature leaves, modify integrity of cell membranes in leaves, and lead to a lesser thickness of the cell wall of sieve and tracheary elements of the stem. It was also reported that triazines absorbed by roots of weeds are translocated to the aboveground parts rapidly, while those in leaves are not translocated readily. The prime mode of action of triazines is the inhibition of the growth of intact plant ascribed to their blockage of photosynthesis. Since the phytotoxic symptoms of triazines mentioned above are not typical of starvation caused by the lack of photosynthate and usually happen unexpectedly fast, it is believed that a secondary phytotoxic substance, the protective carotenoid related reactions, and/or the photooxidative pigment-destructive reactions are also involved in the development of these phytotoxic symptoms (Ashton and Crafts, 1981), as similar to the mode of action of substituted ureas.

Atrazine is irritating to eyes, but only slightly irritating to skin. Rats poisoned by atrazine at a lethal dose showed symptoms of excitation followed by depression with a reduced respiratory rate, motor incoordination, clonic spasm, and hypothermia, and finally died within 1 day, probably due to hemorrhagic pneumonia, hemorrhage in other organs, and dystrophic changes of the kidney tubules. Atrazine alone is not mutagenic, but is a suspected carcinogen. Splashing atrazine on human skin leads to severe pain, skin rash, swelling, and formation of blisters and hemorrhagic bullaes. Depending on the dosage of atrazine which a human exposes, satisfactory recovery of the contacted skin was achieved from 1 week to 1 month.

10.3   Occurrence and Behavior in Natural Systems

Pesticides introduced into natural environments through an application for pest controls, disposal, and spill are not completely broken into benign compounds such as carbon dioxide and water. In fact, the occurrence and behavior or fate of the introduced pesticides in natural environments are complex and continuously affected by many processes. For instance, pesticides sprayed onto the crops may move through the air. Moreover, pesticides applied to the soil may be washed to nearby bodies of surface water or leach to groundwater. Figure 10.13 summarizes the overall fate of pesticides in natural environments. The environmental processes that govern the behavior and fate of a pesticide can be classified into three compartments: (1) phase transfer processes, which control its movement among different environmental compartments, such as water, biota, soils, sediments, and the atmosphere; (2) transport processes, which move the pesticide away from its initial point of introduction to the environment and throughout the environmental system; and (3) transformation processes, which changes its chemical structure through microbial degradation, chemical degradation, and photodegradation (Fishel, 1997; Nowell et al., 1999). In many cases, the phase transfer and transport processes are linked together by volatilization, runoff, leaching, absorption or uptake by animals and plants, and crop removal.

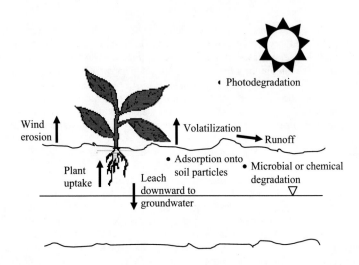

Figure 10.13 A schematic diagram summarizing pesticide fate in natural environments.

10.3.1 Surface Water and Groundwater

Many studies show widespread occurrence of pesticides, with concentrations in many streams at levels that may have effects on aquatic life and fish-eating wildlife (Colborn et al., 1993; Nowell et al., 1999; Gilliom, 2007). Pesticides were detected throughout much of the year in streams (> 90% of the time) and detected in > 50% of wells that sampled shallow groundwater beneath agricultural and urban areas (Gilliom, 2007). Pesticides occurred as mixtures of multiple pesticides much more often than individually. The herbicides atrazine (and its degradate, deethylatrazine), simazine, and prometon were common in mixtures found in streams and groundwater in agricultural areas. The insecticides diazinon, chlorpyrifos, carbaryl, and malathion were common in mixtures found in urban streams. The total combined toxicity of pesticides in aquatic ecosystems may often be greater than that of any single pesticide that is present (Colborn et al., 1993; Nowell et al., 1999; Gilliom, 2007).

Pesticides in surface water can transport a great distance. Pesticide contamination can be traced for hundreds of kilometers in rivers. It was reported that pesticides in the river in Czechoslovakia was transported through Germany and reached the Southern North Sea. Generally, the most soluble pesticides are carried in water, while others can be transported by first attaching to particulates such as planktons, which are then carried in water. These dissolved and particle-bound pesticides can be deposited onto sediments at the bottom of water bodies. Dissolved pesticides can also be taken up by bottom dwelling species. Filter feeders such as shellfish and prawn can take in particle-bound pesticides. Most aquatic organisms can absorb pesticides through the bioconcentration process. In addition, pesticides can reach groundwater from most sources. Because of the limited quantities of organic matters for pesticides adsorption and slow degradation processes in groundwater, pesticides in groundwater are chemically inert and readily available to produce direct toxic effect (Pesticide Action Network UK, 2000).

The movement of pesticides to surface and ground water mainly follows two routes, that is, runoff and leaching. Runoff is the movement of water and pesticides over a sloping surface. It occurs when water from rainfall and irrigation is applied faster than it can enter the soils. Pesticides can be carried in water in dissolved forms or by adsorbing onto eroding soil particles. Leaching is the movement of water and pesticides through the soils rather than over the soil surface. The potential and severity for pesticide movement by runoff and leaching vary with soil properties, pesticide properties, site conditions, and management practices. Table 10.2 illustrates the situations giving low or high risk of groundwater contamination by pesticides. The soils containing high organic matters can retain more water than the soils with low organic matters. In a similar manner, clay and well-structured soils can hold more water than sandy and poorly-structured soils. Thus, pesticide-contaminated water tends to run off rather than leach to groundwater in organic matter-rich and well-structured soils. Soil moisture affects the rate for water traveling through the soil media. Pesticide-contaminated water tends to run off if the soil is wet and saturated before rainfall or irrigation. Soil moisture is also an important factor controlling the rate of microbial pesticide degradation. Furthermore, in coarse-textured soils such as loamy sand and sandy loams which have high soil permeability and infiltration capacity, water tends to leach to groundwater rather than run off (Buttler et al., 2003; Waldron, 1992).

Table 10.2 Effects of soil and pesticide properties, site conditions, and climate on potential groundwater contamination by pesticides.

	Risk of Groundwater Contamination by Pesticides	
	Low Risk	**High Risk**
Soil properties		
Organic matter content	High	Low
Texture	Clay and well-structured	Sandy and poorly-structured
Soil moisture	High	Low
Permeability and infiltration capacity	Low	High
Pesticide properties		
Adsorption coefficient	High	Low
Solubility	Low	High
Half-life	Low	High
Site conditions		
Groundwater depth	Deep (>100 ft)	Shallow (<20 ft)
Sinkholes or cracked bedrocks	Absent	Present
Confined layer	Present	Absent
Climate		
Heavy or sustained rain	Before application	After application

In addition to soil properties, pesticide properties (e.g., the adsorption coefficient, water solubility, and persistence) are also key factors affecting pesticide runoff and leaching. The soil/water adsorption coefficient of a pesticide is a measure of how tightly the pesticide adsorbs onto the soil or soil organic matters. This property is important in regulating the pesticide concentration in soil water. Pesticides with a high adsorption coefficient (>5) bind strongly onto the soil particles, thereby less likely leaching to groundwater. Pesticide solubility is its tendency to dissolve in water. The lower solubility of pesticides, the less likely they are to move with water through the soil media and contaminate groundwater. Generally, pesticides are less likely to leach when their water solubility is less than 30 ppm (Nowell et al., 1999). Pesticide persistence describes the staying power in natural environments. In fact, pesticides in the natural environment are degraded at different rates by soil microorganisms, chemical reactions, and sunlight. Warm and moist conditions in the soil allow microorganisms to effectively use pesticides as food sources and turn them into benign compounds such as carbon dioxide and water. Generally, degradation processes happen in root zones and become slow in deep soils and sediments. Persistent pesticides can maintain their structure for a long time. Customarily, persistence is expressed in terms of half-life. Pesticides with a long half-life are long lasting in the soils and more likely leach to groundwater (Buttler et al., 2003; USEPA, 2007d).

Site conditions influencing the severity of pesticide runoff and leaching involve a depth to groundwater, site geologic conditions, and climate. If the depth to the groundwater table is shallow, there is less soil to act as a filter to degrade and adsorb pesticides. Thus, pesticides have less distance to travel and have a great potential to reach groundwater. Sinkholes, cracked bedrocks, or confining layers in the bedrocks greatly affect the vertical movement of water in the soil media. Pesticide-contaminated groundwater and pesticide runoff can freely move to groundwater through the sinkholes and cracked bedrock as long as there are no confining layers. Once dissolved pesticides enter the sinkholes, they receive little filtration or chance for degradation. Runoff of the dissolved pesticides or particle-bound pesticides is more serious on a sloped ground surface or soil surface that is readily eroded (Waldron, 1992; Buttler et al., 2003).

Heavy or sustained rain immediately after pesticide application results in severe pesticide runoff or pesticide leaching, especially where little runoff occurs. Poor application and irrigation practices also greatly increase the potential of pesticides leaching to groundwater. Such poor practices include selection of pesticides which are susceptible to leach, application of pesticides at a rate higher than the recommended level, spill of pesticides on the ground surface because of careless mixing and loading, application of pesticides immediately prior to irrigation or a heavy rain, over-irrigation, and improper storage and disposal.

One reason for pesticides being ECs is because some pesticides have been proven to be endocrine disrupting chemicals (EDCs), as shown in Table 10.3 (Colborn et al., 1993; Nowell et al., 1999; USEPA, 2001a). Several field studies have shown an association between exposure to pesticides and various adverse effects (e.g., reproductive impairment in field populations). Examples include DDT in western gulls (Larus occidentalis) in southern California (Fry and Toone, 1981); DDE in American alligators from Lake Apopka, Florida (Guillette et al., 1994); Polychlorinated biphenyls (PCBs) in fish-eating birds from the Great Lakes (Giesy et al., 1994) and in common seals from the Wadden Sea (Reijnders, 1986). Effects on steroid hormones on field populations exposed to pesticides also were reported (Nowell et al., 1999). While studies with laboratory animals have reproduced some of the abnormalities observed in field populations, the evidence in humans is much sketchier. Exposure to EDCs has been suggested as contributing to the increased incidence of breast, testicular, and prostate cancers, ectopic pregnancies, and cryptorchidism, and to a decrease in sperm count. However, the role of EDCs in human health trends remains uncertain.

Table 10.3 Pesticides which have been identified as endocrine disrupting chemicals (EDCs).

Insecticides	aldicarb	dieldrin	Heptachlor	oxychlordane
	β-HCH	DDT	Lindane	synthetic pyrethroids
	carbaryl	DDD	Methomyl	parathion
	chlordane	DDE	methoxychlor	toxaphene
	dicofol	endosulfan	Mirex	transnonachlor
Herbicides	2,4-D	alachlor	Atrazine	nitrofen
	2,4,5-T	amitrole	Metribuzin	trifluralin
Fungicides	zineb	mancozeb	Benonmyl	metiram-complex
	ziram	maneb	tributyl tin	hexachlorobenzene
Nematocides	Aldicarb	DBCP		
Biocides	TBT			

DDE, DBCP, and TBT refer to dichlorodiphenyl dichloroethylene, dibromo chloro propane, and tributyltin, respectively.

With complex mechanisms to coordinate and regulate internal communication among cells, endocrine systems release hormones which act as chemical messengers to interact with receptors in cells in order to trigger responses and prompt normal biological functions such as growth, embryonic development, and reproduction. EDCs can permanently and irreversibly impact the development of the endocrine systems and of the organs responding to endocrine signals in organisms indirectly exposed to EDCs during prenatal and/or early postnatal life and directly exposed to EDCs after birth or hatching. Generally, the impacts of EDCs on organisms are not evident and appear until later in life, such as effects on learning ability, behavior, and reproduction, as well as increased susceptibility to cancer and other diseases (Pesticides Action Network UK, 2000). The particular concern about the endocrine disrupting pesticides is their persistence in body fat and ability to bioconcentrate via the food chain. These pesticides can be mobilized and transferred to the developing offsprings via placenta or egg, or to the new-born infants via lactation. It has been reported that tributyltin (TBT), a biocide used to control a broad spectrum of organisms, can make the male whelk sterile. Furthermore, deformity and embryo mortality in birds and fish have been reported due to the exposure to organochlorine insecticides. The presence of endocrine disrupting pesticides in drinking water supplies has also been blamed for falling sperm counts in males (Environment Canada, 1999).

10.3.2 Municipal and Industrial Wastewater

Pesticides are not common constituents in municipal wastewater (Tchobanoglous and Burton, 1991). Generally, only ppb level of pesticides is detected in municipal wastewater (Janssens et al., 1997; Jiries et al., 2002). Pesticides in municipal wastewater primarily result from the pesticide runoff and leaching from the agricultural land, especially from intensive agricultural areas. Urban use of herbicides and insecticides is another nonpoint source in which they may result in runoff into sewers which in turn may expel pesticides into wastewater treatment plants (Müller et al., 2002). The disposal sites and landfills in which industrial or agricultural wastes are buried , and the discharges from hospitals and from industrial effluents of pesticides production plants are the possible point source of pesticides in the municipal wastewater (Hernando et al., 2005). For example, about 45,000 and 56,882 pounds of HCCPD (Hexachlorocyclopentadiene, used in a group of related pesticides) were transferred off-site to landfills and/or to treatment and disposal facilities, and about 900 and 1,580 pounds were discharged to publicly owned treatment works (POTWs) in 1990 and 1996, respectively (http://www.epa.gov/tri, 2008).

Jiries et al. (2002) reported that ppb level of the organochlorine pesticides were detected in the municipal wastewater in Karak City, Jordon, including vinclozoline, tetradifon, trans-heptachlor epoxide, DDT, DDE, dieldrin, endrin, fenpropathrin, and endosulfan. This pesticide-contaminated wastewater consequently contaminated the soil in the Karak raw wastewater disposal site near Al-Lajoun valley. In addition, investigation of the pesticide loads in Zwester Ohm River illustrated that about 65% of the total pesticide loads came from a nearby wastewater treatment plant (Müller, 2002). Rule et al. (2006) found that pesticides in wastewater catchment mainly came from the runoff rather than from residential sewage; they reported that trifluralin was the major pesticide being detected.

Pesticides in industrial effluents of pesticide production plants are mainly generated during manufacturing and formulating/packaging (Nowell et al., 1999). During manufacturing, specific technical grades of the active ingredients of pesticides are made through the process of chemical synthesis, separation, recovery and purification, and product finishing. Chemical synthesis generally includes chlorination, alkylation, nitration, and many other substitution reactions. Separation processes involve filtration, decantation, extraction, and centrifugation. Evaporation and distillation are the common recovery and purification processes used to reclaim solvents or excess reactants, and to purify intermediates and final products. Product finishing involves blending, dilution, drying, pelletizing, packaging, and canning.

Depending on the types of pesticides/active ingredients being manufactured, the possible sources of the pesticide-containing wastewater in the manufacturing are described as follows (Wang et al., 2004):

- vent gas scrubber water from caustic soda scrubbers;
- wastewater from reactors (e.g., chlorinators);
- excess mother liquor from centrifuges;
- scrubber water from dryer units;
- aqueous wastes from filters and decanters;
- distillation vacuum exhaust scrubber wastes;
- washwater from product purification processes;
- process/production area cleanup wastes; and
- washwater from equipment cleanout.

After manufacturing relative pure forms (technical grades) of pesticides or active ingredients, the next step is formulating and packaging. In formulating units, manufactured active ingredients are processed into liquid, granules, dusts, and powders to improve their properties of storage, application, handling, safety, and effectiveness. All formulating systems are batch-mixing operations, involving dry mixing and grinding of solids, dissolving solids, and/or blending. In liquid formulation units, liquid active ingredients or melted solid active ingredients are usually mixed with solvents and/or blending agents such as emulsifiers and synergists in batch-mixing tanks with open-top vessels and standard agitators. Air pollution control devices are used in all exhaust systems of the liquid formulation units. In dry formulation units, technical grade active ingredients are mixed with an appropriate inert carrier using rotary or ribbon blender-type mixers; the mixtures are then ground to obtain correct particle sizes. The dry products include dusts, powders, and granules. After formulating, the finished pesticides are packaged into marketable containers.

During formulating/packaging, the main source of pesticide-contaminated wastewater is from equipment cleaning, such as periodical cleaning of the filling equipment to prevent cross-contamination. The other sources include washdown wastewater generated during formulation area cleaning, natural runoff at formulating/packaging plants, and wastewater from control laboratories. Drum-washing wastewater generated during decontamination and recondition of pesticide drums usually contains caustic solution and washed pesticides in the drums. Wastewater generated from air pollution control devices such as water-scrubbing devices may also be contaminated with pesticides (Wang et al., 2004).

10.3.3 Soils and Sediments

Occurrence and behavior of pesticides in soil and sediment environments are predominantly governed by retention, transport, and transformation processes in which they are influenced by climatic factors, pesticide properties, mode of entry, plant or microbial factors, or soil properties (Cheng et al., 1990). Retention results from interaction between pesticides and soil particle surfaces or soil surface components; it is mainly described as adsorption, but also includes absorption into soil matrices or soil microorganisms. The retention process, which can be either reversible or irreversible, retards the pesticide movement in the soil and affects the availability of pesticides for plant or microbial uptakes or for abiotic or biotic transformations. It also influences the pesticide transport to atmosphere, groundwater, and surface water significantly. Transport processes determine the phase in which pesticides may be present in the soil. For example, volatilization leads to a distribution of pesticides to the gas phase in the soil environment or to atmosphere. Leaching causes the movement of pesticides to groundwater, while runoff brings pesticides to surface water bodies nearby. Transformation processes involve characteristic changes in chemical properties of pesticides. It can be purely chemical in nature catalyzed by soil constituents or induced photochemically. However, most pesticides in the soil or sediments are transformed biologically by soil microorganisms. Through the transformation process, pesticides may be degraded into simpler, and less toxic or less persistence forms, but occasionally the metabolic products formed may be more toxic than the parent pesticides. Effects of retention, transport and transformation processes on pesticides' fate and transport in soil and sediments are briefly discussed below.

10.3.3.1. Retention

Relative soil retention (organic carbon partition coefficient or K_{oc}) and relative soil persistence (half-life) have the greatest influence on pesticide fate. In general, if a pesticide is weakly retained ($K_{oc} < \sim 50$; e.g., sulfonylureas and imidazolinones), its movement will be related to the likelihood of runoff, modeled as the Index Surface Runoff (ISRO; USDA-NRCS, 2003). If a pesticide is very firmly held by soil ($K_{oc} > \sim 2000$; e.g. dinitroanilines), movement in water is greatly reduced and primarily depends on soil displacement, reflected in soil erodibility (K Factor), which is related to the potential of particle-adsorbed pesticide movement and is indirectly related to pesticide retention. The K-factor is a factor in the Universal Soil Loss Equation (USLE; Wischmeier and Smith, 1978) and the revised equation (RUSLE; Renard et al., 1993).

Pesticide retention in the soil and sediments are mainly governed by the organic matter and clay contents of the soil. Organic matter in the soil provides a large number of binding sites for pesticide's retention because it has an extremely high surface area and is very chemically reactive. The higher the organic matter contents of the soil, the greater the amount of the binding sites available for pesticide retention. Usually, hydrophobic pesticides have a stronger tendency to bind with the organic matter of the soil than hydrophilic pesticides. The organic matter content in the soil depends on climate, vegetation, position in the landscape, soil texture, and farming practices. For instance, abundant rainfall combined with lush natural vegetation generally results in organic matter-rich soils. Forest vegetation generally produces less organic matter deeper in the soil than grassland vegetation. Poorly drained soils tend to have more organic matter than better drained soils since organic matter decomposes more slowly in damp soils. Moreover, medium- and fine-textured soils have higher organic matter than the coarse-textured soils since sandy and gravelly soils generally support less vegetation. Farming practices, which involve return of crop residues and animal wastes to soils, maintain a better soil organic matter content than the practices focusing on harvest or destruction of crop residues. Clay content reflects the portion of microscopic plate-shaped grains in the soil. These tiny flat grains are chemically reactive and provide a huge surface area per unit weight of the soil. Clay content is particularly important to subsoil for pesticide retention since its organic matter content is generally much lower than that in the surface soil.

10.3.3.2 Transport

Transport of pesticides in the soil to groundwater through leaching and to surface water via runoff depends on soil properties, pesticide properties, site conditions, and climate, which have been fully mentioned in Section 10.3.1, and thus, will not be discussed in this section except soil permeability. Soil permeability refers to the rate at which water moves through the soil medium. It is highly controlled by the size and continuity of the soil pores. Pesticides in the soil with a higher permeability are more readily to leach to groundwater than the soil with a lower permeability. On the contrary, pesticides on a poorly permeable soil layer are easier to transport to surface water by the runoff in comparison to the highly permeable soil layer. Coarse-textured sandy and gravel soils possess larger pores so that they have a higher permeability than the fine-textured clay soil. Furthermore, soils possessing roots, burrowing insects or animals can have a high permeability because they can create big voids or macropore for water movement (e.g., preferential flow). Contrarily, dense, compact, or cemented soil layers have very low permeability.

Volatilization of pesticides in the soil to atmosphere involves evaporation of pesticide molecules from residues present on soil surfaces into the air, followed by the dispersion of the resulting vapor into the overlying atmosphere by diffusion and turbulent mixing (Taylor and Spencer, 1990). Volatile pesticides or pesticides with a high vapor pressure are easier to transport to the atmosphere than the non-volatile or poorly volatile pesticides. Moreover, pesticide volatilization depends on the amount of pesticides adsorbed onto the soil particles, the temperature, moisture content, and organic matter content of the soil. It was reported that an increase in the residue concentration and temperature in the soil resulted in an increase in the vapor density of DDT, lindane, dieldrin, and trifluralin under controlled laboratory conditions (Spencer et al., 1969; Spencer and Cliath, 1970, 1972, 1974). In addition, retardation of pesticide volatilization in dry soils was also reported. When the soil moisture content drops below a certain level, the adsorption sites on which pesticides are strongly bound become active so that lesser amounts of pesticides transports to the atmosphere (Spencer, 1970). Organic matter in the soil gives detrimental impact on pesticide volatilization. Spencer (1970) observed a decrease in the vapor density of dieldrin over both wet and dry soils with an increase in the soil organic matter content.

10.3.3.3. Abiotic Transformation

Abiotic transformation of pesticides in sediments is probably dominated by hydrolysis and redox reactions since pesticides are exposed to a permanently saturated environment with a proximity close to solid surfaces (Nowell et al., 1999). Hydrolysis reactions involve the cleavage of the pesticide molecule into smaller, more water-soluble portions and in the formation of new C-OH or C-H bonds. They can generally be categorized into acid-catalyzed, alkaline, and neutral (or pH independent) hydrolysis. Functional groups of pesticides which are susceptible to the hydrolysis include carboxylic acid esters, organophosphate esters, amides, anilides, carbamates, organohalides, triazines, oximes, and nitriles. Hydrolysis of the carboxylic acid esters results in carboxylic acid and an alcohol. Alkaline and neutral hydrolysis transfer organophosphate esters into phosphoric acid and alcohol moiety via P-O and C-O bond cleavage, respectively. Cleavage of the carbonyl C-N bond of the amide and anilides through alkaline hydrolysis results in carboxylic acid and amines. Alkaline hydrolysis of carbamates involves a breakdown of both carbonyl C-O and carbonyl C-N bonds, resulting in an alcohol, CO_2, and an amine. Hydrolysis of organohalides leads to the loss of their halogen substituents. In general, the tendency of pesticide hydrolysis is affected in two major ways by the sediments. First, the environment of interstitial water, such as reactants' activities, including water molecules, protons, and hydroxide ions for pesticide hydrolysis, is influenced by the proximity to solid surfaces and the porous nature of the sediments. Second, solid particles directly affect the susceptibility of adsorbed pesticides to hydrolysis. This is because most pesticide hydrolysis reactions in the sediments involve acid-base, metal, or surface-mediated heterogeneous catalysis (Wolfe et al., 1990). Common hydrolysis rate constants and estimated half-lives of selected pesticides can be found in many references (e.g., Watts, 1998; Nowell et al., 1999).

Redox reactions occurring in the sediments may be affected by heterogeneous catalysis and restricted by the oxygen diffusion through the porous medium. (Macalady et al., 1986). Abiotic reduction reactions are the major redox reactions of pesticides occurring in the sediments. Abiotic reduction reactions involved include dehalogenation to alkane, nitroreduction to corresponding amine, azo reduction to a hydrazo or an amino group, and sulfone reduction to the sulfoxide or sulfide. For example, redox transformation of methyl parathion and aldicarb in the sediments has been reported. Inorganic species such as iron and manganese ions are usually the reducing agents responsible for the abiotic pesticide reduction, but bioorganics such as flavins, porphyrins, and extracellular enzymes are also involved (Wolfe et al., 1990).

Abiotic transformation of pesticides in soils and sediments can occur in liquid phase and at solid-liquid interface. In the soil solution, a large variety of chemical transformation of the pesticide can occur; hydrolysis is, perhaps, the most common reaction underwent. The products of hydrolysis are often more polar than the parent pesticides so that they are more soluble in water. In addition, non-biological oxidation is another pesticide transformation reaction in the soil. Pesticides containing oxidizable metal cations can be oxidized abiotically in the soil. Moreover, chemical oxidation of a herbicide amitrole by free radicals such as hydroxyl radicals has been reported. The species in the soil solution can produce free radicals through chemical or photochemical processes, which really implicates in various pesticide transformations in the soil. Under anaerobic conditions, DDT can be transformed to DDD by a chlorine free radical. Trifluraline degradation in soils involves nonbiological reductive processes in which the nitro groups are transformed onto amino groups. The insecticides mirex and toxaphene are reduced by ferrous iron through partial dechlorination. In addition to hydrolysis and redox reactions, chlorotriazine herbicides readily undergo a displacement of the chlorine atom by various nucleophiles. Organophosphate esters in solution can undergo transformations catalyzed by many heavy metals. The transition complexes formed between the pesticide and the heavy metal ion in these transformation reactions are stable complexes, which can affect the mobility, stability, and bioactivity of pesticides (Wolfe et al., 1990).

In general, abiotic transformation of pesticides in the proximity of a charged solid surface is faster than that in the bulk solution. This is because adsorption sites on the solid surface can catalyze many pesticide transformations. The spatial distribution of ions and the charged distribution within polarizable species are greatly influenced by the electric field created from the charged surface. Surface catalysis has been reported during the transformation of trifluralin, diazinon, ciodrin, parathion, and azido-triazine in soil (Wolfe et al., 1990). On clay surfaces, Mortland and Raman (1967) reported that Cu(II) catalyzed the hydrolysis of organophosphate esters in solution and could retain its catalytic capacity as an exchangeable cation on montmorillonite. Transformation of s-triazines in soils (e.g., at momtmorillonitic clay surfaces) has been extensively studied (Brown and White, 1969; Gunther and Gunther, 1970; Erickson and Lee, 1989). Russell et al. (1968) reported the catalytic hydroxylation of atrazine on H-montmorillonite in which the chlorine atom is substituted by a hydroxyl ion. Soil organic matters is capable of promoting abiotic transformation of pesticides because they contain many reactive groups being able to enhance chemical changes of certain families of organic substances. Moreover, humic substances possess a strong reducing capacity, and there are stable-free radicals in fulvic and humic acid fractions. Soil organic matter-enhanced transformation includes hydrolysis of organophosphorous esters, dehydrodechlorination of DDT and lindane, and conversion of aldrin to dieldrin. Hydrolysis of chloro-s-triazines in soils through the interaction of the pesticide with the carboxyl group of the soil organic matter has also been reported (Armstrong and Konrad, 1974). Other than clay and soil organic matters, abiotic and surface-catalyzed degradation of DDT on MgO (Nash et al., 1973) and aluminum oxide-catalyzed hydrolysis of parathion (Mingelgrin and Saltzman, 1979) have also been reported.

Biotic Transformation

Biotic transformation of pesticides involves different pathways, which occur at different rates under aerobic or anaerobic conditions, or at different pHs. Generally, pesticides are degraded more rapidly in the plant rhizosphere than outside of it. Occurrence and the rate of the biotic transformation are highly affected by the population of microorganisms in the soil, the abundance of nutrients to the biodegraders, and other conditions that would affect the activity of microorganisms (e.g., pH and oxygen level). Soil moisture content is also an important factor since only dissolved pesticides in the soil solution are subject to degradation (Gan and Koskinen, 1998).

Biotic transformation of pesticides in the soil can be classified into five processes, that is, biodegradation (mineralization), cometabolic transformation, conjugation or polymerization, microbial accumulation, and nonenzymatic transformation (Bollag and Liu, 1990). Complete biodegradation or mineralization of pesticides in the soil is the most interesting and environmentally valuable pesticide transformation process since there is no generation of potentially hazardous pesticidal intermediates. By degrading pesticides into CO_2 and other inorganic components, microbes can obtain the required substrate(s) and energy for growth. In cometabolism, microorganisms can transform pesticides without deriving any nutrient or energy for growth from the transformation process in which the enzyme involved in catalyzing the initial reaction are often not substrate specific. Complete degradation of the pesticide generally cannot be achieved by a single cometabolism. Thus, cometabolic transformation may lead to an accumulation of more or less toxic intermediate products. Polymerization or conjugation is a microbially-mediated transformation process. Polymerization involves a combination of pesticides or their intermediates themselves or with naturally occurring compounds to form a larger molecule. It has been proven that polymerization plays an important role in incorporation of xenobiotics into soil organic matter (Bollag and Loll, 1983). In conjugation reactions, pesticides and their intermediates are linked together with endogenous substrates to form methylated, acetylated, alkylated compounds, and amino-acid conjugates. While it is out of the scope of this chapter, detailed information on the major biotransformation pathways of a pesticide that may occur in fish and other aquatic organisms and biomagnification of pesticides via food chains of aquatic biota can be found in Nowell et al. (1999). The pesticide in the soil or sediments can also be taken by microbes through a passive physical absorption process. The adsorption rate of pesticides varies with different types of organisms, types and concentration of pesticides. Johnson and Kennedy (1973) reported that DDT and methoxychlor can be accumulated by *Aerobacter aerogenes*. Absorption of lindane and dieldrin in yeast has also been reported (Voerman and Tammes, 1969). Fungi were able to accumulate dieldrin, DDT and pentachloronitrobenzene from soil (Ko and Lockwood, 1968). In many cases, metabolic activity of microorganisms leads to alternation of soil environmental parameters such as pH and redox potential, which are conducive to nonenzymatic transformation of pesticides. For instance, microbial degradation of protein under anaerobic conditions creates an alkaline condition, while metabolism of carbohydrate, oxidation of organic-nitrogen to nitrile or nitrate, sulfide to elemental sulfur, and ferrous to ferric iron result in an acidic environment in the soil. Anaerobic or reduced environments created by microbial activities, particularly in the flooded soil, allow the occurrence of many nonenzymatically reductive degradations of pesticides including DDT, methoxychlor, toxaphene, and heptachlor.

Major pesticide metabolisms occurring in soil environments include oxidation, reduction, hydrolysis, and synthetic reactions (Table 10.4). Oxidation is one of the most important and basic microbial-mediated reactions, which include hydroxylation, N- dealkylation, β-oxidation, decarboxylation, ether cleavage, epoxidation, oxidative coupling, and sulfoxidation. Major microbially reduction reactions of pesticides in soils include reduction of the nitro group, reduction of double or triple bonds, sulfoxide reduction, and reductive dehalogenation. Pesticides possessing ether, ester, or amide linkage are susceptible to be hydrolyzed. A microbially-mediated synthetic reaction involves binding of pesticides or their intermediates among themselves or with other compounds to form larger products. They can be divided into conjugation reactions involving the unions of two substrates and condensation reactions yielding polymeric products (Bollag and Liu, 1990).

10.3.4 Air and Atmosphere

Pesticides enter the atmosphere through a variety of processes both during and after application. Some application methods disperse pesticides directly into the air, by aerial spraying, ground rig broadcast sprays, and orchard mist-blowers. Movement of the sprayed droplets of pesticides from target to non-target sites by air is called drift. High pressure and fine nozzles generate small pesticide droplets, which are more readily to drift. Furthermore, pesticides released close to the ground are not as likely to drift in wind current as those released from aircrafts. Drifting is also less serious in calm weather than windy weather conditions. Once pesticides reach target surfaces, pesticide residue can volatilize by evaporation or be transported into the atmosphere by adhering onto small particulates (e.g., soil particles) or large objects (e.g., leaves) which are caught up by the wind. Movement of the gas vapors of pesticides is called vapor drift, which is invisible. Generally, volatilization from soil and surface waters is a primary pathway for many pesticides to dissipate — as much as 80-90 percent of certain compounds can be lost within a few days of application through this process. Once in the atmosphere, the pesticide droplets, pesticide vapors, and pesticide-adhered objects can be transported by wind currents, undergo photochemical and hydrolytic degradation, or be deposited back to the earth's surfaces (Larson et al., 1997). Wet deposition involves cleaning of the airborne pesticide vapors and pesticide-adhered particles by precipitation, which then deposits them to aquatic and terrestrial surfaces. Dry deposition in the form of gaseous vapor and particulate matters also deposits airborne pesticides to the earth's surface. It has been reported that annual deposition of airborne pesticides by precipitation accounts for less than 1% of the total amount of pesticides applied (USGS, 2007). The movement of pesticides in air not only poses a serious threat to non-target sites such as wetland and other sensitive ecosystems, but also represents a significant economic loss to farmers.

Table 10.4 Summary of microbial-mediated metabolic reactions of pesticides occurring in soil environments (Bollag and Liu, 1990).

Oxidation reactions	
Hydroxylation	$RCH \rightarrow RCOH$
	$ArH \rightarrow ArOH$
N-Dealkylation	$RNCH_2CH_3 \rightarrow RNH + CH_3CHO$
	$ArNRR' \rightarrow ArNH_2$
β-Oxidation	$ArO(CH_2)_nCH_2CH_2COOH \rightarrow ArO(CH_2)_nCOOH$
Decarboxylation	$RCOOH \rightarrow RH + CO_2$
	$ArCOOH \rightarrow ArH + CO_2$
	$Ar_2CH_2COOH \rightarrow Ar_2CH_2 + CO_2$
Ether cleavage	$ROCH_2R' \rightarrow ROH + R'CHO$
	$ArOCH_2R \rightarrow ArOH + RCHO$
Epoxidation	$RCH=CHR' \rightarrow RC\overset{O}{\overset{\diagup\diagdown}{H\text{-}C}}HR'$
Oxidative coupling	$ArOH \rightarrow (Ar)_2(OH)_2$
Sulfoxidation	$RSR' \rightarrow RS(O)R'$
	$RSR' \rightarrow RS(O_2)R'$
Reduction reactions	
Reduction of nitro group	$RNO_2 \rightarrow ROH$
	$RNO_2 \rightarrow RNH_2$
Reduction of double or triple bond	$RC\equiv OH \rightarrow RCH=CH_2$
	$Ar_2C=CH_2 \rightarrow Ar_2CHCH_3$
Sulfoxide reduction	$RS(O)R' \rightarrow RSR'$
Reductive dehalogenation	$Ar_2CHCCl_3 \rightarrow Ar_2CHCHCl_2$
Hydrolysis reactions	
Ether hydrolysis	$ROR' + H_2O \rightarrow ROH + R'OH$
Ester hydrolysis	$RC(O)OR' + H_2O \rightarrow RC(O)OH + R'OH$
Phosphor-ester hydrolysis	$(RO)_2P(O)OR' + H_2O \rightarrow (RO)_2P(O)OH + R'OH$
Amide hydrolysis	$RC(O)NR'R'' + H_2O \rightarrow RC(O)OH + HNR'R''$
Hydrolytic dehalogenation	$RCl + H_2O \rightarrow ROH + HCl$

Note R, R' or R'' refers to organic moiety. Ar means aromatic moiety.

Today, pesticides have been detected in the atmosphere throughout the nation, and a wide variety of pesticides are present in air, rain, snow, and fog. For example, atrazine, 2.4-D, and alachlor levels in rain was reported to exceed the maximum contaminant levels set by the U.S. EPA for drinking water (USGS, 1995). There is significant evidence that pesticides used in one part of the country are transported through the atmosphere and deposited in other parts of the country and beyond, sometimes in places where pesticides are not even used. Even in the Arctic and Antarctic pesticides are found in the air, snow, people, and animals (Nowell et al., 1999). Majewski and Capel (1995) reported that 63 pesticides and pesticide transformation products have been identified in atmospheric matrices.

In general, those pesticides that are more volatile (with a high Henry's law constant), resistant to hydrolysis and photochemical degradation as well as those that are applied aerially have a greater chance of entering the atmosphere and transporting over a great distance. Many of the pesticides observed in bed sediment and tissues (e.g., the organochlorine insecticides) are also routinely observed in the atmosphere (Nowell et al., 1999). The scientists at Environment Canada and Agri-Food Canada reported that Arctic air samples contained the lindane mainly used in India and other countries (Environment Canada, 2002). It is also estimated that up to 30% of the lindane applied in the Canadian prairies enters the atmosphere through volatilization, which corresponds to an atmospheric loading of up to about 190 tons. In addition, atmospheric deposition of organic contaminants such as PCBs from the industrial and urban centers nearby is a source of pollution in the Laurentian Great Lakes (Eisenreich et al., 1992). Toxaphene, which was used on cotton in the Southern United States and has been banned in 1982, is still being transported into the Great Lakes by southerly winds from the Gulf of Mexico. In the United States, organochlorine insecticide such as DDT, αHCH, lindane, heptachlor, and dieldrin have been detected in the air and rain of every state because of their widespread uses during the 1960's and 1970's, and their persistence in environments. The organophosphorous insecticides detected most often in the air, rain, and fog were diazinon, methyl parathion, parathion, malathion, chlorpyrifos, and methidathion. Generally, a high concentration of pesticides in the air and rain occurs in spring and summer months, coinciding with the higher application times and warmer temperature (USGS, 2007).

10.4 Fate and Transformation in Engineered Systems

Pesticides in water, groundwater, and soil can be treated by many different methods, such as oxidation (e.g., with chlorine dioxide), photodegradation, UV-radiation, soil washing, solvent extraction, surfactant washing, critical fluid extraction, air stripping, solidification, *in situ* vitrification, biodegradation, etc. Some of these treatment technologies will be introduced in this section.

10.4.1 Water and Wastewater Treatment

The National Water Quality Assessment implemented by the U.S. Geological Survey (USGS) indicated that there were widespread occurrences of pesticides in potential drinking water sources. It was reported that at least one pesticide was detected in 95% of the 8,500 surface and groundwater samples (Ballard and MacKay,

2005). Miltner et al. (1989) also detected atrazine, alachlor, metolachlor, cyanazine, metribuzin, carbofuran, linuron, and simazine in the stormwater runoff as the influents of three water treatment plants. Since the Food Quality Protection Act (FQPA) of 1996 regulates the risk posed to human beings by the pesticides from all "anticipated dietary exposures", removal/treatment of pesticides in drinking water, which is a potential pathway for the dietary exposure in addition to food crops, is a problem being concerned. Several studies have demonstrated that pesticides in drinking water sources could be partially removed in conventional water treatment systems by coagulation and flocculation, softening, and disinfection/chemical oxidation processes (Ballard and MacKay, 2005). Hansch et al. (1995) reported that coagulation and flocculation processes resulted in 97% removal of DDT and 55% removal of dieldrin from the drinking water, but there was only 10% alachlor removal, most likely due to its low hydrophobicity. Zhang and Emary (1999) reported that neither lime softening nor alum coagulation (conventional/enhanced dosages ranging from 6-18 mg/L) demonstrated atrazine removal. Rebhun et al. (1998) proposed that the removal of hydrophobic pesticides in coagulation and flocculation processes was due to the removal of the pesticide-sorbed humic materials.). Potentially, hydrophobic pesticides with low molecular weight acidic functional groups such as carbonyl and carboxyl, or high molecular weight pesticides can be removed through coagulation and flocculation processes (USEPA, 2001b).

Water softening involves precipitation of calcium (Ca^{2+}) and magnesium (Mg^{2+}) ions as calcium carbonate ($CaCO_3$) and ($Mg(OH)_2$) at pH between 9.3 and 10.5 or ion exchange. Data collection from full-scale water treatment plants showed no removal of atrazine, cyanazine, metribuzin, alachlor, and metolachlor, but complete removal of carbofuran after the water softening process, probably ascribed to an alkaline hydrolysis of carbofuran at high pH (Miltner et al., 1989).

Disinfection or chemical oxidation processes in conventional water treatment systems aim to inactivate or destruct the pathogens in water. Simultaneously, it could remove or degrade pesticides through oxidation (USEPA, 2001). Chemical disinfectants commonly applied in water treatment systems are ozone (O_3), chlorine dioxide (ClO_2), and chlorine (Cl_2). Laboratory studies of chemical oxidation of alachlor conducted by Miltner et al. (1987) showed that ozone was able to remove 75-97% of alachlor in water. Pilot-plant studies conducted by Speth (1993) showed the reduction of glyphosate (herbicide) by chlorine and ozone. In addition, after the chlorination process at full-scale treatment plants, there were approximately 24% and up to 98% removal of carbofuran and metribuzin, respectively (Miltner et al., 1989). Thiobencarb, which is a herbicide, in tap water could be degraded to thiobencarb sulfoxide by free chlorine, as illustrated in Eq. (10.2). In a high free chlorine medium (more than 10 mg/L), thiobencarb sulfoxide was further degraded to p-chlorobenzyl alcohol, p-chlorobenzaldehyde, and p-chlorobenzyl chloride (Kodama et al., 1997).

Eq. (10.2)

Zhang and Pehkonen (1999) reported the oxidation of diazinon to diazoxon by aqueous chlorine in which the oxidation rate increased with decreasing pH [Eq. (10.3)]. Degradation of aldicarb, aldicarb sulfoxide, and aldicarb sulfone by hypochlorite (ClO⁻) had also been demonstrated by Miles (1991).

Diazinon   Diazoxon   Eq. (10.3)

Carbamate insecticides including aldicarb, methhomyl, carbaryl, and propoxur were observed being degraded swiftly by ozone, whereas none of them reacted with ClO_2 (Mason et al., 1990). In addition, sulfur-containing s-triazines including prometryne, terutryne, ametryne, and desmetryne could be degraded quickly by hypochlorous acid (HClO), but their reactions with ClO_2 were extremely slow. After the reactions with HClO, sulfoxide, sulfone, and sulfone's hydrolysis products were the byproducts of the sulfur-containing s-triazines formed, as shown in Eq. (10.4) (Mascolo et al., 1994).

s-triazines   Sulfoxide   Sulfone   Sulfone's
                   product   product   hydrolysis product

R_1 and R_2 are either CH_3, $C(CH_3)_3$, CH_2CH_3, or $CH(CH_3)_2$.

Eq. (10.4)

Chlorpyrifos, which is a kind of organophosphorous pesticide, was found being rapidly oxidized to chloropyrifos oxon in the presence of free chlorine in which HClO was the primary oxidant involved. This oxidation reaction was more rapid at acidic pH (below 7.5). At elevated pH, alkaline hydrolysis of chlorpyrifos and chloropyrifos oxon was susceptible to form 3,5,6-trichloro-2-pyridinol, which is a stable end product (Duirk and Collette, 2006).

Adsorption water treatment processes using activated carbon, which is mainly used for control of taste, odor and disinfection by-products, is also able to remove pesticides in drinking water (USEPA, 2001b). Miltner et al. (1987, 1989) reported 28-87% removal of atrazine and 33-94% removal of alachlor by the powdered activated carbon used in full-scale water treatment plants, and the removal efficiencies was found increasing with increasing the dose of the powered activated carbon (PAC) applied. Zhang and Emary (1999) reported that in their jar tests, adding 16 mg/L PAC with an alum coagulation dosage of 6 mg/L at a pH of 5.8 resulted in an 58% removal of atrazine; increase in either mixing energies or contact time resulted in higher atrazine adsorption on the PAC. Likewise, granular activated carbon is also effective in adsorbing pesticides from water. The granular activated carbon, which had been in operation for 30 months, was found to be capable of removing 47-62% of triazines (e.g., atrazine, cyanazine, metribuzin, and simazine) and 56-99% of acetanilides (e.g., alachlor, metolachlor, and pendimethalin) (USEPA, 2001b).

Membrane technology has been used in reverse osmosis, microfiltration, ultrafiltration, and nanofiltration for desalination, removal of specific ions, color, organics, nutrients, and suspended solids, and has been demonstrated to remove pesticides in water. The membranes which are commonly used include cellulose acetate, polyamide membrane, and thin film composites. In a short-term laboratory test, Chian (1975) found that the thin film composite membrane called cross-linked polyethlenimine and cellulose acetate membrane used in reverse osmosis could remove about 97.8% and 84% of atrazine, respectively. In comparison to cellulose acetate and polyamide membranes, thin film composite membranes generally possess superior performance in removing pesticides. For instance, cellulose acetate and polyamide membrane used in reverse osmosis could only remove 23-59% and 68-85% of triazines from water, whereas thin film composite membrane could reach 80-100% of removal (USEPA, 2001b). Moreover, the thin film composite membrane used in ultrafiltration showed a nearly complete removal of chlordane, heptachlor, methoxychlor, and alachlor (Fronk et al., 1990). A combination of microfiltration or ultrafiltration with adsorbents such as activated carbon is also proven to be effective in removing pesticides. Anselme et al. (1991) demonstrated that an integrated ultrafiltration and powdered activated carbon system could remove 70% of cyanazine and 61% of atrazine. In addition, nanofiltration could achieve 63-93% removal of atrazine and 80-98% removal of simazine (USEPA, 2001b).

Aeration and air stripping processes in water treatment systems, which aims to inject disinfectants in finished water, inject O_2, and to remove ammonia and volatile organic compounds, can remove volatile pesticides. Pesticides with Henry's constant larger than 1×10^{-3} atm m^3 mole^{-1} are susceptible to removal by aeration (McCarty, 1987). Generally, these processes are applied in packed towers, spray towers, or agitated diffused gas vessels for the pesticide removal.

In the past, pesticide wastewater from the manufactures was dumped into sufficient dilution (e.g., the ocean) (Nemerow and Agardy, 1998). Since the establishment of the Clean Water Act (CWA) in 1977, the wastewater discharges generated from all industries including pesticide manufacturing and formulating industries were regulated. Under Section 304 and 306(b) of the CWA, the USEPA was required to establish "effluent guidelines" and "pretreatment standards", respectively, which are applicable to the introduction of wastewater from industry and other nondomestic sources into publicly owned treatment works (POTWs). For the "effluent guidelines," there are 2 types of standards requiring the application of the best practicable control technology (BPT) currently available and effluent limitations requiring application of the best available technology (BAT). Under the BPT regulations, the pesticide industries were divided into 3 subcategories: (1) organic pesticide chemicals manufacturing; (2) metallo-organic pesticide chemical manufacturing; and (3) pesticide chemicals formulating and packaging. For the first subcategory, BPT effluent limitations regulate the amount of chemical oxygen demand (COD), 5-day biochemical oxygen demand (BOD$_5$), total suspended solids (TSS), and organic pesticide chemicals which plants are allowed to discharge during any one day or any thirty consecutive days (see Table 10.5). For the second and third subcategories, the BPT effluent limitations are "no discharges of process wastewater pollutants into navigable waters". In addition, the "pretreatment standards" for existing and new sources for the organic pesticide chemical manufacturing subcategory show the maximum daily and maximum monthly discharge limitations of 24 priority pollutants into POTWs. There are no "pretreatment standards" for the metallo-organic pesticide chemical manufacturing subcategory, whereas the "pretreatment standards" for the pesticide chemicals formulating and packaging subcategory are "no discharges of process wastewater pollutants to POTWs". Details and updated standards or guidelines for the discharges from pesticide industries can be found in "40 CFR Part 455 – Pesticide Chemicals".

Technologies currently available to dispose the pesticide-laden wastewater are generally based on the principles of containment, detoxification/degradation, and volume reduction (Dillon, 1981). These technologies, as summarized in Table 10.6, include lined evaporation beds, chemical oxidation, hydrolysis, and precipitation, reverse osmosis and adsorption as well as trickling filters and activated sludge systems. However, the inert materials such as surfactants, emulsifiers, and petroleum hydrocarbons, which are usually mixed with pesticide active ingredients to formulate pesticide products to achieve specific application characteristics, reduce the performance efficiency of the disposal technologies such as chemical oxidation and activated carbon adsorption because of the formation of emulsion when mixing with water. Hence, emulsion breaking is an important pretreatment step required for the pesticide wastewater treatment in order to facilitate the pesticide removal. Temperature control and acid addition are two common pretreatment methods applied in pesticide manufacturing and formulating industries. Acids such as sulfuric acid dissolve the solid materials holding the emulsions together. After the acid addition, demulsified oil floats and can be skimmed off from water surface, thereby leaving the wastewater ready for subsequent treatment. Increasing the wastewater temperature decreases the viscosity of oil and water, which consequently increases their apparent specific gravity differential. Since oil has significantly lower apparent specific gravity, it readily rises to the water surface. Furthermore, heating the wastewater increases the kinetic energy of the individual molecules in the wastewater, which causes them to collide with each other more frequently and aids in breaking the film in the oil/water interface. Once freed from the water, the oil floats and can be skimmed off from the wastewater. Other pretreatment methods for emulsion breaking include membrane filtration (ultrafiltration), chemically assisted clarification, and settling (USEPA, 1998).

Table 10.5. [a]BPT effluent limitations for the organic pesticide chemicals manufacturing subcategory (Wong, 2004)

Effluent characteristics	Maximum for any 1 day	Average of daily values for 30 consecutive days shall not exceed
COD	13.000	9.0000
BOD5	7.400	1.6000
TSS	6.100	1.8000
Organic pesticide chemicals	0.010	0.0018
pH	6.0-9.0	6.0-9.0

[a]BPT refers to best practical control technology currently available, which is defined as hydrolysis or adsorption technology followed by biological treatment technology.
Note: For COD, BOD5, and TSS, metric units: kilogram/1,000 kg of total organic active ingredients. English units: Pound/1,000 lb of total organic active ingredients. For organic pesticide chemicals — metric units: kilogram/1,000 kg of organic pesticide chemicals. English units: Pound/1,000 lb of organic pesticide chemicals.

Table 10.6. Technologies available for disposal of pesticide-laden wastewater (Felsot et al., 2003)

Physical treatment	Chemical treatment	Biological treatment
- Activated carbon adsorption	- Acid or alkaline hydrolysis	- Activated sludge systems
- Biomass adsorption	- Chemical reduction	- Biobeds
- Cyclodextrin encapsulation	- Fenton reaction	- Composting
- Lined evaporation beds	- Microwave plasma destruction	- Landfarming
- Peat adsorption	- Solar photodecomposition	- Phytoremediation
- Reverse osmosis	- Supercritical water oxidation	- Trickling filters
- Solvent extraction	- Titanium dioxide	
- Synthetic resin adsorption	- Ultraviolet ozonation	
	- Wet-air oxidation	

Lined evaporation beds/basins generally consist of an excavated pit with a plastic liner overlain by soils. Pesticide-laden wastewater is usually pumped into the pits from the bottom. The pits are open to the atmosphere to allow evaporation of water, thereby reducing the volume of the applied pesticide-laden solution. The pesticide left in the pits may simultaneously be degraded by microbial degradation, chemical hydrolysis, or photolysis processes (Dillon, 1981). Researchers at the University of Alabama at Huntsville have designed a stainless steel/glass evaporation basin (2.4x2.1x1.5m); the evaporation basin could evaporate 3400-4500 litres of water per year (Ash et al., 1992). Chemical oxidation of the pesticides in wastewater usually involves direct oxidation with hydrogen peroxide (H_2O_2) or O_3, or advanced oxidation processes in which oxidants are combined with ultraviolent (UV) light, iron salts or titanium dioxide to generate free hydroxyl radicals (Felsot et al., 2003). Kearney et al. (1984) reported that O_3 and UV could degrade paraquat, 2,4-D, and atrazine in wastewater inside a mobile pilot reactor. However, whether alone or in combination with UV or H_2O_2, ozonation could only partially transform pesticides (Hapeman-Somich et al., 1992). Coupling of chemical treatment with biotreatment is needed to fully degrade the pesticides in wastewater. For instance, mineralization of atrazine (Leeson et al., 1993; Hapeman et al., 1995), alachlor (Somich et al., 1988), and bromacil (Acher et al., 1994) could be achieved by passing their ozonation products through soil columns, continuously stirred bioreactors or fixed-film bioreactors. Coumaphos cattle dip waste was not degraded efficiently by UV-ozonation (Kearney et al., 1986). However, treatment of the waste with a culture of *Flavobacterium* sp. prior to the UV-ozonation process could hydrolyze coumaphos to chlorferon, which could then be completely degraded in a UV-ozonation chamber. Further incubation of the resulting mixtures in the soil could lead to release of 70% of the coumaphos as carbon dioxide.

Advanced oxidation processes are also effective in degrading pesticide in wastewater. Eisenhauer (1964) observed significant oxidation of phenolic constituents in the industrial effluent containing phenol and substituted phenols in the presence of Fenton regents. An electrode Fenton treatment (EFT) system involves immersing of sacrificial iron electrodes in aqueous sodium chloride to generate ferrous iron. Subsequent addition of H_2O_2 leads to generation of free hydroxyl radicals. By flowing pesticide-laden wastewater through the EFT system at neutral pH, 50% disappearance times of alachlor, metolachlor, atrazine, cyanazine, and picloram were observed to be less than 30 min (Pratap and Lemley, 1994). In addition, Arnold et al. (1996) observed that introduction of *Rodococcus corallinus* and *Pseudomonas* sp. could further mineralize the reaction products resulting from dark-assisted Fenton reaction with atrazine. Photocatalytic oxidation of carbaryl with titanium dioxide was also observed in which complete mineralization of carbaryl to CO_2 was achieved within an hour and was not inhibited by surfactants (Prevot et al., 1999).

USEPA (1985) reported that activated sludge systems with detention times ranging from 7 to 79 hrs or trickling filter systems in pesticide plants could remove 87.4-98.8% of wastewater BOD and 60.5-89.7% of wastewater COD. Application of a trickling filter prior to an activated carbon adsorption system was successful for the treatment of 2,4-D in wastewater. Sequential batch reactor (SBR), which is a variation of activated sludge system using the sequential steps of fill, aerate, settle, and withdrawal, was reported to be able to achieve over 99% removal of 2,4-D at steady state operation in which the removal rate was influenced by the type of supplementary substrates added (Wong, 2006). Van Leeuwen et al. (1999) reported treatment of a wastewater containing different pesticides (with a chemical oxygen demand (COD) of ca. 1600 mg/l) using biological treatment (biofilters) only and biological granulated activated carbon followed by coagulation and separation with microfiltation; they found that both processes removed most of the pesticides (> 99% removal), but simazine could not be removed effectively.

10.4.2 Soil and Groundwater Remediation Systems

Pesticides in soil and groundwater are amenable to physical, chemical, and biological remediation. In this section, some of these remediation technologies will be introduced.

10.4.2.1. In Situ Soil Flushing and Hydrolysis

In soil flushing processes, solvents, surfactant solution or water with/without co-solvents are first injected into contaminated aquifers to enhance release and movement of contaminants, which are then extracted above ground for further treatment or recovery. Di Palma (2003) reported that *in situ* soil flushing followed by alkaline hydrolysis is an effective approach for the remediation of soil contaminated by phosalone, which is an organophosphorous insecticide widely used on almonds, grapes, and stone fruits. By flushing the contaminated soils with aqueous solution containing 10% of ethanol at a flow rate of 150 mL/hr, up to 99% of the phosalone onto the soils could be extracted after flushing with 7 pore volumes of the solution. Alkaline hydrolysis using 30% sodium hydroxide as an alkaline agent at pH between 10 and 12 could degrade approximately 70-95% of the phosalone in the extracted solution in which the degradation half-lives ranged from 0.57 to 2.96 hrs. The remaining ethanol in the extracted solution was quite stable under the extreme alkaline conditions and thereby did not affect the phosalone hydrolysis.

10.4.2.2. Thermal Desorption

Thermal desorption treatment of pesticide-contaminated soils applies the principles that the pesticide vapor pressure increases with increasing temperature. As the contaminated soils are heated, the pesticides are vaporized and driven from the contaminated soils into a purge gas stream for further destruction and recovery. The USEPA defined that the thermal desorption is a process of

> heating in an enclosed chamber under either oxidizing or non-oxidizing atmospheres at sufficient temperature and treatment time to vaporize hazardous contaminants from contaminated surfaces and surface pores and to remove the contaminants from the heating chamber in an exhaust gas (USEPA, 1991).

Thermal desorbers commonly used can either be directly-heated or indirectly-heated systems. Directly-heated systems utilize an auxiliary-fuel-fired burner to produce a hot gas, which will be in direct contact with the pesticide-contaminated soils. On the other hand, indirectly-heated systems transfer heat through the shell to heat the contaminated soils. The pesticides desorbed from the soils can either be destructed using thermal oxidizers or recovered using wet scrubbers, condensers, or granular activated carbon. Generally, the performance of the thermal desorption processes on the soil remediation depends on the pesticide boiling point, treatment temperature and time applied, soil type, soil pesticide concentration as well as soil moisture content. Pesticides with a low boiling point, high treatment temperature and long treatment time, soils with a low clay content and moisture content as well as a low pesticide concentration in soils usually allow effective soil remediation by thermal desorption processes. Full-scale remediation of pesticide-contaminated clay soil in an industrial site at Albany, Georgia using a thermal desorption system equipped with liquid-phase activated carbon system showed, on average, 95.58 to 99.78% removal of aldrin, HCH, chlordane, DDD, DDE, DDT, endosulfan, endrin, and toxaphene from the soils after the treatment (Troxler, 1998).

10.4.2.3. Incineration

Proper designed incinerators are effective treatment devices for most hazardous waste management problems. Because of the demonstrated effectiveness, safety, low environmental and health risk of the incineration processes, the USEPA has designated incineration as the preferred solution to most waste management problems. Incinerators are devices which apply combustion to treat wastes or contaminated environmental media. When pesticide-contaminated soils are treated in incinerators, pesticides are destructed, and the clean and inert soil particles remain. Incineration treatment of pesticide-contaminated soils is theoretically based on two main mechanisms: desorption and combustion. Generally, an incinerator consists of two chambers in which the primary chamber receiving soil feed is to promote the desorption mechanism and the secondary chamber aims to destruct pesticides liberated in the primary chamber. In most soil remediation application, the primary chamber is a rotating vessel or rotary kiln in which contaminated soils are exposed to a high temperature (> 540 $^{\circ}$C) flowing gas. In the secondary chamber, 99.99 to 99.9999% of the destruction and removal efficiency of the pesticides flowing with the gas stream from the primary chamber is usually achieved at the general rule-of-thumb time (2 seconds) and a temperature parameter of 1000 $^{\circ}$C, respectively. During the incineration, carbon dioxide and water are the simple combustion products of the carbon and hydrogen atoms in pesticides. However, hydrochloric acid and chlorine gas, sulfur dioxide and trioxide, nitrogen oxide and dioxide as well as phosphorous pentoxide are the general combustion products of the chlorine, sulfur, nitrogen, and phosphorous atoms in pesticides, respectively. An incinerator mainly consisting of a rotary kiln primary chamber and secondary combustion chamber in Old Midland Products site at Arkansas had been reported to be able to achieve 99.9999 to 99.999951% and 99.99 to 99.9998% destruction and removal efficiency of trichlorobenzene and naphthalene, respectively (Steverson, 1998).

10.4.2.4. Direct Oxidation and Radical Oxidation Processes

Oxidation processes show a great potential for the treatment of pesticide-contaminated groundwater and soils. This is because oxidative transformation of organic chemicals is primarily limited to those chemicals containing hereroatoms with lone pair electrons such as oxygen, nitrogen, and sulfur, and many pesticides possess these atoms. Common oxidation systems include ozone (O_3) and hydrogen peroxide (H_2O_2) alone or in combination with O_3, ultraviolet (UV) light or iron salts (Fenton's reactions) (Lai et al., 2007). All these systems rely on oxidative degradation involving the organic radicals generated from direct reactions with oxidants or the free radicals, most likely hydroxyl radicals (OH•), generated from a combination of the oxidants. H_2O_2 is a strong oxidant which can oxidize organic contaminants to a less toxic and more biodegradable form. However, H_2O_2 alone is not effective to treat most pesticide-contaminated groundwater and soils at reasonable concentration. Thus, O_3, iron salts or UV is usually applied simultaneously to induce the OH• generation for enhancing the degradation rate. O_3 can oxidize pesticides via direct attack by electrophilic or dipolar cyclic addition and indirect attack by the free radicals generated from the reaction with water. Sites for the direct attack include multiple bonds (C=C, C=C-O-R) or atoms containing a negative charge (N, P, O, S). In general, the presence of electron-withdrawing groups such as halogen, nitro, and carboxyl significantly reduces the rate of direct pesticide ozonolysis. Unlike the direct attack, the indirect attack by the OH•, however, is non-selective. O_3 reacts rapidly with organophosphorous to form more toxic intermediates (oxon) together with the products of cleavage of P-S or P-O bonds. Phenoxyalkyl acids and esters, carbamates, acetanilides, and uracils are also reactive towards ozonation, but complete mineralization is usually not achieved (Hapeman and Torrents, 1998).

10.4.2.5. Photochemical Processes

Photochemical processes rely on oxidative degradation reactions to mineralize the pesticides in groundwater using the free radicals generated by pesticide photolysis or by reactions with photogenerated OH•. Under practical conditions, direct UV photolysis of organic contaminants is quite unusual since it is limited to a number of light-excited organics. Other organic molecules require mediary stages involving participation of organic dyes to undergo decomposition in photolysis. Photogeneration of OH• involves photolysis of H_2O_2 in the UV-H_2O_2 systems or O_3 in the UV-O_3 systems. However, the practical uses of UV-H_2O_2 or UV-O_3 systems may be limited by low light absorption coefficient of H_2O_2 and low water solubility of O_3. Photocatalysis is another photochemical process in which oxidizing species are generated *in situ* from dissolved oxygen or from water on the photocatalytic particles such as titanium oxide (TiO$_2$) after absorbing light. TiO$_2$-photocatalytic oxidation is able to completely mineralize almost all organic compounds dissolved in water. Moreover, the reaction rate is fast and only natural sunlight is needed as an energy source. Field study of the photocatalytic remediation of herbicides in groundwater samples in Israel showed that complete metobromurone and bromoxynil mineralization were achieved after treating the groundwater with TiO$_2$ and H_2O_2 for

3.5 hrs. After treating for 15 hrs, alachlor and chloroxynil were completely mineralized while atrazine, trietazine, and bromacil concentration dropped from 43 to 4 ppb, 23 to 5 ppb, and 115 to 6 ppb, respectively (Muszkat, 1998).

10.4.2.6. Zero-Valent Iron

Zero-valent iron (Fe^0) has been widely used in permeable reactive barrier (PRB) technology for *in situ* remediation of chlorinated organics (Andrea et al., 2005; Lai et al., 2006), heavy metals (Simon et al., 2002; Lo et al., 2005; Lo et al., 2006; Lai et al., 2008), and oxyanions (Alowitz and Scherer, 2002; Westerhoff, 2003; Huang et al., 2003) in groundwater. The degradation of groundwater contaminants by Fe^0 involves a series of reduction and oxidation reactions in which electrons released from Fe^0 oxidation or ferrous iron generated from the Fe^0 oxidation transfer to the contaminants to reduce them into less toxic, non-toxic, or more biodegradable forms. Since the released electrons can only transfer to the contaminants adsorbed onto the Fe^0 surface, the abiotic contaminant reduction by Fe^0 is heterogeneous reactions, and the reduction rate is generally limited by the rate of electron transfer from Fe^0 to the adsorbed contaminants and the rate of mass transfer of contaminants between the bulk solution and the Fe^0 surface via a stagnant fluid layer surrounding the Fe^0 (Lai and Lo, 2007). Agrawal and Tratnyek (1996) reported that nitrobenzene, which is commonly used as an insecticide and herbicide, could be reduced abiotically by Fe^0 under anaerobic conditions to aniline with nitrosobenzene as an intermediate product. The nitrobenzene reduction rate was observed increasing linearly with an increase in the iron surface area, but decreasing with an increase in the bicarbonate concentration.   Huang and Zhang (2006) reported that a limited amount of nitrobenzene could be reduced to aniline by Fe^0 because the formation of a lepidocrocite (γFeOOH) coating could significantly slow down the reaction. However, augmenting Fe^0 with substoichiometric $FeCl_2$ could dramatically accelerate the reaction. Surface-adsorbed Fe(II), not pH nor Cl^-, was found to be responsible for rejuvenating the system.   In the presence of aq. Fe(II), a stratified corrosion coating could develop, with magnetite (Fe_3O_4) as the inner layer and lepidocrocite as the outer layer. Fe^{2+} was not the main reductant for the reactions, but might accelerate the autoreduction of lepidocrocite to magnetite by the underlying Fe^0 (Zhang and Huang, 2006) In addition, abiotic reduction of PCBs by Fe^0 in subcritical water at 250°C and 10 MPa, and by the Fe^0 impregnated with palladium were also reported by Yak et al. (2000) and Korte et al. (2002), respectively. Palladium-impregnated Fe^0 could completely dechlorinate *o*-, *p*-, and *m*-chlorophenol at an initial concentration of 20 ppm within 5 min (Liu et al., 2001). In addition, Fe^0 has also been applied to remediate the soils highly contaminated with pesticides. Shea et al. (2004) and Comfort et al. (2001) treated contaminated soil containing metolachlor (>1400 mg kg^{-1}), atrazine (>250 mg kg^{-1}), alachlor (>90 mg kg^{-1}), pendimethalin (>90 mg kg^{-1}), and chlorpyrifos (>25 mg kg^{-1}) with various Fe^0 treatments in a pilot-scale compost project (in which soil windrows were amended with Fe^0, adjusted to >35% water content, and covered with clear plastic). Within 90 d, pesticide concentrations decreased up to 99% (metolachlor), 70% (atrazine), 99% (alachlor), 90% (pendimethalin), and 96% (chlorpyrifos).   In a second field demonstration, on-site treatment of contaminated soil (ca. 653 mg atrazine, 177 mg metolachlor and 4220 mg nitrate-N per kg) with ZVI and $Al_2(SO_4)_3$ under anaerobic conditions reduced metolachlor, atrazine and nitrate-N concentrations by > 99%, 80% and 40% (unpublished data). These results indicate the potential for using Fe^0 to remediate pesticide-contaminated sediments and soils is very high.

10.4.2.7. Bioremediation

Bioremediation and biodegradation are identical processes in which both of them also rely on microorganisms to transform or metabolize pesticides. The difference is that biodegradation is natural processes; whereas bioremediation involves human intervention including stimulation and augmentation to enhance the biodegradation processes. Biostimulation involves an addition of limiting nutrients such as carbon, nitrogen, phosphorous, and oxygen, acid or base to stimulate the overall activity/growth of indigenous microorganisms or addition of analogous substrates to stimulate their specific enzymes. This strategy assumes that the requisite microorganisms are present on-site and successful bioremediation only requires enhancement of indigenous rates of biodegradation (Shelton and Karns, 1998). Some contaminants at sufficiently high concentrations can inhibit microbial activity. Sometimes, the presence of easily degradable substrates can inhibit the degradation of target pollutants due to preferential substrate utilization (Alvarez and Illman, 2006). Some compounds can be used as N or P sources (e.g., methylphosphonate was used as a P source); the absence of these nutrients may exert selective pressure for the biodegradation of relatively recalcitrant compounds (Alexander, 1999).

It has been reported that biodegradation of pesticides in the sediment of a pit which had been used to store rinsate from pesticide spray equipment could be enhanced by organic and mineral amendment of the sediment (Winterlin et al., 1989). Toxaphene-contaminated soils could be partially decontaminated under anaerobic conditions after addition of cornmeal or cotton gin wastes to the soils (Mirsatari et al., 1987). DDT and lindane were reported being biodegraded much faster in the flooded soils amended with organic matters than the unamended soils (Castro and Yoshida, 1974; Mitra and Raghu, 1986). Besides, the addition of anaerobic sludge was observed to be able to enhance biodegradation of PCP and β-HCH in soils (Mikesell and Boyd, 1988; Van Eekert et al., 1998). Alachlor in the contaminated soils at a level of about 100 mg/kg was biodegraded significantly faster after cornmeal or soybean meal amendment in comparison to unamended soils (Felsot and Dzantor, 1990). Cole et al. (1995) reported that biodegradation of the aged residues of trifluralin, metolachlor, and pendimethalin in the soils from a contaminated field site could be enhanced through compost addition to the soils. Glucose and wheat straw were also observed to be capable of enhancing biodegradation of herbicide metribuzin under aerobic conditions (Pettygrove and Naylor, 1985).

Bioaugmentation involves inoculation of sites lacking appropriate strains with non-indigenous pesticide-degrading microorganisms or microbial consortia in addition to amendment with nutrients and/or substrates (Shelton and Karns, 1998).

Extensive metabolism of DDT by an anaerobic microbial culture followed by an aerobic culture isolated under *in vitro* conditions (i.e., enrichment cultures are used to isolate specific pesticide-degrading microbes or consortia of two or more microbes) has been reported by Pfaender and Alexander (1972). In addition, successful reduction of phytotoxicity of dicamba and pentachlorophenol in soils by inoculating the soils with an isolated culture capable of mineralizing pesticides has been observed (Krueger et al., 1991; Pfender, 1996). White-rot fungi including *Phanerochaete chrysosporium* and other species are the most studied inocula for augmenting pesticide biodegradation. White-rot fungi are unique among other microbiota because they can rapidly depolymerize lignin, which is a complex and irregular nonhydrolyzable wood polymer largely composed of nonrepeating phenylpropane subunits. Their ability in rapidly metabolizing lignin is believed to be ascribed to the proven capability of fungus on metabolizing a variety of recalcitrant halogenated aromatic compounds and PAHs (Barr and Aust, 1994). In fact, metabolism of DDT, many chlorinated cyclodiene insecticides, 2,4,5-T, atrazine, PCP, and creosote constituents (i.e., PAHs) in soils and liquid by the white-rot fungi has been demenstrated (Felsot et al., 2003). Field study of the bioremediation of the soils with PCP- and creosote-contaminated wood preservative sludge through bioaugmentation with the fungus *Phanerochaete sordia* showed 82% reduction in PCP concentration within 60 days of the inoculation (Lamar et al., 1993). Moreover, a maximum of 95% and 72% reduction in 3-ring and 4-ring PAH constituents of creosote, respectively within 56 days of the inoculation have been observed (Davis et al., 1993).

10.4.2.8 Landfarming

Landfarming has been applied for decades to dispose municipal sludges, oily wastes, and municipal wastewater (Felsot, 1998; Surampalli et al., 2007). Overcash and Pal (1979) defined that landfarming is

> the intimate mixing or dispersion of wastes into the upper zone of the soil-plant system with the objective of microbial stabilization, adsorption, immobilization, selective dispersion or crop recovery, leading to an environmentally acceptable assimilation of the waste.

Landfarming of pesticide-contaminated soils and groundwater involves placement of the soils within the plow layer of uncontaminated soils and spraying of the contaminated rinsate or well water on lands, respectively in which dilution will lower the pesticide concentration sufficiently to facilitate both chemical and aerobic microbial degradation. Depending on the type of the pesticides being remediated, nonsensitive crops can be planted on the lands. In this manner, landfarming is a form of bioremediation relying on the microbial and rhizosphere ecology of soils, and physicochemical processes to transform and/or immobile the pesticides into less biologically active materials. However, the loading rate of the contaminated soils and groundwater should not exceed the assimilative capacity of the soils. Otherwise, a high loading rate may result in excessive runoff or leaching of pesticides, and lead to an increase in microbial toxicity, thereby reducing the potential for the pesticide degradation. Field-scale landfarming of trifluralin-contaminated soils from warehouse fire had been conducted at Lexington, Illinois. The contaminated soils were applied by manure spreaders over about 32 ha of farmland. Several months after the application, trifluralin concentration in the collected individual soil cores dropped from an initial average level of 158 ± 247 ppm down to 10-15 ppm. However, trifluralin dissipation in the farmland was slowed down and inhibited as the temperature declined throughout the fall. Another example of landfarming treatment of triazine-contaminated soils from a manufacturing plant was conducted at St. Gabriel, Los Angeles. Approximately 19000 m^3 of the contaminated soils were spread to an 8 ha site. Simultaneously, an acclimated heterogeneous microbial consortium was inoculated into the soils for rapid triazines degradation, and fertilizers were also spread over the area at a rate of 90 g/m^2. Five months after the spreading, the triazine concentration in the soils reduced from 12-500 ppm to 0.034-48.2 ppm (Finklea and Fontenot, 1995). Successful remediation of polycyclic aromatic hydrocarbons (PAHs)- and pentachlorophenol (PCP)-contaminated soils from a lumber mill facility in Montana had also been reported by Piotrowski (1991). Two months after spreading 34200 m^3 of the contaminated soils to a 0.4 ha land treatment unit, the total concentration of carcinogenic PAHs and PCP was reduced below the cleanup standards, and the mutagenic potential of the soils also declined to a background level (Felsot, 1998).

10.4.2.9 Composting

Composting relies on tremendous increases in microbial activities, which occurs when naturally occurring microorganisms start off to decompose readily degradable organic substrates, to remediate pesticide-contaminated soils. The resulting metabolic heat released usually raises the compost temperature as high as 60-70 °C, thereby resulting in a microbial succession of mesophilic organisms to a thermophilic community (Felsot et al., 2003). It has been reported that there were nearly complete degradation of insecticides carbaryl, diazinon, and parathion during composting of sewage sludge (Racke, 1989), dairy manure (Petruska et al., 1985), and cannery wastes (Rose and Mercer, 1968), respectively. Six weeks composting of grass clippings also resulted in complete degradation of chlorpyrifos, isofenphos, diazinon, and pendimethalin therein (Lemmon and Pylypiw, 1992). In addition, extensive mineralization (47%) of the 2,4-D in grass clippings was detected after 50 days of composting, whereas 43% of the 2,4-D formed bound residues (Michel et al., 1995). Michel et al. (1997) observed that the loss of diazinon during composting of deciduous tree leaves and grass clippings was mainly attributed to hydrolysis reactions (36%) and formation of bound or unextractable residues (32%). Composting of the 2,4,6-trichlorophenol-, 2,3,4,6-tetrachlorophenol-, and PCP-contaminated soils at initial chlorinated phenolic level of 437-1108 mg/kg led to about 90% destruction of the contaminants in which approximately 60% of the destruction was due to mineralization (Laine and Jorgensen, 1997). Over 90% removal of DDT from the contaminated soils has also been observed after 60 days of thermally heated composting with green waste in Australia in which DDD was the main metabolite produced. Approximately 90% degradation of PCP in the contaminated soils was also achieved during 15 weeks of composting with the green waste without an external heating supply (Singleton et al., 1998). The amitraz level in the contaminated sludge from cattle dip was reported dropping from 483 mg/kg to less than 0.5 mg/kg after 104 days of composting with saw dust, cow manure, and green waste (Van Zwieten et al., 1998).

10.5 Considerations of Pesticide Application and Pollution Control

To achieve the "fishable/swimmable" goal of the federal Clean Water Act, pesticide contamination must be under control. If a stream or a lake is found to be impaired by a pesticide, a Total Maximum Daily Load (TMDL) will need to be estimated. Development of TMDLs requires information on background pollutant contribution, watershed hydrological information, and seasonal variation, etc., which often is not readily available. For example, even with extensive monitoring (e.g., implementation of the EPA "CORE" monitoring network), no readily available data can be found on the water chlordane level for Cheat River, Pennsylvania (USEPA, 2006) for comparison to the criterion. Sometimes, water column concentration of a pesticide needs to be calculated based on the fish tissue concentration and a bioconcentration factor, which often is not very reliable. Therefore, tracking of pesticide used in agricultural and nonagricultural areas and monitoring new pesticides and others not yet studied needs to be improved. Currently, Integrated Pest Management (IPM) of pesticide, an important part of Best Management Practices (BMPs) is implemented to control the sources of pesticides and minimize their pollution. However, an information gap needs to be filled of evaluating the effects of management practices (e.g., conservation behaviors, IPM and BMP implementation) on concentration and transport of pesticides. Without this information, it will be very difficult to achieve anything simply relying on monitoring programs and the Total Maximum Daily Load (TMDL) recommendations.

Controlling of environmental contamination by pesticides is very complex and full of challenges, requiring (at least) (a) implement of IPM and BMPs, (b) understanding of landscape capacity (or vulnerability) to contamination from leaching and runoff at a watershed and/or regional scale, and (c) knowledge/modeling to evaluate the effects of conservation behaviors and management practices on concentration and transport of pesticides. In this section, IPM for the agricultural pesticide uses will be introduced first in order to highlight the actions which farmers can implement to reduce the potential for pesticides to move into water resources via either surface runoff or leaching. Information and current status of research will then be introduced to link conservation practices, farmer motivations and behaviors with watershed vulnerability to pesticide pollution.

10.5.1 BMPs and IPM

BMPs include both management and operation of facilities. The practices are designed to alleviate the impact of pesticide uses on groundwater and surface water quality. These practices not only protect the water resources, but also should be economically sound. Depending on soils, topography, and the farm operation pattern, suggested practices should be modified or varied accordingly (Czapar, 1996).

Integrated Pest Management (IPM) is an important part of BMPs for protecting the water resources. Under Florida law, IPM is defined as follows:

> *The selection, integration, and implementation of multiple pest control techniques based on predictable economic, ecological, and sociological consequences, making maximum use of naturally occurring pest controls, such as weather, disease agents, and parasitoids, using various biological, physical, chemical, and habitat modification methods of control, and using artificial controls only as required to keep particular pests from surpassing intolerable population levels predetermined from an accurate assessment of the pest damage potential and the ecological, sociological, and economic cost of other control measures (Florida Green Industries, 2002).*

Implementation of IPM can reduce pesticide use to the minimum amount which is, however, still enough to produce high quality food, thereby maximizing farmers' profits (Waskom, 1995). IPM combines cultural, biological, genetic, and chemical controls to form a comprehensive program for managing pests. The cultural component involves proper selection and maintenance of crops. Keeping crops healthy reduces their susceptibility to diseases, nematodes, and insects, thereby minimizing the need of chemical pesticide treatment. Implementation of crop rotation and selection of suitable planting and harvest dates can also minimize the pest damages. The biological component involves release and/or conservation of beneficial organisms in farmlands such as pollinators and pest natural enemies, including parasites, predators, and pathogens. The fields or farmlands can be slightly modified to attract the natural enemies and provide them the habitats. The genetic component focuses on the breeding or genetic engineering of crops which are resistant to key pest. Pests may develop slowly on the partially resistant crops and thereby increase their susceptibility to natural enemies or less toxic pesticides. The chemical component pertains to the mixing of conventional, broad-spectrum pesticides with more selective, newer chemicals such as microbial insecticides and insect growth regulators, thereby promoting the use of the least toxic and the most selective pesticides. In addition to cultural, biological, genetic, and chemical components, pest monitoring is also an important part of IPM. Generally, it includes the understanding of the life cycle of pests, and knowing of the specific plants and conditions the pests may prefer. Monitoring pest population, knowing the historical trends, and finding the place at which the pests most likely occur can enable the control practices to be targeted to specific pests in specific locations (Waskom, 1995; Florida Green Industries, 2002).

In addition to IPM, proper pesticide application, handling and disposal of pesticides, and pesticide selection are also the essential parts of BMPs. Over application of pesticides can result in unnecessary expense to farmers and increase the risk of pesticide contamination of water resources, whereas under application leads to poor pest control. To ensure accurate pesticide application, spray equipment must be properly maintained and calibrated. All sprayer components should be clean and in good working conditions. Moreover, periodic recalibration of the spray equipment is required to compensate for the wear in pumps, nozzles, and metering systems. During calibration, clean water should be used instead of the pesticide, and it should be done in the places which are far away from wells, sinkholes, and surface water bodies (Czapar, 1996; Howard and Thomas, 1998). In addition, unnecessary and poorly timed application of pesticides should be avoided to prevent the need for re-treatment. It is recommended that the pesticide application time should be decided based on the soil moisture, anticipated weather conditions, and irrigation schedules in order to achieve the greatest efficiency and reduce the potential of off-site transport. Pesticide application at the time at which soil moisture status or scheduled irrigation increases the runoff or deep percolation possibility should be avoided. Furthermore, if heavy rains are forecasted for the next few days, pesticide application should be postponed. The other important point is that buffer zones around wells, surface water, and vulnerable areas in which pesticides are not applied should be established (Waskom, 1995).

Proper storage, handling, and disposal of pesticides can greatly reduce the risk of point source contamination to water resources. Spills and improper disposal of pesticide may overload the soil ability to degrade and hold the pesticides, thereby resulting in groundwater contamination. Normally, pesticides have to be stored in a roofed concrete or metal structure with a lockable door to keep them secure and isolated from the surrounding environment. Moreover, the floor of these structure should be constructed with seamless metal or concrete sealed with a chemical-resistant paint to prevent the spills from contaminating soil and groundwater. Different types of pesticides such as insecticides, herbicides, and fungicides should be segregated to prevent cross-contamination and misapplication. Generally, cross-contaminated pesticides cannot be applied in accordance with the labels, and should be disposed off as hazardous wastes. In addition, pesticide mixing and loading activities should be carried out away from groundwater wells and the areas where runoff can readily carry spilled pesticides into surface water bodies. If not possible, every effort has to be made to properly case and cap the wells, and to keep spills out of surface water. During pesticide preparation, back-siphoning of pesticides to water sources (e.g., wells, nurse tank - a tank of clean water transported to the field to fill sprayers, cannel, etc.) should be avoided by leaving an air gap between the water source and the chemicals or applying one-way valve. If possible Excess pesticide spray should not be disposed by dumping and should be re-applied to the field based on the label guidelines. Washwater from pesticide application equipment must be managed properly since it contains pesticide residues. The exterior parts of the application equipment such as tires can be washed at random spots in the field which are not in the vicinity to wells or surface water bodies using the water from a nurse tank. The interior parts of the application equipment, which usually contain more pesticide residues than the exterior parts, should be washed at designated places such as a mix/load pad. Similarly, pesticide containers should be rinsed as soon as they are empty by pressure rinse or triple rinse. The rinsate collected can be re-applied as a pesticide, stored for use as make-up water for next compatible application, or treated as hazardous wastes. After cleaning, the used containers should be punctured to prevent re-use, and stored in a clean and out of weather areas for further disposal or recycling (Howard and Thomas, 1998).

Other BMPs for protecting the water resources from the pesticide contamination include conservation tillage and conservation buffers practices. Conservation tillage involves leaving crop residues (plant materials from past harvests) on soil surface. These residues help reduce sediment loading and slow water runoff, thereby reducing pesticide movement into surface water. Moreover, pesticides which are moved into soil rather than running off can be degraded by microbial activity, chemical reactions, and plant uptake and metabolisms. Conservation buffers involve establishment of grass waterways, grass filter strips or riparian areas to provide additional barrier of protection to surface water bodies by capturing potential pesticides which may move into surface water and reducing runoff flow rate.

10.5.2 Linking Vulnerability with Conservation Behaviors

Watershed vulnerability to pesticide contamination is determined by a combination of physical setting (e.g., soils, topography and climate) and management practices for land (e.g., BMPs) and pesticides applications (e.g., IMP). Since the 1990s, some researchers started to integrate hydrologic models (e.g., Maidment, 1993) and surface and subsurface process modeling (e.g., Moore et al., 1993) within a GIS framework. The integration of spatially and temporally referenced data into a modeling framework would introduce complexities related to data and process scale, data quality, spatial dependency and error analysis (Aspinall and Pearson, 1996). STATSGO data have been used in several GIS modeling efforts (Lytle et al., 1996; Eason et al., 2004; Sinkevich et al., 2005). However, the STATSGO database resolution limits its applicability with respect to making field scale management decisions. Currently, the USDA-NRCS SSURGO data sets, with scales of 1:12,000 to 1:63,360, provide information for exploring regional processes and relationships and reflect what one might expect to find at a field scale.

Substantial research has been conducted into the behavioral dimension of farm conservation and environmental farming (Lynne and Rola, 1988; Lynne et al., 1988; Lockeretz, 1990; Lynne, 1995; Nowak and Korsching, 1998); some theories have been developed in this area, including *Theory of Planned Behavior* (Ajzen, 1991, 2002; Lynne et al., 1995) and *Metaeconomic Theory* (Lynne and Casey, 1998; Lynne, 1999, 2002). Important behavioral areas are those related to implementation of IMP and BMPs for pesticide applications and land conservation (Lynne et al., 1988; Lynne and Rola, 1988; Lynne et al., 1995; Lynne and Casey, 1998; Cutforth et al., 2001). Willock et al. (1999) made efforts to integrate the spatial and temporal data describing the physical attributes of a site along with producer behavioral data also viewed in the spatial and temporal sense, resulting in representing "personological typologies" (i.e. personality characteristics, behavioral expectations, summarized across a resident population in an area) of farmers on a spatial scale. Georeference of the behavioral dimension (i.e. mapping personological typologies) is deemed as essential to effective policy implementation and control of pesticides (or other pollutant) contamination.

Building on the current understanding of landscape processes and pesticide chemistry, Shea et al. (2006) used SSURGO and the USGS/USEPA National Hydrography Datasets (NHD), and pesticide property data as model inputs to assess landscape vulnerability to pesticide leaching and runoff. In their studies, Shea et al. (2006) developed the models by introducing new variables and functions into the existing models, and used models to determine the most vulnerable areas within impaired watersheds in Nebraska, Kansas, Missouri, and Iowa (see Table 10.7 and Fig. 10.14). They validated the model with monitoring data from those areas. In addition, a behavioral assessment model was used in selected areas upstream of Tuttle Creek, Kansas to predict the probability that producers would adopt the technologies and practices associated with the TMDL recommendations, as well as the extent of adoption. In addition, a survey tool was used to determine the practices, motivations and behaviors of producers in vulnerable areas; a statistical model was used to predict their responsiveness to change, and decision typologies were mapped.

Table 10.7. Classification of landscape and pesticide variables used in the vulnerability models (Shea et al., 2006).

Property	Unit	Class		
		Low	Medium	High
Organic matter	%	<1.5	1.5 – 3.0	>3.0
Clay	%	<15	15 - 30	>30
pH		<6.0	6.0 – 7.0	> 7.0
ISRO[†]		Low	Medium	High
Free water depth[‡]	M	<0.5	0.5 - 1.0	>1.0
Flooding frequency		Frequent	Occasional	Rare
K_{oc}[§]		<50	50 - 500	>500
Half-life	D	<15	15 - 30	>30

[†] Index of soil runoff; classes include negligible (N), very low (VL) and very high (VH).
[‡] Free water depths defined only through the soil profile.
[§] Organic carbon partition coefficient.

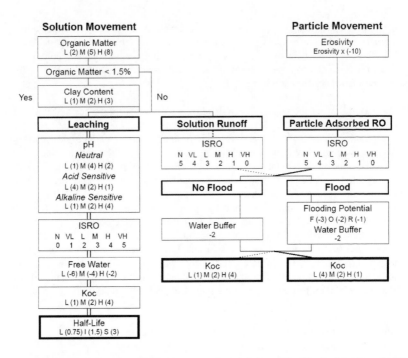

Fig. 10.14. Model schema with classes and weightings. N, VL, L, M, H and VH represent negligible, very low, low medium, high and very high, respectively. Flooding potential classes are frequent (F), occasional (O), and rare (R) while half-life classes are long (L), intermediate (I), and short (S) (Shea et al., 2006).

Currently, Shea P. J. and the co-workers are developing a software module that can be used by educators, consultants, and agency personnel, with ESRI ArcView and Spatial Analyst 9.x to process the SSURGO and NHD input data, allow user selection of whole system modifications (such as herbicide selection), run the models, and display the model output (Shea P. J., personal communication, Jun. 13, 2008). Using ESRI's ArcPublisher 9.x, the output could be exported for viewing with ESRI's free ArcReader 9.x. These results (will) provide insights on how to link conservation practices, personal behaviors with vulnerable landscape areas at a spatial and/or temporal scale that have significant benefits to those tasked with implementing the total maximum daily load (TMDL) recommendations. Similarly, their models could be expanded beyond the management of pesticides to include a variety of water contaminants related to sources of nonpoint pollution (sediment, nutrients, and bacteria).

10.6   Conclusions

Pesticides have been widely used in agriculture, forestry, transportation, urban and suburban areas, lakes and streams, and various industries for pest control. Among various pesticides including insecticides, miticides, fungicides, nematicides, and fumigants, herbicides are most commonly used, which accounts for approximately 65% of the total pesticide consumption in the U.S. Currently, organochlorines including DDT and lindane, organophosphorous such as TEPP and diazinon as well as carbamate esters including aldicarb and carbaryl are the common insecticides being applied. Herbicides can be classified into inorganic and organic categories. Inorganic herbicides include various inorganic acids and salt such as sodium chlorate and sulfuric acid; whereas phenoxyaliphatic acids, substituted amides, substituted ureas, and heterocyclic nitrogens etc. are the common organic herbicides being used.

The mode of action of the insecticides generally involve the impact/inhibition on nervous systems, cuticle production systems, water balance, energy production systems or endocrine systems of insects. In addition, herbicidal actions are usually related to the shrinkage of a plant cell away from its cell wall via water removal from its large central vacuole, the impact on plant's cellular division, activation of plant's phosphate metabolism, modification of plant's nucleic acid metabolism or photosynthesis inhibition. However, potential effects of pesticides on humans and aquatic ecosystems are difficult to evaluate because of inadequate information on effects of low-level mixtures, transformation products, and seasonal exposure.

Pesticide contamination of the environment is in the form of point- and/or non-point-source pollution. Point-source pollution comes from specific and identifiable places or sources. It generally involves small areas with high pesticide concentrations and most likely causes acute incidents. Wash water generated and spills released during clean up of pesticide-sprayed equipment, improper pesticide container rinsing and disposals, leaks and spills at pesticide storage sites, back-siphoning and spills generated during mixing and loading of pesticides, and pesticide residues washed out from impervious areas during storm events are examples of point-source pesticide pollution. Contrarily, non-point source pollution is more widespread and more difficult to identify and quantify than the point-source pollution. It comes from many diffuse sources including pesticides from agricultural and residential lands.

The occurrence and fate of the introduced pesticides in natural environments are complex and continuously affected by transfer, transport, and transformation processes. In many cases, the phase transfer and transport processes are linked together through the way of volatilization, runoff, leaching, absorption or uptake by animals and plants, and crop removal. Pesticides introduced to the soil or the plant can move to surface water and groundwater via runoff and leaching, respectively. Runoff is the movement of water pesticides over a sloping surface, while leaching is the movement of water and pesticides through the soils. Soil properties such as organic matter content, texture, permeability, and moisture content, pesticide properties including soil/water adsorption coefficient, solubility, and degradability as well as site and weather conditions are the key factors affecting pesticide runoff and leaching.

In municipal wastewater, pesticides are not common constituents. Their appearance in municipal wastewater primarily results from the pesticide runoff and leaching from agricultural lands, urban uses of herbicides and insecticides, improper design of pesticide disposal sites and landfills as well as the discharges from hospitals and pesticide production plants. Pesticides in industrial effluents of pesticide production plants are mainly generated during manufacturing and equipment cleaning processes.

Occurrence and fate of pesticides in soil and sediment environments are predominantly governed by retention, transport, and transformation processes. These processes are influenced by climatic factors, pesticide properties, mode of entry, plant or microbial factors, and soil properties. The retention process retards pesticide movement in soils, affects the availability of pesticides for plant or microbial uptakes or for abiotic or biotic transformation, and also influences the pesticide transport to atmosphere, groundwater, and surface water significantly. Transport processes determine the phase in which pesticides may exist in soils. Transformation processes involve abiotically or biotically characteristical changes in chemical properties of pesticides.

Pesticides enter the atmosphere via movement of the sprayed droplets of pesticides called drift from target to non-target sites by air, evaporation/volatilization and/or adhering onto small particulates (e.g., soil particles) or large objects (e.g., leaves) which are then caught up by wind. In general, volatilization from soil and surface waters is a primary pathway for many pesticides to dissipate. Once in the atmosphere, the pesticide droplets, pesticide vapors, and pesticide-adhered objects can be transported by wind currents, undergo photochemical and hydrolytic degradation or be deposited back to the earth's surfaces.

Pesticides in drinking water sources can be partially removed in the coagulation and flocculation, softening, and disinfection/chemical oxidation processes of the conventional water treatment systems. Soil and groundwater contaminated by pesticides can be remediated by *in situ* soil flushing and hydrolysis, thermal desorption, incineration, chemical oxidation, photochemical processes, abiotic reduction using zero-valent iron, bioremediation, landfarming, and composting.

Improved information is needed on long-term trends, pesticides and transformantion products that have not been widely measured, biological effects of typical exposure patterns, and cumulative risk from multiple contaminants or media. It is imperative to establish systems for widespread monitoring of pesticides in air and precipitation. Considerable research is needed for evaluation of the effects of management practices on concentrations and transport of pesticides; a careful and comprehensive assessment of the environmental impacts of pesticides on agriculture and natural ecosystem is needed. Currently, IPM and BMPs are becoming widespread and have been used with success in many countries. New pesticides are being developed, including biological and botanical derivatives and alternatives, to reduce health and environmental risks. Pesticide safety education and pesticide applicator regulation are designed to protect the public from pesticide misuse, but do not eliminate all misuse. Continuous changes in regulations, guidelines, personal behaviors and public awareness of pesticides will be a future tendency.

References

Acher, A. J., Hapeman, C. J., Shelton, D. R., Muldoon, M. T., Lusby, W. R., Avni, A., Warters, R. (1994). "Comparison of formation and biodegradation of bromacil oxidation products in aqueous solutions." J. Agric. Food Chem., 42, 2040-2047.

Agrawal, A., and Tratnyek, P. G. (1996). "Reduction of nitroaromatic compounds by zero-valent iron metal." Environ. Sci. Technol., 30, 153-160.

AGCare (2007). "What is a pesticide?." Agriculture Pesticides Facts, Agricultural Groups Concerned About Resources And The Environment, <http://www.agcare.org/uploadattachments/WhatisPesticide.pdf>, (Jun. 17, 2007).

Ajzen, I. (1991). "The theory of planned behavior." Organizational Behavior and Human Decision Processes, 50, 179-211.

Ajzen, I. (2002). "Perceived behavioral control, self-efficacy, locus of control, and the theory of planned behavior." J. Applied Social Psychology, 32, 1-20.

Alexander (1999). Biodegradation and bioremediation, 2nd Ed., Academic Press, San Diego, CA.

Alowitz, M. J., and Scherer, M. M. (2002). "Kinetics of nitrate, nitrite, and Cr(VI) reduction by iron metal." Environ. Sci. Technol., 36(3), 299-306.

Alvarez, P. J., and Illman, W. A. (2006) Bioremediation and natural attenuation, John Wiley & Sons, Inc., Hoboken, NJ.

Andrea, P. D., Lai, K. C. K., Kjeldsen, P., and Lo, I. M. C. (2005). "Effect of groundwater inorganics on the reductive dechlorination of TCE by zero-valent Iron." J. of Water, Air & Soil Pollut., 162, 401-420.

Anselme, C., Bersillion, J. L., and Mallevialle, J. (1991). "The use of powdered activated carbon for the removal of specific pollutants in ultrafiltration processes." Proc., American Water Works Association Membrane Processes Conference, March 10-13, 1991, Orlando, Florida.

Armstrong, D. E., and Konrad, J. G. (1974). "Nonbiological degradation of pesticides." Pesticides in soil and water, SSSA, Madison, Wisconsin.

Arnold, S. M., Hickey, W. J., Harris, R. F., and Talaat, R. E. (1996). "Integrating chemical and biological remediation of atrazine and s-triazine-containing pesticide wastes." Environ. Toxicol. Chem., 15, 1255-1262.

Ash, D. H., Salladay, D. G., Norwood, V. M., and Guinn, G. R. (1992). "Solar evaporation of aqueous wastes from fertilizers/ag-chemical dealships." Fertilizer Research, 33, 177-185.

Ashton, F. M., and Crafts, A. S. Mode of action of herbicides, 2nd Ed., John Wiley & Sons, Inc., Canada.

Aspinall, R. J., and Pearson, D. M. (1996). Data quality and spatial analysis: Analytical use of GIS for ecological modeling. GIS and environmental modeling: Progress and research issues. GIS World Books, Fort Collins, CO, 35-38.

Agency for Toxic Substances and Disease Registry (ATSDR). (1999). Toxicological profile for hexachloropentadiene (HCCPD), U.S. Department of Health and Human Services, Public Health Service, Atlanta, GA..

Barr, D. P., and Aust, S. D. (1994). "Mechanisms white hot fungi use to degrade pollutants." Environ. Sci. Technol., 28, 78A-87A.

Bollag, J. M., and Loll, M. J. (1983). "Incorporation of xenobiotics into soil humus." Experientia, 39, 1221-1231.

Bollag, J. M., and Liu, S. Y. (1990). "Biological transformation processes of pesticides." Pesticides in the soil environment: Processes, impacts, and modeling – Number 2, Soil Science Society of America, Inc., Wisconsin.

Ballantyne, B., Marrs, T. C., and Syversen, T. (1999). General and applied toxicology, Stockton Press, New York.

Ballard, B. D., and MacKay, A. A. (2005). "Estimating the removal of anthropogenic organic chemicals from raw drinking water by coagulation and flocculation." J. Environ. Eng., 131(1), 108,-118.

Brown, C. B., and White, J. L. (1969). "Reactions of 1,2-s-triazines with soil clays." Soil Sci. Soc. Am. Proc., 33, 863-867.

Buttler, T., Martinkovic, W., and Nesheim, O. N. (2003). Factor influencing pesticide movement to ground water, Agronomy Department, Florida Cooperative Extension Service, Institute of Food and Agricultural Sciences, University of Florida.

Büchel, K. H. (1983) Chemistry of pesticides, John Wiley and Sons Inc., New York.

Castro, T. F., and Yoshida, T. (1974). "Effect of organic matter on the biodegradation of some organochlorine insecticides in submerged soils." Soil Sci. Plant Nutr., 20, 363-370.

Carson, R. (1962). Silent spring, Houghton Mifflin, Boston.

Chen, H. H., Bailey, G. W., Green, R. E., and Spencer, W. F. (1990). Pesticides in the soil environment: Processes, impacts, and modeling – Number 2, Soil Science Society of America, Inc., Wisconsin.

Chian, E. (1975). "Removal of pesticides by reverse osmosis." Environ. Sci. Technol., 9(1), 52-59.

Colborn, T., Saal, F. S, V., and Soto, A. M. (1993). "Review and commentary – Developmental effects of endocrine-disrupting chemicals in wildlife and humans." Environ. Health Perspect., 101(5), 378-384.

Cole, M. A., Zhang, L., and Liu, X. (1995). "Remediation of pesticide contaminated soil by planting and compost addition." Compost Sci. Util, 3, 20-30.

Comfort, S. D., Shea, P. J., Machacek, T. A. (2001). "Field-scale remediation of a metolachlor-contaminated spill site using zero-valent iron." J. Environ. Qual., 30, 1636-1643.

CCG (Cascadia Consulting Group) (2001). Compost tea trials, Office of Environmental Management, City of Seattle.

CPR (2008). California for pesticide reform, <http://www.pesticidereform.org> (Jun. 13, 2008).

Crafts, A. S. (1961). The chemistry and mode of action of herbicides, Interscience Publishers Inc., New York.

Cutforth, L. B., Francis, C. A., Lynne, G. D., Mortensen, D. A., and Eskridge, K. M. (2001). Factors affecting farmers' diversity approach. Amer. J. Alternative Agric. 16, 168-176.

Czapar, G. F. (1996). "Best management practices (BMPs) for pesticides." Proc., Illinois Fertilizer Conference, < http://frec.cropsci.uiuc.edu/1996/report2/> (Dec. 11, 2007).

Davis, M. W., Glaser, J. A., Evans, J. W., Lamar, R. T. (1993). "Field evaluation of the lignin-degrading fungus Phanerochaete sordida to treat creosote-contaminated soil." Environ. Sci. Technol., 27, 2572-2576.

Delaplane, K. S. (2000). Pesticides usage in the United States: Historical, benefit, risks, and trends, Cooperative Extension Service, The University of Georgia College of Agricultural and Environmental Sciences.

Dillon, A. P. (1981). Pesticide disposal and detoxification, Noyes Data Corporation, Park Ridge, New Jersey.

Di Palma, L. (2003). "Experimental assessment of a process for the remediation of organophosphorous pesticides contaminated soils through in situ soil flushing and hydrolysis." Water, Air, and Soil Pollution, 143, 301-314.

Duirk, S. E., and Collette, T. W. (2006). "Degradation of chlorpyrifos in aqueous chlorine solutions: Pathways, kinetics, and modeling." Environ. Sci. Technol., 40(2), 546-551.

Eason, A., Tim, U. S., and Wang, X. (2004). Integrated modeling environment for statewide assessment of groundwater vulnerability from pesticide use in agriculture. Pest. Management Sci. 60, 739-745.

Eisenhauer, H. R. (1964). "Oxidation of phenolic wastes." J. Water Pollut. Control Fed., 36, 1116-1128.

Eisenreich, S. J., Bkaer, J. E., Franz, T., Swanson, M., Rapaport, R. A., Strachan, W. M. J., and Hites, R. A. (1992). "Atmospheric deposition of hydrophobic organic contaminants to the Laurentian Great Lakes." Fate of Pesticides and Chemicals in the Environment, John Wiley & Sons, Canada.

Environment Canada (1999). Endocrine disrupting substances in the environment, Minister of Public Works and Government Services Canada.

Erickson, L. E., and Lee, K. H. (1989) Degradation of atrazine and related s-triazines. Critical Reviews in Environ. Control, 19, 1-14.

Felsot, A. S., and Dzantor, E. K. (1990). "Enhancing biodegradation for detoxification of herbicide waste in soil." Enhanced biodegradation of pesticides in the Environment, ACS Symposium Series 426, American Chemical Society, Washington, DC.

Felsot, A. S. (1998). "Landfarming pesticide-contaminated soils." Pesticide Remediation in Soils and Water, John Wiley & Sons Ltd., West Sussex, England.

Felsot, A. S., Racke, K. D., and Hamilton, D. J. (2003). "Disposal and degradation of pesticide waste." Rev. Environ. Contam. Toxicol., 177, 123-200.

Finklea, H. C., and Fontenot, M. F., Jr (1995). "Accelerated bioremediation of triazines contaminated soils: a practical case study" Bioremediation: Science and Applications, SSSA Speical Publication 43, Soil Sicence Society of American, Madison, Wisconsin.

Fishel, F. (2007). Pesticides and the environment, Agricultural MU Guide, University of Missouri-Colmbia.

Florida Green Industries (2002). Best management practices for protection of water resources in Florida, Department of Environmental Protection, Florida.

Fronk, C. A., Lykins, B. W., and Carswell, J. K. (1990). "Membranes for removing organics from drinking water." Proc., 1990 American Filtration Society Annual Meeting, March 18-22, Washington, D.C.

Fry, F. M., and Toone, C.K. (1981) "DDT-induced feminization of gull embryos." Science, 213 (4510), 922-924.

Gan, J. Y., and Koskinen, W. C. (1998). "Pesticide fate and behavior in soil at elevated concentrations." Pesticide remediation in soils and water, John Wiley & Sons, England.

Giesy, J. P., Ludwig, J. P., and Tillitt, D. E. (1994). "Deformities in birds of the Great Lakes region, assigning causality." Environ. Sci. Technol., 28(3), 128A-135A.

Gilliom, R. (2007) "Pesticides in U.S. streams and groundwater." Environ. Sci. Technol., 41(10), 3408-3414.

Guillette, L. J., Jr., Gross, T. S., Masson, G. R., Matter, J. M., Percival, H. F., and Woodward, A. R. (1994). "Developmental abnormalities of the gonad and abnormal sex hormone concentrations in juvenile alligators from contaminated and control lakes in Florida." Environ. Health Perspect., 102(8), 680-688.

Gunther, F. A., and Gunther, J. D. (1970). "The triazine herbicides." Residue Review, Springer-Verlag, NY.

Hamilton, D., and Crossly, S. (2004). Pesticide residues in food and drinking water: Human exposure and risks, John Wiley & Sons Ltd, England.

Hansch, C., Leo. A., and Hoekman, D. (1995). Exploring QSAR: Hydrophobic, electronic, and steric constants, American Chemical Society, Washington, D.C.

Hapeman-Somich, C. J. (1992). "Chemical degradation of pesticide wastes." Pesticide Waste Management Technology and Regulation, ACS Symposium Series 510, American Chemical Society, Washington, DC.

Hapeman, C. J., Karn, J. S., and Shelton, D. R. (1995). "Total mineralization of aqueous atrazine in the presence of ammonium nitrate using ozone and Klebsiella terragena (strain DRS-I): Mechanistic considerations for pilot scale disposal." J. Agric. Food Chem., 43, 1383-1391.

Hapeman, C. J., and Torrents, A. (1998). "Direct radical oxidation processes." In Pesticide Remediation in Soils and Water, John Wiley & Sons Ltd., West Sussex, England.

Hayes, W. J., and Laws, E. R. (1991). Pesticide toxicology volume 3 – Classes of pesticides, Academic Press Inc., CA.

Hernando, M. D., Ferrer, I., Agüera, A., and Fernandez-Alba, A. R. (2005). "Evaluation of pesticides in wastewaters. A combined (chemical and biological) analytical approach." The Handbook of Environmental Chemistry, Vol. 5, Part O, Springer-Verlag Berlin Heidelberg.

Howard, D., and Thomas, M. (1998). Best management practices for agrichemical handling and farm equipment maintenance, Florida Department of Agriculture and Consumer Services, and Florida Department of Environmental Protection, Florida.

Huang, Y. H., Zhang, T. C., Shea, P. J., and Comfort, S. D. (2003). "Effects of oxide coating and selected cations on nitrate reduction by iron." J. Environ. Qual., 32, 1306-1315.

Huang, Y. H., and Zhang, T. C. (2006). "Reduction of nitrobenzene and formation of corrosion coatings in zerovalent iron system." Wat. Res., 40, 3075-3082.

Huddleston, J. H. (1996). How soil properties affect groundwater vulnerability to pesticide contamination, EM8559, Oregon State University Extension Service.

Janssens, I., Tanghe, T., and Verstraete, W. (1997). "Micropollutants: A bottleneck in sustainable wastewater treatment." Wat. Sci. Tech., 35(10), 13-26.

Jiries, A. G., Al Nasir, F. M., and Beese, F. (2002). "Pesticide and heavy metals residue in wastewater, soil and plants in wastewater disposal site near Al-Lajoun Valley, Karak/Jordon." Water, Air, and Soil Pollution, 133, 97-107.

Johnson, B. T., and Kennedy, J. O. (1973). "Biomagnification of p,p'-DDT and methoxychlor by bacteria. Appl. Microbiol., 26, 66-71.

Kearney, P. C., Zeng, Q, and Ruth, J. M. (1984). "A large scale UV-ozonation degradation unit. Field Trials on soil pesticide waste disposal." Treatment and Disposal of Pesticide Wastes, ACS Symposium Series 259, American Chemical Society, Washington, DC.

Kiely, T., Donaldson, D., and Grube, A. (2004). Pesticides industries sales and usage 2000 and 2001 market estimates, Office of Prevention, Pesticides, and Toxic Substances, U.S. Environmental Protection Agency, Washington, DC.

Ko, W. H., and Lockwood, J. L. (1968). "Accumulation and concentration of chlorinated hydrocarbon pesticides by microorganisms in soil." Can. J. Microbial, 14, 1075-1078.

Kodama, S., Yamamoto, A, and Matsunaga, A. (1997). "S-oxygenation of thiobencarb in tap water processed by chlorination." J. Agric. Food Chem., 45, 990-994.

Krueger, J. P., Butz, R. G., and Cork, D. J. (1991). "Use of dicamba-degrading microorganisms to protect dicamba susceptible plant series." J. Agric Food Chem., 39, 1000-1003.

Lai, K. C. K., Lo, I. M. C., Birkelund, V., and Kjeldsen, P. (2006). "Field monitoring of a permeable reactive barrier for removal of chlorinated organics." J. of Environ. Engrg., ASCE, 132(2), 199-210.

Lai, K. C. K., and Lo, I. M. C. (2007). "Effects of seepage velocity and temperature on the dechlorination of chlorinated aliphatic hydrocarbons." J. of Environ. Engrg., ASCE, 133(9), 859-868.

Lai, K. C. K., Surampalli, R., Tyagi, R. D., Lo, I. M. C., and Yan, S. (2007). "Performance monitoring of remediation technologies for soil and groundwater contamination: Review." Pract. Periodical of Haz., Toxic, and Radioactive Waste Mgmt., ASCE, 11(3), 132-157.

Lai, K. C. K., and Lo, I. M. C. (2008). "Removal of hexavalent chromium using acid-washed zero-valent iron: Removal capacity and types of precipitates," Environ. Sci. Technol., 42(4), 1238-1244.

Laine, M. M., and Jorgensen, K. S. (1997). "Effective and safe composting of chlorophenol-contaminated soil in pilot scale." Environ. Sci. Technol., 31, 371-378.

Larson, S. J., Capel, P. D., and Majewski, M. S. (1997) Pesticides in surface waters: Distribution, trends, and governing factors. Chelsea, Mich. Pesticides in the Hydrologic Systems series, Vol. 3, p373, Ann Arbor Press,

Lamar, R. T., Evans, J. W., and Glaser, J. A. (1993). "Solid-phase treatment of a pentachlorophenol-contaminated soil using lignin-degrading fungi." Environ. Sci. Technol., 27, 2566-2571.

Leeson, A., Hapeman, C. J., and Shelton, D. R. (1993). "Biomineralization of atrazine ozonation products. Application to the development of a pesticide waste disposal system." J. Agric. Food Chem., 41, 983-987.

Lemmon, C. R., and Pylypiw, H. M. (1992). "Degradation of diazinon, chlorpyrifos, isofenphos, and pendimethalin in grass and compost." Bull. Environ. Contam. Toxicol., 48, 409-415.

Liu, Y. H., Yang, F. L., Yue, P. L., and Chen, G. H. (2001). "Catalytic dechlorination of chlorophenols in water by palladium/iron." Wat. Res., 35(8), 1887-1890.

Lo, I. M. C., Lam, C. S. C., and Lai, K. C. K. (2005). "Competitive effects of TCE on Cr(VI) removal by zero-valent iron," J. of Environ. Engrg., ASCE, 131(11), 1598-1606.

Lo, I. M. C., Lam, C. S. C., and Lai, K. C. K. (2006). "Hardness and carbonate effects on the reactivity of zero-valent iron for Cr(VI) removal." Wat. Res., 40(3), 595-605.

Lockeretz, W. (1990). "What have we learned about who conserves soil?." J Soil Water Conservation, 45, 517-523.

Lynne, G. D., and Rola, L. R. (1988). "Improving attitude-behavior prediction models with economic variables." J. Social Psychology, 128, 19-28.

Lynne, G. D., Shonkwiler, J. S., and Rola, L. R. (1988). "Attitudes and farmer conservation behavior." Amer. J. Agric. Econ., 70, 12-19.

Lynne, G. D. (1995). "Modifying the neo-classical approach to technology adoption with behavioral science models." J. Agric. Applied Econ., 27, 67-80.

Lynne, G. D., Casey, C. F., Hodges, A., and Rahmani, M. (1995). "Conservation technology adoption decisions and the theory of planned behavior." J. Econ. Psychology, 16, 581-598.

Lynne, G. D., and Casey, C. F. (1998). "Regulatory control of technology adoption by individuals pursuing multiple utility." J. Socio-Economics, 27, 701-719.

Lynne, G. D. (1999). "Divided self models of the socioeconomic person: The metaeconomics approach." J. Socio-Economics, 28, 267-288.

Lynne, G. D. (2002). "Agricultural industrialization: a metaeconomics look at the metaphors by which we live." Rev. Agri. Econ., 24, 410-427.

Lytle D. J., Bliss, N. B., and Waltman, S. W. (1996). "Interpreting the State Soil Geographic Database (STATSGO)." GIS and environmental modeling: Progress and research issues. GIS World Books, Fort Collins, CO, 49-52.

Macalady, D. L., Tratnyek, P. G., and Grundy, T. J. (1986). "Abiotic reduction reactions of anthropogenic organic chemicals in anaerobic systems: A critical review." J. Contam. Hydrol., 1, 1-28.

Maidment D. (1993). GIS and hydrologic modeling. Environmental modeling with GIS. Oxford University Press, 147-167.

Majewski, M. S., and Capel, P. D. (1995) Pesticides in the atmosphere: distribution, trends, and governing factors. Mich. Pesticides in the Hydrologic Systems series, Ann Arbor Press, Chelsea.

Mascolo, G., Lopez, A., Passino, R., Ricco, G., and Tiravanti, G. (1994). "Degradation of sulfur containing s-triazines during water chlorination." Wat. Res., 28(12), 2499-2506.

Mason, Y., Chohen, E., and Rav-Acha, C. (1990). "Carbamate insecticides: Removal from water by chlorination and ozonation." Wat. Res., 24(1), 11-21.

Matsumura, F. (1985). Toxiciology of insecticides, 2nd Ed., Plenum Press, New York.

McCarty, P. L. (1987). "Chapter 4: Removal of organic substances from water by air stripping." Control of organic substance in water and wastewater, Noyes Data Corporation, Park Ridge, New Jersey.

Michel, F. C. Jr., Reddy, A., and Forney, L. J. (1995). "Microbial degradation and humification of the law care pesticide 2,4-dichlorophenoxyacetic acid during the composting of yard trimmings." Appl. Environ. Microbiol., 61, 2566-2571.

Michel, F. C. Jr, Reddy, C. A., and Forney, L. J. (1997). "Fate of carbon-14 diazinon during the composting of yard trimmings." J. Environ. Qual., 26, 200-205.

Mikesell, M. D., and Boyd, S. A. (1988). "Enhancement of pentachlorophenol degradation in soil through induced anaerobiosis and bioaugmentation with anaerobic sewage sludge." Environ. Sci. Technol., 22, 1411-1415.

Miles, C. J. (1991). "Degradation of aldicarb, aldicarb sulfoxide, and aldicarb sulfone in chlorinated water." Environ. Sci. Technol., 25, 1774-1779.

Miller, G. T. (2004). Sustaining the Earth, 6th Ed., Thompson Learning, Inc. Pacific Grove, California, 211-216.

Miller, T. L. (2008). The EXtension TOXicology NETwork, <http://ace.orst.edu/info/extoxnet> (Jun. 13, 2008).

Miltner, R. J., Fronk, C. A., and Speth, T. F. (1987). "Removal of alachlor from drinking water." Proc., National Conference on Environmental Engineering, ASCE, July 1987, Orlando, Florida.

Miltner, R. J., Baker, D. B., Speth, T. F., and Fronk, C. A. (1989). "Treatment of seasonal pesticides in surface water." J. Am. Water Works Assoc., 81(1), 43-52.

Mingelgrin, U., and Saltzman. S. (1979). "Surface reactions of parathion on clays." Clays Clay Miner., 27, 72-78.

Mirsatari, S. G., McChesney, M. M., Craigmill A. C., Winterlin, W. L., and Seiber J. N. (1987). "Anaerobic microbial dechlorination: An approach to on-site treatment of toxaphene-contaminated soil." J. Environ. Sci. Health B, 22, 663-690.

Mitra, J., and Raghu, K. (1986). "Rice straw amendment and the degradation of DDT in soils." Toxicol. Environ. Chem., 11, 171-181.

Moore I. D., Turner, A. K., Wilson, J. P., Jenson, S. K., and Band, L. E. (1993). "GIS and land-surface-subsurface process modeling." GIS and environmental modeling: Progress and research issues. GIS World Books, Fort Collins, CO, 196-230.

Mortland, M. M., and Raman, K. V. (1967). "Catalytic hydrolysis of some organic phosphate pesticides by copper(II)." J. Agric. Food Chem., 15, 163-167.

Muszkat, L. (1998). "Photochemical processes." Pesticide remediation in soils and water, John Wiley & Sons Ltd., West Sussex, England.

Müller, K., Bach, M., Hartmann, H., Spiteller, M., and Frede, H. G. (2002). "Point- and nonpoint-source pesticide contamination in the Zwester Ohm catchment, Germany." J. Environ. Qual., 31, 309-318.

Nash, R. G., Harris, W. G., and Lewis, C. C. (1973). "Soil pH and metallic amendment effects of DDT conversion to DDE." J. Environ. Qual., 2. 390-394.

Nowak, P., and Korsching, P. (1998). The human dimension of soil and water conservation: A historical and methodological perspective. Advances in soil and water conservation, Ann Arbor Press, Chelsea, MI.

Nemerow, N. L., and Agardy, F. J. (1998) Strategies of industrial and hazardous waste management. Van Nostrand Reinhold.

Nowell, L.H.; Capel, P.D.; Dileanis, P.D. (1999) Pesticides in stream sediment and aquatic biota: Distribution, trends, and governing factors. Lewis Publishers, Inc. Washington, DC.

O'Brien, R. D. (1967). Insecticides - action and metabolism, Academic Press, Inc., New York.

Overcash, M. R., and Pal. D. (1979). Design of land treatment systems for industrial wastes: theory and practice, Ann Arbor Science Publ. Inc., Ann Arbor, Michigan.

PCC (2002). News Bites, <http://www.pccnaturalmarkets.com/sc/0803/newsbites.html> (Jan. 19, 2003).

Perry, A. S., Yamamoto, I., Ishaaya, I., and Perry, R. Y. (1998). Applied agriculture – Insecticides in agriculture and environment, Springer-Verlag, Berlin, Heidelberg.

Pesticides Action Network UK (2000). "Pesticides in water – Costs to health and the environment." Briefing paper, Pesticides Action Network UK, London.

Petruska, J. A., Mullins, D. E., Young, R. W., and Collins, E. R. (1985). "A benchtop system for evaluation of pesticide disposal by composting." Nuclear Chem. Waste Manag., 5, 177-182.

Pettygrove, D. R., and Naylor, D. V. (1985). "Metribuzin degradation kinetics in organically amended soil." Weed Sci., 33, 267-270.

Pfaender, F. K., and Alexander, M. (1972). "Extensive microbial degradation of DDTin vitro and DDT metabolism by natural communities." J. Agric. Food Chem., 20, 842-846.

Pfender, W. F. (1996). "Bioremediation bacteria to protect plants in pentachlorophenol-contaminated soil." J. Environ. Qual., 25, 1256-1260.

Pimentel, D. (2005). "Environmental and economic costs of the application of pesticides primarily in the United States." Environment. development and sustainability, 7, 229-252.

Piotrowski, M. R. (1991). "U.S. EPA-approved, full-scale biological treatment for remediation of a Superfind site in Montana." Hydrocarbon contaminated soils, Lewis Publishers., Chelsea, Michigan.

Pratap, K., and Lemley, A. T. (1994). "Electrochemical peroxide treatment of aqueous herbicide solutions." J. Agric. Food Chem., 42, 209-215.

Prevot, A. B., Pramauro, E., and De La Guardia, M. (1999). "Photocatalytic degradation of carbaryl in aqueous TiO2 suspensions containing surfactants." Chemosphere, 39, 493-502.

Racke, K. D. (1989). "Fate of organic contaminants during sewage sludge composting." Bull. Environ. Contam. Toxicol., 42, 526-533.

Rebhun, M., Meir, S., and Laor, Y. (1998). "Using dissolved humic acid to remove hydrophobic contaminants from water by complexation-flocculation process." Environ. Sci. Technol., 32, 981-986.

Reigart, J. R., and Roberts, J. R. (1999). Recognition and management of pesticide poisonings, 5th Ed., U.S. Environmental Protection Agency, Washington DC.

Reijnders, P. J. H. (1986). "Reproductive failure in common seals feeding on fish from polluted coastal waters." Nature, 324, 456-457.

Renard, K. G., Foster, G. R., Weesies, R. A., and Porter, J. P. (1991). "RUSLE: Revised universal soil loss equation." J. Soil Wat. Cons. 46, 30-33.

Rose, W. W., and Mercer, W. A. (1968). "Fate of insecticides in composting agricultural wastes." Fate of pesticides in composting agricultural wastes, National Canners Association, Washington, DC.

Rule, K. L., Comber, S. D. W., Ross, D., Thornton, A., Makropoulods, C. K., and Rautiu, R. (2006). "Sources of priority substances entering an urban wastewater catchment – tracer organic chemicals." Chemosphere, 63, 581-591.

Russell, J .D., Curz, M., and White, J. L. (1968). "Mode of chemical degradation of s-triazines by morillonite." Science, 160, 1340-1342.

Shea, P. J., Machacek, T. A., and Comfort, S. D. (2004). "Accelerated remediation of pesticide-contaminated soil with zerovalent iron." Environ. Pollution, 132, 183-188.

Shea. P. J., Maribeth, M., Martin, A. R., Lynne, G. D., and Burbach, M. E. (2006). "Targeting watershed vulnerability and behaviors leading to adoption of conservation Management Practice." Research proposal, USDA.

Shelton, D. R., and Karns, J. S. (1998). "Pesticide bioremediation: Genetic and ecological consideration." Pesticide remediation in soils and water, John Wiley & Sons Ltd., West Sussex, England.

Shepard, H. H. (1951). The chemistry and action of insectides, 1st Ed., McGraw-Hill Book Company Inc., New York.

Shrivastava, P. (1987) Bhopal – Anatomy of a crisis. Ballinger Publishing Co., Cambridge, Massachusetts.

Simon, F. G., Meggyes, T., and McDonald, C. (2002). Advanced groundwater remediation: Active and passive technologies, Thomas Telford, London.

Singleton, I. N., McClure, C., Bentham, R., Xie P., Kanatachote, D., Megharaj, M., Dandie, C, France, C. M. M., Oades, J. M., and Naidu, R. (1998). "Bioremediation of organochlorine-contaminated soil in South Australia: A collaborative venture." Seeking agricultural produce free of pesticide residues, Australian Centre for International Agricultural Research, Canberra.

Sinkevich Jr., M. G., Walter, T. W., Lembo Jr., A. J., Richards, B. K., Peranginangin, P., Aburime, S. A., and Steenhuis, T. S. (2005). A GIS-based ground water contamination risk assessment tool for pesticides. Ground Water Monitoring and Remediation, 25, 82–91.

Somich, C. J., Muldoon, M. T., Kearney, P. C. (1990). "Enhanced soil degradation of alachlor by treatment with ultraviolet light and ozone." J. Agic. Food Chem., 36, 1322-1326.

Spencer, W. F., Cliath, M. M., and Farmer, W. J. (1969). "Vapor density of soil-applied dieldrin as related to soil-water content, temperature, and dieldrin concentration." Soil, Sci. Soc. Am. Proc., 33, 509-511.

Spencer, W. F. (1970). "Distribution of pesticides between soil, water, and air." Pesticides in the soil: Ecology, degradation, and movement, Michigan State University, East Lansing.

Spencer, W. F., and Cliath, M. M. (1970). "Desorption of lindane from soil as related to vapor density." Soil, Sci. Soc. Am. Proc., 34, 574-578.

Spencer, W. F., and Cliath, M. M. (1972). "Volatility of DDT and related compounds." J. Agric. Food Chem., 20, 645-649.

Spencer, W. F., and Cliath, M. M. (1974). "Factors affecting vapor loss of trifluralin from soil." J. Agric. Food Chem., 22, 987-991.

Speth, T. F., and Adams, J. Q. (1993). "GAC and air stripping design support for the safe drinking water act." Strategies and technologies for meeting SDWA Requirements, Lewis Publishers, Michigan.

Steverson, E. M. (1998). "Incineration as a pesticide remediation method." Pesticide remediation in soils and water, John Wiley & Sons Ltd., West Sussex, England.

Surampalli, R. Y.; Lai, K. C. K.; Banerji, S. K. and Tyagi, R. D. (2007). "Long-term land application of biosolids, - A case study," Water Sci. Technol., 57(3), 345-352.

Taylor, A. W., and Spencer, W. F. (1990). "Volatilization and vapor transport processes." Pesticides in the soil environment: Processes, impacts, and modeling – Number 2, Soil Science Society of America, Inc., Wisconsin.

Tchobanoglous, G., and Burton, F. L. (1991). Wastewater engineering – Treatment, disposal and reuse, 3rd Ed., Metcalf & Eddy, Inc., Singapore.

Troxler, W. L. (1998). "Thermal desorption." Pesticide remediation in soils and water, John Wiley & Sons Ltd., West Sussex, England.

Tyagi, R. D., Sikati Foko, V., Barnabe, S., Vidyarthi, A. S., Valéro, J. R., and Surampalli, R. (2002). "Simultaneous production of biopesticide and alkaline proteases by Bacillus thuringiensis using sewage sludge as a raw materials." Wat. Sci. Technol., 46(10), 247-254.

USDA-NRCS (2008). "National soil survey handbook (NSSH)." Natural Resources Conservative Service, <http://soils.usda.gov/technical/handbook/> (Jun. 13, 2008).

USEPA (1985). Development document for effluent limitation guidelines for the pesticide point source category, EPA/440/1-85/079, Office of Water and Hazardous Materials, Office of Water, Washington, DC.

USEPA (1991). Thermal desorption treatment engineering bulletin, EPA/540/2-91/008, Cincinnati, Ohio.

USEPA (1998). Pollution prevention (P2) guidance manual for the pesticide formulating, packaging, and repackaging industry: Implementing the P2 alternative, EPA/821/B-98/017Office of Water, Office of Pollution Prevention and Toxics, Washington, DC.

USEPA (2001a). Removal of endocrine disruptor chemicals using drinking water treatment processes, EPA/625/R-00/015, Office Of Research and Development, United States, Environmental Protection Department, Washington, DC.

USEPA (2001b). The incorporation of water treatment effects on pesticide removal and transformations in Food Quality Protection Act (FQPA) Drinking Water Assessments, Office of Pesticides Program, U.S. Environmental Protection Agency, Washington, DC.

USEPA (2006) Final total maximum daily load for Cheat River, Fayette County Chlordane, http://www.epa.gov/owow/tmdl/examples/ pesticides/pa_cheat.html (Feb. 24, 1999).

USEPA (2007a). About pesticides, <http://www.epa.gov/pesticides/about/index.htm>, (Jun. 17, 2007).

USEPA (2007b). History – DDT ban takes effect, U.S. Environmental Protection Agency, < http://www.epa.gov/history/topics/ddt/01.htm>, (Jun. 26, 2007).

USEPA (2007c). An introduction of indoor air quality – Pesticides." Indoor Air Quality (IAQ), <http://www.epa.gov/iaq/pesticid.html> (Jun. 29, 2007).

USEPA (2007d). "Agricultural pesticides." Ag 101, < http://www.epa.gov/oecaagct/ ag101/croppesticideuse.html> (Aug. 31, 2007).

USEPA (2008a). The EPA integrated risk information service (IRIS) database, <http://www.epa.gov/iris> (Jun. 13, 2008).

USEPA (2008b). "Pesticide product information system (PPIS)." Pesticides: regulating pesticides, <http://www.epa.gov/opppmsd1/PPISdata/> (Jun. 13, 2008b).

USEPA (2008c). "Human health risk assessment." Mid-Atlantic risk assessment, <http://www.epa.gov/reg3hwmd/risk/human/index.htm> (Jun. 13, 2008).

USEPA (2008d). Toxics release inventory (TRI) program, <http://www.epa.gov/tri> (Jun. 13, 2008).

USGS (2007). "Pesticides in the atmosphere." U.S Geological Survey, <http://ca.water.usgs.gov/pnsp/atmos/atmos_2.html>, (Dec. 4, 2007).

USGS (1995). Pesticides in the atmosphere. USGS Fact Sheet FS-15295.

Valles, S. M., and Koehler, P. G. (2003). "Insecticides used in the urban environment: Mode of action." University of Florida IFAS Extension, < http://edis.ifas.ufl.edu/IN077> (Jul. 3, 2007).

Van Eekert, M. H. A., Van Ras, N. J. P., Mentink, G. H., Rijnaarts, H. H. M., Stams, A. J. M., Field, J. A., and Schraa, G. (1998). "Anaerobic transformation of β-hexachlorocyclohexane by methanogenic granular sludge and soil microflora." Environ. Sci. Technol., 32, 3299-3304.

Van Leeuwen, J. Edgehill, R. U., and Jin, B. (1999). "Biological treatment of wastewaters from pesticide and starch manufacture." Bioreactor and ex situ biological treatment technologies, Proc., Fifth International In Situ and On-Situ Bioremediation Symposium, 5(5), 203-209, April 19-22 in San Diego, CA, Battelle Press, Columbus.

Van Zwieten, L. V., Ayres, M., and Curran, P. (1998). "Remediation of contaminated soil and fluid at cattle dip sites in Australia." Seeking Agricultural Produce Free of Pesticide Residues, Australian Centre for International Agricultural Research, Canberra.

Voerman, S., and Tammes, R. M. C. (1969). "Absorption and desorption of lindane and dieldrin by yeast." Bull. Environ. Contam. Toxicol., 4, 271-277.

Waldron, A. C. (1992). Bulletin 820 – Pesticides and groundwater contamination, Ohio Cooperative Extension Service, The Ohio State University.

Wang, L. K., Hung, Y., Lo, H. H., and Yapijakis, C. (2004). Handbook of industrial and hazardous wastes treatment, 2nd Ed., Marcel Dekker, inc., New York.

Ware, G. W. (1983). Pesticides – Theory and application, W. H. Freemand and Company, San Francisco.

Ware, G. W., and Whitacre, D. M. (2004). "An introduction to insecticides." The Pesticide Book, 6th Ed., Meister Media Worldwide, Willoughby, Ohio.

Warhurst, M. (1999). "Mechanisms of endocrine disruption." Introduction to hormone disrupting chemicals, <http://website.lineone.net/~mwarhurst/mechanisms.html> (Sep. 4, 2007).

Waskom, R. M. (1995). Best management practices for agricultural pesticide use, XCM-177, Colorado State-University Cooperative Extension.

Watts, R. J. (1998). Hazardous wastes: Sources, pathways, receptors. John Wiley & Sons, Inc., New York.

Westerhoff, P. (2003). "Reduction of nitrate, bromate, and chlorate by zero-valent iron (Fe^0)." J. of Environ. Engrg., ASCE, 129(1), 10-16.

Willock, J., Deary, J. Edwards-Jones, G., Gibson, G., McGregor, M., Sutherland, A., Dent, J., Morgan, O., and Grieve, R. (1999). "The role of attitudes and objectives in farmer decision making: business and environmentally-oriented behaviour in Scotland." J. Agric. Econ., 50, 286-303.

Winterlin, W., Seiber, J. N., Craigmill, A., Baier, T., Woodrow, J., and Walker, G. (1989). "Degradation of pesticide waste taken from a highly contaminated soil evaporation pit in California." Arch. Environ. Contam. Toxicol., 18, 734-747.

Wischmeier W. H., and Smith, D. D. (1978). Predicting rainfall erosion losses: A guide to conservation planning. Department of Agriculture, Agriculture Handbook No. 537, Washington, DC.

Wolfe, N. L., Mingelgrin, U., and Miller, G. C. (1990). "Abiotic transformations inwater, sediments, and soil." Pesticides in the soil environment: Processes, impacts, and modeling – Number 2, Soil Science Society of America, Inc., Wisconsin.

Wong, J. M. (2004). "Chapter 22 – Treatment of pesticide industry wastes." Handbook of industrial and hazardous wastes treatment, Marcel Dekker, Inc., New York.

Wong, J. M. (2006). "Chapter 11 – Treatment of pesticide industry wastes." Waste treatment in the process industries, CRC Press, Hoboken.

Yak, H. K., Lang, Q. Y., and Wai, C. M. (2000). "Relative resistance of positional isomers of polychlorinated biphenyls toward reductive dechlorination by zero-valent iron in subcritical water." Environ. Sci. Technol., 34, 2792-2798.

Zhang, Q., and Pehkonen, S. O. (1999). "Oxidation of diazinon by aqueous chlorine: Kinetics, mechanisms, and product studies." J. Agric. Food Chem., 47, 1760-1766.

Zhang, T. C., and Emary, S. C. (2000). "Jar tests for evaluation of atrazine removal at drinking water treatment plants." Environ. Eng. Sci., 16, 417-432.

Zhang, T. C., and Huang, Y.H. (2006). "Profiling iron corrosion coating on iron grains in a zero-valent iron system under the influence of dissolved oxygen." Wat. Res., 40, 2311-2320.

CHAPTER 11

Nanoparticles

Satinder K. Brar, Mausam Verma, R.D. Tyagi, Rao Y. Surampalli

11.1 Introduction

Nanotechnology is a broad interdisciplinary area of research, development and industrial activity which has been growing rapidly world wide for the past decade. It has been estimated that products utilizing nanotechnology could reach $1 trillion per year by 2015 (Program #5261 of the Earth & Sky Radio Series, 2007). Nanoparticles are the end products of a wide variety of physical, chemical and biological processes with some being novel and radically different, while others quite commonplace. Nanoparticle products include nanotubes, nanowires, quantum dots and "other" nanoparticles. Today, nanoscale materials find use in a variety of different areas such as electronic, biomedical, pharmaceutical, cosmetics, energy, environmental, catalytic and material applications with increased investment in the time to come (Guzman et al., 2006). Nanoparticle properties differ from the bulk material as their novel properties typically develop at a critical length scale of under 100 nm. A nanometre is 1×10^{-9} m or one millionth of a millimeter as seen in Figure 11.1 which demonstrates the relative size spectrum of nano-range versus micro-range. To give a sense of this scale, a human hair is of the order of 10,000 to 50,000 nm, a single red blood cell has a diameter of around 5000 nm, viruses typically have a maximum dimension of 10 to 100 nm and a DNA molecule has a diameter of $2 - 12$ nm (The National Science and Technology Council, 2005.). Worldwide, governments and major industrial companies are committing significant resources for research into the development of nano-scale processes, materials and products. In Europe, the Sixth Framework Program places nanotechnology as one of its seven main thematic programmes (CORDIS, 2007). The programme, *"Nanotechnology and nanosciences, knowledge-based multifunctional materials and new production processes and devices"* has a budget of €1,300 million for the period 2002-2006 (CORDIS, 2007). A similar large scale program, the National Nanotechnology Initiative (NNI), is running in the USA with a budget of approximately over $1.44 billion (for 2008), more than triple the estimated $464 million spent when the initiative started in 2001, and an increase of 13% over the 2007 request (The National Science and Technology Council, 2005.). In the UK, the Department of Trade and Industry (DTI) has recently launched the "Micro and nanotechnology manufacturing initiative" with a budget of more than £90 million (The Royal Society and The Royal Academy of Engineering, UK, 2004).

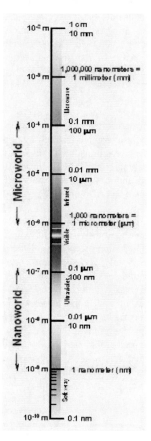

Figure 11.1 Relative particle size spectrum (Reference : US EPA, 2007)

Despite the current interest, nanoparticles are not a new phenomenon, with scientists being aware of colloids and sols, for more than 100 years. The scientific investigation of colloids and their properties was reported by Faraday (1857) in his experiments with gold. He used the term "divided metals" to describe the material which he produced. Zsigmondy (1905) described the formation of a red gold sol which is now understood to comprise particles in the 1-10 nm size range. Throughout the last century the field of colloid science has developed to a great deal and led to the production of many materials including metals, oxides, organics and pharmaceutical products.

Many other well known industrial processes produce materials which have dimensions in the nanometre size range. One example is the synthesis of carbon black by flame pyrolysis which produces a powdered form of carbon with a very high surface to mass ratio. This is usually highly agglomerated but has a primary particle size of the order of 100 nm. Worldwide production of carbon black was approximately 6 million tonnes in 1993 (IARC,1996). Other common materials produced by flame pyrolysis or similar thermal processes include fumed silica (silicon dioxide), ultrafine titanium dioxide (TiO_2) and ultrafine metals such as nickel.

Most of these nanomaterials are made directly as dry powders, and it is a common myth that these powders will stay in the same state when stored (Koehler et al., 2007). In fact, they rapidly aggregate through a solid bridging mechanism in as little as a few seconds. The adverse impacts of these aggregates will depend entirely on the application of the nanomaterial. If the nanoparticles need to be kept separate, then they must be prepared and stored in a liquid medium designed to facilitate sufficient interparticle repulsion forces to prevent aggregation.

There are four main groups of nanoparticle production processes (gas-phase, vapour deposition, colloidal and attrition) all of which may potentially result in exposure by inhalation, dermal or ingestion routes. Types of particles include spheres, sometimes known as nanocrystals, rods, carbon nanotubes and other geometric shapes. They may exist bound to surfaces by covalent bonds, bound by van der Waals forces, in an adherent matrix, such as a plastic polymer, carried in solution, or handled as a dry powder and find wide applications as given in Table 11.1. The most likely pathway of free nanoparticles from various sources like consumer products (cosmetics) is via the aquatic environment. It is essential to understand the transport behavior in natural waters to predict the fate in surface and ground waters.

Table 11.1 Applications of selected nanomaterials.

Nanomaterials	Applications
Metal oxides such as Ceria, Zinc oxide,Alumina, Zirconia.	Car catalysts, fuel cells, transparent UV absorbers, antibacterial functions, structural ceramics, Sunscreens.
Carbon	Electrical and thermal conductors.
Aluminosilicate (imogolite)	Catalyst support, ceramics filter, and humidity controlling building materials
Calcium phosphates (hydroxyapatite)	Implants such as eyes, knees, and hips

Nanoparticles entering the environment may not initially be toxic to living species in the environment, but they could in their lifecycle become toxic. The nanoparticles could react with other substances in the environment, break-down in the environment, provide a catalyst for the reactions already taking place, or prevent essential reactions from actually taking place. Measuring these effects would require a complete understanding of the nanoparticles themselves, the reactions taking place in the environments encountered by the particles, and the lifecycle of the nanoparticle. A simple schematic of possible fate of nanoparticles during different developmental phases is given in Figure 11.2. For example, waste nanoparticles from a manufacturing plant entering a stream could alter the pH of the stream. Altering the pH of a stream can lead to metals that are not normally soluble, such as, aluminium. Aluminium in the water supply would in turn be toxic to living things in the stream. To minimize possible adverse and unintended effects during production, use and disposal, the development and application of nanotechnology must be accompanied by risk assessment (Colin, 2003; Royal Society, 2004; Maynard et al., 2006). The toxicity of different nanomaterials has been tested on some organisms and studies revealed that they may have ecotoxicological effects which depend on the characteristic properties of engineered nanomaterials and test conditions (Jia et al., 2005; Hardman, 2006; Hunde-Rinke and Simon, 2006; Thill et al., 2006).

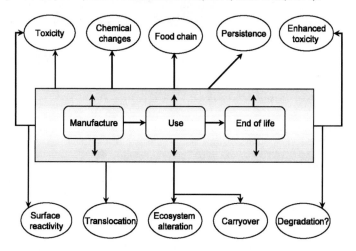

Figure 11.2 Probable environmental fate of nanoparticles during different development phases.

Nano-materials, as of today are credited for their current uses and dreamed of possibilities, but a critical question remains with regard to the environmental impact of manufactured nano-sized materials. Thus, there is a need to understand the fate of the nanoparticles in different environmental compartments as well as their effects on human health and other fauna. The chapter begins with a brief introduction of type of nanomaterials and production methods.

11.2 Classification of Nanomaterials

There are many types of engineered nanomaterials, and a variety of others are expected to appear in the future. For the purpose of this chapter, most current nanomaterials could be organized into four types: a) carbon based materials; b) metal based materials; c) dendrimers; and d) composites. Figure 11.3 demonstrates different commonly encountered structures of nanomaterials.

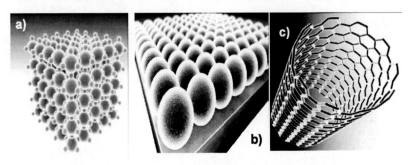

Figure 11.3 Different structures of nanomaterials: a) nanocrystal/nanoparticle; b) nanolayer and; c) nanotube.

11.2.1 Carbon Based Materials

These nanomaterials are composed mostly of carbon, most commonly taking the form of a hollow spheres, ellipsoids, or tubes. Spherical and ellipsoidal carbon nanomaterials are referred to as fullerenes, while cylindrical ones are called nanotubes. These particles have many potential applications, including improved films and coatings, stronger and lighter materials, and applications in electronics (Nesper et al., 2006). A list of products that utilize nanotechnology has been appended in Table 11.2.

Carbon black is another type of carbon based nanomaterial that is used in car tyres to increase the life of the tyre and provide the black color. The discovery of the buckyball in 1985 led to a new class of carbon-based materials, the fullerenes. Just as diamond and graphite are forms of carbon, so are fullerenes and nanotubes. The buckyball is the most basic fullerene, comprised of 60 carbon atoms arranged in a pattern resembling a soccer ball. As this structure is elongated, it forms a nanotube, which is about a nanometer (10^9 m) in diameter. Multi-walled nanotubes were discovered in 1991. Single-walled carbon nanotubes are only 1/50,000th the diameter of a human hair, but are known to have extraordinary mechanical, electrical, and thermal properties. Structurally, a nanotube is like a single graphitic sheet wrapped into a cylinder and capped at the ends (Figure 11.3).

As individual molecules, single-walled nanotubes are believed to be mostly defect- free, leading to high strength despite their low density. The tensile strength of single-walled nanotubes is about 50 times higher than some common high-strength materials like, graphite fibres, Kevlar and stainless steel. Nanotubes can be either electrically conductive or semiconductive, depending on their helicity, leading to nanoscale wires and electrical components. These one-dimensional fibers exhibit electrical conductivity as high as copper, thermal conductivity as high as diamond, strength 100 times greater than steel at one sixth the weight, and high strain to failure. If utilized to its promising potential, the field of nanotechnology will yield smaller and more lightweight components to revolutionize next generation spacecraft. Applications which are currently being investigated include; polymer composites (conductive and structural filler), electromagnetic shielding, electron field emitters (flat panel displays), super capacitors, batteries, hydrogen storage and structural composites (Maynard et al., 2004).

Table 11.2 Products that use nanotechnology (potential contributors of nanoparticles) and timeline of nanomaterials

Appliances	Automotive	Goods for Children	Electronics and Computers	Food and Beverage	Health and Fitness	Home and Garden
Heating	Exterior	Basics	Audios	Cooking	Clothing	Leaning
Cooling	Maintenance	Toys	Cameras and film	Food	Cosmetics	Construction materials
Air	Accessories	Games	Computer hardware	Storage	Filtration	Home furnishings
Large kitchen			Display	Supplements	Personal care	Luxury
Laundry and clothing			Mobile and communication devices		Sporting goods	Paint
			Television		Sunscreen	
			Video			

First generation (passive nanostructures) – 2001 – in coatings, nanoparticles, bulk materials (metals, polymers, ceramics)

Second generation (active nanostructures) – 2005 – transistors, amplifiers, targeted drugs and chemicals, actuators, adaptive structures

Third generation (3D nanosystems) – 2010 – heterogenous nanocomponents and various assembling techniques; bio-assembling; networking at the nanoscale and new architectures

Fourth generation (molecular nanosystems) – 2020 – heterogenous molecules, based on biomimetics and new designs

Source: Adapted from: (The project on Emerging nanotechnologies, 2007; The Promise of Nanotechnology, 2007)

11.2.2 Metal Based Materials

These nanomaterials include quantum dots, nanogold, nanosilver and metal oxides, such as titanium dioxide. A quantum dot is a closely packed semiconductor crystal comprised of hundreds or thousands of atoms, and whose size is on the order of a few nanometers to a few hundred nanometers (Jolivet et al., 2000). Changing the size of quantum dots changes their optical properties.

Typical examples of such nanomaterials include zinc oxide and cerium oxide. Under normal conditions, the oxides form crystals that can be as large as a cubic millimetre in size, but often these nanoparticles are less than 100 nm in length. They are not amorphous, as the atoms are still arranged in discrete crystals.

Nanowires are small conducting or semi-conducting nanoparticles with a single crystal structure and a typical diameter of a few 10s of nanometres and a large aspect ratio. They are used as interconnectors for the transport of electrons in nanoelectronic devices. Various metals have been used to fabricate nanowires including cobalt, gold and copper. Silicon nanowires have also been produced. Van Zant (2000) provides a comprehensive review of microchip fabrication which is an excellent background reading material in this field. Typically, they involve the manufacture of a template followed by the deposition of a vapour to fill the template and grow the nanowire. The template may be formed by various processes including etching, or the use of other nanoparticles, in particular, nanotubes (Figure 11.3).

11.2.3 Dendrimers

These nanomaterials are nanosized polymers built from branched units. The surface of a dendrimer has numerous chain ends, which can be tailored to perform specific chemical functions. This property could also be useful for catalysis (Cloninger, 2004). Welding fume is the best known example of this. Additionally, as three-dimensional dendrimers contain interior cavities into which other molecules could be placed, rendering them highly useful for drug delivery. Nanoparticles of this type may be formed from many materials including, metals, oxides, ceramics, semiconductors and organic materials. The particles may be composites comprising, for example, a metal core with an oxide shell or alloys in which mixtures of metals are present. This group of particles may be categorized as being less well defined in terms of size and shape, generally larger (however, respecting the norm so that they could be considered as nanoparticles), and likely to be produced in larger bulk quantities than other forms of nanoparticles. From an occupational hygiene perspective, the likelihood of aerosol generation and their availability in bulk quantities makes these nanoparticles of particular interest.

11.2.4 Composites

Composites constitute combination of nanoparticles with other nanoparticles or with larger, bulk-type materials. Nanoparticles, such as nanosized clays, are already being added to products ranging from auto parts to packaging materials, to enhance mechanical, thermal, barrier, and flame-retardant properties (Ushakov et al., 2006).

The unique properties of these engineered nanomaterials give them novel electrical, catalytic, magnetic, mechanical, thermal, or imaging features that are highly desirable for applications in commercial, medical, military, and environmental sectors. These materials may also find their way into more complex nanostructures and systems. As new uses for materials with these special properties are identified, the number of products containing such nanomaterials and their possible applications continues to grow.

11.2.5 Aluminosilicate Nanomaterials (Imogolite)

Some naturally occurring or synthetic clays such as imogolite are inorganic nanotubes with mesopore. Imogolite forms tubular structures. The external tube diameter of imogolite has been shown to be approx 2.5 nm and the tubes are several micrometers long. Electron micrographs have shown that the tubes exist in a high degree of order as self-aligned bundles. Adsorption has been used to measure the internal diameter of the intra-tube pore. Nitrogen studies of the pore volume indicate that the intra tube pore diameter is 1.5 nm. By controlling chemical compositions, pore and gain size distributions, surface properties, imogolite can find wide application in the field of chemical reaction, catalysis support, ceramic filter, humidity controlling building materials.

11.2.6 Calcium Phosphate Nanomaterials (Hydroxyapatite)

Current artificial eyes although widely used encounter various problems due to the motility and fail to deliver natural movement. They also cause sagging of the lids due to unsupported weight of the prosthesis. It is expected that application of a porous bioceramic such as the hydroxyapatite can generate good mechanical and chemical bonding to the tissue and hence a life- like eye movement. Nanostructural material, hydroxyapatite, and other related calcium phosphates have been studied as implant materials in orthopaedics and dentistry, because of their: excellent soft and/or hard tissue attachment, biocompatibility and ease of formation (Hu et al., 2001). The list of the nanomaterials is not limited to the applications discussed herein. The coming decade will engender new nanomaterials with advances in nanoparticles as listed in Table 11.3.

Table 11.3 Major types of nanoparticles anticipated to be commercially available in 2006-2014.

Product	2006-07	2008-10	2011-14
Nickel (carbon-coated) (Ni-C) powders	3 500	7 500	15 000
Poly (L-lactic acid) (PLLA) nanofibres	500	2 500	5 000
Yttrium oxide (Y_2O_3) nanopowders	2 500	7 000	7 500
Ceria (CeO2) nanoparticles, coatings	N/A	10 000	N/A
Fullerenes	N/A	300	N/A
Graphite particles	1 000 000	N/A	N/A
Silica (SiO_2) nanoparticles, coatings	100 000	100 000	> 100 000
Titania (TiO_2) nanoparticles, thin layers	5 000	5 000	> 10 000
Zinc oxide (Zn)) nanopowders, thin films	20	N/A	N/A
		USD/year	
Carbon black	≈ 8 billion	10 billion	12 billion
Carbon nanotubes	700 million	3.6 billion	13 billion

Source: Nanoroad SME, European Commission, 2006.

11.3   Fate of Nanomaterials in Air, Water and Soil

Nanoparticles and other nano-structures will be released into the air, soil and water in the form of environmental remediation products; through waste streams from factories and research laboratories; as fixed or unfixed nanoparticles in composite products and particularly after nanoproducts have been disposed of; in the form of nano-chemical pesticides and fertilisers; accidental releases during handling or transport; as components of military weapons; and through the explosion of nanopowders. Domestic nano waste discharge will also expand as large quantities of cosmetics, sunscreens and personal care products containing nanomaterials are washed off in the shower and find their way into waste water streams, or are washed off directly into oceans and lakes, due to use by the swimmers and sunbathers. The eminent pathways of the transport of nanoparticles in the environment have been summed up in Figure 11.4.

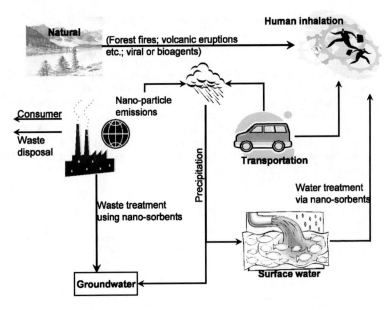

Figure 11.4 Eminent pathways of transport of nanoparticles via various emissions in the environment.

11.3.1 Air

Nanoparticles are bigger than air molecules (< 0.3 nm), but are smaller than the upper limits regulated by ambient air quality standards. U.S. National Ambient Air Quality Standards regulate the mass of particles with diameters less than 2500 and 10,000 nm, $PM_{2.5}$ and PM_{10}, respectively. Nanoparticles are produced by condensation of hot vapors in fresh combustion emissions. They are also formed from natural and anthropogenic gases such as, secondary aerosol by photochemical oxidation of gaseous compounds. Nanoparticles may contain transition metals, organic material, sulfuric acid, and free radicals. Owing to their small size and high mobility, they diffuse rapidly and may combine with each other, with larger particles, get deposited on nearby surfaces. Owing to their short lifetimes and low mass concentrations, nanoparticles are not difficult to measure in source emissions and ambient air. Airborne nanoparticles, whether engineered or as byproducts of combustion, attach quickly to ambient airborne particles. Nanoparticles in gas suspension often exhibit fractal-like shapes formed by random or interaction assisted agglomeration of much smaller primary solid particles. The agglomerates possess holes as well as internal and external surfaces. A surrogate concept for defining a size of such an object is the mobility diameter. This diameter is defined as the diameter of a sphere with one elementary charge having the same electrical mobility as the agglomerate. Many individual shapes and densities are classified as one specific mobility diameter. However, when such nanoparticles contain liquid matter or when water condenses on them, the surface tension of the liquid tends to contract the structure to a spherical particle with little hollow areas inside (Weingartner et al., 1995). Thus, nanoparticles may change size, shape, and density when released into the atmosphere, or when transported into the moist respiratory tract.

These particles have been well studied and their behavior can be accurately assessed. They are amenable to conventional aerosol emission controls when needed. The transport and deposition does not differ from background ambient aerosols. Nanoparticles are often considered to be deleterious when they are inhaled or ingested into the human body (Oberdoster et al., 1995). They penetrate deeply into the lung (Daigle et al., 2003), where their large numbers overwhelm defensive mechanisms. Thereafter, they can be transported through the bloodstream or lymphatic system to vital organs (Oberdoster et al., 2006). Today, the evidence is increasing that nanoparticles suspended in ambient air are toxic and have a severe impact on public health, shortening the life expectancy even in allegedly clean cities such as the city of Zurich by as much as several years (Leuenberger et al., Sapalida Team, 1995; Godleski et al., 1996).

The exposure scoping study suggests that the greatest exposure for humans to nanoparticles via air at the moment is likely to be to those who manufacture, process or use nanoparticles in the workplace or research laboratory. Exposure of the general population to nanoparticles is likely to be very small. However, to place this in context, the main form of human exposure to nanoparticles via air is from non-engineered nanoparticle combustion products, in particular from diesel engines, which accounted for half the total of nanoparticle emissions in 2001.

The atmosphere is a major route of human and environmental exposure to particulates, in particular through inhalation. While there are data on the overall atmospheric exposure to non-engineered nanoparticles, none of this is specific to nanoparticles. Most exposure data is from non-engineered nanoparticles (often referred to as ultrafine particles) largely from combustion processes.

11.3.2 Water

There is no direct information about water as a potential source of exposure to nanoparticles and very little is known about their behaviour in aquatic environments.

It is, however, reasonable to assume that the primary route of human exposure would be through drinking water, i.e. abstraction. Remediation of contaminated groundwater using nanoparticles presents an immediate opportunity to remediate polluted aquifers but also presents a potential pathway of exposure to the environment (and humans if there is abstraction). An understanding of the fate and behaviour of nanoparticles used for remediation purposes (for example, zero valent iron proposed for use in groundwater remediation of chlorinated solvents) is considered to be a priority research area. As said earlier, little is known about the behavior of nanoparticles in the aquatic environment. Effects on fish are being studied, with some adverse effects noted in high dose aquarium studies as discussed later in the section on fauna. This appears to be related to the surface chemistry of the particles.

Industrial products and wastes tend to end up in waterways (e.g., drainage ditches, rivers, lakes, estuaries and coastal waters) despite safeguards (Moore, 2002; Daughton, 2004; Moore et al., 2004). Consequently, as the nanotechnology industries enter large scale production sector, it is inevitable that nanoscale products and by-products will enter the aquatic environment (Moore, 2002; Daughton, 2004; Howard, 2004; Moore et al., 2004; Royal Society and Royal Academy of Engineering, 2004). This makes it imperative to carry out effective risk assessment procedures in place as soon as possible to deal with potential hazards. In developing a risk strategy for manufactured nanoparticles, much can probably be learnt from our past experience with conventional industrial materials and pollutant chemicals such as lipophilic organic xenobiotics (e.g., PAHs, heterocyclics and organohalogens). For instance, fullerenes are lipophilic while many inorganic and polymeric nanoparticles will be hydrophilic (Oppenheim, 1981; Brigger et al., 2002; Sayes et al., 2004).

Nanomaterials possess characteristic properties by virtue of their high size surface to mass ratio, facilitating detailed studies on the aggregation behaviour in the environment. It has been estimated that under certain conditions, e.g. sedimentation processes, in the aquatic environment the particles tend to aggregate restricting transport in the environment (Degushi et al., 2001; Brant et al., 2005; Hyung et al., 2007). Parameters such as pH, presence of anions and cations (ion type and concentration) and presence of humic acids are influencing the surface properties and chemical reaction of engineered nanoparticles.

11.3.3 Soil

Exposure to nanoparticles in soils will result from a number of activities, including deliberate releases via soil and water remediation technologies (see section on water above), potential agricultural uses (e.g. fertilisers) and unintentional releases via air, water and from sewage sludge applied to land. There may be risks of contamination of groundwater as a result of transport of nanoparticles through the soil profile.

Very little is known about the behavior, transfer and fate of nanoparticles in soils. For example, nanoparticles may be taken up and degraded by soil organisms, but little data exists on this. The data that is available is for nanoparticles used in the remediation of contaminated land. Much of this has focused on nanoparticle transport in the soil, since effective remediation requires movement of particles through the soil (Greg, 2004; Zhang, 2004).

Nanoparticles also have a demonstrated ability to bind to sediments and soil particles. Rice University's Center for Biological and Environmental Nanotechnology has pointed out the tendency for nanoparticles to bind to contaminating substances already pervasive in the environment like cadmium and petrochemicals. This tendency would make nanoparticles a potential mechanism for long range and widespread transport of pollutants in groundwater (Colvin, 2002). Substances such as nano-formulated pesticides and fertilizers that may be applied regularly and widely are therefore a concern. Nanoformulated pesticides are already on the market (ETC group, 2004). These pesticides have been developed precisely because they are more toxic to their target pests, and their effects are longer lasting, but they may also potentially become more toxic to other living organisms as well.

However, a recent study on the environmental impact of manufactured nanoparticles on ordinary soil showed no negative effects, which is contrary to concerns voiced by some that the microscopic particles could be harmful to organisms. Scientists at Purdue University added both dry and water-based forms of manufactured fullerenes - nanosized particles also known as buckyballs - to soil. The nanoparticles didn't change the soil constitution and the functioning of microorganisms (Tong et al., 2007). In this study, dry buckyballs and buckyballs suspended in water were added to the soil in levels of one part per million parts of soil and 1,000 parts per million parts of soil. Over a six-month period, the scientists monitored the size, composition and function of the bacterial community in the soil samples. Carbon dioxide levels in the soil, or soil respiration, the soil microbes' response to added nutrients, and enzyme activities in the soil were measured. No significant differences were found in soil containing no added nanoparticles and soil samples with either the low-level or high-level of buckyballs, the researchers reported. The data confirmed no adverse effects on the soil microbiology in concordance with another study carried out by Lubick (2007). Meanwhile, the global forecast of nanotechnology based products, as presented in Figure 11.5, raises alarms on their potential compartmentalization in different environmental streams.

11.4   Health Effects

Like chemical pollution, the concerns over nano-pollution are based on the persistence, bioaccumulation and toxicity of nanoparticles and other nano-structures and products. Remarkably little information exists on the potential of nanomaterials to cause environmental harm. There is no body of literature equivalent to that which exists for the potential of nanomaterials to cause harm to humans that examine the impacts of nanotoxicity on non-human animals, micro-organisms and plants (Royal Society, 2004). Preliminary study in this area has begun, however it has received even less funding than the relatively small amount available for the examination of implications of nanotoxicity on human health (National Science and Technology Council, 2005).

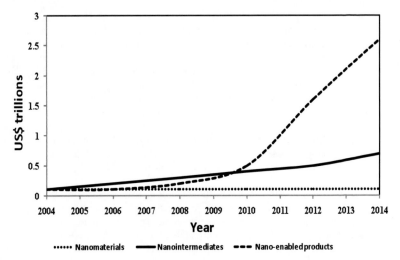

Figure 11.5 Global forecasts of nanotechnology based products (Source: Modified from: Lux Research Report, 2004)

11.4.1 Flora

Research on interaction of nanoparticles with plants is almost non-existent. However, some studies suggested that microorganisms and plants may be able to produce, modify and concentrate nanoparticles that can then bioaccumulate (or even biomagnify) through the food chain. Researchers have noted that although most people are concerned with the impacts of nanoparticles on large wildlife, the basis of many food chains depends on soil flora and fauna which could be dramatically affected by nanoparticle exposure through waste streams or deliberate release (eg for environmental remediation)(Oberdorster et al., 2005). The antimicrobial properties of nanoparticles have led to concerns that they may shift into microbial populations and disrupt signalling between nitrogen-fixing bacteria and their plant hosts (Oberdorster et al., 2005). Any significant disruption of nitrogen fixing could halt plant growth and have serious negative impacts on the functioning of entire ecosystems. This would have significant ecological and economic impacts. High levels of exposure to nanoscale aluminium (currently used in face powders and sunscreen) have been also found to cause stunt root growth in five commercial crop species (Yang and Watts, 2004).

Aluminum oxide nanoparticles were found to affect root elongation in hydroponic studies (Yang and Watts, 2005). A slight reduction in root elongation was found in the presence of uncoated alumina nanoparticles but not with nanoparticles coated with phenanthrene. It was proposed that the surface characteristics of the alumina played an important role in phytotoxicity. However, according to Murashov (2006), the author did not take into account the root toxicity of soluble Al^{3+} which is known to inhibit root growth. The solubility of aluminum oxide is known to increase with decreasing particle size and modification of the surface by adsorbed compounds is known to affect the dissolution rate making the early study baseless.

On the contrary, some studies pointed to the positive impact of nanoparticles on the flora. Hong et al. (2005a,b), Zheng et al. (2005), Gao et al. (2006) and Yang et al. (2006a) showed that nano-sized TiO_2 had a positive effect on growth of spinach when administered to the seeds or sprayed onto the leaves. Nano-TiO_2 was shown to increase the activity of several enzymes and to promote the adsorption of nitrate and accelerate the transformation of inorganic into organic nitrogen. Normal-sized TiO_2 did not have these effects.

Few data that are available indicate that at least inorganic, oxidic nanoparticles are able to interact with plant cells or green algae with a similar cell wall structure. To date no information about cellular internalization has become available. Possible interactions of nanoparticles with plant roots are adsorption onto the root surface, incorporation into the cell wall, and uptake into the cell. The nanoparticles could also diffuse into the intercellular space, the apoplast, and be adsorbed or incorporated into membranes there. Plant cells carry a negative surface charge, which allows the transport of negatively charged compounds into the apoplast as shown in Figure 11.6. The Casparian strip poses a barrier to the apoplastic flow and transport and only symplastic transport is possible into the xylem. However, this barrier is not perfect and compounds can enter the xylem through holes or damaged cells without ever crossing a cell membrane and be further transported to the shoots.

If the roots of plants were to absorb nanoparticles, the nanoparticles could enter the food-chains. nanoparticles with colloidal characteristics ideal to carry toxic material, such as water-repelling pollutants and heavy metals As they tend to be more reactive, due to the size, they may react with other substances in the environment and lead to new and possible toxic compounds. In the long term it is a possibility for a wide exposure of the entire ecosystem to engineered nanomaterials through the water and soil.

11.4.2 Fauna

Literature is abundant with studies that show that nano-particles are taken up by a wide variety of mammalian cell types, can traverse the cell membrane and become internalized (Lynch et al., 2006; Rothen-Rutishauser et al., 2006; Smart et al., 2006). The uptake of nanoparticles is usually size-dependent (Limbach et al., 2005; Chithrani et al., 2006). Aggregation and size-dependent sedimentation onto the cells or diffusion towards the cell were the main parameters determining uptake (Limbach et al., 2005). The uptake occurs via endocytosis or by phagocytosis in specialized cells. One hypothesis stated that the coating of the nanoparticles by protein in the growth medium resulted in conformational changes of the protein structure, which triggers the uptake into the cell by specialized structures, limiting uptake below 120 nm (Lynch et al., 2006).

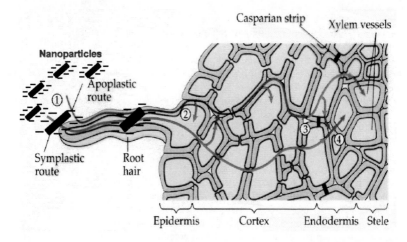

Figure 11.6 Longitudinal section of plant endodermis showing different parts: cells of the endodermis possess cell walls that are ringed by the Casparian Strip, a waxy layer (composed of suberin). The Casparian Strip is a wax and therefore prevents the apoplastic flow of water. Water must pass through the plasma membrane and enter the symplast. The plasma membrane of the endodermal cells contain many transport proteins to actively transport some molecules in and others to pump other molecules out (in this case nanoparticles are carried along this way). Once water passes under the Casparian Strip in the endodermal cells, it is free to enter the apoplast again on its way to the xylem causing accumulation of nanoparticles.

Carbon fullerenes (buckyballs), already used in several face creams and moisturisers, have been found to cause brain damage in largemouth bass (Oberdorster, 2004), a species accepted by regulatory agencies as a model for defining ecotoxicological effects. Fullerenes have also been found to kill water fleas and have bactericidal properties (Oberdorster et al., 2005).

Nanoparticles are stored in certain locations (e.g. inside vesicles, mitochondria) within the cells and are able to exert a toxic response. The small particle size, a large surface area and the ability to generate reactive oxygen species are key players in the toxicity of NPs (Nel et al., 2006). Inflammation and fibrosis are effects observed on an organism level, whereas oxidative stress, antioxidant activity and cytotoxicity are observed on a cellular level (Oberdorster et al., 2005). Several respiratory and cardiovascular diseases in humans are caused by nanoparticles (Avakian et al., 2002; Morawska and Zhang, 2002; Armstrong et al., 2004). Ultrafine soot globules migrate deep into the lungs and carry very toxic, often carcinogenic compounds such as polycyclic aromatic hydrocarbons (PAH) (Avakian et al., 2002; Morawska and Zhang, 2002; Armstrong et al., 2004). Air pollution related illnesses causes "premature death" among especially asphalt-, coke plant-, and gasworkers, chimney sweeps, and carbon electrode manufacturers(Armstrong et al., 2004).

In environmental conditions, the uptake and accumulation of these nanoparticles by aquatic biota will be of major concern. Potential uptake routes include direct ingestion or entry across epithelial boundaries such as gills or body wall. At the cellular level, prokaryotes like bacteria may be largely protected against the uptake of nanoparticles since they do not have mechanisms for transport of colloidal particles across the cell wall (Moore, 2006). However, for eukaryotes, e.g. protists and metazoans, the situation is different since they possess processes for the cellular internalization of nanoscale or microscale particles, namely endocytosis and phagocytosis (Moore, 2006). It has been reported that carbon nanotubes were taken up by a unicellular protozoan and were localized on the mitochondria of the cells (Zhu et al., 2006c). The eggs of the fish *Oryzias latipes* and adult fish accumulated the nanoparticles in gills and intestine and also brain, testis, liver and blood (Kashiwada, 2006). Another study reported that C60 adsorption onto the gram-negative *E. coli* was 10 times higher than on gram-positive *Bacillus subtilis* (Lyon et al., 2005). Nano-sized ZnO, for example, was internalized by bacteria (Brayner et al., 2006). Further, nano-sized CeO_2 particles were adsorbed onto the cell wall of *E. coli*, but the microscopic methods were not sensitive enough to discern whether or not internalization had taken place (Thill et al., 2006).

Ecotoxicological studies have also shown wide ranging toxicity to other aquatic organisms: growth inhibition in protozoans (Zhu et al., 2006c); respiratory toxicant in rainbow trout (Smith et al., 2007); mortality of copepods (Templeton et al., 2006); acute toxicity of *Daphnia magna* (Roberts et al., 2007). C60 has been found to affect several aquatic organisms, e.g. bacteria (Lyon et al., 2005, 2006), *Daphnia* (Lovern and Klaper, 2006) and fish (Oberdorster, 2004; Oberdorster et al., 2006b; Zhu et al., 2006a).

Even inorganic nanoparticles are not immune from causing toxic effects on aquatic organisms. Inorganic nanoparticular TiO_2, SiO_2 and ZnO had a toxic effect on bacteria, and the presence of light was a significant factor increasing the toxicity (Adams et al., 2006). Although, bulk TiO_2 is considered to have no health effects on aquatic organisms, still nano-sized TiO_2 can produce acute toxic effects (Lovern and Klaper, 2006). NPs that damage bacterial cell walls have been found to be internalized while those without this activity were not taken up (Stoimenov et al., 2002).

11.4.3 Humans

It has been shown that nanomaterials can enter the human body through several ports. Accidental or involuntary contact during production or use is most likely to happen via the lungs from where a rapid translocation through the blood stream is possible to other vital organs (Nemar et al., 2001). A number of recent studies indicated that the inflammatory response depends on the surface area of particles deposited in the proximal alveolar region of the lungs (Tran *et al*; 2000; Faux *et al;* 2003). Inhalation of the nanoparticles in the respiratory tract via different pathways is shown in Figure 11.7. On the cellular level an ability to act as a gene vector has been demonstrated for nanoparticles (Xiang et al., 2003). Carbon black nanoparticles have been implicated in interfering with cell signalling (Brown et al., 2004). A study demonstrated uses of DNA in the size separation of carbon nanotubes (Zheng et al., 2004). The DNA strand just wrapped around it if the tube diameter was right. Although the methods are excellent for the separation purposes, it raised some concerns over the consequences of carbon nanotubes entering the human body.

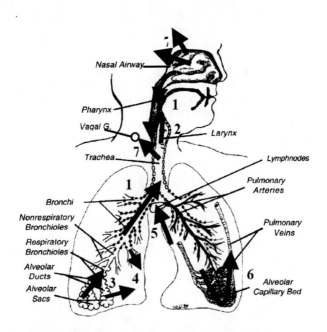

Figure 11.7 Respiratory tract particle clearance pathways (probable zones of inhalation of nanoparticles). (1) Mucociliary escalator; (2) gastro-intestinal tract; (3) Alveolar macrophage -mediated clearance; (4) Interstitium (*via* epithelium); (5) Lymphat. circulation; (6) Blood circulation; (7) Sensory neurons (olfactory, trigeminus, t-bronchial).

Nanotubes have been reported to show signs of toxicity (Service et al., 2003), confirmed in two independent publications by Warheit et al. (2003) and Lam et al. (2003), which demonstrated the pulmonary effects of single walled carbon nanotubes *in vivo* after intratracheal instillation, in both rats and mice.

Exposure to nanoparticles is primarily known to lead to pulmonary diseases. The precise mechanisms by which these materials exhibit higher levels of toxicity, at smaller particle sizes, have yet to be elucidated, although it is now understood that oxidative stress plays an important role in initiating the chain of events, at the molecular and cellular level, leading to the observed health effects. This was demonstrated in a study in which *in vitro* experiments with a range of poorly soluble particles were conducted (Faux et al; 2003). The data generated by this study demonstrated a relationship between the surface area dose and inflammatory response *in vitro* and the earlier *in vivo* results. This discovery has important practical and theoretical implications. At the theoretical level, it has highlighted the role of the epithelial cells in the proximal alveolar region, their interaction with the deposited particles and the ensuing molecular events leading to inflammation. At the practical level, it has offered a new approach to interpreting *in vitro* data, a cost effective way of screening new materials for toxicity and an alternative way to animal testing. But particles also exist in a wide range of surface reactivity. Upon deposition in the pulmonary region, such reactivity exerts an oxidative stress on the cells *via* contact between particle/cell surface area. Consequently, both highly reactive particles with a low surface area and low toxicity particles with a high surface area can exert the same oxidative stress level on the cells which they come into contact. When this inhaled dose reaches a critical level, an inflammatory response occurs. For low toxicity dusts, inflammation will diminish with the cessation of exposure, while for high toxicity dusts, inflammation will persist (Tran et al, 2000).

Persistent inflammation is likely to lead to diseases such as fibrosis and cancer. Thus, it is important to control inflammation. This can be done if we can (i) determine the critical dose of particles that initiates inflammation and (ii) set exposure limits, according to the relevant metric, so that such a dose cannot be reached within a lifetime exposure scenario. However, it has also been suggested that due to their small diameter, nanoparticles are capable of penetrating epithelial cells, entering the bloodstream from the lungs (Gilmour et al, 2004), and even translocating to the brain via the olfactory nerves (Oberdorster et al, 2004) as seen in Figure 7. The ability of inhaled nanoparticles to disperse beyond the lungs suggests that the health effects of nanoparticles may not be confined to the lungs. Also, nanoparticles cleared by macrophages via the mucociliary escalator, can be swallowed and therefore available for transfer to other body organs via the gastro-intestinal compartment. There is also some evidence that smaller particles can be transferred more readily than their larger counterparts across the intestinal wall (Behrens *et al;* 2002). Little is currently known about the health effects of nanoparticles on the liver and kidneys as well as the correct metric for describing the nanoparticle dose in these organs. Another area which merits further research is the transfer of nanoparticles across the placenta barrier. Exposure to nanoparticles during the critical window of foetal development may lead to developmental damage in the offspring.

A review of dermal exposure issues concluded that there was no evidence to indicate that specific health problems are currently arising from dermal penetration of ultrafine particles (HSE, 2000). However, the review conceded that dermal absorption of ultrafine particles (nanoparticles) has not been well investigated and suggested that ultrafine particles may penetrate into hair follicles where constituents of the particles could dissolve in the aqueous conditions and enter the skin. Direct penetration into the skin has been reported by Tinkle et al (2003) for particles with a diameter of 1000 nm, much larger than nanoparticles. It is reasonable to postulate that nanoparticles are more likely to penetrate, but this has not yet been demonstrated. Several pharmaceutical companies are believed to be working on dermal penetration of nanoparticles as a drug delivery route.

In conclusion, scientific evidence, so far, has demonstrated that particle surface area and surface reactivity is likely to be the metric of choice to describe the inflammatory reaction to deposited particles in the proximal alveolar region of the lung. For nanoparticles, their potential dispersal to other organs as well as the possibility of exposure by other routes such as dermal or ingestion mean that possible health risks beyond the lung cannot be ruled out. Further research to generate vital data on the possible mode action of nanoparticles in the extra-pulmonary system is needed in order to assess realistically the health risks to nanoparticle exposure.

Accidental spillages or permitted release of industrial effluents in waterways and aquatic systems may result in direct exposure to nanoparticles of humans via skin contact, inhalation of water aerosols and direct ingestion of contaminated drinking water or particles adsorbed on vegetables or other foodstuffs (Daughton, 2004; Howard, 2004).

More indirect exposure could also arise from ingestion of organisms such as fish and shellfish (i.e., molluscs and crustaceans) as part of the human diet. Surface sediment- and filter-feeding molluscs are prime candidates for uptake of manufactured nanoparticles from environmental releases. It can be postulated that some of these nanomaterials will associate with natural particulates; since the molluscs are already known to accumulate suspended particleand sediment-associated conventional pollutants (Livingstone, 2001; Galloway et al., 2002).

11.5. Regulations

The growing toxicity effects of nanomaterials have stirred United States (US) and the European Union (EU) to develop environmental regulatory programs for nanomaterials and nanoproducts. Meanwhile, independent voluntary standards-setting entities, such as, the American National Standards Institute (ANSI) and the International Standards Organization (ISO), propose to develop standards for measuring and evaluating toxicity effects, environmental impact, risk assessment, metrology, and methods of analysis for nanomaterials. This section will present an overview on current and future environmental regulations on release of nanoparticles in the US and the EU.

11.5.1 Environmental Regulations in the US

In the United States, the manufacture, use, transport and disposal of engineered nanomaterials is currently unregulated. It is difficult to develop new nanoparticle based regulations at this point. Instead, existing regulatory programs such as the Toxic Substances Control Act (TSCA) and the Occupational Health and Safety Act (OSHA) likely will be applied, or adapted to apply, to the regulation of nanomaterials (ANSI-NSP, 2004). However, environmental and health-based regulation of nanomaterials is unlikely to progress significantly until a standardized nomenclature for nanomaterials is developed. The key objectives of TSCA are to ensure that adequate risk assessment data are developed with respect to a given substance, and appropriate actions are taken, using regulations based on those data, to mitigate human and environmental exposure to any unreasonable risk. Prior to manufacturing or using a "new chemical substance," TSCA requires chemical manufacturers and importers to submit a pre-manufacture notification (PMN) and risk assessment information to the EPA.

Further, the standards developed by voluntary standards-developing entities such as ANSI and the American Society for Testing and Materials International (ASTM) are quasi-regulatory in nature, and are readily incorporated into regulatory programs. This pattern is not uncommon in the United States, especially, in the context of environmental regulatory programs. It should be assumed that any ANSI NSP and ASTM standards that are developed for the purpose of measuring the toxicity effects of nanomaterials, as well as assessing the potential environmental impacts and risks associated with the manufacture, use, and disposal of nanomaterials, will be used in a similar manner (www.epa.gov).

Consider an example, if a carbon nanotube (CNT) manufacturer applied to EPA for a low volume exemption (LVE) to manufacture CNTs. If EPA determined that the CNTs are covered by an existing TSCA inventory listing (*e.g.*, for carbon), the LVE application would be denied as unnecessary. In essence, all data related to health and environmental effects in the possession, custody or control—defined broadly to include several categories of related corporate entities if they are associated with the research, development, and marketing of the substance of the company must be submitted to EPA.

11.5.2 Environmental Regulations in the EU

Similar to the situation in the US, the manufacture, use, transport and disposal of engineered nanomaterials is also currently unregulated in the EU. Unlike the US, however, the novelty and unique features of these materials make it likely that the European Union will develop new environmental regulatory directives that are specifically tailored to nanomaterials. Environmental issues concerning the EU fall within the jurisdiction of the European Environmental Agency (EEA). The EEA recently developed its Sixth Action Programme for the Environment, which established the environmental priorities for the EU for the period from 2001 through 2010 (Europian Union, 2004).

In the context of nanomaterials and nanoproducts, the management of waste materials generated and the development of "end-of-life" vehicles is of primary importance. In December 2004, the European Commission (EC) released a formal communication entitled "Towards a European Strategy for Nanotechnology." The EC stressed via the communication that nanotechnology must be developed in a safe and responsible manner. As such, the EC urged that any potential public health, safety, environmental and consumer risks be addressed up front by generating the data needed for risk assessment, integrating risk assessment into every step of the lifecycle of nanotechnology-based products, and adapting existing methodologies (and, as necessary, developing new ones) for the regulation of nanomaterials and nanoproducts. Meanwhile, until the strategy is implemented, the EC has advised member states to make maximum possible use of their existing regulatory programs to address public health, worker and consumer safety, and environmental protection issues that may arise as a result of the manufacture and use o   f nanomaterials and nanoproducts. Thus, regulations on release of nanoparticles from nanomaterial products is still in its natal stage with strenuous efforts in future on the amends to the existing environmental regulations or incorporating new ones.

11.6. Socioeconomic Issues

Nanotechnology has encroached human entity in all spheres of life. The wider use of nanotechnologies in sensing and surveillance devices, for example, could both deliver increased security, but also impact on people's sense of privacy. Other concerns centre on the need for adequate management and controls around the development and use of nanotechnologies, and the equitable distribution of the benefits from their exploitation.

Nanotechnology research is dominated by the military and the first nonmilitary nanoproducts to be released commercially are targeted squarely at wealthy consumers in the Global North. In 2006, the US government, which is the world's biggest funder of nano research, allocated a third of its US$1.3 billion nanotechnology research budget to the US defence program, which was a greater share than that received by the entire National Science Foundation. In stark contrast, research into the environmental and health impacts of nanotechnology received less than 4% of the budget. The first non-military nanoproducts to be released commercially include: anti-ageing cosmetics; odour-eating socks; superior display screens for computers, televisions and mobile phones; premium coatings for luxury cars; and self-cleaning windows and bathrooms. In 2004, the United Kingdom's Royal Society noted that of the engineered nanomaterials in commercial production, the majority were being produced for use in the cosmetics industry (The Royal Society and The Royal Academy of Engineering, 2004). Experience tells us that technological innovation in and of itself will not be enough to deliver environmentally positive and socially just outcomes. Industrial-technological solutions alone cannot fix problems stemming from flawed economic ideologies, a failure to value the natural world, socioeconomic inequity or the unequal distribution of power.

At this stage, evidence does not suggest that nanotechnologies raise economic issues that differ significantly from other cases of technological innovation. However, it would contribute greatly to the wider societal debate and to decisions about the introduction of nanotechnologies if appropriate economic analysis of developments with widespread societal impacts, including an assessment of the advantages and disadvantages, is undertaken at an appropriate stage. Since this cannot be done on a systematic basis in advance of the development of new technologies (for reasons given at the start of this chapter), such analysis would need to proceed on a case by-case basis, as developments and applications come closer to market. Such analysis should also take full account of the uncertainties involved, of the case for relying on alternative technologies and which may lead to potential economic shocks and surprises.

11.7 Conclusion

Nanotechnology is mushrooming fast and represents the most powerful attempt to date to deconstruct the world into the most basic elements or units and to reshape it to meet our requirements. Thus, nanotechnology opens up new avenues for the exploitation of the earth's resources, as ever more parts of the earth become mere putty to be reconstructed and harnessed to the goals of commodity production. Far from leading us towards more environmentally sustainable and benign systems of production and consumption, nanotechnology will more likely facilitate the radical expansion of current levels of resource and energy consumption as well as pollution and waste emissions. Nanoparticles and devices which are non-biodegradable and are released en masse for 'environmental' purposes - such as nano-scale sensors, or nanoparticle iron oxide used already in the world for remediation - may also simply introduce new set of environmental pollutants and hazards intensifying resources for clean-up.

More recent uses of nanotechnology means that more and more man-made nanoparticles could in their life-time enter our atmosphere, soil or water environments. The risk of nanoparticles entering these environments, as well as the adverse effects on human health, needs to be assessed and researched in the following areas: detection of the particles; measurement of emissions of nano-particles; environmental life-cycle of the particles; toxicity of the particles to the environment and; impact on the immediate and longer range environment. The probable risk zones could be: waste streams from industrial plants being discharged into streams and rivers; accidental releases from production and during transportation of nanoproducts into streams, rivers and the atmosphere and domestic nano waste discharge from the use of nanotechnology in cosmetics, toiletries and sun creams, among many others.

Present day research lacks exact information on the nanoparticles entering the environment, the health risks and consequences to the environment. The main concern will be if any of the nanoparticles entering the environment are toxic or could become toxic to living species in the environment. For example, there is the possibility of nanoparticles being toxic to microorganisms in the soil and groundwater. There would be possible hazards from the nanoparticles or from consuming the microorganisms affected by the nanoparticles to the fish, insects or mammals. The risk to plants from nanoparticles could also have a follow-on effect on the food chain. For example the deposition of atmospheric particles on crops could provide another route for toxic or reactive nanoparticles into the food chain.

Ample information is lacking on reliable and comprehensive risk assessment of nanoparticles. Nanoparticles should be assigned to classes characterized by similar effects and by defining suitable reference parameters (e.g. mass, particle count, surface area) in order to achieve a comparable evaluation of results. Particles are seldom present as a single size (monodisperse) but rather can be represented by a distribution of sizes which is commonly characterized by median (in terms of mass or number) and a geometric standard deviation. This simplistic definition (less than 100 nm) fails to take account of size distribution. It is not clear, for example, whether the definition implies all particles less than 100 nm, 95% of particles less than 100 nm, a mean (or median) of less than 100 nm or any particles less than 100 nm. Suitable methods of measurement are to be developed and optimized to determine exposure in order to be able to record the parameters required for an assessment. Methods to establish ecotoxicity and toxicity for humans and testing strategies have to be examined and optimized as to their suitability for the evaluation of nanoparticles regarding their special properties and possible new endpoints. Like the opportunities, the risks associated with this technology have to be paid attention to. Given the very dynamic development of technology and indications of very serious implications and possible risks for human health and the environment that may arise from nanotechnology, there is an urgent need to identify and assess such risks. In spite of increasing numbers of scientific studies, there are still considerable gaps as far as information is concerned, so that there is a great need for further research.

Acknowledgements
We would like to express special thanks to INRS-ETE, NSERC and Canada research chair for giving us the opportunity to present the research work.

References

Adams, L.K., Lyon, D.Y., Alvarez, P.J.J., 2006. Comparative eco-toxicity of nanoscale TiO_2, SiO_2, and ZnO water suspensions. Water Res. 40, 3527-3532.

Air Quality Expert Group (2001) Particulate Matter in the UK. London: Defra. See:http://www.defra.gov.uk/environment/airquality/aqeg/particulate-matter/index.htm.

ANSI-NSP. Priority Recommendations Related to Nanotechnology Standardization Needs (Nov. 2004).

Armstrong, B., Hutchinson, E., Unwin, J., Fletcher, T., 2004. Lung cancer risk after exposure to polycyclic aromatic hydrocarbons: a review and meta-analysis. Environ. Health Perspect. 112, 970-978.

Avakian, M.D., Dellinger, B., Fiedler, H., Gullet, B., Koshland, C., Marklund, S., Oberdorster, G., Safe, S., Sarofim, A., Smith, K.R.,Schwartz, D., Suk, W.A., 2002. The origin, fate, and health effects of combustion by-products: a research framework. Environ. Health Perspect. 110, 1155-1162.

Behrens, I., Pena, A.I., Alonso, M.J., Kissel, T. 2002. Comparative uptake studies of bioadhesive and non-bioadhesive nanoparticles in human intestinal cell lines and rats: the effect of mucus on particle adsorption and transport. Pharmaceutical Research: 19(8); 1185-93.

Brant, J., Lecoanet, H., Wiesner, M.R. 2005. Aggregation and deposition characteristics of fullerene nanoparticles in aqueous systems. Journal of Nanoparticle Research, 7, 545- 553.

Brayner, R., Ferrari-Illiou, R., Brivois, N., Djediat, S., Benedetti, M.F., Fievet, F., 2006. Toxicological impact studies based on Escherichia coli bacteria in ultrafine ZnO nanoparticles colloidal medium. Nano Lett. 6, 866-870.

Brigger, I., Dubernet, C., Couvreur, P. 2002. Nanoparticles in cancer therapy and diagnosis. Adv Drug Deliv Rev; 54:631–51.

Brown, D.M., Donaldson, K., Borm, P.J., Schins, R.P., Dehnhardt, M., Gilmour, P., Jimenez, L.A., Stone, V. 2004. Calcium and ROS-mediated activation of transcription factors and TNF-alpha cytokine gene expression in macrophages exposed to ultrafine particles. Am J Physiol Lung Cell Mol Physiol. 286:L344–353. doi: 10.1152/ajplung.00139.2003.

Chithrani, B.D., Ghazani, A.A., Chan, W.C.W., 2006. Determining the size and shape dependence of gold nanoparticle uptake into mammalian cells. Nano Lett. 6, 662-668.

Cloninger, M. 2004. Dendrimers and protein cages as nanoparticles in drug delivery. Drug Discovery Today, 9 (3), 111-112.

Colvin, V. 2002. Responsible Nanotechnology: Looking Beyond the Good News. Nature Biotechnology, 21, 1166-1170.

Colvin, V.L. 2003. The Potential Environmental Impact of Engineered Nanomaterials.

CORDIS, 2007. Nanotechnologies and nanosciences, knowledge-based multifunctional materials and new production processes and devices. http://cordis.europa.eu/nmp/home.html. (24 August, 2007).Daigle, C.C., Chalupa, D.C., Gibb, F.R., Morrow, P.E., Oberdorster, G., Utell, M.J., Frampton, M.W. 2003. Ultrafine Particle Deposition in Humans During Rest and Exercise, Inhal. Toxicol. 15, 539–552.

Daughton, C.G. 2004. Non-regulated water contaminants: emerging research. Environ Impact Asses Rev;24:711–732.

Degushi, S., Alargova, R.G. and Tsujii, K. 2001. Stable dispersions of fullerenes, C60 and C70, in water. Preparation and Characterization. Langmuir, 17, 6013-6017.

ETC Group (2004). Down on the farm. Available at: http://www.etcgroup.org.

EurekAlert!
http://www.eurekalert.org/context.php?context=nano&show=essays&essaydate=1
102.

Faux, S.P., Tran, C.L., Miller, B.G., Jones, A.D., Monteiller, C., Donaldson, K. 2003.
In vitro determinants of particulate toxicity: the dose metric for poorly soluble
dusts. HSE Research Report 154.

Galloway, T.S., Sanger, R.C., Smith, K.L., Fillmann, G., Readman, J.W., Ford, T.E.
2002. Rapid assessment of marine pollution using multiple biomarkers and
chemical immunoassays. Environ Sci Technol; 36:2219–2226.

Gao, F.Q., Hong, F.H., Liu, C., Zheng, L., Su, M.Y., Wu, X., Yang, F., Wu, C.,
Yang, P., 2006. Mechanism of nano-anatase TiO2 on promoting photosynthetic
carbon reaction of spinach e inducing complex of Rubisco-Rubisco activase. Biol.
Trace Elem. Res. 111, 239-253.

Gilmour PS, Ziesenis A, Morrison ER, Vickers MA, Drost EM, Ford I, Karg E,
Mossa C, Godleski, J.J., et al., 1996. Proceedings of the 2nd Colloquium on
Particulate Air Pollution and Human Health 4, 136-143.

Greg, W. 2004. Demonstration of In Situ Dehalogenation of DNAPL Through Injection
of Emulsified Zero-Valent Iron at Launch Complex 34 in Cape Canaveral Air Force
Station, FL, Battelle Conference on Nanotechnology Applications for Remediation:
Cost-Effective and Rapid Technologies Removal of Contaminants From Soil,
Ground Water and Aqueous Environments, September 10, 2004.

Guzman, K.A.D., Taylor, M.R., Banfield, J.F., 2006. Environmental risks of
nanotechnology: national nanotechnology initiative funding, 2000e2004. Environ.
Sci. Technol. 40, 1401-1407.

Hardman, R. 2006. A Toxicologic Review of Quantum Dots: Toxicity Depends on
Physcochemical and Environmental Factors. Environmental Health Perspect, 114,
165- 172.

Hong, F.H., Yang, F., Liu, C., Gao, Q., Wan, Z.G., Gu, F.G., Wu, C., Ma, Z.N.,
Zhou, J., Yang, P., 2005a. Influences of nano-TiO2 on the chloroplast aging of
spinach under light. Biol. Trace Elem. Res. 104, 249-260.

Hong, F.H., Zhou, J., Liu, C., Yang, F., Wu, C., Zheng, L., Yang, P., 2005b. Effect of
nano-TiO2 on photochemical reaction of chloroplasts of spinach. Biol. Trace
Elem. Res. 105, 269-279.

Howard C.V. 2004. Small particles — big problems. Int Lab News; 34(2):28–29.

HSE. 2000. MDHS 14/3 General methods for sampling and gravimetric analysis of
respirable and total inhalable dust. HSE Books, Sudbury.

Hu, J., Russell, J. Ben-Nissan, B. and Vago, R. 2001. Production and analysis of
hydroxyapatite derived from Australian corals via hydrothermal process, Journal
of Materials Science Letters, 20, 85.

Hund-Rinke, K. and Simon, M. 2006. Ecotoxic Effect of Photocatalytic Active
Nanoparticles (TiO2) on Algae and Daphnids. Environmental Science and
Pollution Research, 13, 225.

Hyung, H., Fortner, J.D., Hughes, J.B. and Kim, J.H. 2007. Natural organic matter
stabilizes carbon nanotubes in the aqueous phase. Environ. Sci. Technol., 41, 179-
184.

IARC. 1996. IARC Monographs on the evaluation of carcinogenic risks to humans.

Volume 65. Printing processes and printing inks, carbon black and some nitro compounds. International Agency for Research on Cancer. World Health Organisation, Lyon, France.

Jia, G.,Wang, H., Yang, L.,Wang, X., Pei, R., Yan, T., Zhao, Y. and Guo, X. 2005. Cytotoxicity of Carbon Nanomaterials: Single-Wall Nanotube, Multi-Wall Nanotube, and Fullerene. Environmental Science and Technology, 39, 1378-1383.

Jolivet, J.P., Tronc, E., Chanéac, C. 2000. Synthesis of iron oxide- and metal-based nanomaterials. Eur. Phys. J. AP 10, 167-172.

Kashiwada, S., 2006. Distribution of nanoparticles in the see-through medaka (Oryzias latipes). Environ. Health Perspect. 114, 1697-1702.

Koehler, A., Som, C., Helland, A., Gottschalk, F. 2007. Studying the potential release of carbon nanotubes throughout the application life cycle. J. Clean. Prod., in press, doi:10.1016/j.jclepro.2007.04.007.

Lam, C.W., James, J.T., McCluskey, R., Hunter, R.L.2003. Pulmonary Toxicity of Single-Wall Carbon Nanotubes in Mice 7 and 90 Days after Intratracheal Instillation. Toxicol Sci.;77:126–134. doi: 10.1093/toxsci/kfg243.

Leuenberger, P. et al., Sapalida Team, 1995. Swiss study on air pollution and lung diseases in adults. Final Report to the Swiss Nat. Res. Foundation, Lausanne/Basel.

Limbach, L.K., Li, Y., Grass, R.N., Brunner, T.J., Hintermann, M.A., Muller, M., Gunther, D., Stark, W.J., 2005. Oxide nanoparticle uptake in human lung fibroblasts: effects of particle size, agglomeration, and diffusion at low concentrations. Environ. Sci. Technol. 39, 9370-9376.

Livingstone, D.R., Chipman, J.K., Lowe, D.M., Minier, C., Mitchelmore, C.L., Moore, M.N. 2000. Development of biomarkers to detect the effects of organic pollution on aquatic invertebrates: recent molecular, genotoxic, cellular and immunological studies on the common mussel (Mytilus edulis L.) and other mytilids. Int J Environ Pollut; 13:56–91.

Lovern, S.B., Klaper, R., 2006. Daphnia magna mortality when exposed to titanium dioxide and fullerene (C60) nanoparticles. Environ. Toxicol. Chem. 25, 1132-1137.

Lubick, N. 2007. Microbes survive in soil with fullerenes Environ. Sci. Technol. A; 41(08), 2658-2659.

Lux Research Report. 2004 Sizing nanotechnology's value chain. www.luxresearchinc.com. (05 Oct., 2007).

Lynch, I., Dawson, K.A., Linse, S., 2006. Detecting cryptic epitopes created by nanoparticles. Sci. STKE, pe14.

Lyon, D.Y., Adams, L.K., Falkner, J.C., Alvarez, P.J.J., 2006. Antibacterial activity of fullerene water suspensions: effects of preparation method and particle size. Environ. Sci. Technol. 40, 4360-4366.

Lyon, D.Y., Fortner, J.D., Sayes, C.M., Colvin, V.L., Hughes, J.B., 2005. Bacterial cell association and antimicrobial activity of a C60 water suspension. Environ. Toxicol. Chem. 24, 2757-2762.

Maynard AD, Baron PA , Foley M, Shvedova AA, Kisin ER, Castranova V. 2004. Exposure to carbon nanotubes material: aerosol release during the handling of unrefined single walled carbon nanotube material. Journal of Toxicology and Environmental Health, Part A; 67: 87–107.

Maynard, A.D., Aitken, R.J., Butz, T., Colvin, V., Donaldson, K., Oberdörster, G., Philbert, M.A., Ryan, J., Seaton, A., Stone, V., Tinkle, S.S., Tran, L., Walker, N.J. and Warheit, D.B. 2006. Safe Handling of Nanotechnology. Nature, 444, 267.

Moore, M.N., and Allen, J.I. 2002. A computational model of the digestive gland epithelial cell of the marine mussel and its simulated responses to aromatic hydrocarbons. Mar Environ Res; 54:579–584.

Moore, M.N., and Noble, D. 2004. Editorial: computational modelling of cell and tissue processes and function. J Mol Histol; 35:655–658.

Moore, M.N., 2006. Do nanoparticles present ecotoxicologiocal risks for the health of the aquatic environment? Environ. Int. 32, 967-976.

Morawska, L., Zhang, J.F., 2002. Combustion sources of particles. 1. Health relevance and source signatures. Chemosphere 49, 1045-1058.

Murashov, V., 2006. Comments on ''Particle surface characteristics may play an important role in phytotoxicity of alumina nanoparticles'' by Yang, L., Watts, D.J. Toxicology Letters, 2005, 158, 122e132. Toxicol. Lett. 164, 185-187.

Nel, A., Xia, T., Madler, L., Li, N., 2006. Toxic potential of materials at the nanolevel. Science 311, 622-627.

Nemmar A, Vanbilloen H, Hoylaerts MF, Hoet PH, Verbruggen A, Nemery B. 2001. Passage of intratracheally instilled ultrafine particles from the lung into the systemic circulation in hamster. Am J Respir Crit Care Med.;164:1665–1668.

Nesper, R. Ivantchenko, A. Krumeich, F. 2006. Synthesis and characterization of carbon-based nanoparticles and highly magnetic nanoparticles with carbon coatings. Advanced Functional Materials, 16(2), 296-305.

Oberdörster, E 2004. Manufactured nanomaterials (fullerenes, C60) induce oxidative stress in the brain of juvenile largemouth bass. Environmental Health Perspectives 112:1058-1062.

Oberdörster, G., Oberdörster, E. and Oberdörster, J. 2005. Nanotoxicology: an emerging discipline from studies of ultrafine particles. Environmental Health Perspectives 113(7):823-839

Oberdorster G, Sharp Z, Atudorei V, Elder A, Gelein R, Kreyling W, Cox C. 2004. Translocation of inhaled ultrafine particles to the brain. Inhalation Toxicology; 16: 437-445.

Oberdorster, E., 2004. Manufactured nanomaterials (fullerenes, C60) induce oxidative stress in the brain of juvenile largemouth bass. Environ. Health Perspect. 112, 1058-1062.

Oberdorster, E., Zhu, S., Blickley, T.M., McClellan-Green, P., Haasch, M.L., 2006b. Ecotoxicology of carbon-based engineered nanoparticles: effects of fullerene (C60) on aquatic organisms. Carbon 44, 1112-1120.

Oberdorster, G., Gelein, R.M., Ferin, J., Weiss, B. 1995. Association of particulate air pollution and acute mortality: involvement of ultrafine particles? Inhal. Toxicol. 7(1), 111–124.

Oberdorster, G., Sharp, Z., Atudorei, V., Elder, A., Gelein, R., Kreyling, W., Cox, C. (2004). Translocation of inhaled ultrafine particles to the brain. Inhal. Toxicol., 16(6 –7), 437–445.

Oppenheim, R.C. 1981. Solid colloidal drug delivery systems: nanoparticles. Int J Pharm; 8:217–234.

Program #5261 of the Earth & Sky Radio Series.2007 http://www.earthsky.org/radioshows/51591/nano-products-could-top-1-trillion-by-2015. (1 Oct, 2007)

Roberts, A.P., Mount, A.S., Seda, B., Souther, J., Qiao, R., Lin, S., Ke, P.C., Rao, A.M., Klaine, S.J., 2007. In vivo biomodification of lipid-coated carbon nanotubes by Daphnia magna. Environ. Sci. Technol. 41, 3025-3029.

Rothen-Rutishauser, B.M., Schu"rch, S., Haenni, B., Kapp, N., Gehr, P., 2006. Interaction of fine particles and nanoparticles with red blood cells visualized with advanced microscope techniques. Environ. Sci. Technol. 40, 4353-4359.

Royal Society and Royal Academy of Engineering. 2004. Nanoscience and nanotechnologies: opportunities and uncertainties. RS policy document 19/04. London: The Royal Society; p. 113.

Sayes, C.M., Fortner, J.D., Guo, W., Lyon, D., Boyd, A.M., Ausman, K.D., 2004. The differential cytotoxicity of water soluble fullerenes. Nanoletters; 4:1881–1887.

Schroeppel, A., Ferron, G.A., Heyder, J., Greaves, M., MacNee, W., Donaldson, K. (2004). Pulmonary and systemic effects of short-term inhalation exposure to ultrafine carbon black particles. Toxicology and Applied Pharmacology; 195: 35-44.

Service, R.F. 2003. Nanomaterials show signs of toxicity. Science. 300:243. doi: 10.1126/science.300.5617.243a

Smart, S.K., Cassady, A.I., Lu, G.Q., Martin, D.J., 2006. The biocompatibility of carbon nanotubes. Carbon 44, 1034-1047.

Smith, C.J., Shaw, B.J., Handy, R.D., 2007. Toxicity of single walled carbon nanotubes to rainbow trout, (Oncorhynchus mykiss): respiratory toxicity, organ pathologies, and other physiological effects. Aquat. Toxicol. 82, 94-109.

Stoimenov, P.K., Klinger, R.L., Marchin, G.L., Klabunde, K.J., 2002. Metal oxide nanoparticles as bactericidal agents. Langmuir 18, 6679-6686.

Templeton, R.C., Ferguson, P.L., Washburn, K.M., Scrivens, W.A., Chandler, G.T., 2006. Life-cycle effects of single-walled carbon nanotubes (SWNTs) on an estuarine meiobenthic copepod. Environ. Sci. Technol. 40, 7387-7393.

The National Science and Technology Council 2005. The National Nanotechnology Initiative: Research and development leading to a revolution in technology and industry. A supplement to the President's FY 2006 budget. Available at: http://www.nano.gov/NNI_06Budget.pdf. (07 Sept., 2007)

The Royal Society and The Royal Academy of Engineering, UK 2004. Chapter 4: Nanoscience and nanotechnologies. Available at http://www.royalsoc.ac.uk/.

The project on Emerging nanotechnologies. 2007. www.nanotechproject.org. (27 August, 2007).

The Promise of Nanotechnology 2007. www.wilsoncenter.org (30 August, 2007).

Thill, A., Zeyons, O., Spalla, O., Chauvat, F., Rose, J., Auffan, M. and Flank, A.M. 2006. Cytotoxicity of CeO$_2$ Nanoparticles for Escherchia Coli. Physico-Chemical Insight of the Cytotoxicity Mechanism. Environmental Science and Technology, 40, 6151-6156.

Tinkle, S.S., Antonini, J.M., Rich, B.A., Roberts, J.R., Salmen, R., DePree, K., Adkins, E.J. 2003. Skin as a route of exposure and sensitisation in chronic beryllium disease. Environmental Health Perspectives; 11: 1202-1208.

Tong, Z., Bischoff, M., Nies, L., Applegate, B., Turco, R. F. 2007. Impact of fullerene (C60) on a soil microbial community. Environ. Sci. Technol., 41, 2985-2991.

Towards A European Strategy for Nanotechnology. Communication from the Commission of European Communities (Dec. 2004).

Toxic Substances Control Act, 15 U.S.C. §§ 2601-2692;40 CFR Parts 700-789.

Tran, C., Donaldson, K., Stones, V., Fernandez, T., Ford, A., Christofi, N., Ayres, J., Steiner, M., Hurley, J., Aitken, R., Seaton, A. 2005. A scoping study to identify hazard data needs for addressing the risks presented by nanoparticles and nanotubes. Research Report. Institute of Occupational Medicine, Edinburgh.

Tran, C.L., Buchanan, D., Cullen, R.T., Searl, A., Jones, A.D., Donaldson, K. 2000. Inhalation of poorly soluble particles II. Influence of particle surface area on inflammation and clearance. Inhalation Toxicology, 12: 1113-1126.

US EPA 2007. U.S. Environmental Protection Agency Nanotechnology White Paper. Science Policy Council U.S. Environmental Protection Agency Washington, DC 20460.

Ushakov, N. M., Yurkov, G. Yu., Baranov, D. A., Zapsis, K. V., Zhuravleva, M. N., Kochubeï, V. I., Kosobudskiï, I. D. and Gubin, S. P. 2006. Optical and photoluminescent properties of nanomaterials based on cadmium sulfide nanoparticles and polyethylene. Optics and Spectroscopy, 101(2), 248-252.

Van Zant. 2000. Microchip fabrication. A practical guide to semiconductor processing (fourth edition). McGraw Hill Book company, ISBN 0-07-135636-3.

Warheit, D.B., Laurence, B.R., Reed, K.L., Roach, D.H., Reynolds, G.A., Webb, T.R. 2003. Comparative pulmonary toxicity assessment of single wall carbon nanotubes in rats. Toxicol Sci. 77:117–125. doi: 10.1093/toxsci/kfg228.

Weingartner, E., Baltensperger, U., Burtscher, H., 1995. Environmental Science and Technology 29, 2982-2986.

Xiang JJ, Tang JQ, Zhu SG, Nie XM, Lu HB, Shen SR, Li XL, Tang K, Zhou M, Li GY. 2003. IONP-PLL: a novel non-viral vector for efficient gene delivery. J Gene Med.;5:803–817. doi: 10.1002/jgm.419.

Yang, L., Watts, D.J. 2005. Particle surface characteristics may play an important role in phytotoxicity of alumina nanoparticles. Toxicol Lett. Volume 158(2):122-32.

Yang, F., Hong, F.S., You, W.J., Liu, C., Gao, F.Q., Wu, C., Yang, P., 2006a. Influences of nano-anatase TiO$_2$ on the nitrogen metabolism of growing spinach. Biol. Trace Elem. Res. 110, 179-190.

Zhang, W-X. 2004. Nanoscale Iron Particles for Environmental Remediation: An Overview. J. Nanoparticle Research, 5(3-4), 323-332.

Zheng M, Jagota A, Strano MS, Santos AP, Barone P, Chou CG, Diner BA, Dresselhaus MS, Mclean RS, Onoa GB, Samsonidze GG, Semke ED, Usrey M, Walls DJ. 2003. Structure-based carbon nanotube sorting by sequence-dependent DNA assembly. Science. 302:1543–1548. doi: 10.1126/science.1091911.

Zheng, L., Hong, F.S., Lu, S.P., Liu, C., 2005. Effect of nano-TiO_2 on strength of naturally and growth aged seeds of spinach. Biol. Trace Elem. Res. 104, 83-91.

Zhou, Q.X., Xiao, J.P., Wang, W.D., 2006c. Using multi-walled carbon nanotubes as solid phase extraction adsorbents to determine dichlorodiphenyltrichloroethane and its metabolites at trace level in water samples by high performance liquid chromatography with UV detection. J. Chromatogr. A 1125, 152-158.

Zhu, Y., Ran, T., Li, Y., Guo, J., Li, W., 2006b. Dependence of the cytotoxicity of multi-walled carbon nanotubes on the culture medium. Nanotechnol. 17, 4668-4674.

CHAPTER 12

Molecular Biology Techniques for CoEEC Degrading Organisms

Bala Subramanian S., Song Yan, R.D. Tyagi, Rao Y. Surampalli

12.1 Introduction

Chemicals of emerging environmental concern (CoEECs) are suspected to cause environmental threat and may be potentially harmful to ecosystem and human health. Many CoEECs are highly carcinogenic and mutagenic, causing major health disorders such as acute, chronic, and systemic allergic reactions. They, in turn, affect the nervous, respiratory and circulatory systems, liver, organ, kidney and reproductive system of human beings and impact the surrounding ecosystem as well (Hovander et al., 2006; Park et al., 2007; She et al., 2007). Some of these CoEECs, such as plasticizers, fire retardants, pesticides and surfactants are well known, while newer ones like hormones, antimicrobials, antibiotics, pharmaceuticals and personal care products are being introduced all the time into the environment. These compounds have emerged as environmental contaminants concomitantly with the development of modern technologies and many toxic chemical industries. The CoEECs are reaching the environment (soil and natural waters) from consumer products through the use of sewage sludge on land, effluents from wastewater treatment plants (WWTP) and industrial discharges into freshwater and marine sites (Scott and Jones, 2000). Sources, types and harmful effects and variety of CoEECs were highlighted and explained in detail in the previous chapters. Several types of CoEEs are summarized in Table 12.1.

A hazardous waste site may contain hundreds of these contaminants of concern; therefore, it is important to remove/degrade/remediate these CoEECs to produce a safe environment for human life. Physiochemical methods are typically employed to remediate these compounds, which only render a temporary solution and do not eliminate the potential to further contaminate the environment by producing new pollutants (intermediate metabolites) or adding chemical reagents that may be more toxic than the parental compounds. Most often physiochemical methods are expensive, and not eco-friendly. Hence we need alternative methods such as biodegradation/bioremediation, which can be ecofriendly and cost-effective (Hattan

446

et al., 2003). Biodegradation/bioremediation is a process in which degradation of toxic compounds results in their conversion into non–toxic substances such as CO_2 and H_2O. This process can be facilitated either at contaminated sites (*in-situ* bioremediation) or in bioreactors (*ex-situ* bioremediation) using microorganisms to achieve complete detoxification of toxic compounds (Hwang and Cutright, 2002). It is found to be highly useful even in large-scale degradation of pollutants in the field.

Microorganisms play an important role in degradation of hydrocarbon contaminants from polluted aquifers and landfills by aerobic and anaerobic processes. Bacterial strains, besides diverse range of metabolic activities, also possess the ability to degrade a variety of organic compounds including numerous aliphatic (n-alkanes), polycyclic aromatic hydrocarbons (PAHs) and xenobiotics such as polychlorinated bi-phenyls (Olivera et al., 1997; Whyte et al., 1998).

The microbial degradation of two- and three-ring PAHs has been extensively reviewed (Atlas, 1981; Cerniglia, 1984 & 1992; Gibson and Subramanian, 1984) and, more recently, a variety of microorganisms have been isolated and shown to metabolize PAHs with up to four rings (Sutherland et al., 1995). Although individual species of bacteria and bacterial consortia have been shown to metabolize PAHs in laboratory culture, identifying such potential in a community of microorganisms *in situ* is more difficult.

Identification and biochemical characterization of such bacterial species using conventional methods are time consuming and laborious (Coenye et al., 2001). Hence, advanced molecular biology techniques have been employed, due to expediency and accuracy in identifying CoEEC degrading bacterial strains and relevant genes (Ringelberg et al., 2001). This degradative microorganism produced degradative/catabolic enzymes to degrade/metabolize the pollutants. These enzymes are encoded by the degradative genes present in their genome or plasmid DNA (Table 12.2). Therefore, identifying these degradative genes can lead to the identification of potential CoEEC degraders from the environment rapidly by using advanced molecular biology tools/techniques (Smits et al., 1999). In this chapter we discuss CoEEC degrading microorganisms and molecular biology techniques used to detect these microorganisms and their genes. Selected examples are presented to illustrate these techniques.

Table 12.1 A brief summary of different types of CoC and its compounds, along with their environmental occurrence.

S. No	Types of CoC	Examples of CoC	Sources	References
1	Plasticizers	1-Methylnaphthalene, 2,6-Dimethylnaphthalene, 2-Methylnaphthalene, Isophorone, Pyrene, Tetrachloroethylene, Tributyl phosphate, Tris(2-chloroethyl) phosphate, Triphenyl phosphate	Waste water Drinking water Surface water Ground water Soils Indoor Outdoor Air Biota	Lietz and Meyer, 2006, Andresen et al., 2007; Cho et al., 2007; et al., 2007; Mendell, 2007
2	Fire retardants	Tris(dichloroisopropyl) phosphate, Tris(2-butoxyethyl) phosphate	Waste water, Surface water, Ground water, Soils, Indoor and outdoor Air	Lietz and Meyer, 2006, Birnbaum and Staskal, 2004; Santillo and Johnston, 2003; Ebert and Bahadir, 2003
3	Pesticides	Carbaryl, Carbazole, Chlorpyrifos, Diazinon, Dichlorvos, Indole, Metalaxyl, Metolachlor, N,N-diethyl-*meta*-toluamide (DEET), Naphthalene, Pentachlorophenol, Prometon	Waste water, Drinking water, Surface water, Ground water, Soils, Biota	Lietz and Meyer, 2006, Posecion et al., 2006; Bielawski et al., 2005
4	Surfactants	Monoethoxyoctylphenol (OPEO1) 4-Cumylphenol 4-Nonylphenol (total, NP) 4-*n*-Octylphenol 4-*tert*-Octylphenol Diethoxynonylphenol (total NPEO2) Diethoxyoctylphenol (OPEO2)	All waters	Lietz and Meyer, 2006; Nielsen et al., 2007
5	Hormones	17 beta-estradiol (E2), Ethenyl estradiol (EE2), Estrone (E1)	Waste water, Drinking water, Surface water, Ground water, Soils, Indoor Air, Biota	Lietz and Meyer, 2006; Waring and Harris, 2005
6	Antimicrobials	3-*tert*-Butyl-4-hydroxyanisole (BHA) Anthraquinone Benzo[*a*]pyrene *d*-Limonene Phenol Triclosan	Waste water, Soils, Biota	Lietz and Meyer, 2006; Peng et al., 2006; Heidler and Halden, 2007
7	Antibiotics	Anhydro-erythromycin, Ciprofloxacin,	Waste water, Soils,	Lietz and Meyer, 2006; Pauwels and

		Clinafloxacin, Doxycycline, Erythromycin, Lincomycin, Ofloxacin, Oxolinic acid, Sulfadiazine, Sulfadimethoxine, Sulfamethoxazole, Sulfathiazole, Tetracycline, Trimethroprim, Tylosin	Biota	Verstraete, 2006; Kim and Carlson, 2007
8	Pharmaceuticals	1,7-dimethylxanthine, Carbamazapine, Codeine, Diltiazem, Diphenhydramine, Metformin, Trimethoprim	Waste water, soil	Lietz and Meyer, 2006; Jose Gomez et al., 2007
9	Personal care products	1,4-Dichlorobenzene 3-Methyl-1H-indole (skatol) Acetophenone Acetyl-hexamethyl-tetrahydro-naphthalene (AHTN), Benzophenone Camphor Cholesterol Cotinine Hexahydrohexamethylcyclopentaben zopyran (HHCB), Isoborneol Isoquinoline Menthol Methyl salicylate Triethyl citrate (ethyl citrate)	Waste water, soil	Lietz and Meyer, 2006; Rice and Mitra, 2007
10	Common CoC (Other hydrocarbons)	Beryllium, Chiral compounds, Chromium VI, Dichlorobenzenes, Dioxin, Di-nitrotoluenes (DNT), Methyl-tertiary-butyl ether (MTBE), Nanomaterials, Naphthalene, N-Nitrosodimethylamine (NDMA), Organotins, Perchlorate, Perfluorooctanoic acid (PFOA), Polybrominated diphenyl ethers (PBDEs) and polybrominated biphenyls (PBBs), Royal Demolition Explosive (RDX) /cyclotrimethylenetrinitramine, Tetrachloroethylene, Trichloroethylene (TCE), Tungsten and alloys, 1,4 dioxane, 1,2,3-trichloropropane (TCP), Isophorone, Isopropylbenzene (cumene), Phenanthrene, Tribromomethane	Waste water, Drinking water, Surface water, Ground water, Soils, Outdoor Air, Biota	Lietz and Meyer, 2006, Phillips et al., 2007; Cai et al., 2007

12.2 CoEEC Degrading Microbes and Their Genes

Many researchers have reported that microorganisms in the environment can efficiently adapt to use xenobiotic chemicals as novel growth and energy substrates. Their specialized enzyme metabolic pathways for the degradation of xenobiotic pollutants have been found in microbial strains that metabolize them completely and at considerable rates. These types of microbial strains have been isolated from natural systems, such as soil, wastewater, lakes and hydrocarbons (oil) polluted aquatic and terrestrial environments. The degradative enzymes include intracellular and extracellular, which are responsible for bioaccumulation, biotransformation, and co-metabolism, and can degrade low molecular weight compounds and a variety of macromolecules (Ju, 1997; Davis, 2002; Evans, 2003).

Bacterial genera like *Rhodococcus* (Smits et al., 1999), *Pseudomonads* (van Beilen et al., 1994 & 2001; Ratajczak et al., 1998) and *Acinetobacter* (Tani et al., 2001) are reported to degrade alkanes of petroleum hydrocarbons. These bacteria are known to degrade petroleum hydrocarbons by their cell bound enzymes in multistep biochemical reactions generating harmless primary and secondary metabolites for growth and metabolism. These microbial metabolisms can reduce hydrocarbon concentrations from polluted sites to the levels that no longer remain a threat to the environment and/or human health (Nalkes and Linz, 1997). Among these microorganisms, *Pseudomonas* species are the most important for biodegradation of toxic pollutants. Many aromatic compounds such as benzoate, p-hydroxybenzoate, maldelate, tryptophan, phthalate, and salicylate have been transformed into a common intermediate beta-ketoadipate by 1,2-dioxygenase produced by *Pseudomonas* species (Cafaro et al., 2004). Furthermore, some recalcitrant compounds, such as PAHs (polyaromatic hydrocarbons), can also be utilized by microbes and synthesize useful byproducts such as biosurfactants (Cooper and Paddock, 1984). Aliphatic hydrocarbons are assimilated by a wide variety of microorganisms, but not all microbial species are capable of utilizing them as a growth substrate. Many details of microbial hydrocarbon metabolism have been elucidated (Ratledge, 1984) including enzymology (Providenti, 1993), regulation and genetics (Witholt et al., 1990). The biodegradation process may be affected by physicochemical conditions (such as temperature, pH, redox potential, salinity, oxygen concentration) or by the availability of the substrates (solubility, dissociation from adsorbed materials, etc.). Little attention has been paid to the primary interaction of microorganisms and the pollutants involved.

Various microorganisms such as bacteria, cyanobacteria, fungi and algae are able to efficiently remove heavy metals from the environment via bioaccumulation and biosorption processes. Microorganisms can concentrate metals to levels that are largely higher than those existed in the environment. Biosorption is an important approach for the bioremediation of metal-contaminated environments by interactions between metals and live or dead microbial biomass. Microbes adsorb the toxic metals on their cell wall and recover them from polluted sites (Nandakumar et al., 1995).

Chlamydomonas reinhardtii grown in laboratory media were found to fractionate radioisotope selenium compound (Johnson, 2004). Microorganisms are also found to produce metalothiones, a metal accumulating protein secreted by bacterial species to adsorb metal ions from the surrounding environment.

Similarly, various plant species are also known to remove toxic metals from the soil and water through biosorption and bioassimilation (Hong et al., 2000; Zumriye, 2005). Researchers have found that plants also have the capability to degrade the toxic pollutants through enzymatic degradation. The use of certain plants to clean-up contaminated sites with a variety of pollutants such as heavy metals, radionuclides, chlorinated solvents and pesticides, petroleum hydrocarbons, polychlorinated biphenyls, polynuclear aromatic hydrocarbons, and explosives, is known as phytoremediation (Lasat, 2002; Singh and Labana, 2003). Elegant transgenic approaches have been designed for the development of mercury or arsenic phytoremediation technologies (Krämer, 2005).

Bioremediation of toxic pollutants by microorganisms is fast due to its easy acclimatization and adaptation to the adverse environmental conditions, fast growth (doubling/generation time) and ability to degrade various emerging toxic substances. Microorganisms capable of degrading toxic pollutants and their degradative genes are summarized in Table 12.2. Detection of these degradative genes using molecular biology techniques can facilitate easy identification of potential CoEEC degraders from mixed microbial communities.

12.3. Molecular Biology Techniques

To isolate potential CoEEC degraders from the environment one must know the degradative genes. Using this valuable information, we can make use of available advanced molecular biology tools to detect the degraders. Routine molecular biology techniques used for these purposes are highlighted and summarized in Figure 12.1.

There are two different approaches to investigate the diversity of catabolic genes in environmental samples: culture-dependent and culture-independent methods. In culture-dependent methods, bacteria are isolated from environmental samples with culture medium. Nucleic acid is then extracted from the bacterial culture. In contrast, culture-independent methods employ direct extraction of nucleic acids from environmental samples (Lloyd-Jones et al., 1999; Okuta et al., 1998; Watanabe et al., 1998). The description of catabolic gene diversity by culture-independent molecular biological methods often involves the amplification of DNA or cDNA from RNA extracted from environmental samples by polymerase chain reaction (PCR), and the subsequent analysis of the diversity of amplified molecules (community fingerprinting). Alternatively, the amplified products may be cloned and sequenced to identify and enumerate bacterial species present in the sample.

Table 12.2 A summary of degradative genes and its plasmid bearing CoC degrading microbial strains.

Name of plasmid	Compounds Degraded	Plasmid Size (kb)	Organisms	Reference
Plasmids carrying genes for degradation of (alkyl-substituted) aromatic compounds and related compounds via meta-cleavage pathway TOU/NAH/SAL plasmids				
pWW0	Xylene and toluene (*xyl* upper, lower, *xylS, xylR*)	117	*Pseudomonas putida mt-2*	Assinder et al., 1990; Genka et al., 2002; Tsuda and Iino, 1988.
pWW53	Xylene and toluene (*xyl* upper, lower, 1, 2, *xylS1, S2(incomplete), S3, xylR*)	107	*P. putida*	Keil et al., 1985a; Osborne et al., 1988; Tsuda and Genka, 2001
pWW15	Xylene and toluene (*xyl* upper 1, 2, lower 1, 2 *(incomplete), xylS, xylR*)	250	*Pseudomonas fluorescens MT15*	Keil et al., 1985b; O'Donnell et al., 1991; Williams and Worsey, 1976
pDK1	Xylene and toluene (*xyl* upper, lower, *xylS1, S2, S3, xylR*)	125	*P. putida HSI*	Assinder et al., 1993.
NAH7	Naphthalene, salicylate (*nah, sal* genes, *nahR*)	83	*P. putida G7*	Tsuda and Iino, 1990.
SAL	Salicylate (*sal* genes)	68	*Pseudomonas aeruginosa PAC (AC165)*	Chakrabarty, 1972; Lehrbach et al., 1983.
pVI150	Phenol, (Di)methylphenol (dmp genes)	>200	*P. putida CF600*	Powlowski and Shingler, 1994.
pWW100	Biphenyl, 4chlorobiphenyl (catechol meta-cleavage genes)	200	*Pseudomonas sp. Cl3406*	Lloyd-Jones et al., 1994.
pNL1	Xylene, naphthalene, biphenyl, etc. (catechol meta-cleavage genes)	184	*Sphingomonas arornaticivorans FI 99*	Romine et al., 1999.
Plasmids carrying genes for degradation of chlorinated aromatic compounds via chlorocatechol ortho-cleavage pathway 2,4-Dichlorophenoxyacetic acid (2,4-D)				
pJP4	2,4-D, 3-CBA, MCPaD (tfdA-S-R- tfd module II and tfd module I)	80	*Ralstonia eutropha JMP134*	Don and Pemberton, 1981 & 1985.
pMAB1	2,4-D, MCPA (tfdA-S- tfd module 1)	90	*Alcaligenes sp. CSV9O*	Bhat et al., 1994.
pRC10	2,4-D, 3-CBA, MCPA	45	*Flavobacterium sp. 50001*	Chaudhry and Huang, 1988.
pKA2	2,4-D	42.9	*Alcaligenes paradoxus 281 1 P*	Ka and Tiedje, 1994.
pEMT1	2,4-D	84	*Unidentified soil bacterium*	Top et al., 1995.
pEMT3	2,4-D	ca 60	*Unidentified soil bacterium*	Top et al., 1995.
pTV1	2,4-D (tfd genes)	200	*Variovorax paradoxus TV1*	Vallaeys et al., 1998.
pEST4011	2,4-D (tfd genes derived from pEST4002, 78 kb)	70	*Achromobacter xylosoxidans subsp.*	Mäe et al., 1993; Vedler et al., 2002a,b

pIJB1	2,4-D (tfd genes)	102	denitrificans EST4002 Burkholderia cepacia 2a	Poh et al., 2002; Xia et al., 1998.
3-chlorobenzoic acid (3-CBA) and others				
clc element (pB13)	3-CBA (clcRABDE)	105	Pseudomonas sp. P13	Ravatn et al., 1998; Chatterjee and Chakrabarty, 1983.
pAC27	3-CBA (clcRABDE)	110	P. putida AC866	Chatterjee and Chakrabarty, 1982; Coco et al., 1993.
pP51	1,2-Di-, 1,4-di and 1,2,4-trichlorobenzenes (tcbAB, tcbRCDEF)	110	Pseudomonas sp. P51	Van de Meer et al., 1991
pENH91	3-CBA (cbnRABCD)	78	R. eutropha NH9	Ogawa et al., 1999.
pPH111	3,5-Dichlorobenzoate (derived from pPB111)	120	P. putida P111D	Brenner et al., 1993.
Plasmids carrying (only) genes of aromatic ring-hydroxylating monooxygenase or dioxygenase system				
pEST1226	Phenol (pheBA)	<100	P. putida PaW85	Kasak et al., 1993.
pHMT112	Benzene (bedDC1 C2BA)	112	P. putida ML2 (NCIB 12190)	Fong et al., 2002; Tan and Mason, 1990
pPS12-1	Chlorobenzene (tecA1AZA3A4B)	<50	Burkholderia PSI2	Beil et al., 1999.
pBAH1	2-Chlorobenzoate (cbdABC)	70	Burkholderia cepacia 2CBS	Haak et al, 1995.
pPB111	2-Chlorobenzoate (chlorobenzoate 1,2-dioxygenase genes)	75	P. putida P111	Nakatsu et al., 1995.
pBRC60	3- and 4-Chlorobenzoates (cbaRABC)	88	Comamonas testosteroni BR60	Fukumori and Saint, 1997.
pTDN1	Aniline, m-, and p- toluidine (tdnQTAIA2B R)	79	P. putida UCC22	Suen and Spain, 1993.
pJS1	2,4-Dinitrotoluene (dntABD)	<180	Pseudomonas sp. DNP	Shields et al., 1995.
TOM	Toluene (toluene orthomonooxygenase and catechol 2,3-dioxygenase genes)	108	B. cepacia G4	Maeda et al., 2003.
pCAR1	Carbazole (carAaAaBaBbCAcAd)	199	Pseudomonas resinovorans CA10	Mäe et al., 1993.
pRHL1 and pRHL2	Polychlorinated biphenyls, ethylbenzene (bphA1AZA3A4CB and etbD1 on pRHL1, and bphB2, C2, C4, DEF, and etbC, D2 on pRHL2)	1100, 450	Rhodococcus sp. RHA1	Shimizu et al., 2001.
pBD2	Isopropylbenzene, trichloroethene (ipbA1A2A3A4C)	208	Rhodococcus erythropolis BD2	Cowles et al., 2000; Kesseler et al., 1996.

Plasmids carrying genes for other types of conversions or modifications of aromatic compounds				
NIC	Nicotine		*Pseudomonas convexa 1*	Thacker et al., 1978.
pADP-1	Atrazine *(atzA,* B, C, DEF)	109	*Pseudomonas sp. ADP*	Martinez et al., 2001.
pTSA	p-Toluenesulfonate (tsaRMBCD)	72	*Comamonas testosteroni T-2*	Junker et al., 1997.
pNB1 and pNB2	Nitrobenzene (nbzA,nbzCDE on pNB1, nbzB on pNB2)	59.1, 43.8	*P. putida HS12*	Park and Kim, 2000.
pPDL11	Carbofuran (mcd genes)	>100	*Achromobacte r sp. WM111*	Tomasek and Karns, 1989.
pRC1 and pRC2	Carbaryl via 1-naphrhol (by pRC1) to gentisate (by pRC2)	130, 120	*Arthrobacter sp. RC100*	Hayatsu et al., 1999.
Plasmids carrying genes of dehalogenases				
pASU1	4-Chlorobenzoate (fcbABC)	120	*Arthrobacter sp. SU*	Schmitz et al., 1992.
pUO1	Monohaloacetate (dehH1 and dehH2)	65	*Delftia acidovorans B*	Sota et al., 2002.
pXAU1	1,2-Dichloroethane (dhlA, ald genes)	ca. 200	*Xanthobacter autotrophicus GJ10*	Tardif et al., 1991.
Plasmids carrying degradative genes for alkanes and other compounds				
CAM	D-Camphor (camRDCAB)	ca. 500	*P. putida PpG1 (ATCC 17453)*	Unger et al., 1986.
OCT	n-alkanes (C5 to C12) (alkST-BFGHJKL)	ca. 500	*P. putida GPo1 (ATCC 29347) (Pseudomonas oleovorans GPo1)*	Van Beilen et al., 2001; Glick and Pasternak, 2003
pOAD2	Nylon oligomer (nylABC)	45.5	*Flavobacteriu m sp. KI72*	Kato et al., 1995; Negoro et al., 1983.

To date, more than 300 catabolic genes involved in catabolism of aromatic compounds have been cloned and identified from culturable bacteria. Several approaches have been used to discover catabolic genes for aromatic compounds from various bacteria; these include shotgun cloning by using indigo formation (Ensley et al. 1983; Goyal and Zylstra 1996), clearing zone formation (de Souza et al. 1995), or meta-cleavage activity (Sato et al. 1997) as screening methods for cloning; applying proteomics (two dimensional gel electrophoresis analysis) of xenobiotic-inducible proteins to obtain genetic information (Khan et al. 2001), transposon mutagenesis to obtain a defective mutant (Foght and Westlake 1996), transposon mutagenesis using a transposon-fused reporter gene (Bastiaens et al. 2001), applying a degenerate primer to generate a probe (Saito et al. 2000), and applying a short probe from a homologous gene (Moser and Stahl 2001).

Biodegradation requires the presence of microbes that can synthesize appropriate enzymes, target compounds at levels not lethal to the microorganisms and environment containing sufficient nutrients, cofactor, etc. The detection of organisms able to grow on toxic substances in samples from contaminated sites has been used to predict intrinsic bioremediation potential. Enrichment on a medium containing a toxic substance and probe hybridization of bacteria growing on the contaminants are common approach for detection (Aislabie et al., 1989; Villarreal et al., 1991; Lobus et al., 1992; Fries et al., 1994; Russ et al., 1994; Binks et al., 1995; Spiess et al., 1995; Sutton et al., 1996). These approaches assumed that the pressure exerted by the presence of toxic substance enriches the bacteria to develop the degradative pathway. Gene probes detect degradative microbes in enriched cultures and DNA extracts of soil by binding with catabolic genes (Walia et al., 1990; Holben et al., 1992; Sanseverino et al., 1993; Erb and wagner-dobler, 1993; Richard et al., 1996; Ka et al., 1996; Pellizari et al., 1996). The assumption is that the growth or positive hybridization with a probe indicates the presence of degradative microorganisms.

Identification can be carried out using culture dependent and culture independent methods. The first step of identification is deciding the degradative genes to be detected, followed by isolation of genomic or plasmid DNA, designing PCR primers and/or development of DNA probe. Subsequently potential degraders can be detected using different PCR techniques and/or hybridization techniques.

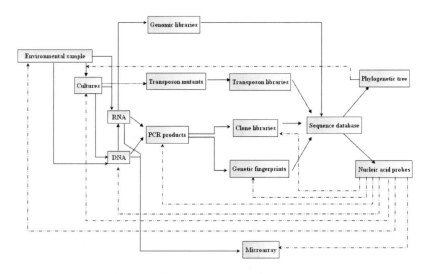

Figure 12.1 Molecular approaches for detection and identification of xenobiotic-degrading bacteria and their catabolic genes from environmental samples.

Different PCR techniques more often used in detection of microorganisms are quantitative PCR (qPCR), RT-PCR (reverse transcription PCR), real-time PCR and multiplex PCR. PCR techniques are used to detect the degraders using specific and/or degenerate primers. Similarly, different hybridization techniques can be employed for detection and quantification purposes, such as slot/dot blot hybridization, colony hybridization and southern blot.

Many novel molecular biology techniques are emerging for detection and quantification of degradative microbes such as DNA microarray and fibre optic biosensors. These techniques are explained with few examples in the following sections.

12.4. Detection of CoEEC Degraders

12.4.1 Conventional Methods

Traditional culture-dependent methods are generally based on differential morphological, physiological and metabolic traits (Leahy and Colwell, 1990; Rosenberg 1992; Atlas and Cerniglia 1995). These include microscopical analysis, isolation and cultivation of degradative microbes on selective solid media, most probable number (MPN) assays, and more recently, BIOLOG substrate utilization patterns. An example of the difficulties of estimating microbial populations using specific substrates or cultural requirements is the estimation of petroleum degrading microorganisms (Mesarch and Nies, 1997). Petroleum-degrading bacteria are abundant in soil and numerous plating techniques have been used for their estimation, including plates containing petroleum, polyaromatic hydrocarbons and crude oil. However, these have been shown to be difficult to use as some of the substrates are volatile and the results variable. It has been shown that a medium containing benzoate was better at discriminating petroleum-degrading microorganisms and could show difference in the microbial populations between clean and petroleum contaminated soil (Mesarch and Nies, 1997; Scragg, 1999). Enumeration and monitoring of xenobiotic-degrading bacterial populations in contaminated environments using traditional microbiological methods can take an inordinate length of time, and often underestimates numbers as a result of our inability to cultivate the majority of soil microorganisms (Lloyd-Jones et al., 1999).

Molecular approaches are now being used to characterize the nucleic acids of microorganisms contained in the microbial community from environmental samples (Figure 12.1). The major benefit of these molecular techniques is the ability to study microbial communities without culturing of bacteria and fungi, whereas analyses using incubation in the laboratory (classic microbiology) are indirect and produce artificial changes in the microbial community structure and metabolic activity. In addition, direct molecular methods preserve the in situ metabolic status and microbial community composition. Recent advances in molecular biology have extended our understanding of the metabolic processes related to the microbial transformation of CoEECs and their possible role in bioremediation.

12.4.2 Molecular Biology Techniques for Detection and Quantification of Pollutant Degraders

12.4.2.1. Polymerase Chain Reaction (PCR)

Based on degradative genes we can design and utilize a variety of PCR primers for PCR amplification of catabolic genes and thereby we can identify the degradative microbes from environment. The specific oligonucleotide sequences of these primers are available for a variety of hydrocarbon degradative genes (Whyte et al., 2002; Margesin et al., 2003; Luz et al., 2004) and dehalogenation of chlorinated organics (Fortin et al., 1998). Generally PCR primers can be designed as follows. Gene sequences for key enzymes from known bacterial biodegradative pathways are identified and searched for in databases such as the nucleotide database (GenBank) at the NCBI website (www.ncbi.nlm.nih.gov). The DNA sequences of all corresponding genes encoding the key enzyme are retrieved for comparative DNA and protein alignments using appropriate molecular biology sequence software. PCR forward and reverse primers for each catabolic gene sequence are then selected from alignments by PCR primer software and/or manually by identifying homologous regions shared by the selected DNA sequences. For construction of the oligonucleotide primers, long sequences originating from within the coding region of the catabolic genes and having a high guanine-cytosine (GC) content are preferred to ensure specificity. In addition, general criteria for designing catabolic gene primers must be followed as mentioned in literature (Sambrook and Russell, 2001; Margesin and Schinner, 2005).

Detection and Quantification of Degradative Microbes by qPCR. PCR is now often used for sensitive detection of specific DNA in environmental samples. Sensitivity can be enhanced by combining PCR with DNA probes, by running two rounds of amplification using nested primers (Moller et al., 1994), or by using real-time detection systems (Widada et al. 2001). Detection limits vary for PCR amplification, but usually between 10^2 and 10^3 cells/g soil can routinely be detected by PCR amplification of specific DNA segments (Fleming et al., 1994b; Moller et al. 1994). Despite its sensitivity, until recently it has been difficult to use PCR quantitatively to calculate the number of organisms (gene copies) present in a sample. Three techniques have now been developed for quantification of DNA by PCR, namely: most probable number-PCR (MPN-PCR), replicative limiting dilution-PCR (RLD-PCR), and competitive PCR (cPCR) (Chandler 1998). MPN-PCR is carried out by running multiple PCR reactions of samples that have been serially diluted, and amplifying each dilution in triplicate. The number of positive reactions is compared to the published MPN tables for an estimation of the number of target DNA copies in the sample (Picard et al. 1996). In MPN-PCR, DNA extracts are serially diluted before PCR amplification and limits can be set on the number of genes in the sample by reference to known control dilutions.

RLD-PCR, an alternate quantitative PCR for environmental application, is based on RLD analysis and the pragmatic tradeoffs between analytical sensitivity and practical utility (Chandler, 1998). This method has been used to detect and quantify specific biodegradative genes in aromatic-compound-contaminated soil. The catabolic genes *cdo*, *nahAc*, and *alkB* were used as target genes (Chandler, 1998). Quantitative cPCR is based on the incorporation of an internal standard in each PCR reaction. The internal standard (or competitor DNA) should be as similar to the target DNA as possible and be amplified with the same primer set, yet still be distinguishable from the target, for example, by size (Diviacco et al., 1992). A standard curve is constructed using a constant series of competitor DNA added to a dilution series of target DNA. The ratio of PCR-amplified DNA yield is then plotted versus initial target DNA concentration. This standard curve can be used for calculation of unknown target DNA concentrations in environmental samples. The competitive standard is added to the sample tube at the same concentration as used for preparation of the standard curve (Diviacco et al., 1992). Since both competitor and target DNAs are subjected to the same conditions that might inhibit the performance of DNA polymerase (such as humic acid or salt contaminants), the resulting PCR product ratio is still valid for interpolation of target copy number for the standard curve. Recently, Alvarez et al. (2000) have developed a simulation model for cPCR, which takes into account the decay in efficiency as a linear function of product yield. Their simulation data suggested that differences in amplification efficiency between target and standard templates induced biases in quantitative cPCR. Quantitative cPCR can only be used when both efficiencies are equal (Alvarez et al., 2000).

Detection of Novel Catabolic Genes. The emergence of methods using PCR to amplify catabolic sequences directly from environmental DNA samples now appears to offer an alternative technique to discover novel catabolic genes in nature. Most research focusing on analysis of the diversity of the catabolic genes in environmental samples has employed PCR amplification using a degenerate primer set (a primer set prepared from consensus or unique DNA sequence), and the separation of the resultant PCR products either by cloning or by gel electrophoresis (Allison et al. 1998; Hedlund et al. 1999; Lloyd-Jones et al. 1999; Watanabe et al. 1998; Wilson et al. 1999). To confirm that the proper gene has been PCR-amplified, it is necessary to sequence the product, after which the resultant information can be used to reveal the diversity of the corresponding gene(s). Over the last few years, these molecular techniques have been systematically applied to the study of the diversity of aromatic-compound-degrading genes in environmental samples (Table 12.3).

Application of a degenerate primer set to isolate functional catabolic genes directly from environmental samples has been reported (Okuta et al., 1998). Fragments of catechol 2,3-dioxygenase (C23O) genes were isolated from environmental samples by PCR with degenerate primers, and the gene fragments were inserted into the corresponding region of the *nahH* gene, the structural gene for C23O encoded by the catabolic plasmid NAH7, to reconstruct functional hybrid genes reflecting the diversity in the natural gene pool. In this approach, the only information necessary is knowledge of the conserved amino acid sequences in the protein family from which the degenerate primers should be designed. This method is generally applicable, and may be useful in establishing a divergent hybrid gene library for any gene family (Okuta et al., 1998).

When degenerate primers cannot be used for amplification of DNA or RNA targets, PCR has limited application for investigating novel catabolic genes from culture collections or from environmental samples. Dennis and Zylstra (1998) have developed a new strategy for rapid analysis of genes for Gram-negative bacteria. For this purpose, they constructed a minitransposon containing an origin of replication in an *Escherichia coli* cell. These artificially derived transposons are called plasposons (Dennis and Zylstra, 1998). Once a desired mutant has been constructed by transposition, the region around the insertion point can be rapidly cloned and sequenced. Mutagenesis with these plasposons can be used as an alternative tool to investigate novel catabolic genes from culture collections, although such approaches cannot be taken for environmental samples. The in vitro transposon mutagenesis by plasposon containing a reporter gene without a promoter will provide an alternative technique to search for desired xenobiotic-inducible promoters from environmental DNA samples.

Table 12.3 Molecular biology techniques have been systematically used to detect the aromatic-compound-degrading genes from environmental samples.

Target Gene	Molecular Approach	Source	Reference
Nah, nahAc, phnAc, glutathione-*S*-transferase, Phenol hydroxylase, (LmPH), RHD, *tfdC* PAH dioxygenase and catechol dioxygenase	RT-PCR with degenerate primers, PCR with several primers, PCR with several degenerate primers and PCR-DGGE with degenerate primers	Groundwater, Soil samples, Activated sludge and Prestine- and aromatic hydrocarbon-contaminated soils (culture-independent) & PAH soil bacteria, Marine sediment bacteria, Wastewater and soil bacteria, River water, sediment, and *Rhodococcus* sp. strain RHA1 (culture-dependent)	Allison et al. 1998 Cavalca et al. 1999 Hamann et al. 1999 Hedlund et al. 1999 Kitagawa et al. 2001 Lloyd-Jones et al. 1999 Meyer et al. 1999 Watanabe et al. 1998 Widada et al. 2002 Wilson et al. 1999 Yeates et al. 2000

Reverse Transcription PCR (RNA-Based Detection of Degradative Genes). One disadvantage of DNA-based methods is that they do not distinguish between living and dead organisms, which limits their use for monitoring purposes. The mRNA level may provide a valuable estimate of gene expression and/or cell viability under different environmental conditions (Fleming et al., 1993). Retrieved mRNA transcripts can be used to compare the expression level of individual members of gene families in the environment. Thus, when properly applied to field samples, mRNA-based methods may be useful in determining the relationships between the environmental conditions prevailing in a microbial habitat and particular in situ activities of native microorganisms (Wilson et al., 1999). Extraction of RNA instead of DNA, followed by reverse transcription-PCR (RT-PCR), gives a picture of the metabolically active microorganisms in the system (Nogales et al., 1999; Weller and Ward 1989). RT-PCR adds an additional twist to the PCR technique. Before PCR amplification, the DNA in a sample is destroyed with DNase. Reverse transcriptase and random primers (usually hexamers) are added to the reaction mixture, and the RNA in the sample – including both mRNA and rRNA – is transcribed into DNA. PCR is then used to amplify the specific sequences of interest. RT-PCR gives us the ability to detect and quantify the expression of individual structural genes. In a recent study, the fate of phenol-degrading *Pseudomonas* was monitored in bioaugmented sequencing batch reactors fed with synthetic petrochemical wastewater by using PCR amplification of the *dmpN* gene (Selvaratnam et al. 1995, 1997). In addition, RT-PCR was used to measure the level of transcription of the *dmpN* gene. Thus, not only was the presence of organisms capable of phenol degradation detected, but the specific catabolic activity of interest was also measured. A positive correlation was observed between the level of transcription, phenol degradation, and periods of aeration. In a similar study, transcription of the *tfdB* genes was measured by RT-PCR in activated-sludge bioreactors augmented with a 3-chlorobenzoate-degrading *Pseudomonas* (Selvaratnam et al. 1997), and the expression of a chlorocatechol 1,2-dioxygenase gene (*tcbC*) in river sediment was measured by RT-PCR (Meckenstock et al., 1998). Similarly, with this approach Wilson et al., (1999) isolated and characterized in situ transcribed mRNA from groundwater microorganisms catabolizing naphthalene at a coal-tar-waste-contaminated site using degenerate primer sets. They found two major groups related to the dioxygenase genes *ndoB* and *dntAc*, previously cloned from *Pseudomonas putida* NCIB 9816-4 and *Burkholderia* sp. strain DNT, respectively. Furthermore, the sequencing of the cloned RT-PCR amplification product of 16S rRNA generated from total RNA extracts has been used to identify presumptive metabolically active members of a bacterial community in soil highly polluted with PCB (Nogales et al., 1999).

Differential display (DD), an RNA-based technique that is widely used almost exclusively for eukaryotic gene expression, has been recently optimized to assess bacterial rRNA diversity (Yakimov et al., 2001). Double stranded cDNAs of rRNAs were synthesized without a forward primer, digested with endonuclease, and ligated with a double-stranded adapter. The fragments obtained were then amplified using an adapter-specific extended primer and a 16S rDNA universal primer pair, and displayed by electrophoresis on a polyacrylamide gel (Yakimov et al., 2001). In addition, the DD technique has been optimized and used to directly clone actively expressed genes from soil-extracted RNA (Fleming et al., 1998). Using this approach, Fleming et al., (2001) successfully cloned a novel salicylate-inducible naphthalene dioxygenase from *Burkholderia cepacia* (Fleming et al., 1998), and identified the bacterial members of a 2,4,5-trinitrophenoxyacetic acid-degrading consortium.

12.4.2.2. Nucleic Acid Probes (DNA-Based Methods)

DNA Probes and Colony and Southern Hybridization. A probe DNA may detect genes or gene sequences in total DNA isolated and purified from environmental samples by a variety of methods. DNA hybridization techniques, using labeled DNA as a specific probe have been used in the past for identification of specific microorganisms in environmental samples (Atlas, 1992; Sayler and Layton, 1990). Although these techniques are still useful for monitoring a specific genome in nature, they have some limitations. Colony hybridization can only be used for detection of culturable cells, and slot blot and Southern blot hybridization methods are not adequately sensitive for the detection of low cell numbers. On the other hand, greater sensitivity of detection, without reliance on cultivation, can be obtained using PCR (Jansson, 1995).

One of the earliest studies to use direct hybridization techniques for monitoring xenobiotic degraders, monitored the TOL (for toluene degradation) and NAH (for naphthalene degradation) plasmids in soil microcosms (Sayler et al. 1985). Colonies were hybridized with entire plasmids as probes to quantify the cells containing these catabolic plasmids. A positive correlation was observed between plasmid concentrations and the rates of mineralization. Exposure to aromatic substrates caused an increase in plasmid levels (Sayler et al. 1985). Similar techniques were used recently to monitor the *xylE* and *ndoB* genes involved in creosote degradation in soil microcosms (Hosein et al., 1997). Standard Southern blot hybridization has been used to monitor bacterial populations of naphthalene-degraders in seeded microcosms induced with salicylate (Ogunseitan et al., 1991). In this study, probes specific for the *nah* operon were used to determine the naphthalene-degradation potential of the microbial population. Dot-blot hybridizations with isolated polychlorinated biphenyl (PCB) catabolic genes have been used to measure the level of PCB-degrading organisms in soil microbial communities (Walia et al., 1990).

Molecular probing has been used in conjunction with traditional most-probable-number (MPN) techniques in several studies. A combination of MPN and colony hybridization was used to monitor the microbial community of a flow-through lake microcosm seeded with a chlorobenzoate-degrading *Alcaligenes* strain (Fulthorpe and Wyndham, 1989). This study revealed a correlation between the size and activity of a specific catabolic population during exposure to various concentrations of 3-chlorobenzoate. In another study, Southern hybridization with *tfdA* and *tfdB* gene probes was used to measure the 2,4-dichlorophenoxyacetic acid (2,4-D)-degrading populations in field soils (Holben et al. 1992). It was shown that amendment of the soil with 2,4-D increased the level of hybridization and that these changes agreed with MPN analyses.

pcpB Gene Probes. While there are many reports of many genes for biodegradation in the presence of contaminants, a few reports have recently appeared on the presence of genes for biodegradation of toluene in the absence of contaminants (Walia et al

1990; Hallier-soulier et al., 1996). This suggests the location of the genes may not depend solely on the presence of the contaminants (although the quantity, or frequency of the genes may be modified by the concentration of the compounds). Possible limiting usefulness of these studies was their use of target genes that may not be unique in terms of participation in biochemical pathways. The degradation pathways of PCP by various aerobic and anaerobic microorganisms have been extensively reviewed by (McAllistr et al., 1996). The pentachlorophenol monooxygenases is a very rare enzyme involved in the degradation of PCP under aerobic conditions by removing the parachlorine. PCP is a restricted use pesticide currently used only for the treatment of wood products for use in drainage ditch, drive ways and fence rows. Previous use was more extensive, but negative ecological effect on animals, especially bird and fish, led to severe limitation. However, it still enters the environment from treated wood, industrial discharges, spills and agriculture runoffs. The toxicity of PCP is apparently due to its ability to act as an uncoupler of oxidative phosphorylation. Several isolates can degrade PCP aerobically (Watanabe, 1973; Radehaus and Schmidt, 1992; Resnick and chapman., 1984). These were described as *Flavobacterium*, *Pseudomonas* SR3, *Pseudomonas* RA2 and *Arthrobacter*. All are now classified as *Sphingomonas chlorophenolica* (Karlson et al., 1995; Nohynek et al., 1995). The PCP-4 monooxygenease activity had been found in these *S. chlorophenolica* strains. Detection of PCP degraders was achieved by detecting *pcpB* genes from the soil microbes by developing DNA probes after amplification of *pcpB* gene using specific gene primers. Developed probes were evaluated using colony hybridization of environmental isolates to screen the PCP degraders (Ahmed et al., 2001).

Dioxygenase Gene Probes Identification of novel PAH and/or CoEEC degrading genes can be accomplished using a standard dioxygenase gene probes. This is mainly useful during development of DNA probes or PCR primers to detect the microbes degrading emerging pollutants, which metabolism is unknown (Zylstra et al., 1997; Singh and Ward, 2004).

dcmA Gene Probes. Similarly, detection of dichloromethane (DCM) degrading microbes using molecular biology techniques such as PCR, FISH and hybridization procedures can be achieved. DCM dehalogenase is an inducible enzyme encoded by the *dcm*A gene (La Roche and Leisinger, 1991). By targeting this gene detection we can detect all the DCM degrading microbes from the environment.

12.4.2.3. Environmental Microarrays

Microarrays are considered an emerging technology with tremendous potential in the field of environmental genomics (Greer et al., 2001; Fennell et al., 2004; Stahl, 2004; Zhou and Thomson, 2000). Successful application of microarrays technology, which uses high-density, high-microbial diversity and microbial ecology, as thousands of potential gene probes can be printed on an array and hybridized to labeled total nucleic acids extracted from environmental samples (Bowtell and Sambrook, 2003).

DNA microarray technology has a bright future, and a great deal of potential. This is due to the fact that there are a number of features of DNA microarray technology that are advantageous to its use. These include:

1. The great advantage of being able to study the behavior of many genes simultaneously.
2. The speed of using the technique is one of its main benefits. There can be as many as 150 copies of an array of 12,000 genes printed in only 1 day.
3. DNA microarray technology is relatively cheap to use. As previously described in Equipment the initial cost of constructing an arrayer is approximately $60,000. After this, the cost per copy of a microarray is small, usually less than $100. This has the added advantage of allowing research to take place much more freely; if investigation must be carried out, it can be done readily without much agonizing over the cost.
4. The technique of DNA microarrays is very user-friendly (the technique is neither radioactive nor toxic, the microscope slide is a convenient base for the technique, and arrays are cheap and easily replaced).
5. DNA microarrays are both adaptable and comprehensive. This accelerates the pace at which information is being generated and efficient ways for transforming this information into methods of exploration are very desirable.

A major advantage of DNA microarrays is that the information about the sequence of the DNA is not required to construct and use the DNA microarrays.

Major disadvantages of DNA microarrays include:

1. Quality and quality-control is highly variable
2. Quantity of data often overwhelms most users
3. Analysis and interpretation is difficult
4. Unknown significance of RNA. The final product of gene expression is protein, not RNA.
5. Relatively expensive.

Applications of Environmental Microarray. Environmental microarray technology is at a developmental stage where significant problems regarding specificity, sensitivity and quantification remain to be solved (Eyers et al., 2004; Fennell et al., 2004; Stahl 2004). Nevertheless, application-specific environmental microarrays were recently used to detect sulphate-reducing bacteria (Loy et al, 2002), methanotrophs (Bodrossy et al, 2003) and biodegradative populations (Fennell et al., 2004) in environmental samples. Two types of environmental microarrays are presently being developed. Functional gene microarrays (FGMA) contain variety of catabolic, biogeochemical cycling, heavy metal transformation genes, etc., as gene targets. Phylogenetic gene microarrays (PGMA) contain taxonomic gene targets, usually the 16S rDNA genes representing most genera of bacteria and Archaea. Environmental microarrays will be increasingly used to detect and characterize complex microbial communities in contaminated soils as well as to monitor degradative populations during bioremediation. In addition, to isolate the novel degradative microbes can be achieved using FGMA based approach. This will lead to a better understanding of important processes such as biogeochemical cycles and bioremediation in soils that are associated with microbial populations in natural environments (Margesin and Schinner, 2005).

12.4.2.4. Fibre Optic Biosensors

Nucleic acid probes are immobilized onto the surface of optical fibers and undergo hybridization with complementary nucleic acids introduced into the local environment of the sensor. Hybridization events are detected by the use of fluorescent compounds which bind into nucleic acid hybrids. The biosensor has many applications in detection and screening of genetic disorders, viruses, and pathogenic microorganisms. Biotechnology applications include monitoring of gene cultures and gene expression and the effectiveness (e.g. dose-response) of gene therapy pharmaceuticals. The invention includes biosensor systems in which fluorescent molecules are connected to the immobilized nucleic acid molecules. The preferred method for immobilization of nucleic acids is by in situ solid phase nucleic acid synthesis. Control of the refractive index of the immobilized nucleic acid is achieved by the support derivatization chemistry and the nucleic acid synthesis. The preferred optical fiber derivation yields a DNA coating of higher refractive index than the fiber core onto the fiber surface (US Patent # 6503711).

In 2003, Almadidy et al., was developed a fibre-optic biosensor for detection of genomic target sequences from *Escherichia coli*. A small portion of the LacZ DNA sequence is the basis for selection of DNA probe molecules that are produced by automated nucleic acid synthesis on the surface of optical fibres. Fluorescent intercalating agents are used to report the presence of hybridization events with target strands. This work reviewed the fundamental design criteria for development of nucleic acid biosensors and reports a preliminary exploration of the use of the biosensor for detection of sequences that mark the presence of *E. coli*. The biosensors were able to detect genomic targets from *E. coli* at a picomole level in a time of a few minutes, and dozens of cycles of use have been demonstrated (Almadidy et al., 2003). Much work on biosensor development was done in medical field (gene therapy pharmaceuticals) and detection of pathogens. Similar approach can be conducted to detect the potential CoC degraders, based on their specific oligonucleotide probes.

12.4.2.5. Bioreporter Technology to Detect GMO to Degrade Pollutants

Bioreporter refers to intact, living cells that have been genetically engineered/genetically modified (GMO) to produce a measurable signal transcriptionally induced in response to a specific chemical and physical agent in their environment. Bioreporters contain three essential genetic elements, a promoter sequence, a regulatory gene, and a reporter gene. In the wild type cell, the promoter gene is transcribed upon exposure to an inducing agent, leading to subsequent transcription of downstream genes encode for proteins that aid in the cell in either adapting or combating the agent to which it has been exposed. In the bioreporter, the downstream genes, or portions thereof, have been removed and replaced with a reporter gene. Consequently, transcription of the promoter gene activates the reporter gene. Reporter proteins are produced, and some type of measurable signal is generated. These signals can be categorized as colorimetric, florescent, luminescent, chemiluminescent, electrochemical or amperometric. These reporter genes includes, β-galactosidase (lacZ gene), green fluorescent protein (GFP gene), Luciferases (insect luciferase (luc gene) and bacterial luciferase (lux gene) and mini-transposons. Minitransposons used as genetic tools in bioreporter construction, which poses heavy metals resistance gene it make antibiotic selection obsolete. In 1994, a method for constructing and using mini transposons was well described in literature (de Lorenzo and Timmis, 1994; Margesin and Schinner, 2005). This is only useful to detect the genetically engineered microorganisms used in bioremediation studies.

12.4.2.6. Recent Development of Methods Increasing Specificity of Detection

A new approach that permits culture-independent identification of microorganisms responding to specified stimuli has been developed (Borneman, 1999). This approach was illustrated by the examination of microorganisms that respond to various nutrient supplements added to environmental samples. A thymidine nucleotide analog, bromodeoxyuridine (BrdU), and specified stimuli were added to environmental samples and incubated for several days. DNA was then extracted from an environmental sample, and the newly synthesized DNA was isolated by immunocapture of the BrdU-labelled DNA. Comparison of the microbial community structures obtained from total environmental sample DNA and the BrdU-labelled fraction showed significantly different banding patterns between the nutrient supplement treatments, although traditional total DNA analysis revealed no notable differences (Borneman, 1999). This approach provides a new strategy to permit identification of DNA from a stimulus or substrate-responsive organism in environmental samples. Application of such an approach in bioremediation by using the desired xenobiotic as a substrate or stimulus added to an environmental sample may provide a robust strategy to discover novel catabolic genes involved in xenobiotic degradation.

12.5 Conclusion

The use of molecular biology techniques in detection and quantification of degradative microorganisms was found useful. These techniques have provided useful information for improving biodegradation approach and monitoring the impact of biodegradation process on ecosystems. Many new developments in molecular biology techniques provided fast, sensitive and specific methods of detecting/identifying bacterial strains and their degradative genes from the complex environment. Also, these molecular techniques have been used for designing active biological containment systems to prevent the potentially undesirable spread of released microorganisms, mainly genetically engineered microorganisms. On the other hand, a complete understanding of the limitations of these techniques is essential.

Acknowledgements: The authors are sincerely thankful to Natural Sciences and Engineering Research Council of Canada (Grants A4984, Canada Research Chair) for financial support. Thanks to FQRNT (Québec) for providing Ph.D. Scholarship to S. Balasubramanian.

References

Ahmed, N., Qureshi, FM. And Khan, OY. 2001. Industrial and environmental biotechnology. Horizon Scientific Press. Wymondham, England.

Aislabie J, S Rothenburger, and R M Atlas. 1989. Isolation of microorganisms capable of degrading isoquinoline under aerobic conditions. Appl. Envir. Microbiol. 55: 3247-3249.

Allison DG, Ruiz B, San-Jose C, Jaspe A, Gilbert P (1998) Analysis of biofilm polymers of Pseudomonas fluorescens B52 attached to glass and stainless steel coupons. In: Abstracts of the General Meeting of the American Society for Microbiology, Atlanta, Georgia 98:325

Almadidy A., Watterson ., Piunno P.A.E., Foulds I.V., Horgen P.A., and Krull U. 2003. A fibre-optic biosensor for detection of microbial contamination1. Can. J. Chem. 81: 339–349.

Alvarez MJ, Depino AM, Podhajcer OL, Pitossi FJ (2000) Bias in estimations of DNA content by competitive polymerase chain reaction. Anal Biochem 28:87–94

Andresen JA, Muir D, Ueno D, Darling C, Theobald N, Bester K (2007) Emerging pollutants in the North Sea in comparison to Lake Ontario, Canada, data. Environ Toxicol Chem. 26(6):1081-9.

Assinder, S. J., P. De Marco, D. J. Osborne, C L. Poh, L. E. Shaw, M. K. Winson, and P. A. Williams. 1993. A comparison of the multiple alleles of xylS carried by TOL plasmids pWW53 and pDK1 and in implications for their evolutionary relationship.]. Gen. Microbiol. 139: 557-568.

Atlas M (1992) Molecular methods for environmental monitoring and containment of genetically engineered microorganisms. Biodegradation 3:137–146

Atlas, R. M. 1981. Microbial degradation of petroleum hydrocarbons: an environmental perspective. Microbiol. Rev. 45:180–209.

Atlas, RM and Cerniglia, CE. 1995. Bioremediation of petroleum pollutants: diversity and environmental aspects of hydrocarbon biodegradation. BioScience. 45: 332-338.

Bastiaens L, Springael D, Dejonghe W, Wattiau P, Verachtert H, Diels L (2001) A transcriptional luxAB reporter fusion responding to fluorine in Sphingomonas sp LB126 and its initial characterizations for whole-cell bioreporter purposes. Res Microbiol 152:849–859.

Beil, S., K. N. Timmis, and D. H. Pieper. 1999. Genetic and biochemical analyses of the tec operon suggest a route for evolution of chlorobenzene degradation genes. J. Baceriol. 181:341-346.

Bhat, M. A., M. Tsuda, K. Horiike, M. Nozaki, C. S. Vaidyanathan, and T. Nakazawa. 1994. Identification and characterization of a new plasmid carrying genes for degradation of 2,4dichlorophenoxyacefaie from Pseudomonas cepacia CSV90. Appl. Environ. Microbiol. 60:307-312.

Bielawski D, Ostrea E, Posecion N, Corrion M, Seagraves J. (2005) Detection of Several Classes of Pesticides and Metabolites in Meconium by Gas Chromatography-Mass Spectrometry. Chromatographia. 62(11-12): 623-629.

Binks PR, S Nicklin, and NC Bruce. 1995. Degradation of hexahydro-1,3,5-trinitro-1,3,5-triazine (RDX) by Stenotrophomonas maltophilia PB1. Appl. Envir. Microbiol. 61: 1318-1322.

Birnbaum LS, Staskal DF. (2004) Brominated flame retardants: cause for concern? Environ Health Perspect. 112(1): 9-17.

Bodrossy L., Nancy Stralis-Pavese, J. Colin Murrell, Stefan Radajewski, Alexandra Weilharter and Angela Sessitsch (2003) Development and validation of a diagnostic microbial microarray for methanotrophs. Environ Microbiol,5: 566–582.

Bollag, J. M and Dec, J. (1995) Detoxification of aromatic pollutants by fungal enzymes. Hinchee, R. E., Brockman, F. J., and Vogel, C. M. (Eds) Microbial process for bioremediation. Battelle press, Columbus, USA. pp. 67-74.

Borneman J (1999) Culture-independent identification of microorganisms that respond to specified stimuli. Appl Environ Microbiol 65:3398–3400.

Bowtell, D. and Sambrook, J. (2003) DNA Microarrays: A molecular cloning manual. CSHL press, New York, USA.

Brenner, V., B. S. Hernandez, and D. D. Focht. 1993. Variation in chlorobenzoate catabolism by Pseudomonas putida P111 as a consequence of genetic alterations. Appl. Environ. Microbiol. 59: 2790-2794.

Cafaro V., V. Izzo, R. Scognamiglio, E. Notomista, P. Capasso, A. Casbarra, P. Pucci, and A. L. Donato. 2004. Phenol hydroxylase and toluene/o-xylene monooxygenase from Pseudomonas stutzeri OX1: Interplay between two enzymes. Appl. Env. Microbiol. 70(4): 221-2219.

Cai QY, Mo CH, Wu QT, Zeng QY, Katsoyiannis A. (2007) Occurrence of organic contaminants in sewage sludges from eleven wastewater treatment plants, China. Chemosphere. 68(9):1751-62.

Cavalca L, Hartmann A, Rouard N, Soulas G (1999) Diversity of tfdC genes: distribution and polymorphism among 2,4-dichlorohenoxyacetic acid degrading soil bacteria. FEMS Microbiol Ecol 29:45–58

Cerniglia, C. E. 1984. Microbial metabolism of polycyclic aromatic hydrocarbons. Adv. Appl. Microbiol. 30:31–71; 8.

Cerniglia, C. E. 1992. Biodegradation of polycyclic aromatic hydrocarbons. Biodegradation 3:351–368.

Chakrabarty, K. M. 1972. Genetic basis of the biodegradation of salicylate in Pseudomonas. J. Bacteriol. 112: 815-823.

Chandler DP (1998) Redefining relativity: quantitative PCR at low template concentrations for industrial and environmental microbiology. J Ind Microbiol Biotechnol 21:128–140.

Chatterjee D. K.. S. T. Kellogg, S. Hamada, and A. M. Chakrabarthy. 1981. Plasmid specifying total degradation of 3 - chlorobenzoate by a modified ortho pathway. J. Bacteriol. 146: 639-646.

Chatterjee, D. K., and A. M. Chakrabarty. 1982. Genetic rearrangements in plasmids specifying total degradation of chlorinated benzoic acids. Mol. Gen. Genet. 188: 279-285.

Chatterjee, D. K., and A. M. Chakrabarty. 1983. Genetic homology between independently isolated chlorobenzoate degradative plasmids. J. Bacteriol. 153: 532-534.

Cho CW, Choi JS, Shin SC. (2007) Controlled release of pranoprofen from the ethylene-vinyl acetate matrix using plasticizer. Drug Dev Ind Pharm. 33(7):747-53.

Coco, W. M., R. K. Rothmel, S. Henikoff, and A M. Chakrabarty. 1993. Nucleotidc sequence and initial functional characterization of the clcR gene encoding a LysR family activator of the clcABD chlorocatechol operon in Pseudomonas putida . J. Bacteriol. 175: 417-427.

Coenye, T., P. Vandamme, J. R. Govan, and J. J. LiPuma. 2001. Taxonomy and identification of the Burkholderia cepacia complex. J. Clin. Microbiol. 39, 3427-3436.

Cooper, D. G. and Paddock, D. A. 1984. Production of a biosurfactants from Torulopsis bombicola. Appl. Env. Microbiol. 47: 173-176.

Cowels, C E, N. N. Nichols, and C. S. Harwood. 2000. BenR, a XylS homologue, regulates three different pathways of aromatic acid degradation in Pseudomonas putida. J. Bacteriol. 182: 6339-6346.

Davis, J. W., Odom, J. M., DeWeed, K. A., Stahl, D. A., Fishbain, S. S., West, R. J. 2002. Natural attenuation of chlorinated solvents at area 6, Dover Air force base: characterization of microbial community structure. Jounal of contaminant hydrology, 57(1-2), 41-59.

de Lorenzo V, Timmis KN (1994) Analysis and construction of stable phenotypes in gram-negative bacteria with Tn5- and Tn10-derived minitransposons. Methods Enzymol. 235: 386-405

de Souza, M. L., L. P. Wackett, K. L. Boundy-Mills, R. T. Mandelbaum, and M. J. Sadowsky. 1995. Cloning, characterization, and expression of a gene region from Pseudomonas sp. strain ADP involved in the dechlorination of atrazine. Appl. Environ. Microbiol. 61:3373–3378.

Dennis JJ, Zylstra GJ (1998) Plasposon: modular self-cloning mini-transposon derivatives for the rapid genetic analysis of Gram-negative bacterial genomes. Appl Environ Microbiol 64:2710–2715.

Diviacco S, Norio P, Zentilin L, Menzo S, Clementi M, Biamonti G, Riva S, Falaschi A, Giacca M (1992) A novel procedure for quantitative polymerase chain reaction by coamplification of competitive templates. Gene 122:313–320.

Don, R. H. and J. M. Pemberton. 1981. Properties of six pesticide degradation plarmids isolated from Alcaligenes eutrophus. J. Bacteriol. 145: 681-686.

Don. R H., and J. M. Pemberton. 1985. Generic and physical map of the dichlorophenoxyacetic acid-degradative plamid pJP4. J. Bacteriol. 161: 466-468.

Ebert J, Bahadir M. (2003) Formation of PBDD/F from flame-retarded plastic materials under thermal stress. Environ Int. 29(6):711-6.

Ensley BD, Ratzkin BJ, Osslund TD, Simon MJ, Wackett LP, Gibson DT (1983) Expression of naphthalene oxidation genes in Escherichia coli results in biosynthesis of indigo. Science 222:167–169.

Erb RW, Wagner-Doebler I (1993) Detection of polychlorinated biphenyl degradation genes in polluted sediments by directed DNA extraction and polymerase chain reaction. Appl Environ Microbiol 59:4065–4073.

Evans, G. M. and J. C. Furlong. 2003. Environmental Biotechnology: theory and applications. West Sussex, UK, John Wiley and Sons.

Eyers L., I. George[1], L. Schuler[1], B. Stenuit[1], S. N. Agathos[1] and Said El Fantroussi. (2004) Environmental genomics: exploring the unmined richness of microbes to degrade xenobiotics. Applied Microbiology and Biotechnology, 66: 123-130.

Fennell, D. E., S. K. Rhee, et al. (2004). "Detection and characterization of a dehalogenating microorganism by terminal restriction fragment length polymorphism fingerprinting of 16S rRNA in a sulfidogenic, 2-bromophenol-utilizing enrichment." Applied and Environmental Microbiology 70(2): 1169-1175.

Fleming CA, Leung KT, Lee H, Trevors JT, Greer CW (1994) Survival of lux-lac-marked biosurfactant-producing Pseudomonas aeruginosa UG2L in soil monitored by nonselective plating and PCR. Appl Environ Microbiol 60:1606–1613.

Fleming JT, Nagel AC, Rice J, Sayler GS (2001) Differential display of prokaryote messenger RNA and application to soil microbial communities. In: Rochelle PA (ed) Environmental molecular microbiology: protocol and applications. Horizon, Norfolk, England, pp 191–205.

Fleming JT, Sanseverino J, Sayler GS (1993) Quantitative relationship between naphthalene catabolic gene frequency and expression in predicting PAH degradation in soil at town gas manufacturing sites. Environ Sci Technol 27:1068–1074. Fleming JT, Yao W-H, Sayler GS (1998) Optimization of differential display of prokaryotic mRNA: application to pure culture and soil microcosms. Appl Environ Microbiol 64:3698–3706.

Foght JM, Westlake DWS (1996) Transposon and spontaneous deletion mutants of plasmid-borne genes encoding polycyclic aromatic hydrocarbon degradation by a strain of Pseudomonas fluorescens. Biodegradation 7:353–366.

Fons K. P. Y.. C B. H. Goh and K. M. Tan. 2000. The genes for benzene catabolism in Pseudomonas putida ML2 are flanked by two copies of the insertion element IS1489, forming a class-I-type catabolic transposon, Tn5542. Plasmid 43:103-110.

Fortin N., Roberta R. Fulthorpe, D. Grant Allen, and Charles W. Greer. (1998) Molecular analysis of bacterial isolates and total community DNA from kraft pulp mill effluent treatment systems. Can J of Microbiol, 44:537-546

Fries M R, J Zhou, J Chee-Sanford, and J M Tiedje. 1994. Isolation, characterization, and distribution of denitrifying toluene degraders from a variety of habitats. Appl. Envir. Microbiol. 60: 2802-2810.

Fukumori, F, and C. P. Saint 1997. Nucleotide sequences and regulatory analysis of genes involved in conversion of aniline to catechol in Pseudomonas putida UCC22 (pTDN1). J. Bacteriol. 179: 399-408.

Fulthorpe RR, Wyndham RC (1989) Survival and activity of a 3-chlorobenzoate-catabolic genotype in a natural system. Appl Environ Microbiol 55:1584-1590.

Genka, H.,Y. Nagata, and M. Tsuda. 2002. Site-specific recombination system encoded by toluene catabolic transposon Tn-4651. J. Bacteriol. 184: 4757-4766.

Gibson, D. I., and V. Subramanian. 1984. Microbial degradation of aromatic hydrocarbons, p. 181-252. In D. T. Gibson (ed.), Microbial degradation of organic compounds—1984. Mercel Dekker, Inc., New York, N.Y.

Glick, B. R. and Pasternak, J. J. (2003) Molecular biotechnology, Principles and applications of recombinant DNA. Chapter 13: bioremediation and biomass utilization. 3rd Ed. ASM press, Washington, D.C. pp. 378-415.

Goyal AK, Zylstra GJ (1996) Molecular cloning of novel genesfor polycyclic aromatic hydrocarbon degradation from Comamonas testosteroni GZ39. Appl Environ Microbiol 62:230-223.

Greer CW, Whyte LG, Lawrence JR, Masson , L., Brousseau R (2001) current and future impact of genomics-based technologies on environmental science. Environ Sci Technol, 35: 360A-366A.

Haak, B.. S. Fetzner , and F. Lingens. 1995. Cloning Nucleotides sequences and expression of the plasmid encoded genes for the two-component 2-halobenzene 1,2-dioxygenase from Pseudomonas cepacia 2CBS. J. Bacteriol. 177: 667-675.

Hallier-Soulier S, Ducrocq V, Mazure N, Truffaut N (1996) Detection and quantification of degradative genes in soils contaminated by toluene. FEMS Microbiol Ecol 20:121-133.

Hamann C, Hegemann J, Hildebrandt A (1999) Detection of polycyclic aromatic hydrocarbon degradation genes in different soil bacteria by polymerase chain reaction and DNA hybridization. FEMS Microbiol Lett 173:255-263

Hattan, G., Wilson B.and Wilson. J. T. (2003) Performance of conventional remedial technology for treatment of MTBE and benzene at UST sites in Kansas. Bioremediation Winter. 85-94.

Hayatsu, M., M. Hirano and T. Nagata. 1999. Involvement of two plasmids in the degradation of carbaryl by Arthrobacter sp. strain RC100. Appl.. Environ. Microbiol. 65: 1015-1019.

Hedlund BP, Geiselbrecht AD, Timothy JB, Staley JT (1999) Polycyclic aromatic hydrocarbon degradation by a new marine bacterium, Neptunomonas napthovorans, sp. Nov. Appl Environ Microbiol 65:251-259.

Heidler J, Halden RU. Mass balance assessment of triclosan removal during conventional sewage treatment. Chemosphere. 66(2):362-9.

Holben WE, Schroeter BM, Calabrese VGM, Olsen RH, Kukor JK, Biederbeck UD, Smith AE, Tiedje JM (1992) Gene probe analysis of soil microbial populations selected by amendment with 2,4-dichlorophenoxyacetic acid. Appl Environ Microbiol 58:3941-3948.

Hong H. B., Hwang S. H., and Chang Y. S. 2000. Biosorption of 1,2,3,4-tetrachlorodibenzo-p-dioxin and polychlorinated dibenzofurans by Bacillus pumilus. Water Res., 34: 349-353.

Hosein SG, Millette D, Butler BJ, Greer CW (1997) Catabolic gene probe analysis of an aquifer microbial community degrading creosote-related polycyclic aromatic and heterocyclic compounds. Microb Ecol 34:81–89.

Hovander L, Linderholm L, Athanasiadou M, Athanassiadis I, Bignert A, Fängström B, Kocan A, Petrik J, Trnovec T, Bergman A. (2006) Levels of PCBs and their metabolites in the serum of residents of a highly contaminated area in eastern Slovakia. Environ Sci Technol., 40(12), 3696-703.

Hwang, S. and Cutright. T. J. (2002) Biodegradability of aged pyrene and phenenthrene in a natural soil. Chemosphere. 47: 891-899.

Jansson JK (1995) Tracking genetically engineered microorganisms in nature. Curr Opin Biotechnol 6:275–283.

Johnson T. M.. 2004. A review of mass-dependent fractionation of selenium isotopes and implications for other heavy stable isotopes. Chemical geogology, 204: 201-214.

Jose Gomez M, Malato O, Ferrer I, Aguera A, Fernandez-Alba AR. (2007) Solid-phase extraction followed by liquid chromatography-time-of-flight-mass spectrometry to evaluate pharmaceuticals in effluents. A pilot monitoring study. J Environ Monit. 9(7): 718-29.

Ju, Y. H., Chen T. and Liu, J. C. 1997. A study on the biosorption of lindane. Colloids Surf B. 9: 187-196.

Junker, F., R. Kiewitz, and A. M. Cook. 1997. Characterization of the p – toluenesulfonates Operon tsaMBCD and tsaR in Comamonas testosteroni T-2. J. Bacteriol. 179: 919-927.

Ka J O, W E Holben, and J M Tiedje. 1994. Genetic and phenotypic diversity of 2,4-dichlorophenoxyacetic acid (2,4-D)-degrading bacteria isolated from 2,4-D-treated field soils. Appl. Envir. Microbiol. 60: 1106-1115.

Ka, J. O., and J. M. Tiedje. 1994. Integration and excision of a 2,4-dichlorophenoxyacetic acid degradative plasmid in Alcaligenes paradoxus and evidence of its natural intergenic transfer. J. Bacteriol. 176:5284-5289.

Karlson, U., Rojo, F., van Elsas, J. D. and Moore, E. 1995. Genetic and serological evidence for the recognition of four pentachlorophenol-degrading bacterial strains as a species of the genus Sphingomonas. Systematic and Applied Microbiology. 18: 539-548.

Kasak, L.,Horak, A. Nurk, K. Talvik, and M. Kivisaar. 1993. Regulation of the catechol 1,2-dioxygenase- and phenol Monooxygenase encoding pheBA operon in Pseudomonas putida PaW85 J. Bacteriol. 175:8038-8042.

Kato, K., K. Ohtsuki, Y. Koda, T. Maekawa, T. Yomo, S Negoro and 1. Urabe. 1995. A plasmid encoding enzymes for nylon oligomer degradation: nucleotide sequence and analysis of pOAD2. Microbiology. 141 : 2585-2590.

Keil. H.. S. Keil, R W. Pickup, and P. A. Williams. 1985a. Evolutionary conservation of genes coding for meta pathway enzymes within TOL plasmids pWW0 and pWW53. J. Bacteriol. 164: 887-895.

Keil. H.. S. Keil, R W. Pickup, and P. A. Williams. 1985b. TOL plasmid pW15 contains two nonhomologous, independently regulated catechol 2,3 –oxygenase genes. J. Bacteriol. 163: 248-255.

Kesseler, M., E. R. Dabbs, B, Averhoff, G. Gottschalk. 1996. Studies on the isopropylbenzene 2,3-dioxygenase genes encoded by the linear plasmid of Rhodococcus erythropolis BD2. Microbiology, 142 : 3241-3251.

Khan AA, Wang R-F, Cao W-W, Doerge DR, Wennerstrom D, Cerniglia CE (2001) Molecular cloning, nucleotide sequence and expression of genes encoding a polycyclic aromatic ring dioxygenase from Mycobacterium sp. strain PYR-1. Appl Environ Microbiol 67:3577–3585.

Kim SC and Carlson K. (2007) Temporal and spatial trends in the occurrence of human and veterinary antibiotics in aqueous and river sediment matrices. Environ Sci Technol. 41(1): 50-7

Kitagawa W, Suzuki A, Hoaki T, Masai E, Fukuda M (2001) Multiplicity of aromatic ring hydroxylation dioxygenase genes in a strong PCB degrader, Rhodococcus sp. strain RHA1 demonstrated by denaturing gel electrophoresis. Biosci Biotechnol Biochem 65:1907–1911.

Krämer, U. Phytoremediation: novel approaches to cleaning up polluted soils, Current Opinions in Biotechnology, 2005, 16:133-141

La Roche SD. and Leisinger T. (1991) identification of dcmR, the regulatory genes governing expression of dichloromethane dehalogenase in Methylobacterium sp. strain DM 4. J bacterial, 173; 6714-6721.

Lasat, M. M. 2002. Phytoextraction of toxic metals: a review of biological mechanisms. J. Environ. Qual. 31, 109-20

Leadbetter, J. R. (2005) Methods in ENZYMOLOGY, Volume 397: Environmental Microbiology, Elsevier Academic press, San Diego, California, USA.

Leahy JG. and Colwell, RR. 1990. Microbial degradation of hydrocarbons in the environment. Microbiol. Rev. 54 : 305-315.

Lehrbach, P. R.. I McGregor, J. M. Ward, and P. Broda. 1983. Molecular relationships between Pseudomonas INC p-9 degradative plasmids TOL, NAH, and SAL. Plasmid. 10: 164-174.

Lietz, A.C., and Meyer, M.T., (2006) Evaluation of emerging contaminants of concern at the South District Wastewater Treatment Plant based on seasonal sampling events, Miami-Dade County, Florida, 2004: U.S. Geological Survey Scientific Investigations Report 2006-5240, p 1-44.

Lloyd-Jones G, Laurie AD, Hunter DWF, Fraser R, (1999) Analysis of catabolic genes for naphthalene and phenanthrene degradation in contaminated New Zealand soils. FEMS Microbiol Ecol, 29, 69–79.

Llyod-Jones, G. C., de Jong, R. C. Orden, W. A. Duetz, and P. A. Williams. 1994. recombination of the bph (biphenyl) catabolic genes from plasmid pWW100 and their deletion during growth on benzoate. Appl. Environ. Microbiol. 60 691-696.

Lobos J H, T K Leib, and T M Su. 1992. Biodegradation of bisphenol A and other bisphenols by a gram-negative aerobic bacterium. Appl. Envir. Microbiol. 58: 1823-1831.

Loy A., Angelika Lehner, Natuschka Lee, Justyna Adamczyk, Harald Meier, Jens Ernst, Karl-Heinz Schleifer, and Michael Wagner (2002) Oligonucleotide

Microarray for 16S rRNA Gene-Based Detection of All Recognized Lineages of Sulfate-Reducing Prokaryotes in the Environment. Appl. Environ. Microbiol. 68: 5064-5081.

Luz A.P., V.H. Pellizari, L.G. Whyte, and C.W. Greer (2004) A survey of indigenous microbial hydrocarbon degradation genes in soils from Antarctica and Brazil, Can J of Microbiol, 50: 323-333.

Mäe, A. A., R. O. Marits, N. R. Ausmees, V. M. Kôiv, and A. L. Heinaru. 1993. Characterization of a new 2.4-dichlorophenoxyacetic acid degrading plasmid pEST4011 : physical map and localization of catabolic genes. J. gen. Microbiol. 139: 3165-3170.

Maeda. K., H Nojiri, M. Shintani, T. Yoshida, H. Habe, and T. Omori. 2003. Complete nucleotide sequence of carbozole/dioxin-degrading plasmid pCAR1 in Pseudomonas resinovorans strain CA10. J. Mol. Biol. 326: 21-33

Margesin R., D. Labbé, F. Schinner, C. W. Greer, and L. G. Whyte. (2003) Characterization of Hydrocarbon-Degrading Microbial Populations in Contaminated and Pristine Alpine Soils. Appl. Envir. Microbiol. 69: 3085-3092.

Margesin, R. and Schinner, F. (2005) Soil biology, Manual of soil analysis, monitoring and assessing soil bioremediation. Volume: 5. Springer publications, Heidelberg.

Martinez, B., J. Tomkins, L P. Wackett, R. Wing, and M. J. Sadowsky. 2001. Complete nucleotide sequence and organization of the attrazince catabolic plasmid pADP-1 from Pseudomonas sp. strain ADP. J. Bacteriol. 183: 5684-5697.

McAllistr, K. A., Lee, H. And Trevors, J. T. 1996. Microbial degradation of pentachlorophenol. Biodegradation. 7: 1-40.

Meckenstock R, Steinle P, van der Meer JR, Snozzi M (1998) Quantification of bacterial mRNA involved in degradation of 1,2,4-trichlorobenzene by Pseudomonas sp. strain P51 from liquid culture and from river sediment by reverse transcriptase PCR (RT/PCR). FEMS Microbiol Lett 167:123–129.

Mendell MJ. (2007) Indoor residential chemical emissions as risk factors for respiratory and allergic effects in children: a review. Indoor Air. 17(4): 259-77.

Mesarch M B., Cindy H. Nakatsu, and Loring Nies. 2000. Development of Catechol 2,3-Dioxygenase-Specific Primers for Monitoring Bioremediation by Competitive Quantitative PCR. Appl. Envir. Microbiol. 66: 678-683.

Meyer S, Moser R, Neef A, Stahl U, Kämpfer P (1999) Differential detection of key enzymes of polyaromatic-hydrocarbondegrading bacteria using PCR and gene probes. Microbiol. 145:7131–1741.

Moller A, Gustafsson K, Jansson JK (1994) Specific monitoring by PCR amplification and bioluminescence of firefly luciferase gene-tagged bacteria added to environmental samples. FEMS Microbiol Ecol 15:193–206.

Moser R, Stahl U (2001) Insights into the genetic diversity of initial dioxygenases from PAH-degrading bacteria. Appl Microbiol Biotechnol 55:609–618.

Nakatsu. C H., R. R. Fulthrope, B. A. Holland, M. C Peel, and R C Wyndham. 1995. Thephylogenetic distribution of a transposable dioxygenase from the niagara River water shed. Mol. Ecol. 4: 593-603.

Nalkes, D. V., and D. G. Linz. 1997. Environmentally acceptable endpoints in soil. American Academy of Environmental Engineers, Annapolis, Md.

Nandakumar, B. A., V. Dushenkov, H. Motto, I. Raskinse. 1995. Phytoextraction: Use of plants to remove heavy metals from soils, Environ. Sci. Technol,. 29(5): 1232-1235.

Negoro, S., T. Taniguchi, M. Kanaoka, H. Kimura, and H. Okada. 1983. Plasmid-determined enzymatic degradation of nylon oligomers. J. Bacteriol. 155: 22-31.

Nielsen GD, Larsen ST, Olsen O, Lovik M, Poulsen LK, Glue C, Wolkoff P (2007) Do indoor chemicals promote development of airway allergy? Indoor Air. 17(3): 236-55.

Nogales B, Moore ERB, Abraham W-R, Timmis KN (1999) Identification of the metabolically active members of a bacterial community in a polychlorinated biphenyl-polluted moorland soil. Environ Microbiol 1:199–212.

Nohynek, L. J., Suhonen, E. L., Nurmiaho-Lassila, E. L., Hantula, J. and Salkinnoja-Salonen, M. 1995. Description of four pentachlorophenol- degrading bacterial strains as Sphingomonas chlorophenolica sp. nov. Systematic and Applied Microbiology. 18: 527-538.

O'Donnell, K. J., and P. A Williams. 1991. Duplication of both xyl catabolic Operons on TOL plasmid pWW15. J. Gen. Microbiol. 137: 2831-2838.

Ogawa, N., Chakrabarty, A. M. and Zaborina, O. (2004) Plasmid biology. Chapter 16: Degradative plasmids. Funnell, B. E. and Philips, G. J. (Eds.). ASM press, Washington, D.C. pp. 341-344.

Ogawa, N.. and K. Miyashita 1999. The chlorocatechol catabolic transposon Tn5707 of Alcaligeens eutrophus NH9, carrying gene cluster highly homologous to that in the 1,2,4-trichlorobenzene degrading bacterium Pseudomonas sp. strain P51, confers the ability to grow on 3-chlorobenzoate. Appl. Environ. Microbiol. 65: 724-731.

Ogunseitan OA, Delgado IL, Tsai YL, Olson BH (1991) Effect of 2-hydroxybenzoate on the maintenance of naphthalene-degrading pseudomonads in seeded and unseeded soil. Appl Environ Microbiol 57:2873–2879

Okuta A, Ohnishi K, Harayama S (1998) PCR isolation of catechol 2,3-dioxygenase gene fragments from environmental samples and their assembly into functional genes. Gene 212: 221–228.

Olivera, N. L., Esteves J. L. and Commendatore, M.G. (1997). Alkane biodegradation by a microbial community from contaminated sediments in Patagonia, Argentina. Int.Biodet and Biodegrad. 40(1):75-79.

Osborne, D. J., R W. Pickup and P. A Williams. 1988. The presence of two complete homologous meta pathway operons on TOL plasmid pWW53. J. Gen. Microbiol. 134: 2965-2975.

Park JS, Linderholm L, Charles MJ, Athanasiadou M, Petrik J, Kocan A, Drobna B, Trnovec T, Bergman A, Hertz-Picciotto I. (2007) Polychlorinated biphenyls and their hydroxylated metabolites (OH-PCBS) in pregnant women from eastern Slovakia. Environ Health Perspect., 5(1), 20-7.

Park, H.S., and H.S. Kim. 2000. Identification and characterization of nitrobenzene catabolic plasmids pNB1 and pNB2 in Pseudomonas putida HS12. J. Bacteriol. 182: 573-580.

Pauwels B, Verstraete W. (2006) The treatment of hospital wastewater: an appraisal. J Water Health. 4(4): 405-16.

Pellizari VH, S Bezborodnikov, JF Quensen, 3rd, and JM Tiedje. 1996. Evaluation of strains isolated by growth on naphthalene and biphenyl for hybridization of genes to dioxygenase probes and polychlorinated biphenyl-degrading ability. Appl. Envir. Microbiol. 62: 2053-2058.

Peng X, Wang Z, Kuang W, Tan J, Li K (2006) A preliminary study on the occurrence and behavior of sulfonamides, ofloxacin and chloramphenicol antimicrobials in wastewaters of two sewage treatment plants in Guangzhou, China. Sci Total Environ. 371(1-3):314-22.

Phillips BM, Anderson BS, Hunt JW, Tjeerdema RS, Carpio-Obeso M, Connor V. (2007) Causes of water toxicity to Hyalella azteca in the New River, California, USA. Environ Toxicol Chem. 26(5):1074-9.

Picard C, Nesme X, Sinomet P (1996) Detection and enumeration of soil bacteria using the MPN-PCR technique. In: van Elsas JD (ed) Molecular ecology manual, vol 2. Kluwer, Dordrecht, pp 1–9.

Poh R. P.-C., A. R W. Smith, and 1. J. Bruce. 2002. Complete characterization of Tn5530 from Burkholderia cepacia strain 2a (pIJB1) and studies of 2,4-dichlorophenoxyacetate uptake by the organism. Plasmid. 48: 1-12.

Posecion N, Ostrea E, Bielawski D, Corrion M, Seagraves J, Jin Y. (2006) Detection of Exposure to Environmental Pesticides During Pregnancy by the Analysis of Maternal Hair Using GC-MS. Chromatographia. 64(11-12): 681-687.

Powlowski, J., and V. Shingler. 1994. Genetics and biochemistry of phenol degradation by Pseudomonas sp. CF600. Biodegradation. 5: 219-236.

Providenti, M.A., Lee, H. and Trevors J. C. 1993. Selected factors limiting the microbial degradation of recalcitrant compounds. J.Industrial Microbiol. 12,379-395.

Radehaus P M and S K Schmidt. 1992. Characterization of a novel Pseudomonas sp. that mineralizes high concentrations of pentachlorophenol. Appl. Envir. Microbiol. 58: 2879-2885.

Ratajczak. A., G. Walter, and H. Wolfgang. 1998. Alkane hydroxylase from Acinetobacter sp. Strain ADP1 is encoded by alk M and belongs to a new family of bacterial integral membrane hydrocarbon hydroxylases. Appl. Env. Microbiol. 64(4): 1175-1179.

Ratledge, C. 1984. Microbial conversion of alkanes and fatty acids. JAOCS 61,447-453.

Ravatn, R., S. Studer. A. J. B. Zehnder, and J. R. Van der Meer. 1998. Int-B13, an unusual site-specific recombinase of the bacteriophage P4 integrase family, responsible for chromosaomal insertion of the 105-kilobase clc element Pseudomonas sp. strain B13. J. Bacteriol.. 180: 5505-5514.

Resnick, S. M. and Chapman, P. J. 1984. Physiological properties and substrate specificity of a pentachlorophenol-degrading Pseudomonas species. Biodegradation. 5: 47-54.

Rice SL, Mitra S. (2007) Microwave-assisted solvent extraction of solid matrices and subsequent detection of pharmaceuticals and personal care products (PPCPs) using gas chromatography-mass spectrometry. Anal Chim Acta. 589(1):125-32.

Richards N.K., H.K. Mahanty, and J. Aislabie. 1994. Construction of a DNA probe to detect isoquinoline-degrading bacteria. Canadian Journal of Microbiology 40: 561-566

Ringelberg, D. B., Talley, J. W., Perkins, E. J., Tucker, S. G., Luthy, R. G., Bouwer, E. J. and Fredrickson. H. L. (2001). Succession of phenotypic, genotypic and metabolic community characteristics during in vitro bioslurry treatment of polycyclic aromatic hydrocarbon-contaminated sediments. Appl. Environ. Microbiol. 67(4): 1542-1550.

Romine M. F., Lisa C. Stillwell, Kwong-Kwok Wong, Sarah J. Thurston, Ellen C. Sisk, Christoph Sensen, Terry Gaasterland, Jim K. Fredrickson, and Jeffrey D. Saffer 1999. Complete Sequence of a 184-Kilobase Catabolic Plasmid from Sphingomonas aromaticivorans F199. J. Bacteriol. 181: 1585-1602.

Rosenberg, E. 1992. The ydrocarbon-oxidizing bacteria. In : Balows A (ed) The prokaryotes : a handbook on the biology of bacteria : ecophysiology, isolation, identification, applications. Springer, Berlin Heidelberg NewYork, pp 446-459.

Russ, R., Muller, C., Knackmuss, H. J. And Scotz, A. 1994. Aerobic biodegradation of 3-aminobenzoate by gram negative bacteria involves intermediate formation of 5-amino salicylate as ring cleavage substrate. FEMS Microbiol Letters. 122: 137-143.

Saboo, V., Sullivan, M., Sobhon, V. and Gealt, M. A. (2001) Industrial and environmental biotechnology. Chapter 3: Genes for pentachlorophenol degradation and the bacteria that contain them. Ahmed, N., Qureshi, F. M. and Khan, O. Y. (Eds). Horizon press, Wymondham, England. pp. 21-32.

Saito A, Iwabuchi T, Harayama S (2000) A novel phenanthrene dioxygenase from Nocardiodes sp. strain KP7: expression in Escherichia coli. J Bacteriol 182:2134–2141.

Sambrook, J. and Russell, D. W. (2001) Molecular cloning, A laboratory Manual, Volume 2. 3rd Ed. CSHL press, New York, USA.

Sanseverino J. , C. Werner, J. Fleming, B. Applegate, J. M. H. King and G. S. Sayler. 1993. Molecular diagnostics of polycyclic aromatic hydrocarbon biodegradation in manufactured gas plant soils. Biodegradation. 4: 303-321.

Santillo D, Johnston P. (2003) Playing with fire: the global threat presented by brominated flame retardants justifies urgent substitution. Environ Int. 29(6):725-34.

Sato S, Nam J-W, Kasuga K, Nojiri H, Yamane H, Omori T (1997) Identification and characterization of the gene encoding carbazole 1,9a-dioxygenase in Pseudomonas sp. strain CA10. J Bacteriol 179:4850–4858.

Sayler GS, Layton AC (1990) Environmental application of nucleic acid hybridization. Annu Rev Microbiol 44:625–648.

Sayler GS, Shields MS, Tedford ET, Breen A, Hooper SW (1985) Application of DNA-DNA colony hybridization to the detection of catabolic genotypes in environmental samples. Appl Environ Microbiol 49:1295–1303.

Schmitz, A., K H Gartemann, J Fiedler, E Grund, and R Eichenlaub. 1992. Cloning and sequence analysis of genes for dehalogenation of 4-chlorobenzoate from Arthrobacter sp. strain SU. Appl. Envir. Microbiol. 58: 4068-4071.

Scott, M. J. and Jones, M. N. (2000) The biodegradation of surfactants in the environment. Biochimica et Biophysica Acta., 1508, 235-251

Scragg, A. (1999) Environmental biotechnology. Chapter 1 & 2: Overview & Environmental monitoring. Addison Wesley Longman Singapore (Pte) Ltd., Singapore. pp. 1-45.

Selvaratnam S, Schoedel BA, McFarland BL, Kulpa CF (1995) Application of reverse transcriptase PCR for monitoring expression of the catabolic dmpN gene in a phenol-degrading sequencing batch reactor. Appl Environ Microbiol 61:3981–3985.

Selvaratnam S, Schoedel BA, McFarland BL, Kulpa CF (1997) Application of the polymerase chain reaction (PCR) and reverse transcriptase/PCR for determining the fate of phenol-degrading Pseudomonas putida ATCC 11172 in a bioaugmented sequencing batch reactor. Appl Microbiol Biotechnol 47:236–240.

She J, Holden A, Sharp M, Tanner M, Williams-Derry C, Hooper K. (2007) Polybrominated diphenyl ethers (PBDEs) and polychlorinated biphenyls (PCBs) in breast milk from the Pacific Northwest. Chemosphere., 67(9), S307-17.

Shields MS, MJ Reagin, RR Gerger, R Campbell, and C Somerville. 1995. TOM, a new aromatic degradative plasmid from Burkholderia (Pseudomonas) cepacia G4. Appl. Environ. Microbiol. 61: 1352-1356.

Shimizu S, Hiroyuki Kobayashi, Eiji Masai, and Masao Fukuda. 2001. Characterization of the 450-kb Linear Plasmid in a Polychlorinated Biphenyl Degrader, Rhodococcus sp. Strain RHA1. Appl. Envir. Microbiol. 67: 2021-2028.

(2007) Tripodal chelating ligand-based sensor for selective determination of Zn(II) in biological and environmental samples. Anal Bioanal Chem. 388(8):1867-76.

Singh, A. and Ward, O. P. (2004) Soil biology, Biodegradation and bioremediation. Volume: 2. Springer publications, Heidelberg.

Singh, O. V, and S. Labana. 2003. Phytoremediation: an overview of metallic ion decontamination from soil. Appl. Microbiol. Biotechnol. 61, 405-12.

Smits, T. H. M., Rothlisberger, M., Witholt. B. & Van Beilen. J. B. (1999). Molecular screening for alkane hydroxylase genes in Gram-negative and Gram-positive strains. Environ. Microbiol. 1(4), 307-317.

Sota M, Masahiro Endo, Keiji Nitta, Haruhiko Kawasaki, and Masataka Tsuda. 2002. Characterization of a Class II Defective Transposon Carrying Two Haloacetate Dehalogenase Genes from Delftia acidovorans Plasmid pUO1 Appl. Envir. Microbiol. 68: 2307-2315.

Spiess E, C Sommer, and H Gorisch. 1995. Degradation of 1,4-dichlorobenzene by Xanthobacter flavus 14p1. Appl. Envir. Microbiol. 61: 3884-3888.

Stahl, DA., (2004) High-throughput techniques for analysing complex bacterial communities. Adv Exp Med Biol, 547: 5-17.

Suen W C and J C Spain. 1993. Cloning and characterization of Pseudomonas sp. strain DNT genes for 2,4-dinitrotoluene degradation. J. Bacteriol. 175: 1831-1837.

Sutherland, J B.; Rafii, F.; Khan, A A.; Cerniglia, C E. (1995) Mechanisms of polycyclic aromatic hydrocarbon degradation. In: Young L Y, Cerniglia C E. , editors; Young L Y, Cerniglia C E. , editors. Microbial transformation and degradation of toxic organic chemicals—1995. New York, N.Y: Wiley-Liss, Inc. pp. 269–306.

Sutton SD, SL Pfaller, JR Shann, D Warshawsky, BK Kinkle, and JR Vestal. 1996. Aerobic biodegradation of 4-methylquinoline by a soil bacterium. Appl. Envir. Microbiol. 62: 2910-2914.

Tan. H.-M, and J. R. Mason. 1990. Cloning and expression of the plasmid encoded benzene dioxygenase genes from Pseudomonas putida ML2. FEMS Microbiol. Lette. 72: 259-264.

Tani, A., T. Ishige, Y. Sakai, and N. Kato. 2001. Gene structures and regulation of the alkane hydroxylase complex in Acinetobacter sp. Strain M-1. J. Bacteriol. 183: 1819-1823

Tardif G, C W Greer, D Labbé, and P C Lau. 1991. Involvement of a large plasmid in the degradation of 1,2-dichloroethane by Xanthobacter autotrophicus. Appl. Envir. Microbiol. 57: 1853-1857.

Thacker R, O Rørvig, P Kahlon, and I C Gunsalus. 1978. NIC, a conjugative nicotine-nicotinate degradative plasmid in Pseudomonas convexa. J. Bacteriol. 135: 289-290.

Tomasek P H and J S Karns. 1989. Cloning of a carbofuran hydrolase gene from Achromobacter sp. strain WM111 and its expression in gram-negative bacteria. J. Bacteriol. 171: 4038-4044.

Top EM, WE Holben, and LJ Forney. 1995. Characterization of diverse 2,4-dichlorophenoxyacetic acid-degradative plasmids isolated from soil by complementation. Appl. Envir. Microbiol. 61: 1691-1698.

Tsuda M., and Iino, T. 1988. Identification and characterization of Tn4653 on TOL plasmid pWW0. Mol. Gen. Genet. 213: 72-77.

Tsuda M., and Iino, T. 1990. Napthalene degrading genes on plasmid NAH7 are on a defective transposon. Mol. Gen. Genet. 223: 33-39.

Tsuda Masataka and Hiroyuki Genka . 2001. Identification and Characterization of Tn4656, a Novel Class II Transposon Carrying a Set of Toluene-Degrading Genes from TOL Plasmid pWW53. J. Bacteriol. 183: 6215-6224.

Unger BP, IC Gunsalus, and SG Sligar. 1986. Nucleotide sequence of the Pseudomonas putida cytochrome P-450cam gene and its expression in Escherichia coli. J. Biol. Chem. 261: 1158-1163.

United States Patent # 6503711, http://www.patentstorm.us/patents/6503711.html.

Vallaeys T, Lionel Albino, Guy Soulas, Alice D. Wright and Andrew J. Weightman. 1998. Isolation and characterization of a stable 2,4-dichlorophenoxyacetic acid degrading bacterium, Variovorax paradoxus, using chemostat culture. Biotechnol. Letters. 20: 1073-1076.

van Beilen, J. B., M. G. Wubbolts, and B. Witholt. 1994. Genetics of alkane oxidation by Pseudomonas oleovarans. Biodegradation. 5: 161-174.

van der Meer J R, A C Frijters, J H Leveau, R I Eggen, A J Zehnder, and W M de Vos . 1991. Characterization of the Pseudomonas sp. strain P51 gene tcbR, a LysR-type transcriptional activator of the tcbCDEF chlorocatechol oxidative operon, and analysis of the regulatory region. J. Bacteriol. 173: 3700-3708.

Vedler, E., Koiv, V. and Heinaru, A. 2000. Analysis of the 2,4-dichlorophenoxyacetic acid-degradative plasmid pEST4011 of Achromobacter xylosoxidans subsp.denitrificans strain pEST4011. Gene, 255 (2): 281-288.

Vedler, E., Koiv, V. and Heinaru, A. 2000. TfdR, the LysR-type transcriptional activator, is responsible for the activation of the tfdCB operon of Pseudomonas.putida.2,4-dicholorophenoxyacetic acid degradative plasmid pEST4011. Gene, 245 (1):.161-168.

Villarreal D T, R F Turco, and A Konopka. 1991. Propachlor degradation by a soil bacterial community. Appl. Envir. Microbiol. 57: 2135-2140.

Walia S, Khan A, Rosenthal N (1990) Construction and applications of DNA probes for detection of polychlorinated biphenyl-degrading genotypes in toxic organic-contaminated soil environments. Appl Environ Microbiol 56:254–259.

Waring RH, Harris RM. (2005) Endocrine disrupters: a human risk? Mol Cell Endocrinol. 244(1-2): 2-9.

Watanabe K, Teramoto M, Futamata H, Harayama S (1998) Molecular detection, isolation, and physiological characterization of functionally dominant phenol-degrading bacteria in activated sludge. Appl Environ Microbiol 64:4396–4402.

Watanabe, I. 1973. Isolation of pentachlorophenol decomposing bacteria from soil. Soil Sci. Plant Nutrit. 19: 109-116.

Weller R, Ward DM (1989) Selective recovery of 16S ribosomal RNA sequences from natural microbial communities in the form of complementary DNA. Appl Environ Microbiol 55:1818–1822.

Whyte, L. G., Hawari, J., Zhou, E., Bourbonniere, L., Inniss, W. E. and Greer, C.W. (1998). Biodegradtion of variable–chain-length alkanes at low temperatures by a psychrotrophic Rhodococcus sp. Appl. Environ. Microbiol. 64: 2578-2584.

Whyte, L.G., A. Schultz, J.B. van Beilen, A.P. Luz, V. Pellizari, D. Labbé, C.W. Greer. 2002. Prevalence of alkane monooxygenase genes in Arctic and Antarctic hydrocarbon-contaminated and pristine soils. FEMS Microbiology Ecology. 41(2):141-150.

Widada J, Nojiri H, Kasuga K, Yoshida T, Habe H, Omori T (2001) Quantification of carbazole 1,9a-dioxygenase gene by real-time competitive PCR combined with co-extraction of internal standards. FEMS Microbiol Lett 202:51–57.

Widada J, Nojiri H, Kasuga K, Yoshida T, Habe H, Omori T (2002) Molecular detection and diversity of polycyclic aromatic hydrocarbon-degrading bacteria isolated from geographically diverse sites. Appl Microbiol Biotechnol 58:202–209

Williams P A and M J Worsey. 1976. Ubiquity of plasmids in coding for toluene and xylene metabolism in soil bacteria: evidence for the existence of new TOL plasmids. J. Bacteriol. 125: 818-828.

Wilson MS, Bakerman C, Madsen EL (1999) In situ, real-time catabolic gene expression: extraction and characterization of naphthalene dioxygenase mRNA transcripts from groundwater. Appl Environ Microbiol 65:80–87.

Witholt, B., de Smet, M.J., Kingma, J., van Beilen, J.B., Kok, M., Lageveen, R.G. and Eggink, G. 1990. Bioconversion of aliphatic compounds by Pseudomonas putida in multiphase bioreactors; Background and economic potential. Trends Biotechnol. 8, 46-52

Xia, X S., Aathithan, S., Kamilla Oswiecimska, Anthony R. W. Smith and Ian J. Bruce. 1998. A Novel Plasmid pIJB1 Possessing a Putative 2,4-Dichlorophenoxyacetate Degradative Transposon Tn5530 in Burkholderia cepacia Strain 2a. Plasmid. 39 (2): 154-159.

Yakimov MM, Giuliano L, Timmis KN, Golyshin PN (2001) Upstream- independent ribosomal RNA amplification analysis (URA): a new approach to characterizing the diversity of natural microbial communities. Environ Microbiol 3:662–666.

Yeates C, Holmes AJ, Gillings MR (2000) Novel forms of ring hydroxylating dioxygenases are widespread in pristine and contaminated soils. Environ Microbiol 2:644–653

Zhou J. and Thomson, DK., (2000) challenges in applying microarrays to environmental studies. Curr Opin Biotechnol, 13: 204-207.

Zumriye Aksu. 2005. Application of biosorption for the removal of organic pollutants: a review. Process Biochemistry 40: 997–1026.

Zylstra GJ, Kim E, Goyal AK. (1997) Comparative molecular analysis of genes for polycyclic aromatic hydrocarbon degradation. Genet Eng, 19: 257-269.

Editor Biographies

Craig D. Adams, P.E., F.ASCE, is Chair of the Department of Civil, Environmental and Architectural Engineering, and the J. L. Constant Distinguished Professor at the University of Kansas. He received his B.S. in Chemical Engineering, and an M.S. and Ph.D. in Environmental Health Engineering at the University of Kansas. Professor Adams was formerly on the faculty of Missouri S&T (formerly UMR) (1995-2008) and Clemson University (1991-1995). He worked in research and development in industry (1983-1987). Dr. Adams serves as a member of the International Water Association (IWA) USA National Committee, and the Chair of the IWA Specialist Group on Adsorption, and as Associate Editor for the ASCE *Practice Periodical for Hazardous, Toxic, and Radioactive Waste Management*. He was also the 2003 winner of the ASCE Rudolph Hering Medal, and the 2005 winner of the ASCE State-of-the-Art Civil Engineering Award. Dr. Adams contributes extensively to professional organizations and government regulatory agencies including the American Water Works Association and the U.S. Environmental Protection Agency, respectively. Dr. Adams' primary research and publications focus on the emerging contaminants including antibiotics, endocrine disrupting chemicals, estrogens, and disinfection byproducts in drinking water, wastewater and food with emphasis on analysis, treatment and fate modeling. Dr. Adams has over 50 peer-reviewed journal publications.

Alok Bhandari, P.E., M.ASCE, is an Associate Professor of Environmental Engineering in the Department of Agricultural and Biosystems Engineering at Iowa State University. He received his B.Tech. in Civil Engineering from J N Tech University, India, and M.S. and Ph.D. degrees in Civil and Environmental Engineering from Virginia Tech. Dr. Bhandari conducted post-doctoral research at the University of Michigan, Ann Arbor (1995-1997) and was formerly on the faculty of civil engineering at Kansas State University (1998-2007). He is a recipient of the NSF CAREER Award, Kansas's K-STAR First Award, and Virginia Tech Civil and Environmental Engineering's Outstanding Young Alumni Achievement Award. Dr. Bhandari serves as an Associate Editor of ASCE's *Journal of Environmental Engineering*, and *Practice Periodical of Hazardous, Toxic and Radioactive Waste Management*. His research focuses on understanding the environmental fate and transport of agrichemicals and contaminants of emerging concern in natural and engineered systems. Dr. Bhandari has authored over 80 technical and scholarly publications including 30 in peer-reviewed journals.

Pascale Champagne, P.Eng., M.ASCE, is an Associate Professor in the Department of Civil Engineering and the Department of Chemical Engineering at Queen's University, Canada. She received a B.Sc. in Biology from McGill University and a B.Eng. in Water Resources Engineering from the University of Guelph, and M.A.Sc. and Ph.D. degrees in Environmental Engineering at Carleton University, Canada. Dr. Champagne was formerly on the faculty of Carleton University in the Department of Civil and Environmental Engineering (1998-2005). She is the recipient of the MRI

Early Researcher Award (2007), Journal of Solid Waste Technology and Management Russell Ackoff Best Paper Award (2005) and the Petro-Canada Young Innovator Award (2002). Dr. Champagne serves as an Associate Editor of ASCE's *Practice Periodical of Hazardous, Toxic and Radioactive Waste Management* and is on the Editorial Board of the *Journal of Applied Sciences in Environmental Sanitation*. Her research focuses on the development and understanding of alternate water and waste management technologies, and sustainable environmental approaches. Dr. Champagne has authored over 85 technical and scholarly publications including 36 in peer-reviewed journals and 6 book chapters.

Say-Kee Ong, P.E., M.ASCE, is a Professor of Environmental Engineering and the Director of the Environmental Engineering program at Iowa State University. Prior to his academic career, he worked as an environmental engineer with Battelle Memorial Institute focusing on the development of innovative remediation technologies. Say-Kee Ong's publication record includes over 45 archival papers, principally in the development of treatment technologies and contaminant fate of pollutants in natural systems (soil and groundwater) and engineered systems (wastewater treatment facilities). He is a recipient of the 2003 Distinguished Service Award by the Association of Environmental Engineering & Science Professors (AEESP), the 2004 Outstanding Achievement in Teaching Award by the Iowa State University Foundation, and the 2005 AEESP Outstanding Teaching Award. Say-Kee Ong received his Ph.D. in Environmental Engineering from Cornell University.

Rao Surampalli, P.E., F.ASCE, is an Engineer Director with the U.S. Environmental Protection Agency. He received his M.S. and Ph.D. degrees in Environmental Engineering from Oklahoma State University and Iowa State University, respectively. He is an Adjunct Professor at Iowa State University, University of Missouri-Columbia, University of Nebraska-Lincoln, Missouri University of Science and Technology-Rolla, University of Quebec-Sainte Foy, and Tongji University-Shanghai and an Honorary Professor in Sichuan University-Chengdu. He is a recipient of the ASCE National Government Civil Engineer of the Year Award for 2006, ASCE State-of-the Art of Civil Engineering Award, ASCE Rudolph Hering Medal (twice), ASCE Wesley Horner Medal and ASCE Best Practice Oriented Paper Award. His awards and honors also include NSPE's Founders Gold Medal for 2001, National Federal Engineer of the Year Award for 2001, WEF's Philip Morgan Award, USEPA's Scientific and Technological Achievement Award, EPA Engineer of the Year Award (thrice), American Society of Military Engineer's Hollis Medal, Federal Executive Board's Distinguished Military Service Award, the U.S. Public Health Service's Samuel Lin Award, and Distinguished Engineering Alumnus Awards from Oklahoma State University and Iowa State University. Dr. Surampalli is a Fellow of the American Association for the Advancement of Science, and a Member of the European Academy of Sciences and Arts. Dr. Surampalli is an Editor of *Water Environment Research* and ASCE's *Practice Periodical of Hazardous, Toxic and Radioactive Waste Management*. He has authored over 400 technical publications, including 8 books, 38 book chapters and 140 peer-reviewed journal articles.

R. D. Tyagi is an internationally recognized Professor with 'Institut national de la recherché Scientifique – Eau, terre, et environement', (INRS-ETE), University of Québec, Canada. He holds Canada Research Chair on, 'Bioconversion of wastewater and wastewater sludge to value added products. He conducts research on hazardous/solids waste management, water/wastewater treatment, sludge treatment/disposal, and bioconversion of wastewater and wastewater sludge into value added products. He has developed the novel technologies of simultaneous sewage sludge digestion and metal leaching, bioconversion of wastewater sludge (biosolids) into *Bacillus thuringiensis* based biopesticides, bioplastics, biofertilisers and biocontrol agents. Dr. Tyagi has published/presented over 430 papers in refereed journals, conferences proceedings and is the author of three books, twenty-six book chapters, ten research reports and nine patents.

Tian C. Zhang, P.E., M.ASCE is a Professor in the department of Civil Engineering at the University of Nebraska-Lincoln. He received his Ph.D. in environmental engineering from the University of Cincinnati in 1994; after a few months of post-doctoral research, he joined the UNL faculty in August 1994. Professor Zhang teaches water and wastewater treatment, biological wastes treatment, senior design, remediation of hazardous wastes, among others. He has authored or co-authored over 50 peer-reviewed journal publications since 1994. Professor's Zhang's research involves water, wastewater, and stormwater treatment and management, remediation of contaminated environments, and detection and control of emerging contaminants in aquatic environments. Professor Zhang is a member of the American Society of Civil Engineers, American Water Works Association, Water Environmental Federation, Association of Environmental Engineering and Science Professors and the International Water Association. He has been the Associate Editor of *Journal of Environmental Engineering* and *Practice Periodical of Hazardous, Toxic, and Radioactive Waste Management* since 2007. He has been a registered professional engineer in Nebraska since 2000.

Index